Coastal Sea Levels, Impacts and Adaptation

Special Issue Editors

Thomas Wahl
Jan Even Øie Nilsen
Ivan Haigh
Sally Brown

MDPI • Basel • Beijing • Wuhan • Barcelona • Belgrade

MDPI

Special Issue Editors
Thomas Wahl
University of Central Florida
USA

Jan Even Øie Nilsen
Ocean and Coastal Remote Sensing, Nansen Environmental and Remote Sensing
Center, and Bjerknes Centre for Climate Research
Norway

Ivan Haigh   Sally Brown
University of Southampton   University of Southampton
UK   UK

Editorial Office
MDPI AG
St. Alban-Anlage 66
Basel, Switzerland

This edition is a reprint of the Special Issue published online in the open access
journal *Journal of Marine Science and Engineering* (ISSN 2077-1312) from 2017–2018
(available at: http://www.mdpi.com/journal/jmse/special_issues/sea_level).

For citation purposes, cite each article independently as indicated on the article
page online and as indicated below:

Lastname, F.M.; Lastname, F.M. Article title. *Journal Name* **Year**, *Article number,*
page range.

First Edition 2018

ISBN 978-3-03842-847-3 (Pbk)
ISBN 978-3-03842-848-0 (PDF)

Cover photo courtesy of Sissel Kvernes.

Table of Contents

About the Special Issue Editors

Thomas Wahl, PhD, is an Assistant Professor for Coastal Risks and Engineering at the University of Central Florida, where he is affiliated with the Civil, Environmental, and Construction Engineering Department and the National Center for Integrated Coastal Research. Through his research, he connects engineering to various science disciplines (e.g., oceanography, hydrology, meteorology, climatology) to better understand the vulnerability of coastal societies, built infrastructures, and fragile ecosystems under climate change conditions. He studies the changes in coastal sea levels (mean and extreme), ocean waves, and freshwater flows and the associated socioeconomic impacts to support the development of sustainable and resilient adaptation strategies.

Jan Even Øie Nilsen, PhD, is a physical oceanographer whose research concerns climatic processes in the Atlantic, Nordic, and Arctic Seas, including sea level change (SLC). Nilsen has worked at the Nansen Environmental and Remote Sensing Center (NERSC) since 2002, also as part of the Bjerknes Centre of Climate Research (BCCR). Nilsen has co-authored several Norwegian reports with social relevance concerning SLC , including the official projections for Norway. He is the contact point of NERSC regarding SLC for governmental agencies and municipalities in Norway and is also co-chair of the European Climate Research Alliance (ECRA) collaborative programme on SLC and Coastal Impacts. Nilsen has led BCCR's projects on "Northern European and Arctic Sea Level" (iNcREASE), "Sea level change and ice sheet dynamics" (SEALEV), and "Changes in the past, present, and future sea level on the coast of Norway" funded by the City of Bergen.

Ivan Haigh, PhD, is an Associate Professor in coastal oceanography at the University of Southampton. In the last 15 years, he has worked on a wide range of projects in both industry and academia, covering many different aspects of coastal oceanography, with a particular focus on sea-level rise and coastal flooding. From 2012 to 2015, he led the UK Natural Environment Research Council (NERC)-funded iGlass consortium project, which defined the high-impact range of future sea-level rise. From 2012 to 2015, he led the University of Southampton component of the Engineering and Physical Sciences Research Council-funded Flood MEMORY project, which investigated the temporal clustering of flood events. He currently leads the NERC-funded E-Rise project, which focuses on the detection of sea-level rise accelerations. He established and leads the development of UK's coastal flood database (www.surgewatch.org).

Sally Brown, PhD, is a coastal engineer specializing in geomorphology, the impacts and adaptation to the effects of sea-level rise at local to global scales, and the long-term sustainability of coastal zones. Within this, she is particularly interested in the physical–engineering–natural–social interactions of changing coastlines. She also has an interest in coastal policies relating to UK flood risk and erosion and enjoys disseminating her research to a wide range of audiences. Sally's research has involved a range of geomorphic settings, including cliffs, beaches, wetlands, deltas, and small islands. She has also worked in a range of environments, such as cities, ports, and heritage sites.

Journal of
Marine Science and Engineering

MDPI

Editorial

Coastal Sea Levels, Impacts, and Adaptation

Thomas Wahl [1,*], Sally Brown [2], Ivan D. Haigh [3] and Jan Even Øie Nilsen [4]

1   Department of Civil, Environmental, and Construction Engineering and National Center for Integrated Coastal Research, University of Central Florida, Orlando, FL 32816-2450, USA; t.wahl@ucf.edu
2   Faculty of Engineering and the Environment and Tyndall Centre for Climate Change Research, University of Southampton, Highfield, Southampton SO17 1BJ, UK; sb20@soton.ac.uk
3   Ocean and Earth Science, National Oceanography Centre, University of Southampton, European Way, Southampton SO14 3ZH, UK; I.D.Haigh@soton.ac.uk
4   Nansen Environmental and Remote Sensing Center, and Bjerknes Centre for Climate Research, 5006 Bergen, Norway; jan.even.nilsen@nersc.no
*   Correspondence: t.wahl@ucf.edu

Received: 9 February 2018; Accepted: 11 February 2018; Published: 21 February 2018

Keywords: mean sea level; storm surges; waves; coastal zone management; impacts; coastal climate services; adaptation; communication; cross-sectorial collaboration; ECRA

1. Introduction

Sea-level rise (SLR) poses a great threat to approximately 10% of the world's population residing in low-elevation coastal zones (i.e., land located up to 10 m of present-day mean sea-level (MSL)) [1], as well as to the human and natural systems supporting these communities. In its Fifth Assessment Report (AR5), the Intergovernmental Panel on Climate Change (IPCC) projected, based on process-based model studies, the upper end of the likely range of global mean SLR, to be 98 cm at the end of the century with respect to 1986-2005 [2]. However, high-end scenarios of 2 m SLR, and more, by 2100 have been developed assuming unmitigated greenhouse gas emissions, and are recognized as a realistic possibility by the scientific community and many stakeholders and decision-makers (e.g., references [3–5] for the Netherlands, United Kingdom (UK), and the United States, respectively). It has been shown in multiple recent studies [6–8] that even moderate changes in MSL can lead to a significant increase in the number of extreme water level exceedances. These arise as a combination of mean sea-level, astronomical tides, storm surges (driven by tropical or extra-tropical storms), and a dynamic wave component (especially at open coastlines) leading to run-up at beaches and overtopping of built and natural coastal (defence) structures, such as dikes, sea walls, or dunes. Changes in the frequency of extreme sea levels will adversely impact coastal communities by increasing the risk of flooding and/or erosion of beaches and cliffs, and it will also impact ecologically and economically valuable marine ecosystems (such as productive estuaries, coastal wetlands, and coral reefs). Extensive adaptation plans and efforts are already underway in some parts of the world, particularly in high risk or urban areas (such as the Delta Works in the Netherlands or the Thames Barrier in the UK) and more will be necessary to mitigate the increasing risks.

By recognizing that combating the negative impacts of SLR presents a multidisciplinary challenge that requires cooperation of scientists from various fields, with stakeholders and decision-makers, the European Climate Research Alliance (ECRA) launched a Collaborative Program on Sea Level Changes and Coastal Impacts in 2012. During a series of workshops, scientists, practitioners, and stakeholders from across Europe came together to discuss solutions to the challenges outlined above. As a result, a white paper was released [9], where the following five themes were identified as research focus areas to facilitate improved sea-level predictions and projections, impact and adaptation assessments, and communication with stakeholders, policy makers, and the public:

1. Observations of MSL change and a better understanding of the contributing processes;
2. Modelling and projections of regional MSL;
3. Changes in extreme sea levels;
4. Potential impacts of, and adaptation strategies to, extreme sea levels and MSL change; and
5. Improved communication and collaboration.

In this Special Issue, 15 papers are published that can be grouped into these topical themes. For simplicity, topics 1 and 2 are merged into the broader topic "MSL changes". Each topic is discussed below in the context of the papers that have been submitted.

2. Mean Sea-Level Changes

Mean sea level changes, ranging from seasonal through multi-decadal variations to linear and non-linear long-term trends, have been studied extensively in the recent past. This is to better understand the processes involved in causing these changes at different temporal and spatial scales, develop models capable of simulating them, and ultimately derive more robust future projections to assess risks and adaptation needs. Four papers in the Special Issue address different aspects of MSL changes at various spatial and temporal scales.

Chafik et al. [10] analyse how sea levels, observed along northern European coastlines, are affected and modulated by large-scale climate features. They detect a non-stationary sea-level response to the North Atlantic Oscillation (NAO), which is explained by the influence of the East Atlantic (EAP) and Scandinavian (SCAN) teleconnection patterns on the NAO. Importantly, they find that coastal sea levels along different coastline stretches respond differently, depending on the phases of the NAO and teleconnection patterns (i.e., whether they are in a positive or negative state), but with variations that can reach the same magnitude as the observed MSL rise during the 20th century. Hence, it is crucial for climate models to be able to reproduce the relevant driving mechanisms acting at large spatial scales, in order to use this information to infer potential changes at the regional and local scale, in terms of both sea-level and flood risk.

Also at the continental scale, but focusing on Australia, Taylor and Brassington [11] introduce a new sea-level forecasting system that aggregates information from heterogeneous operational systems, including gridded ocean and atmosphere models and tidal predictions, to provide 7-day sea-level forecasts at different locations along the coast. Model bias is assessed and corrected for by including information from in-situ sea-level measurements. The system has the ability to provide meaningful forecasts under non-extreme conditions (i.e., storm surges are not captured), to support routine coastal decision processes, and offers a benchmark for future developments in MSL forecasting.

Breili et al. [12] undertake a regional study. They use a range of different data sets derived from tide gauges, satellite altimetry, Global Navigation Satellite System (GNSS), levelling campaigns, metrological stations, and model results, to investigate MSL changes along the Norwegian coast since the mid-20th century. They pay particular attention to spatial variability along the coast and its drivers, mainly vertical land movement caused by glacial isostatic adjustment (GIA) or anthropogenic impacts, and the inverse barometer effect. For example, they show that MSL trends along the Norwegian coast, when adequate corrections are applied for regional factors, are consistent with estimated global MSL trends for the last 30 to 50 years amounting to approximately 2 mm/year but increase to up to 3.5 mm/year for the more recent period from 1993 to 2016, again in accordance with an observed acceleration in global MSL rise.

Finally, the study by Watson [13] has a local focus and includes a detailed analysis of the long sea-level record from the Battery tide gauge in New York City. Observed rates of MSL rise over the last decade (again, after applying adequate corrections to filter out location specific temporal fluctuations) are compared with those derived from climate model projections from the Climate Model Intercomparison Project—Phase 5 (CMIP5) over the same time period (2007 to 2016). They find that the projected rates of rise are larger than the observed rates at the tide gauges. Such differences may be site specific for the particular location, but this highlights the need for in-depth validation of the model projections with in-situ measurements from tide gauges and remote sensing data from satellite altimetry.

3. Extreme Sea Levels

While the slow rise in MSL will eventually threaten many low lying and unprotected areas, the majority of the impacts to people, property, infrastructure, and the environment, will be felt through extreme sea-level events. Extreme sea levels arise as a combination of four factors: MSL, astronomical tides, storm surges, and waves. Three contributions in this Special Issue analyse extreme sea-level changes, including the dynamic wave component. Slangen et al. [14] undertake a global assessment of extreme sea-level allowances (i.e., the height a coastal structure needs to be elevated to keep the same frequency and likelihood of sea-level extremes under a certain sea-level rise scenario). Simpson et al. [15] and Malagon Santos et al. [16] undertake regional assessments of mean and extreme sea levels around Norway and extreme waves around the UK, respectively.

Slangen et al. [14] calculate sea-level allowances at the global scale using the Global Extreme Sea Level Analysis (Version 2) tide gauge database. In particular, they address one of the major uncertainties in future sea-level projections: the contribution of the ice sheets in Greenland and Antarctica to sea-level rise. Their results show that allowances increase significantly for ice sheet dynamics' uncertainty distributions that are more skewed, due to the increased probability of a much larger ice sheet contribution to sea-level rise. They find that allowances are largest in regions where a relatively small observed variability in the extremes is paired with relatively large magnitude and/or large uncertainty in the projected sea-level rise. This typically occurs around the equator. Finally, for the Representative Concentration Pathway (RCP) 8.5 sea-level rise projections, they show that the likelihood of extremes increases by more than a factor 10,000 at the majority of tide gauges analysed.

In their article, Simpson et al. [15] present new relative sea-level projections for Norway for the 21st century. The region is commonly perceived as being at low risk from sea-level rise, because the coastline is characterised by steep rocky topography. However, as the authors point out, most of Norway's major cities and numerous towns and villages are located in low-lying coastal areas and, hence, are vulnerable to sea-level rise and coastal flooding. To create their new projections, they use findings from the IPCC AR5 and CMIP5 model outputs and scale them to take into account spatial variations in ocean density, ocean mass redistribution, ocean mass changes and associated gravitation effects, and vertical land motion. Then, they calculate return heights for extreme sea levels around the coast using the average conditional exceedance rate (ACER) statistical method. Finally, they adapt the ACER method to also calculate sea-level allowances.

Malagon Santos et al. [16] carry out a spatial footprint and temporal clustering analysis of extreme storm-wave events around the coast of the UK using measurements from wave buoys. As the authors point out, economical, societal, and environmental impacts from extreme events are often correlated spatially. Furthermore, temporal clustering of extreme wave events may have important consequences for coastal structures as there may be insufficient time to properly repair structures between storms. These two issues have important financial and practical implications for the flood risk management sector and, yet, recognition and analysis of spatial and temporal wave characteristics is lacking. The authors identify six categories of spatial footprints of wave events and the distinct storm tracks that generated them. They find that the majority of large wave events occurred between November and March, with large inter-annual differences in the number of events per season associated with the West Europe Pressure Anomaly (WEPA).

4. Impacts and Adaptation

As outlined above, both observations and modelling studies point toward changes in mean and extreme sea levels on decadal to centennial time scales. Over the last few years, the scientific community has responded to this by focusing on sector specific studies or local scale studies to better understand potential impacts. Increasingly, adaptive response is becoming part of that process, with an emphasis on engineering (particularly encouraging natural methods) and societal response, including the role of decision-making. Six of the papers in this Special Issue reflect this range of advancements.

Some of the earliest effects of SLR are seen on beaches, which includes shifts in shoreline position and sediment redistribution. Kinsela et al. [17] describe an approach to estimate potential beach erosion and shoreline change on wave dominated sandy beaches in New South Wales, Australia. They find that sediment compartments help quantify sediment redistribution, including the source and sediment pathways. Exposure to coastal erosion is expected to increase, primarily due to SLR driven shoreline recession. This indicates that thousands of properties may be at risk from coastal erosion over this century. Similarly, Van De Lageweg and Slangen [18] assess changes in deltaic systems as a result of a combination of tides, waves, river-flow, and SLR. They use a set of models to quantify how different types of deltas respond to the abovementioned forcing factors and they evaluate related impacts, in the form of flooding, shoreline recession, and habitat change. Park et al. [19] investigate the effects of SLR and associated implications further inland. Focusing on Florida, they find that large amounts of land could be inundated unless adaptation is undertaken, including areas of marshland. SLR could have significant implications for coastal infrastructure and national parks in the region.

Whilst beaches erode, innovative methods are required to consider how to accrete sediment without causing knock-on problems down-coast. This need is particularly acute after rapid erosion events, as infrastructure is left exposed. Goreau and Prong [20] describe their findings on biorock electric reefs, which encourage damaged reefs to grow, thus having secondary impacts on coastal protection. From a case study in Indonesia, they find that after storm conditions, biorock reefs have allowed beaches to grow. If beach erosion is projected to increase with SLR, biorock reefs may encourage sedimentation along vulnerable shorelines.

Understanding how to respond to shoreline change is important, and the study by Hirschfield and Hill [21] is an example of how adaptation and decision-making, at the local scale in urban areas, is addressed. They analyse the San Francisco Bay shoreline and estimate unit costs for raising current infrastructure, taking into account the shoreline position and design heights, as well as the range of shoreline infrastructure. They conclude that defending the shortest length of shoreline might not be the best option. Although costs to protect a shorter shoreline stretch are lower, longer shoreline (which represents a boundary between saltwater habitat and freshwater habitat) protection could bring multiple ecosystem benefits. This paper has important implications for engineering, as it indicates that a wide range of options need to be considered when planning future defences, not just cost.

Being situated directly on the coast, ports are on the forefront of impacts from adverse weather conditions, including extreme events and SLR. However, even as a commercial business ports and harbours infrequently consider long-term (>100 years) climate change adaptation as they are more focused on day-to-day delivery. Becker et al. [22] recognise the need for long-term adaptation in port environments, such as raising infrastructure as sea-levels rise. Despite there being thousands of ports in need of long-term land raising, this paper is one of the first to estimate the volume of fill cost of materials to raise land. Focusing on 100 major ports in the United States, they estimate that 704 million m^3 of fill is required to raise land and infrastructure by 2 m at a cost of US\$57–US\$78 billion. For a large industry, these costs are achievable and suggestive that when the time comes, ports will be capable (from a technical and financial standpoint) to adapt to SLR.

5. Communication and Collaboration

The practical use of SLR information for coastal adaptation by governance and managers, is a challenge in its own right. Success requires extensive collaboration between all involved parties, from users to providers of SLR information. Separately, neither the specific requirements for, nor the usefulness of the provided climate information can be known a priori. For numerous places of the world, particularly in built up areas or those of high risk, the time for taking measures is now or in the near future. Hence, there is a dire need for establishing practices that facilitate the efficient use of SLR information for adaptation.

Two of the papers in this Special Issue address the challenge of improved communication (including the assessment of uncertainties inherent to coastal hazard assessments) and collaboration. Le Cozannet et al. [23] assess the translation of SLR information for efficient adaptation, while Stephens et al. [24] offer a framework for uncertainty identification and management and show, by practical example, how flexibility in decision-making for adaptation to future hazards can be supported by maps of the degree of hazard exposure.

Le Cozannet et al. [23] review the practices of coastal climate services (CCS) in France, the US, and Australia, by identifying current barriers and offering recommendations to overcome these. They find that coastal climate services based on sea-level projections are emerging in a scattered manner, and, overall, too slowly to meet the diversity of challenges. All the while the demand is there, driven by the user need to analyse the benefits of mitigation, to highlight research needs, and to support the many aspects of adaptation. The more technical barriers are the need for topical research into, for example, near- and long-term regional, relative sea-level projections and the uncertainties involved, as well as the gap between what sea-level science can provide and the methods of coastal engineers. In conclusion, the authors recommend and propose a framework involving all stakeholders, addressing issues of user interaction, decision-making, and uncertainties, as well as topical research on sea-level science, hydro- and morpho-dynamics, biology, demography, and economy.

Stephens et al. [24] explore the matter of uncertainty management for decision-making, with respect to coastal hazards and adaptation, addressing that near-term decisions need to build in flexibility, both in order to reduce exposure and to enable changes to actions, or pathways that can accommodate higher sea levels over longer timeframes. They outline a logical framework, starting from the land use situation, through the level of uncertainty, hazard scenarios in question, and complexity of the hazard modelling, to the decision (accept, adapt, or avoid). They also demonstrate enhancements to coastal flood exposure mapping by isolating both flooding depth and frequency, showing the degree of exposure and likelihood and how it will change with SLR, in an incremental manner. This gives flexibility in planning and helps inform when intolerable risks emerge. Together, the uncertainty framework and mapping techniques may improve the identification of trigger points for adaptation pathway planning and their expected time range, compared to traditional coastal flooding hazard assessments.

6. Conclusions

This Special Issue has emerged from the ECRA Collaborative Program on Sea Level Changes and Coastal Impacts and it provides a snapshot of the current state of research in the broad field of sea-level rise, its impacts, and adaptation. The 15 papers highlight the different challenges and pathways that can be taken to address either only one or multiple of the topical areas listed in the introduction, and how changes in both the physical drivers of coastal risks and our responses to it can happen at multiple temporal and spatial scales. Rising sea-levels will continue to pose challenges in many different coastal environments. In some coastal zones, lack of information is a key barrier, whereas in others, it is the lack of understanding of how to respond that leaves coastal communities, built infrastructure, and ecosystems vulnerable to changes in mean and extreme sea levels. Communicating these issues to decision-makers, stakeholders, and the general public is a major challenge, particularly when the threat is not always obvious to see, or a short-term solution is favoured but may reduce long-term sustainability that could be otherwise achieved or improved. This compendium shows that the community, including scientists from multiple fields, practitioners, and stakeholders, have made important advances, working side-by-side in order to make progress, but as sea levels continue to rise we will be facing major challenges in managing the coast for years to come.

Acknowledgments: The Collaborative Program on Sea Level Changes and Coastal Impacts thank ECRA, the Bjerknes Centre for Climate Research, and Nansen Environmental and Remote Sensing Center for support of the workshop where this Special Issue was conceived. All contributing authors and reviewers are thanked for their efforts.

Conflicts of Interest: The authors declare no conflict of interest.

References

1. McGranahan, G.; Balk, D.; Anderson, B. The rising tide: Assessing the risks of climate change and human settlements in low elevation coastal zones. *Environ. Urban.* **2007**, *19*, 17–37. [CrossRef]
2. Church, J.A.; Clark, P.U.; Cazenave, A.; Gregory, J.M.; Jevrejeva, S.; Levermann, A.; Merrifield, M.A.; Milne, G.A.; Nerem, R.S.; Nunn, P.D.; et al. Sea level change. In *Climate Change 2013: The Physical Science Basis. Contribution of Working Group I to the Fifth Assessment Report of the Intergovernmental Panel on Climate Change*; Stocker, T.F., Qin, D., Plattner, G.-K., Tignor, M., Allen, S.K., Boschung, J., Nauels, A., Xia, Y., Bex, V., Midgley, P.M., Eds.; Cambridge University Press: Cambridge, UK; New York, NY, USA, 2013.
3. Katsman, C.A.; Sterl, A.; Beersma, J.J.; van den Brink, H.W.; Church, J.A.; Hazeleger, W.; Kopp, R.E.; Kroon, D.; Kwadijk, J.; Lammersen, R.; et al. Exploring high-end scenarios for local sea level rise to develop flood protection strategies for a low-lying delta—The Netherlands as an example. *Clim. Chang.* **2011**, *109*, 617–645. [CrossRef]
4. Nicholls, R.J.; Hanson, S.E.; Lowe, J.A.; Warrick, R.A.; Lu, X.; Long, A.J. Sea-level scenarios for evaluating coastal impacts. *Wiley Interdiscip. Rev. Clim. Chang.* **2014**, *5*, 129–150. [CrossRef]
5. Sweet, W.V.; Kopp, R.E.; Weaver, C.P.; Obeysekera, J.; Horton, R.M.; Thieler, E.R.; Zervas, C. *Global and Regional Sea Level Rise Scenarios for the United States*; NOAA Technical Report NOS CO-OPS 083; NOAA/NOS Center for Operational Oceanographic Products and Services: Silver Spring, MD, USA, 2017.
6. Buchanan, M.K.; Oppenheimer, M.; Kopp, R.E. Amplification of flood frequencies with local sea level rise and emerging flood regimes. *Environ. Res. Lett.* **2017**, *12*, 064009. [CrossRef]
7. Vitousek, S.; Barnard, P.L.; Fletcher, C.H.; Frazer, N.; Erikson, L.; Storlazzi, C.D. Doubling of coastal flooding frequency within decades due to sea-level rise. *Sci. Rep.* **2017**, *7*, 1399. [CrossRef] [PubMed]
8. Wahl, T.; Haigh, I.D.; Nicholls, R.J.; Arns, A.; Dangendorf, S.; Hinkel, J.; Slangen, A. Understanding extreme sea levels for coastal impact and adaptation analysis. *Nat. Commun.* **2017**, *8*, 16075. [CrossRef]
9. Nilsen, J.E.Ø.; Sannino, G.; Bordbar, M.; Carrasco, A.R.; Dangendorf, S.; Haigh, I.D.; Hinkel, J.; Haarstad, H.; Johannessen, J.A.; Madsen, K.S.; et al. *White Paper: Sea Level Related Adaptation Needs in Europe*; Collaborative Programme: Sea Level Change and Coastal Impacts (CP SLC); European Climate Research Alliance (ECRA): Brussels, Belgium, 2016.
10. Chafik, L.; Nilsen, J.E.Ø.; Dangendorf, S. Impact of North Atlantic Teleconnection Patterns on Northern European Sea Level. *J. Mar. Sci. Eng.* **2017**, *5*, 43. [CrossRef]
11. Taylor, A.; Brassington, G.B. Sea Level Forecasts Aggregated from Established Operational Systems. *J. Mar. Sci. Eng.* **2017**, *5*, 33. [CrossRef]
12. Breili, K.; Simpson, M.J.R.; Nilsen, J.E.Ø. Observed Sea-Level Changes along the Norwegian Coast. *J. Mar. Sci. Eng.* **2017**, *5*, 29. [CrossRef]
13. Watson, P.J. Integrating Long Tide Gauge Records with Projection Modelling Outputs. A Case Study: New York. *J. Mar. Sci. Eng.* **2017**, *5*, 34. [CrossRef]
14. Malagon Santos, V.; Haigh, I.D.; Wahl, T. Spatial and Temporal Clustering Analysis of Extreme Wave Events around the UK Coastline. *J. Mar. Sci. Eng.* **2017**, *5*, 28. [CrossRef]
15. Simpson, M.J.R.; Ravndal, O.R.; Sande, H.; Nilsen, J.E.Ø.; Kierulf, H.P.; Vestøl, O.; Steffen, H. Projected 21st Century Sea-Level Changes, Observed Sea Level Extremes, and Sea Level Allowances for Norway. *J. Mar. Sci. Eng.* **2017**, *5*, 36. [CrossRef]
16. Slangen, A.B.A.; van de Wal, R.S.W.; Reerink, T.J.; de Winter, R.C.; Hunter, J.R.; Woodworth, P.L.; Edwards, T. The Impact of Uncertainties in Ice Sheet Dynamics on Sea-Level Allowances at Tide Gauge Locations. *J. Mar. Sci. Eng.* **2017**, *5*, 21. [CrossRef]
17. Kinsela, M.A.; Morris, B.D.; Linklater, M.; Hanslow, D.J. Second-Pass Assessment of Potential Exposure to Shoreline Change in New South Wales, Australia, Using a Sediment Compartments Framework. *J. Mar. Sci. Eng.* **2017**, *5*, 61. [CrossRef]

J. Mar. Sci. Eng. **2018**, *6*, 19

18. Van de Lageweg, W.I.; Slangen, A.B.A. Predicting Dynamic Coastal Delta Change in Response to Sea-Level Rise. *J. Mar. Sci. Eng.* **2017**, *5*, 24. [CrossRef]

19. Park, J.; Stabenau, E.; Redwine, J.; Kotun, K. South Florida's Encroachment of the Sea and Environmental Transformation over the 21st Century. *J. Mar. Sci. Eng.* **2017**, *5*, 31. [CrossRef]

20. Goreau, T.J.F.; Prong, P. Biorock Electric Reefs Grow Back Severely Eroded Beaches in Months. *J. Mar. Sci. Eng.* **2017**, *5*, 48. [CrossRef]

21. Hirschfeld, D.; Hill, K.E. Choosing a Future Shoreline for the San Francisco Bay: Strategic Coastal Adaptation Insights from Cost Estimation. *J. Mar. Sci. Eng.* **2017**, *5*, 42. [CrossRef]

22. Becker, A.; Hippe, A.; Mclean, E.L. Cost and Materials Required to Retrofit US Seaports in Response to Sea Level Rise: A Thought Exercise for Climate Response. *J. Mar. Sci. Eng.* **2017**, *5*, 44. [CrossRef]

23. Le Cozannet, G.; Nicholls, R.J.; Hinkel, J.; Sweet, W.V.; McInnes, K.L.; Van de Wal, R.S.W.; Slangen, A.B.A.; Lowe, J.A.; White, K.D. Sea Level Change and Coastal Climate Services: The Way Forward. *J. Mar. Sci. Eng.* **2017**, *5*, 49. [CrossRef]

24. Stephens, S.A.; Bell, R.G.; Lawrence, J. Applying Principles of Uncertainty within Coastal Hazard Assessments to Better Support Coastal Adaptation. *J. Mar. Sci. Eng.* **2017**, *5*, 40. [CrossRef]

Journal of
Marine Science and Engineering

MDPI

Article

The Impact of Uncertainties in Ice Sheet Dynamics on Sea-Level Allowances at Tide Gauge Locations

Aimée B. A. Slangen [1,2,*], Roderik S. W. van de Wal [2], Thomas J. Reerink [2], Renske C. de Winter [3], John R. Hunter [4], Philip L. Woodworth [5] and Tamsin Edwards [6]

1   Royal Netherlands Institute for Sea Research (NIOZ), Department of Estuarine and Delta Systems (EDS), Utrecht University, Korringaweg 7, Yerseke 4401 NT, The Netherlands
2   Institute for Marine and Atmospheric research Utrecht (IMAU), Princetonplein 5, Utrecht 3584 CC, The Netherlands; r.s.w.vandewal@uu.nl (R.S.W.v.d.W.); tjreerink@gmail.com (T.J.R.)
3   Department of Physical Geography, Utrecht University, Heidelberglaan 2, Utrecht 3584 CS, The Netherlands; r.c.dewinter@uu.nl
4   Antarctic Climate & Ecosystems Cooperative Research Centre, 20 Castray Esplanade, Hobart TAS 7000, Australia; jrh@johnroberthunter.org
5   National Oceanography Centre, 6 Brownlow Street, Liverpool L3 5DA, UK; plw@noc.ac.uk
6   School of Environment, Earth & Ecosystem Sciences, Faculty of Science, Technology, Engineering & Mathematics, The Open University, Milton Keynes MK7 6AA, UK; tamsin.edwards@open.ac.uk
*   Correspondence: aimee.slangen@gmail.com; Tel.: +31-0-113-577-300

Academic Editors: Thomas Wahl, Jan Even Øie Nilsen, Ivan Haigh and Sally Brown
Received: 10 March 2017; Accepted: 16 May 2017; Published: 23 May 2017

Abstract: Sea level is projected to rise in the coming centuries as a result of a changing climate. One of the major uncertainties is the projected contribution of the ice sheets in Greenland and Antarctica to sea-level rise (SLR). Here, we study the impact of different shapes of uncertainty distributions of the ice sheets on so-called sea-level allowances. An allowance indicates the height a coastal structure needs to be elevated to keep the same frequency and likelihood of sea-level extremes under a projected amount of mean SLR. Allowances are always larger than the projected SLR. Their magnitude depends on several factors, such as projection uncertainty and the typical variability of the extreme events at a location. Our results show that allowances increase significantly for ice sheet dynamics' uncertainty distributions that are more skewed (more than twice, compared to Gaussian uncertainty distributions), due to the increased probability of a much larger ice sheet contribution to SLR. The allowances are largest in regions where a relatively small observed variability in the extremes is paired with relatively large magnitude and/or large uncertainty in the projected SLR, typically around the equator. Under the RCP8.5 (Representative Concentration Pathway) projections of SLR, the likelihood of extremes increases more than a factor 10^4 at more than 50–87% of the tide gauges.

Keywords: sea-level rise; allowances; sea-level extremes

1. Introduction

Sea level is projected to rise in the coming century and beyond as a result of a warming climate, with major contributions from warming of the ocean water and increasing land ice mass loss [1]. To protect against or mitigate the effects of sea-level rise (SLR) in highly populated coastal zones, it is important to understand the processes causing SLR, and to reduce the uncertainties for each of the projected contributions. Sea-level rise here is *relative* sea-level change, i.e., the difference as measured by tide gauges attached to the Earth's surface. Presently, one of the main uncertainties in SLR is the contribution of the Antarctic and Greenland ice sheets [2,3]. Particularly the timing and magnitude of dynamical changes associated with the ice sheets (e.g., marine ice sheet instability) are still uncertain. This is therefore an active area of research.

In the Fifth Assessment Report (AR5) of the Intergovernmental Panel on Climate Change (IPCC), the projected ice dynamical contribution to SLR was based on observations and a range of scenarios from models ([1], Table 13.5). Around the same time, Bamber and Aspinall [4] presented an estimate based on expert judgement, which had a strongly skewed uncertainty distribution rather than the symmetric distribution presented in IPCC. With the same data but a different way to process the expert opinions, De Vries and Van de Wal [5] presented an estimate with a lower high-end estimate than Bamber and Aspinall [4], though still a strongly skewed uncertainty distribution. Although expert elicitations document the level of consensus on a certain topic, they are sensitive to, for instance, the level of expertise of the interviewees and the way experts are selected. Therefore, process-based estimates are generally a preferable method. Until recently, process-based estimates were only available for specific parts of the ice sheets (e.g., Pine Island Glacier [6] or Thwaites Glacier [7]). However, Ritz et al. [8] recently presented an estimate for the Antarctic ice dynamical contribution to SLR in the 21st century using a process-based approach. They too found a non-normal distribution of the uncertainties, though less skewed than the De Vries and Van de Wal [5] distribution.

In this study, we will compare and contrast three different types of uncertainty distributions for the mass loss contribution to SLR due to ice dynamics [1,5,8]. We will specifically look at the effect on sea-level allowances rather than focusing on the sea-level projections. A sea-level allowance indicates the height a coastal structure needs to be elevated to maintain the same frequency and likelihood of extreme sea-level events under future SLR [9,10]. An allowance gives more information than just a 90% or 95% confidence level of projected mean SLR because it also accounts for the local variability in the sea-level extremes. An important assumption for the allowances in general is that the variability in the extremes does not change under the influence of climate change. For the 20th century, this is a reasonable assumption, as Menéndez and Woodworth [11] found that the historical change in extremes is mostly due to changes in mean sea level rather than changes in the storminess contributions to the extremes. For the 21st century, model projections of changes in the wave climate and storm surges are subject to large uncertainties, and it is therefore difficult to say if the assumption will hold in the future [1].

We use the regional relative sea-level projections of Slangen et al. [12], which describe the change of the ocean surface *relative* to the ocean floor. The projections follow the high-end RCP8.5 (Representative Concentration Pathway [13]) climate change scenario based on CMIP5 model output (5th phase of the Climate Model Intercomparison Project [14]). While these projections assume a normal (Gaussian) uncertainty distribution, more recent publications have incorporated skewed ice sheet dynamics distributions into their projections to provide a more complete uncertainty distribution (e.g., [15–18]) which can be used for risk-assessment and policy purposes. We follow this line and combine the Slangen et al. [12] sea-level projections with three different ice sheet dynamics scenarios (Section 2.1) using the framework of De Winter et al. [18]. Whereas De Winter et al. [18] discussed the impact and sensitivity of the high-end regional sea-level projections to different ice sheet uncertainty distributions, here we focus on the effect of the shape of the skewed ice sheet uncertainty distributions on the projected allowances. De Winter et al. [18] showed that the 90th, 95th and 97.5th percentiles of regional SLR projections are highly sensitive to the uncertainty distribution of ice sheet dynamics. In contrast to De Winter et al. [18], all the ice sheet dynamics scenarios will be shifted to the same median (as described in Section 2.1) to focus on the effects of the different shapes rather than differences in their magnitude. Previously, regional allowances were mainly computed using normal (Gaussian) uncertainty distributions for the projected SLR [10]. Buchanan et al. [19] used full probability distributions in combination with allowances to compute an "Average Annual Design-Life Level" for the United States coastline. Here, we present allowances for skewed uncertainty distributions for tide gauge stations across the world and compare the different shapes of the distributions (Section 2.3).

There are various ways to express sea-level extreme events statistically, depending on the application and the quality of the available data. Methods such as fitting Generalised Extreme Value distributions (of which the Gumbel distribution is a special case) to annual maxima in water levels and fitting Generalised Pareto distributions to peaks over a threshold [20] are often used to represent the statistics of extreme water levels. There is no consensus as to a universal best approach [21], and the method chosen depends on the purpose of the analysis and the data availability and quality. Here, we use a Gumbel parameterisation, which will allow us to compute the allowances following Hunter [9] and Hunter et al. [10]. We use the GESLA-2 tide gauge dataset (Global Extreme Sea Level Analysis Version 2 [22,23], Section 2.2), which significantly increases the number of locations with respect to previous studies of allowances [9,10]. Finally, we show the increase in the frequency of sea level extremes under the RCP8.5 scenario if the allowances are not applied (Section 3.3).

2. Data and Methodology

2.1. Uncertainty Distributions of Sea-Level Change

We use the regional sea-level projections for the RCP8.5 scenario based on CMIP5 model output as presented in Slangen et al. [12], for the period 2010–2100. These projections are similar to the results presented in IPCC AR5 [1] and include the contributions of steric/dynamic change, glacier and ice sheet mass change, groundwater extraction and Glacial Isostatic Adjustment. The global mean projected change for 2010–2100 is 0.72 ± 0.28 m, of which the Antarctic dynamics mass loss contribution is 0.08 ± 0.06 m and the Greenland dynamics mass loss contribution 0.05 ± 0.06 m. All uncertainty distributions in Slangen et al. [12] are assumed to be normal, with means and standard deviations as provided in their Table 1.

The two dynamical ice sheet contributions will be removed from the total projections of Slangen et al. [12], and replaced by three different scenarios for the ice sheet dynamically-driven mass loss. For clarity, this means that the ice sheet surface mass balance (SMB) contribution to SLR remains unaltered with respect to Slangen et al. [12]. The three scenarios used are:

(1) The IPCC scenario (Figure 1, green, from Church et al. [1]). We use the cumulative ice sheet dynamical contributions for the Antarctic Ice Sheet (AIS) and Greenland Ice Sheet (GRIS) as presented in IPCC AR5 Table 13.5 (0.05 ± 0.03 m for Greenland, 0.07 ± 0.08 for Antarctica), assuming a normal distribution. As Church et al. [1] present the change for 1986–2005 to 2081–2100, we correct the values linearly to estimate the change for 2010 to 2100. The Antarctic contribution is split into 11% for the East Antarctic Ice Sheet (EAIS) and 89% for the West Antarctic Ice Sheet (WAIS), which is the split for the year 2010 as presented in the IMBIE project (Ice sheet Mass Balance Inter-comparison Exercise, [24]).

(2) The VW15 scenario (Figure 1, blue) is from De Vries and Van de Wal [5]. We use the skewed uncertainty distribution of rates of SLR in 2100 for WAIS, EAIS and GRIS contributions, based on expert elicitation estimates. VW15 presented a reanalysis of the data in Bamber and Aspinall [4], which more rigorously accounts for the lack of consensus in the expert estimates. To convert mm/year in 2100 into cumulative SLR for 2010–2100, we assume a rate of change in 2010 of 0.6 mm/year for WAIS, 0.0 mm/year for EAIS and 0.3 mm/year for GRIS following VW15 and Bamber and Aspinall [4], and assume a linear change in the rate for the period 2010–2100.

(3) The R15 scenario for the Antarctic dynamical ice sheet contribution (Figure 1, black) is from Ritz et al. [8]. This is an observation and model-based estimate of the ice dynamical mass loss for 2010–2100. The shape of the Antarctic distributions differs slightly from the original R15 study due to the use of a narrower bandwidth (less smoothing) in the kernel density estimation. For Greenland dynamics, this scenario is the same as the IPCC scenario, as R15 did not calculate the Greenland contribution.

All the ice dynamics scenarios are shifted to match the median values of the IPCC scenario (Figure 1, dashed lines), in order to make sure that the differences in the allowances are only due to changes in the *shape* of the distribution rather than the *magnitude*. We use the median (i.e., the point where the cumulative density function is 0.5) for this shift, rather than the mean, as the median is less sensitive to extremely large values in the non-normal distributions. Although the published VW15 estimates are for SMB and ice sheet dynamics combined, we here assume that the skewed *shape* of the distribution is primarily a result of the uncertainty in the dynamical processes and correct for the differences in *magnitude* by the aforementioned shift in the median.

The regional SLR for each of the ice sheet dynamics contributions is computed by multiplying each contribution as presented in Figure 1 (Greenland, East Antarctica, West Antarctica) with its respective sea-level "fingerprint" ratio (e.g., [25]). These fingerprints are computed with a gravitationally consistent sea-level model. The resulting projected ice dynamical contribution to the relative regional SLR was shown in Slangen et al. [12] in their Figure 1e. Typically, sea level falls near the sources of ice mass loss, and rises above average in the far field. The individual fingerprint ratios of each ice sheet may be scaled with the global mean SLR contribution of each ice sheet if the location of the mass change remains the same as the location used to compute the fingerprint. This assumption is most significant in proximity to the ice sheet, where gravitational gradients decrease relatively sharply from the centres of mass loss.

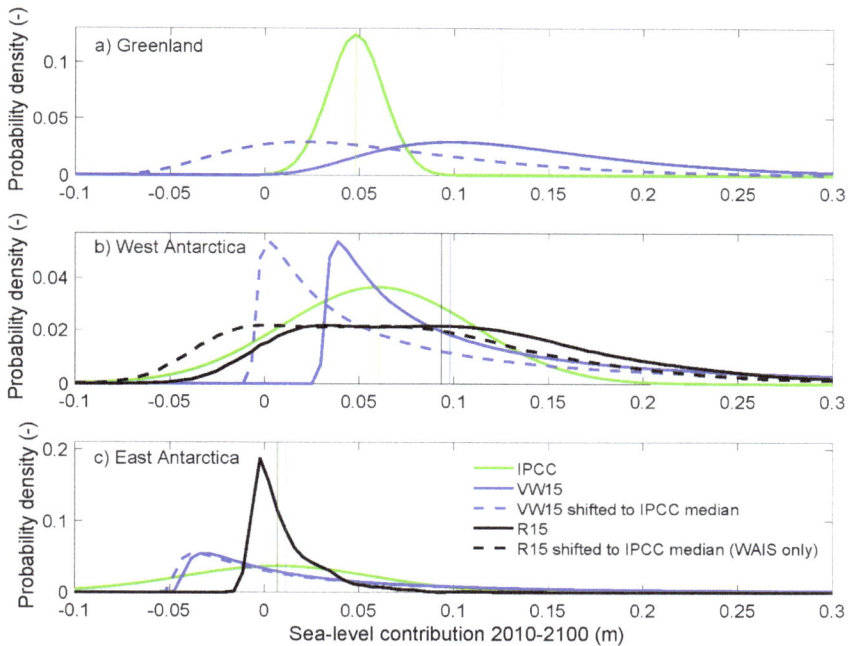

Figure 1. The three scenarios for ice sheet dynamics contributions to sea-level rise used in this study, for (**a**) Greenland, (**b**) West Antarctica and (**c**) East Antarctica (cumulative m sea-level change between 2010–2100). Original uncertainty distributions in solid lines, medians in vertical lines, shifted uncertainty distributions (where all medians match the IPCC medians) in dashed lines.

We use the SEAWISE model [18] to combine the skewed SLR distributions for Greenland, West-Antarctica and East-Antarctica one by one with the projected non-ice sheet dynamics SLR from Slangen et al. [12] (for which a normal distribution is assumed) at each tide gauge location (Appendix, Figure A1). SEAWISE first scales the ice dynamic uncertainty distributions with the individual fingerprints to obtain the ice sheet uncertainty distribution at each location. Then, SEAWISE combines multiple SLR distributions, e.g., for West-Antarctica and Greenland (these can be normal or skewed). The model can combine independent distributions or assign a correlation coefficient to assume a dependency between distributions. Here, all distributions are assumed to be independent. Details of the SEAWISE methodology are presented in the Appendix and in De Winter et al. [18].

2.2. The Statistics of Sea-Level Extremes

In addition to the regional sea-level projections, we require location-specific information on the statistics of sea-level extremes resulting from tides, storm surges and other (high-frequency) sea-level variability. For this, we use Gumbel scale parameters. The Gumbel scale parameter for a certain location can be derived from historical records of high-frequency sea-level measurements (see Figure 2 for four examples). Here, we use data from the GESLA-2 database (Global Extreme Sea Level Analysis Version 2 [22,23]), which contains high-frequency records (at least hourly) from 1300 tide gauge stations as of February 2016. Gumbel distributions were fitted to the annual maxima (expressed relative to the linear trend in mean sea level over the record) of each of the 658 tide gauges, using the Matlab® *evfit* (Mathworks, Natick, MA, USA) function, as described in Hunter et al. [26]. In a comparison of the Gumbel scale parameters obtained from the *evfit* method to the *ismev* method as used in Hunter [9] (both methods are extensively described in Hunter et al. [26]), we found that the results are very similar to the extent required for the present analysis and we therefore use the values from the Matlab® *evfit* method only. We select all tide gauge records that contain at least 20 years of data and where each year with data is more than 75% complete (Figure 3), which yields a data set of 658 records. Following Hunter et al. [26], we reject four stations that show significant non-Gumbel behaviour, and also leave out two Hudson Bay tide gauges (Canada) as we do not have sea-level projections available there. This leaves 652 records, of which 448 records have $\geq$30 years of data available, 319 records $\geq$40 years, 164 records $\geq$50 years and 94 records $\geq$60 years.

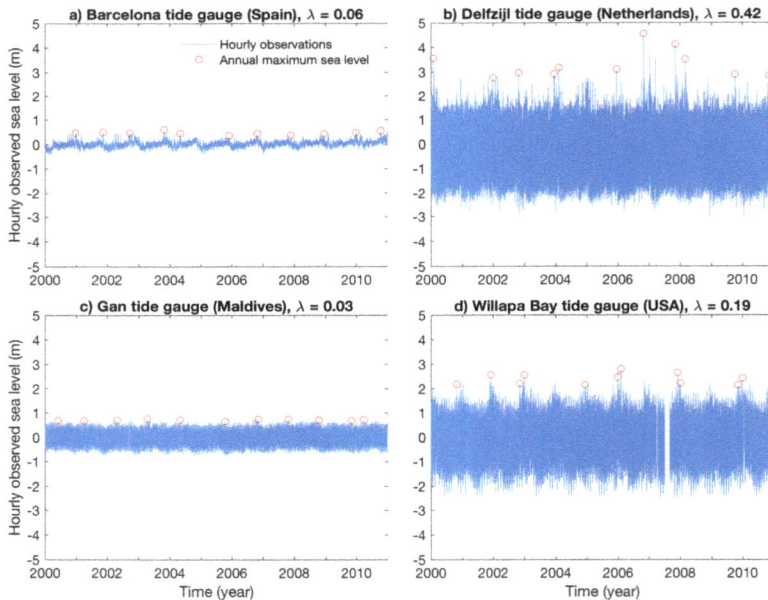

Figure 2. Four examples of tide gauge records: (**a**) Barcelona, (**b**) Delfzijl, (**c**) Gan, (**d**) Willapa Bay. Hourly data (in blue) and their annual maximum values (red) for 2000–2010 (m), where the mean of each time series has been removed. λ = the Gumbel scale parameter (m). Locations are indicated in Figure 5a by their first letter.

The Gumbel scale parameter describes the variability of the extremes; i.e., a smaller parameter means that the *variability in the extremes* is small (e.g., Figure 2a,c) — this does not necessarily mean that the average *magnitudes* of the extremes are small. A larger scale parameter indicates larger variability in the extremes (e.g., Figure 2b,d). Sea-level extremes can be a combination of high tide and extreme storm surge, but also a medium tide and an extreme surge, or a high tide and a medium surge. Other factors such as interannual or seasonal variability can have significant effects as well: Barcelona (Figure 2a), for instance, has a strong seasonal cycle, leading to a more regular occurrence of the annual maximum than for instance in Delfzijl (Figure 2b). For a detailed discussion of contributing factors to extreme sea levels, we refer the reader to Merrifield et al. [27].

Historically, there have been more tide gauge stations in the Northern Hemisphere, resulting in a larger number of estimates of Gumbel scale parameters in the Northern Hemisphere than in the Southern Hemisphere. The Gumbel scale parameters are typically in the range of tenths of meters (0.12 ± 0.06 m, mean $\pm 1\sigma$). Values are generally lower around the equator, increasing towards the mid-latitudes (specifically in the Northern Hemisphere), with the highest values between 40° and 70° N (Figure 3), where both the tidal range and the storm surge activity tend to be larger [27]. For instance, Gumbel scale parameters are high at the Dutch coast (e.g., Figure 2b, Delfzijl tide gauge), due to a large tidal range and a large variability in storm surge heights. This is the result of the geometry of the semi-enclosed and relatively shallow North Sea basin, resulting in the development of larger storm surges under north western storm wind conditions, while the storm surge levels are considerably lower for other wind directions (e.g., [28]). This in contrast to the Barcelona tide gauge (Figure 2a), which has a much smaller tidal range since it is located in the Mediterranean and experiences less extreme storm surge heights.

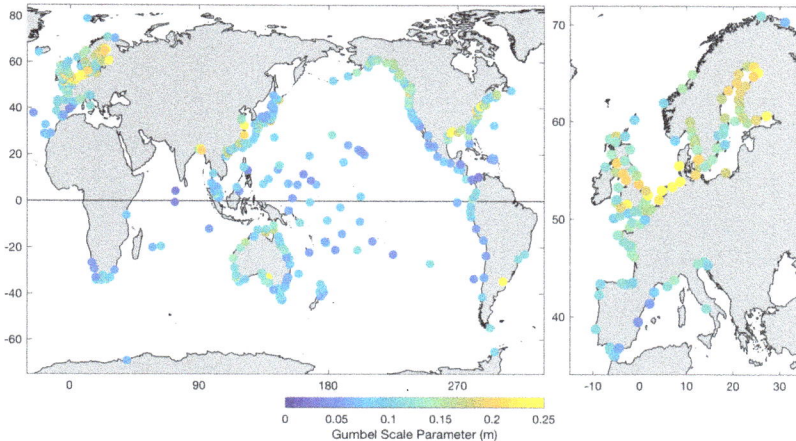

Figure 3. (**left**) Gumbel scale parameters (m) at 652 tide gauge stations and (**right**) a zoom on the European region.

2.3. Allowances Methodology

We assume a projected sea-level rise by an amount z', where z' has an uncertainty distribution function $P(z')$. If a coastal structure is raised by an allowance a to preserve the frequency of extreme (flooding) events in the future, a is given by

$$a = \lambda \ln \left(\int_{-\infty}^{\infty} P(z') \exp \left(\frac{z'}{\lambda} \right) dz' \right) \tag{1}$$

where λ is the scale parameter of the Gumbel distribution (Section 2.2), which describes the sea level extreme events and which is assumed to remain unchanged under sea-level rise. For the full derivation of the allowance, the reader is referred to Hunter [9] and Hunter et al. [10]. Note that, in these earlier papers, z' was defined relative to a projected mean sea level, whereas here we use z' as the total projected sea level to simplify the equations.

For a normal distribution, the allowance can be reduced to a simple analytical form:

$$a_{normal} = \mu + \frac{\sigma^2}{2\lambda} \tag{2}$$

(Equation *viii* in Supplementary Material of Hunter [9]), where μ is the mean SLR and σ the standard deviation of SLR when $P(z')$ follows a normal distribution.

However, Equation (2) is only valid for simple normal distributions of SLR. As we want to test different types of (non-normal) uncertainty distributions, the approach needs to be modified. We first redefine the skewed distributions of the projected SLR (from Section 2.1) in such a way that they can be used in the allowance computation. This is done by using a nonlinear fitting procedure (Powell's method, Press et al. [29]), which fits up to four normal distributions (see Figure 4 for an example) to optimally represent the (skewed) sea-level distribution. Then, the equation for the allowances only needs to be modified such that the allowance represents this group of normal distributions rather than just one normal distribution.

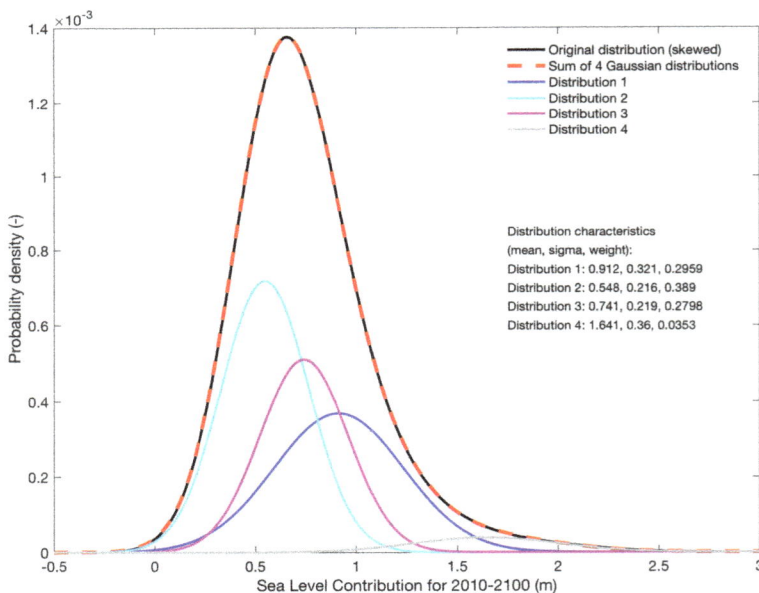

Figure 4. Example of a skewed sea-level uncertainty distribution (black), broken down into a set of four normal distributions (blue), which together describe the original distribution best (red dash).

We start from the description of a single normal distribution with a zero mean and standard deviation σ, where $P(z')$ is expressed as:

$$P(z') = \frac{1}{\sigma\sqrt{2\pi}} \exp\left(-\frac{(z')^2}{2\sigma^2}\right) \tag{3}$$

(Eq. *ii* from Supplementary Material of Hunter [9]). Then, the distribution $P(z')$ is modified such that it represents a group of normal distributions rather than one normal distribution:

$$P(z') = \frac{1}{\sqrt{2\pi}} \sum_{i=1}^{n} \frac{w_i}{\sigma_i} \exp\left(-\frac{(z' - \mu_i)^2}{2\sigma_i^2}\right) \tag{4}$$

where each distribution i in the group of normals has a weight w_i, a standard deviation σ_i, and is centered on μ_i, such that $P(z')$ is described by a group of n normal distributions. The sum of all weights w equals 1, where individual weights may be negative. Now, Equation (4) is inserted in Equation (1):

$$\int_{-\infty}^{\infty} P(z') \exp\left(\frac{z'}{\lambda}\right) dz' = \sum_{i=1}^{n} w_i \exp\left(\frac{\mu_i}{\lambda} + \frac{\sigma_i^2}{2\lambda^2}\right) \tag{5}$$

This results in the following definition of the allowance for a group of normal distributions:

$$a_{skewed} = \lambda \ln\left(\sum_{i=1}^{n} w_i \exp\left(\frac{\mu_i}{\lambda} + \frac{\sigma_i^2}{2\lambda^2}\right)\right) \tag{6}$$

This equation will be used to compute allowances for non-normal distributions in the remainder of the paper. For one normal distribution ($n = 1$) with weight 1, Equation (6) reduces again to Equation (2).

3. Results

3.1. Sea-Level Projections for 2010–2100

Sea-level projections are computed for each grid point of the RCP8.5 projections of Slangen et al. [12] in conjunction with the regional patterns of the three scenarios for ice sheet dynamics (Section 2.1), which leads to three sea-level projection scenarios (Figure 5). In the three scenarios, the median values (i.e., the point where the cumulative density function reaches 0.5) of the uncertainty distributions of WAIS, EAIS and GRIS are set to match the IPCC median values (Figure 1, dashed lines) in order to study only the effect of the change in the shape of the uncertainty distributions, rather than the shifts in magnitude. This is a different approach compared to the work presented in De Winter et al. [18], who did not shift the medians, as the purpose of that study was to look at the sensitivity of the high-end percentiles of regional sea-level projections to different ice sheet scenarios, considering both the shape and the magnitude of the distributions.

As a result of the differences in the shape of the ice sheet dynamics distributions, the final median SLR is not exactly the same for the three scenarios, even though the medians of the individual contributions are the same. Therefore, the VW15 scenario has a higher median (over the global ocean area) of 0.81 m compared to 0.71 m for IPCC and 0.73 m for R15 (Figure 5, left column). The maps show low or negative SLR around the glaciers and ice sheets. As a result of projected changes in the ocean heat content and circulation, there is a dipolar band associated with a shift in the Antarctic Circumpolar Current stretching south of South Africa and Australia [30] and generally higher values in a wide band around the equator in the Indian and Pacific Oceans.

The 95th percentile figures of the sea-level projections (Figure 5, right column) show significantly larger values for the VW15 scenario compared to the other two scenarios, reaching projected SLR over 1.75 m. The global mean 95th percentile of the VW15 scenario is 1.56 m, which is higher than the other two scenarios (1.16 m IPCC and 1.18 m R15). This is a result of the skewness of the VW15 dynamical ice sheet uncertainty distribution, which has a heavier tail that leads to much higher values for the 95th percentile of all three dynamical ice sheet contributions in the VW15 scenario (Figure 1). Although the R15 scenario has a higher 95th percentile contribution for the West-Antarctic ice dynamics (+0.12 m compared to IPCC) and a slightly longer tail to higher values, the shape of the distribution is less skewed than that of the VW15 scenario. The R15 East Antarctic 95th percentile contribution is smaller (−0.04 m compared to IPCC) due to the distribution being much sharper than the IPCC distribution. Combined, the 95th percentile total SLR of R15 is only marginally larger than that of IPCC (+0.02 m).

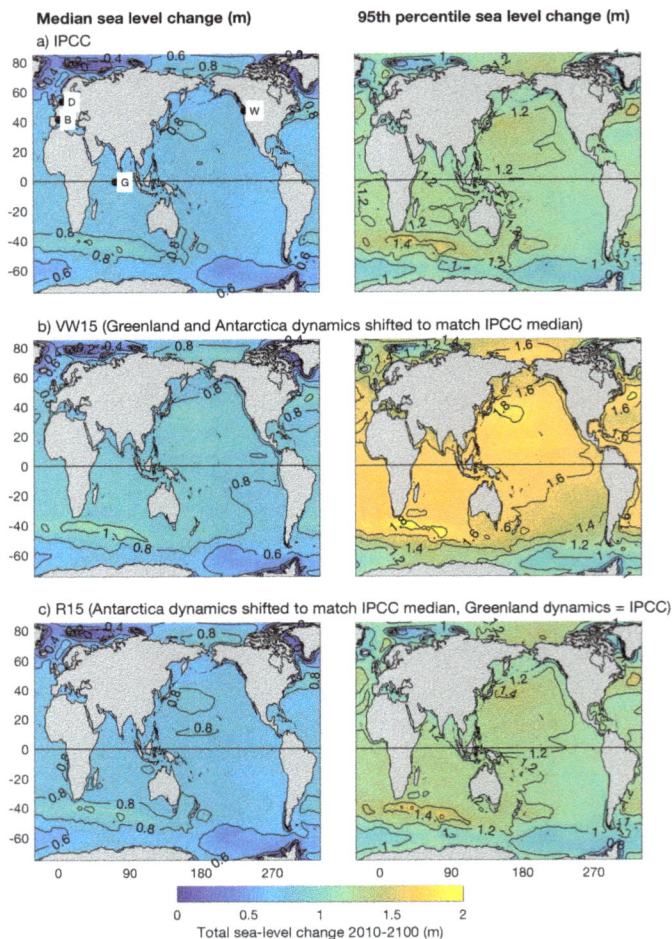

Figure 5. Projected cumulative sea-level change (m) for 2010–2100; median (left column) and 95th percentile (right column). All projections have the same median for the projected individual dynamical ice sheet contributions to SLR but a different shape of the uncertainty distribution. (**a**) IPCC; (**b**) VW15; (**c**) R15. The locations of the tide gauges in Figure 2 are indicated in (**a**) by their first letter.

3.2. Allowances for Different Uncertainty Distributions

From the sea-level projections and the Gumbel scale parameters, we compute sea-level allowances for the year 2100 (with respect to 2010) as outlined in Section 2.3. The allowances can only be computed at tide gauge locations, where Gumbel scale parameters are available. The projected SLR for each location is taken from the regional sea-level maps using a nearest neighbour approach. In some regions with multiple tide gauge stations, this means that the same sea-level projection grid point will be used for different stations.

Similar to the 95th percentile sea-level projections (Figure 5), the allowances show a distinct effect from the skewed distribution of the VW15 scenario, and less so from the more mildly skewed R15 distribution (Figure 6, left column). On average, the VW15 allowances are more than twice as large as the allowances in the other two scenarios (2.19 m vs. 1.01 (IPCC) and 1.03 (R15) m). For the normal and close to normal distributions, it appears that the allowance is slightly less than the 95th percentile SLR at the tide gauge locations (which is 1.08 m for IPCC and 1.12 m for R15). For the skewed VW15 distribution, the allowance is much larger than the 95th percentile SLR at the tide gauges (1.50 m), due to the heavy upper tails of the uncertainty distributions of the dynamical ice sheet contributions.

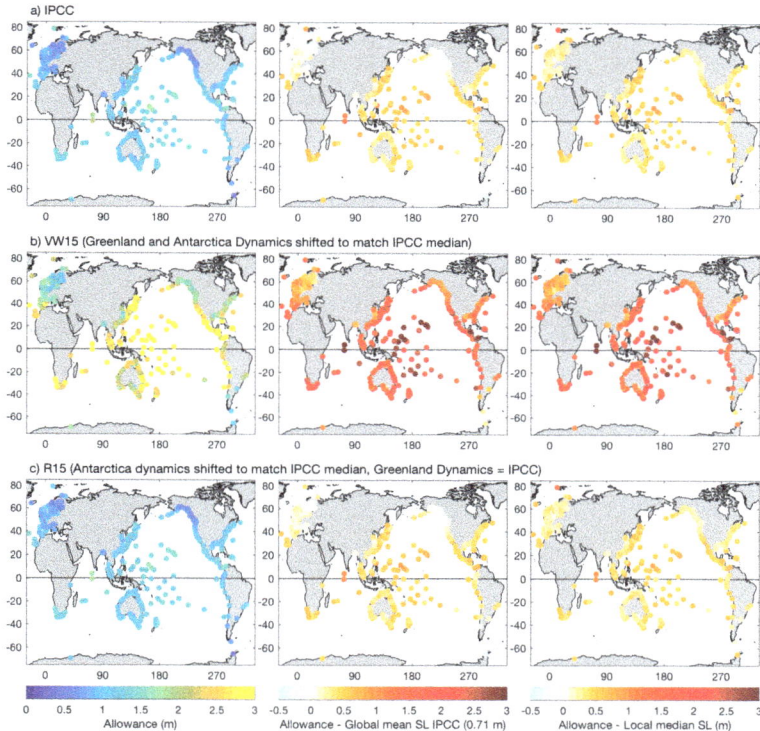

Figure 6. Allowances (m) for 2100 (left column), Allowances minus global mean SLR (middle column, m) and Allowances minus local SLR (right column, m), using 2010–2100 sea-level change projections, all projections have the same median for the projected dynamical ice sheet contributions to SLR but a different shape of the uncertainty distribution. (**a**) IPCC; (**b**) VW15; (**c**) R15.

When we compare the allowance to the global mean SLR (Figure 6, middle column) and to the local projected SLR (Figure 6, right column), this shows that an allowance is more than a simple addition of the change in the mean sea level: it accounts for the local distribution of extremes through the Gumbel scale parameter as well. In some regions, the projected local SLR is less than the global mean change in the IPCC and R15 scenarios, for instance in the Baltic Sea and the northwest of the United States. Although the allowance may be less than the *global mean* SLR in some places, it is always larger than the *local* SLR (compare middle to right columns, Figure 6). The largest differences between the projected SLR (global and regional) and the allowances are in and along the Central Pacific ocean basin, where Gumbel scale parameters (Figure 3) tend to be lower (indicating smaller variability in the extremes) and projected SLR and its uncertainty tend to be higher (due to large distance from ice mass loss sources and large steric/dynamic changes).

A focus on Europe (Figure 7) shows that the allowances in this region fall between 0.5 and 1 m for the IPCC and R15 scenarios, while the allowances are larger for the VW15 scenario, being at least 1 m and as much as 2.5 m. The allowances pattern generally is opposite to the Gumbel scale parameter (Figure 3): regions with higher Gumbel scale parameters tend to have lower allowances and vice versa. Due to Glacial Isostatic Adjustment (ongoing response of the solid Earth following ice sheet unloading after the Last Glacial Maximum [31]), the projected SLR is lower in the Baltic, leading to lower allowances in all three scenarios. Around the Mediterranean basin, projected SLR is substantial and the present sea-level variability is relatively small (small Gumbel scale parameters), which leads to the highest allowances in Europe (Figure 2a).

There is no immediately obvious relation between the Gumbel scale parameter and the projected mean SLR, apart from a tendency for locations with a smaller Gumbel scale parameter (<0.1 m) to be in locations with an average to above-average SLR projection, for all three scenarios (Figure 8). However, if the allowance is included (colours in Figure 8), a pattern emerges in the allowance that results from the combination of the projections and the scale parameters. It shows that tide gauges with smaller Gumbel scale parameters and larger projected SLR (uncertainties) require larger allowances to maintain the same likelihood of flooding (following Equation (6)).

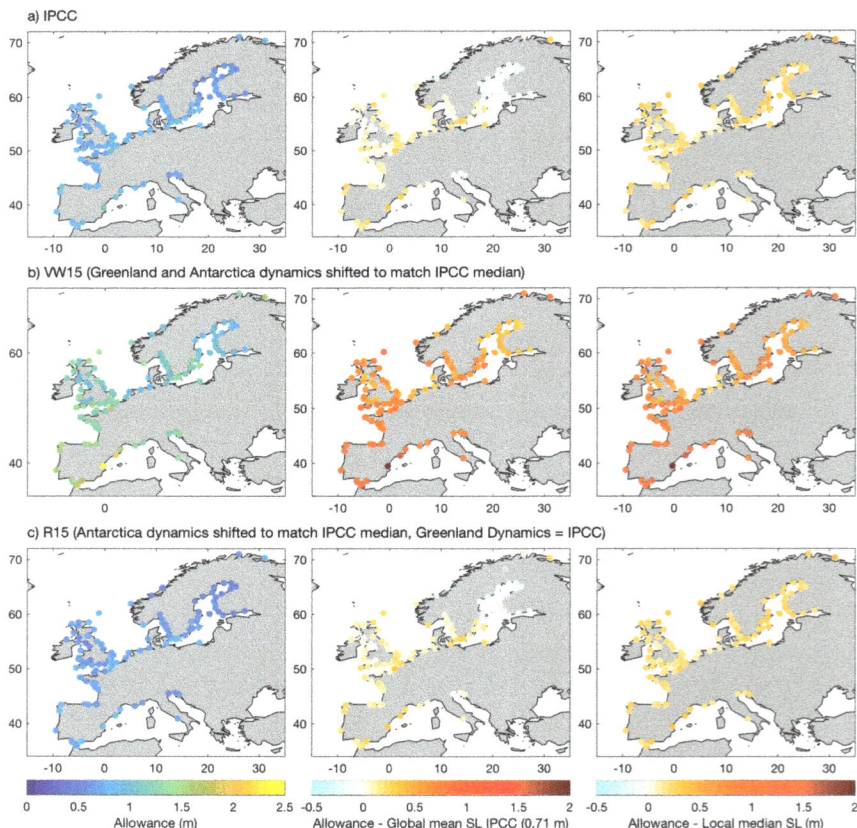

Figure 7. Same as Figure 6, but zoom on European region. Scenarios (**a**) IPCC; (**b**) VW15; (**c**) R15.

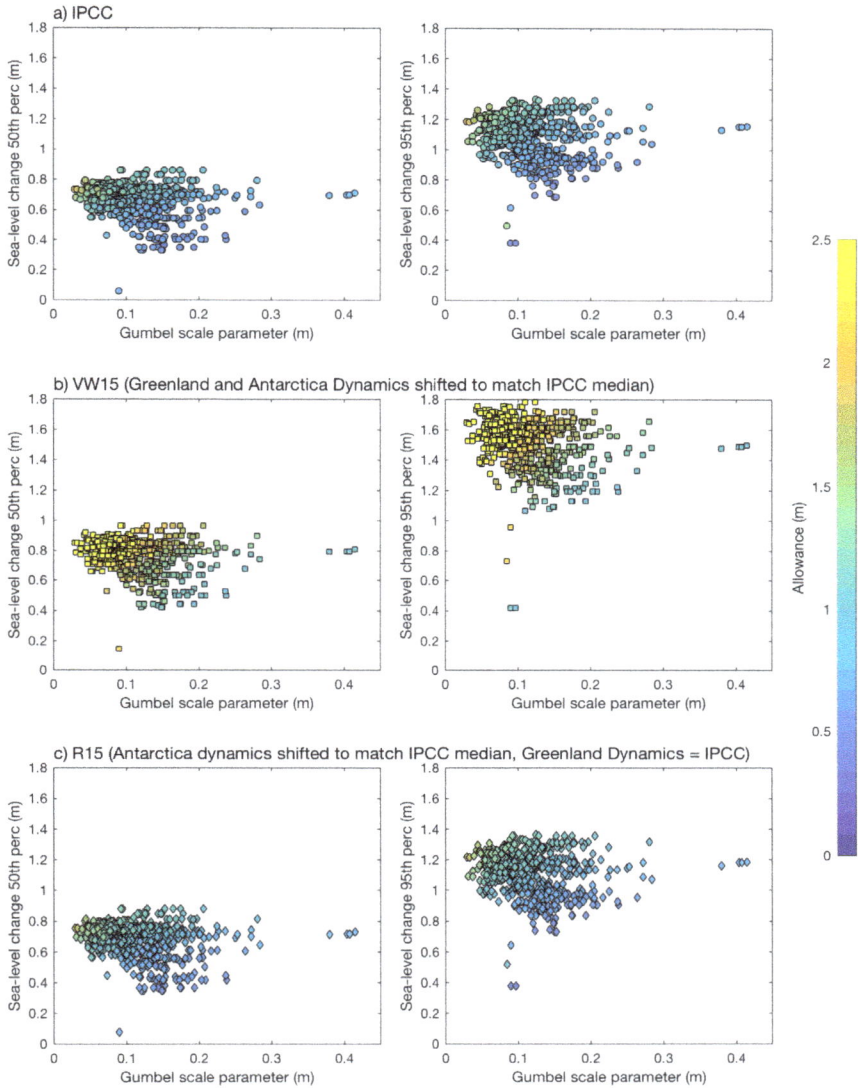

Figure 8. Scatter of 2010–2100 change in sea level (m) median (left column) and 95th percentile (right column) vs. the Gumbel scale parameter (m), colourscale indicating allowance (m), for the three scenarios (**a**) IPCC; (**b**) VW15; (**c**) R15.

3.3. Changes in the Frequency of Extreme Events

The frequency of sea-level extremes at any given location will change when mean sea level changes. Looking at Figure 2, the same amount of SLR would have a larger effect on the number of extreme events in places with small variability (both in sense of total range of variability and in the small variability in the extremes, such as Barcelona and Gan, Figure 2a,c). In these places, even 0.5 m of SLR would bring the mean sea level outside the range of current sea-level variability. In places such as Delfzijl and Willapa Bay (Figure 2b,d), which have a larger range already, a mean SLR of 0.5 m would not be out of the ordinary range of variability. However, it would mean that the extremes, which are already of considerable magnitude, would become even larger.

The sea-level allowance (Equation (6)) is designed such that a location maintains the same frequency of extremes when the infrastructure is raised by the amount of the allowance. To quantify the effect of not applying the allowance (i.e., the infrastructure is not moved), we compute the change in the frequency of extremes N following

$$N = \exp\left(a/\lambda\right) \tag{7}$$

where a is the (skewed or normal) allowance and λ the Gumbel scale parameter. If the frequency of the extremes increases by a factor 500 this means that a 1 in 10,000 year extreme event will by 2100 occur once every 20 years. For a factor 5, the 1 in 10,000 year event would become a 1 in 2000 year event. It is therefore important to consider regional sea-level projections and their uncertainties, and sea-level extreme statistics, which all differ significantly between different locations.

The changing frequencies of extreme events for the IPCC and R15 scenarios (Figure 9), computed using Equation (7), show larger increases than in (Church et al. [1] their Figure 13.25), due to the use of the RCP8.5 scenario here rather than the RCP4.5 scenario in Figure 13.25b. Another difference with respect to Church et al. [1] is that there are more stations with updated time series included in Figure 9, which can lead to small variations in the Gumbel scale parameter λ. Since the allowances for the VW15 scenario are on average twice as large as the allowances in the other two scenarios due to the skewed uncertainty distribution, the increase in frequencies is even larger due to the exponent in Equation (7) (Figure 9, Table 1). Overall, the largest increases in frequencies are projected in the Equatorial Pacific and along the Southern and Eastern coasts of the U.S., where projected SLR and uncertainties are relatively large and Gumbel scale parameters are small. If Equation (7) results in ratios of more than 10^4, this means that a 1 in 10,000 year event would occur at least once every year by 2100, and a 1 in 100 year event would occur at least 100 times per year by 2100. Factors over 10^4 therefore effectively mean that a coastal structure, designed for any normal range of return periods (i.e., 100–10,000 years), would be lost unless the allowance is applied. For IPCC and R15, we find factors over 10^4 for 50% and 52% of the tide gauge locations, respectively, while, for VW15, this increases to 87% due to the skewed ice sheet contribution (Table 1).

Table 1. Percentage (number) of tide gauge stations for every 10^n increase in the frequency of extreme events by 2100 per ice sheet scenario IPCC, VW15 and R15).

10^n	IPCC	VW15	R15
0–1	1% (6)	0% (2)	1% (6)
1–2	11% (73)	1% (9)	10% (67)
2–3	21% (139)	6% (37)	21% (136)
3–4	17% (109)	6% (39)	16% (105)
>4	50% (325)	87% (565)	52% (338)

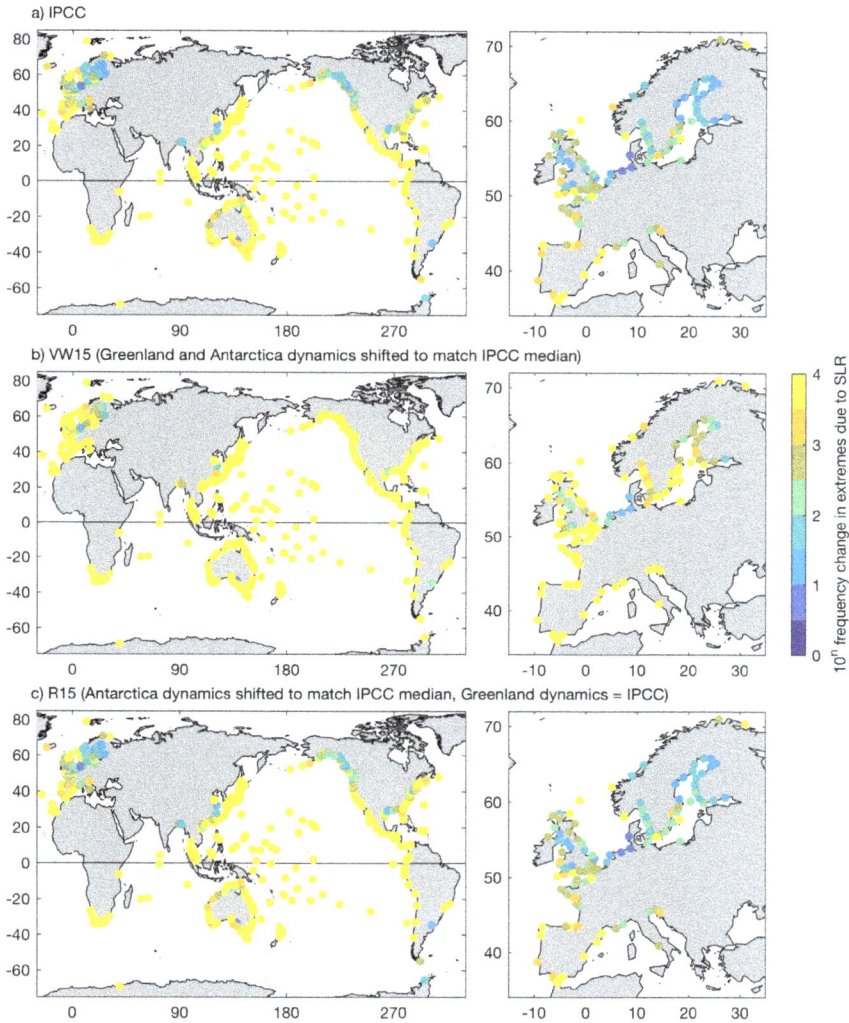

Figure 9. Ratio of the change in the frequency of extreme events (2100 vs. 2010) if allowances are not applied, based on the (**a**) IPCC (**b**) VW15 and (**c**) R15 sea-level change scenario. Note the logarithmic scale. A limit of 10^4 has been imposed, as frequency increases above this level mean that a coastal structure effectively would be lost if no allowance is applied.

4. Discussion

In computing the allowances, several choices and assumptions were made for the projected mean SLR and for the representation of extreme sea levels.

Regarding the projections, we used the RCP8.5 scenario as presented in Slangen et al. [12]. The RCP8.5 scenario is chosen because it is currently the more realistic option among the available scenarios, which is therefore relevant for coastal protection. Compared to the projections presented in [1], there are small differences (for the period 1986–2005 vs. 2080–2100, RCP8.5 scenario) in the individual contributions: glaciers (+6 cm), Greenland SMB (−1 cm), Antarctic SMB (−4 cm), Greenland

Dynamics (+1 cm), Antarctic Dynamics (+2 cm) and groundwater depletion (+4 cm). The main reason is that Slangen et al. [12] used one model or approach for each contribution to sea level, while the IPCC AR5 projections are based on an assessment of multiple estimates for each contribution. As a result, the global mean SLR under the RCP8.5 scenario is 71 cm in Slangen et al. [12], compared to 63 cm in IPCC AR5 (for 2081–2100 relative to 1986–2005). However, the methodology used for computing regional patterns is the same for the IPCC AR5 and the Slangen et al. [12] projections, and therefore any differences are mainly in the magnitude of the signal rather than the spatial distribution.

For the contributions of the ice sheet dynamics, we used three different shapes for the uncertainty distributions, from three different sources. The IPCC estimate is a consensus estimate, based on a multitude of models and publications, the VW15 estimate is based on an expert elicitation, and the R15 estimate is based on physical and statistical models. We chose to use these published estimates of the actual ice sheet dynamical contributions rather than completely artificial uncertainty distributions, which probably would have not included a distribution with the shape of the R15 distribution. A systematic sensitivity study could be part of future work. Nevertheless, the current set of distributions shows that a highly skewed distribution leads to much higher estimates of the allowances. It is therefore very important to better understand the shape of the uncertainty distribution using physical understanding and modelling, as the uncertainties are critical to estimating future SLR and to decide on the necessary coastal protection.

In order to focus on the influence of the shape of the uncertainty distributions rather than the actual magnitude, we shifted all distributions such that they matched the median values of IPCC. If we compare the differences in allowances before and after the shift of the medians, we find that for the VW15 scenario the shift in the median SLR projections is −12 cm, but the impact on the allowances (comparing allowances with and without median shift) is −8 cm. For the R15 scenarios, the medians were shifted by only −3 cm, and the impact on the allowances is −4 cm.

With regard to the use of the Gumbel scale parameter, which is based on annual maximum sea levels, it is considered to be a method that is relatively straightforward, but perhaps not the ideal representation of sea-level extremes. It is relatively robust for temporal and spatial variations and therefore widely used for global applications [32], but, on the other hand, does not use all the information that is available. Some other potential methods, such as the Generalised Extreme Value distribution or a Generalised Pareto distribution, include more information, as they not only provide a slope, but also a shape of the extreme value distribution [33]. However, these more complicated distributions cannot (yet) be used to compute allowances. In a potential follow-up of this work, it would be useful to examine the effect of different types of extreme value distributions on sea-level allowances, but here we chose to focus on the sensitivity to the shape of the sea-level projection uncertainties alone.

Generally, the longer the time period used, the better the estimate of the extreme sea-level variability [34]. Here, we use a minimum of 20 years of data in the tide gauge record as a constraint, but find that the average allowance only varies with a maximum of 0.06 m for the VW15 scenario when selecting longer time series (which results in less tide gauge records): 2.19 m for ≥20 years vs. 2.13 for records ≥50 years. For IPCC and R15, the differences are even smaller and the selection of longer time series has a negligible effect on the allowances (respectively, 1.01 and 1.03 m for ≥20 years and, respectively, 1.00 and 1.02 m for ≥50 years). One caveat for all extreme value statistics is that they depend on the occurrence of extreme events in the observational period, which becomes a more relevant restriction for longer return periods (e.g., 1 in 10,000 years) [35]. As a result, the statistics may change significantly when new extreme events are included in the analysis and it is therefore important to keep them up to date.

A final assumption regarding the statistics of the extremes is that they do not change over time, i.e., that both the magnitude with respect to the mean sea level and the frequency not change. This assumption is supported by observations and analysis of past extreme sea levels, which show that the observed changes in the frequency of flooding events is mainly caused by the increase in mean sea levels (e.g., [11]). However, the statistics of storms, and therefore of sea-level extremes, could well change significantly in the future. It is important to keep in mind that sea level can also rise due to local land subsidence, for instance due to groundwater extraction. In some locations, the subsidence even exceeds the climate-driven sea-level rise and therefore amplifies the flooding frequency [36] and changes the location of the coastline due to submergence of the coast. However, information on vertical land motion (VLM, measured with GPS) is not available all over the world and projections are difficult since they directly depend on human decisions. Therefore, VLM and changing coastlines could not yet be taken into account in the sea-level projections.

5. Conclusions

We have compared and contrasted three different scenarios for the projected dynamical ice sheet contribution to SLR. For each scenario, we combined information on sea-level extremes at more than 650 tide gauge stations with regional SLR projections in 2100 with respect to 2010, to compute so-called sea-level allowances. An allowance is a guide for the expected height a coastal structure needs to be elevated to keep the same flooding frequency. Allowances depend not only on the mean SLR, but also on the uncertainty in the mean sea-level projection and on the variability and magnitude of sea-level extremes. Each of the three sea-level projection scenarios has a different shape for the uncertainty distributions in the ice sheet dynamics contributions, leading to differences in the allowances.

We found that the allowances change significantly for our skewed scenario, doubling their values compared to allowances based on a normal distribution of the ice sheet dynamics. This is due to a heavier upper tail associated with higher contributions from ice sheet dynamics (the low-probability/high-impact events). The IPCC-based 95% SLR at the tide gauge stations is 1.08 m, with an average allowance of 1.01 m, while the skewed VW15 scenario leads to a projected 95% SLR of 1.50 m at the tide gauges and a much larger allowance of 2.19 m. For the R15 scenario, which has a less skewed but still non-normal distribution, the 95% SLR at the tide gauges is 1.12 m and the average allowance is 1.03 m. This sensitivity to the skewness of the uncertainty distribution means that it is very important to improve estimates of ice sheet dynamics contributions to SLR, and in particular their uncertainties. As we focused on the shape of the uncertainty distribution here rather than the magnitude, computations of allowances should be repeated with improved ice sheet uncertainty distributions to deliver allowances that are suitable for decision making and coastal protection.

The sea-level allowances are largest in regions with a relatively small variability in sea-level extremes and SLR projections of large magnitude and/or uncertainty. These are also the regions where frequencies in extremes will increase most if the allowances are not applied (i.e., coastal structure is not elevated), especially for the skewed probabilities in the VW15 scenario. Most of these vulnerable locations are in the tropics, where the largest regional sea-level changes (and uncertainties) are projected due to increasing thermal expansion and above-average regional contributions due to land ice mass loss [12]. This means that these regions would also be particularly vulnerable to larger contributions from the ice sheets.

Here, the computation of allowances could only be done at tide gauge locations, as this is where high-resolution observational time series are available for reasonable lengths of time. It would, however, be interesting to try and fill in the gaps for the entire coastline, using modelling approaches. Recently, Muis et al. [32] published a global set of extreme sea levels and storm surges (the GTSR data set), including Gumbel scale parameters along the global coastline, which showed good agreement between observations and the model data set. This set could possibly be used in future work to provide sea-level allowances at all coastal locations, which would be helpful for coastal planning purposes.

Acknowledgments: We gratefully acknowledge the contributors to the GESLA-2 database which are listed on www.gesla.org. A.B.A.S. and R.C.d.W. were funded by the NWO Netherlands Polar Programme, and T.J.R. was funded by the Netherlands Earth Science Center. P.L.W. was funded partly by the Leverhulme Trust.

Author Contributions: A.B.A.S. and R.S.W.v.d.W. conceived and designed the experiments; A.B.A.S. performed the experiments, analysed the data and wrote the paper; and all authors contributed to the writing of the paper. R.S.W.v.d.W., R.C.d.W. and T.J.R. designed the SEAWISE model and T.J.R. developed the code. J.R.H., P.L.W. and T.E. contributed with data and analysis tools.

Conflicts of Interest: The authors declare no conflict of interest.

Abbreviations

The following abbreviations are used in this manuscript:

AR5	Fifth Assessment Report
CMIP5	5th phase of the Climate Model Intercomparison Project
EAIS	East Antarctic Ice Sheet
GESLA-2	Global Extreme Sea Level Analysis Version 2
GRIS	Greenland Ice Sheet
IPCC	Intergovernmental Panel on Climate Change
RCP	Representative Concentration Pathway
SLR	Sea-Level Rise
WAIS	West Antarctic Ice Sheet

Appendix A. SEAWISE Methodology

SEAWISE is used to combine different uncertainty distributions, e.g., Greenland (P_1) and West-Antarctica (P_2), into one uncertainty distribution (e.g., $P_{combined1}$). The composed distribution $P_{combined1}(z)$ (where z is the projected SLR) exists for each z of all contributions of two independent distributions $P_1(z_1)$ and $P_2(z_2)$ for which the summed values z_1 and z_2 add to z. $P_{combined1}(z)$ is the resulting combined uncertainty distribution of $P_1(z_1)$ and $P_2(z_2)$:

$$P_{combined1}(z) = \sum_{m=c}^{d} P_1(z - m\Delta z)P_2(m\Delta z) \tag{A1}$$

All data within 6σ-boundaries around the mode of each distribution is selected, which usually covers over 99.9% of the data. The distribution is normalized based on this selected part. In Equation (A1), the interval counter m runs in steps of Δz between c and d corresponding with z_c and z_d, respectively the left and right 6σ-boundaries of P_2, while taking $z_1 = z - m\Delta z$ and $z_2 = m\Delta z$. To obtain the entire distribution $P_{combined1}(z)$, Equation (A1) is calculated for all z between z_a at the left 6σ-boundary of P_1 and $z_b + z_d$, where z_b is the right 6σ-boundary of P_1 and z_d the right 6σ-boundary of P_2. In this way, the uncertainty distribution for each SLR contribution can be added, until all uncertainty distributions are combined (Figure A1). In addition to providing the combined uncertainty distributions for single points, SEAWISE can produce global maps of the SLR mode, SLR median and a number of SLR nth percentiles or SLR nth σ-intervals [18].

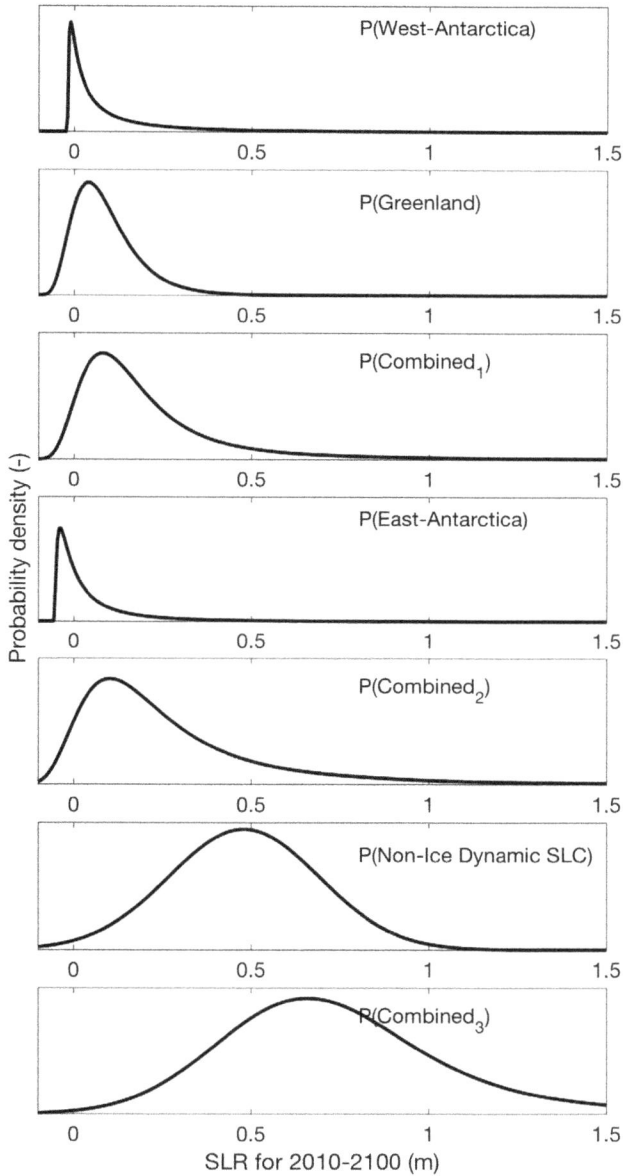

Figure A1. An example showing the SEAWISE methodology which step-wise combines different uncertainty distributions into a final uncertainty distribution $P_{combined3}$. All y-axes are probability density (dimensionless). This example is for a grid point off the Canadian coast (290W-67N), using the VW15 scenario for the ice sheet dynamics contributions. P(Non-Ice Dynamic SLR) contains the contributions from glaciers, ocean density variations and ocean dynamics, ice sheet surface mass balance, groundwater extraction and glacial isostatic adjustment [12].

References

1. Church, J.; Clark, P.; Cazenave, A.; Gregory, J.; Jevrejeva, S.; Levermann, A.; Merrifield, M.; Milne, G.; Nerem, R.; Nunn, P.; et al. Sea Level Change. In *Climate Change 2013: The Physical Science Basis. Contribution of Working Group I to the Fifth Assessment Report of the Intergovernmental Panel on Climate Change*; Stocker, T., Qin, D., Plattner, G.K., Tignor, M., Allen, S., Boschung, J., Nauels, A., Xia, Y., Bex, V., Midgley, P., Eds.; Cambridge University Press: Cambridge, UK; New York, NY, USA, 2013.
2. Clark, P.U.; Church, J.A.; Gregory, J.M.; Payne, A.J. Recent Progress in Understanding and Projecting Regional and Global Mean Sea Level Change. *Curr. Clim. Chang. Rep.* **2015**, *1*, 224–246.
3. Slangen, A.B.A.; Adloff, F.; Jevrejeva, S.; Leclercq, P.W.; Marzeion, B.; Wada, Y.; Winkelmann, R. A Review of Recent Updates of Sea-Level Projections at Global and Regional Scales. *Surv. Geophys.* **2016**, *38*, 385–406.
4. Bamber, J.L.; Aspinall, W.P. An expert judgement assessment of future sea level rise from the ice sheets. *Nat. Clim. Chang.* **2013**, *3*, 424–427.
5. De Vries, H.; van de Wal, R.S.W. How to interpret expert judgment assessments of 21st century sea-level rise. *Clim. Chang.* **2015**, *130*, 87–100.
6. Favier, L.; Durand, G.; Cornford, S.L.; Gudmundsson, G.H.; Gagliardini, O.; Gillet-Chaulet, F.; Zwinger, T.; Payne, A.J.; le Brocq, A.M. Retreat of Pine Island Glacier controlled by marine ice-sheet instability. *Nat. Clim. Chang.* **2014**, *4*, 117–121.
7. Joughin, I.; Smith, B.E.; Medley, B. Marine Ice Sheet Collapse Potentially Under Way for the Thwaites Glacier Basin, West Antarctica. *Science* **2014**, *344*, 735–738.
8. Ritz, C.; Edwards, T.L.; Durand, G.; Payne, A.J.; Peyaud, V.; Hindmarsh, R.C.A. Potential sea-level rise from Antarctic ice-sheet instability constrained by observations. *Nature* **2015**, *528*, 115–118.
9. Hunter, J. A simple technique for estimating an allowance for uncertain sea-level rise. *Clim. Chang.* **2012**, *113*, 239–252.
10. Hunter, J.R.; Church, J.A.; White, N.J.; Zhang, X. Towards a global regionally varying allowance for sea-level rise. *Ocean Eng.* **2013**, *71*, 17–27.
11. Menéndez, M.; Woodworth, P.L. Changes in extreme high water levels based on a quasi-global tide-gauge dataset. *J. Geophys. Res.* **2010**, *115*, C10011.
12. Slangen, A.B.A.; Carson, M.; Katsman, C.; van de Wal, R.; Koehl, A.; Vermeersen, L.; Stammer, D. Projecting twenty-first century regional sea-level changes. *Clim. Chang.* **2014**, *124*, 317–332.
13. Moss, R.H.; Edmonds, J.A.; Hibbard, K.A.; Manning, M.R.; Rose, S.K.; van Vuuren, D.P.; Carter, T.R.; Emori, S.; Kainuma, M.; Kram, T.; et al. The next generation of scenarios for climate change research and assessment. *Nature* **2010**, *463*, 747–756.
14. Taylor, K.; Stouffer, R.J.; Meehl, G.A. An overview of CMIP5 and the experiment design. *Bull. Am. Meteorol. Soc.* **2012**, *93*, 485–498.
15. Kopp, R.E.; Horton, R.M.; Little, C.M.; Mitrovica, J.X.; Oppenheimer, M.; Rasmussen, D.J.; Strauss, B.H.; Tebaldi, C. Probabilistic 21st and 22nd century sea-level projections at a global network of tide-gauge sites. *Earth Future* **2014**, *2*, 383–406.
16. Jackson, L.P.; Jevrejeva, S. A probabilistic approach to 21st century regional sea-level projections using RCP and High-end scenarios. *Glob. Planet. Chang.* **2016**, *146*, 179–189.
17. Jevrejeva, S.; Jackson, L.P.; Riva, R.E.M.; Grinsted, A.; Moore, J.C. Coastal sea level rise with warming above 2 °C. *Proc. Natl. Acad. Sci. USA* **2016**, *113*, 13342–13347.
18. De Winter, R.; Reerink, T.J.; de Vries, H.; Slangen, A.B.A.; van de Wal, R.S.W. Impact of asymmetric uncertainties in ice sheet dynamics on regional sea level projections. *Nat. Hazards Earth Syst. Sci. Discuss.* **2017**, In Review.
19. Buchanan, M.K.; Kopp, R.E.; Oppenheimer, M.; Tebaldi, C. Allowances for evolving coastal flood risk under uncertain local sea-level rise. *Clim. Chang.* **2016**, *137*, 347–362.
20. Coles, S. *An Introduction to Statistical Modeling of Extreme Values*; Springer: London, UK, 2001.
21. Arns, A.; Wahl, T.; Haigh, I.D.; Jensen, J.; Pattiaratchi, C. Estimating extreme water level probabilities: A comparison of the direct methods and recommendations for best practise. *Coast. Eng.* **2013**, *81*, 51–66.
22. GESLA. Available online: www.gesla.org (accessed on 23 May 2017).
23. Woodworth, P.; Hunter, J.; Marcos, M.; Caldwell, P.; Menendez, M.; Haigh, I. Towards a global higher-frequency sea level data set. *Geosci. Data J.* **2016**, *3*, 50–59.

24. Shepherd, A.; Ivins, E.R.; Geruo, A.; Barletta, V.R.; Bentley, M.J.; Bettadpur, S.; Briggs, K.H.; Bromwich, D.H.; Forsberg, R.; Galin, N.; et al. A Reconciled Estimate of Ice-Sheet Mass Balance. *Science* **2012**, *338*, 1183–1189.

25. Mitrovica, J.X.; Tamisiea, M.E.; Davis, J.L.; Milne, G.A. Recent mass balance of polar ice sheets inferred from patterns of global sea-level change. *Nature* **2001**, *409*, 1026–1029.

26. Hunter, J.R.; Woodworth, P.L.; Wahl, T.; Nicholls, R.J. Using Global Tide Gauge Data to Validate and Improve the Representation of Extreme Sea Levels in Flood Impact Studies. *Glob. Planet. Chang.* **2017**, Under Review.

27. Merrifield, M.A.; Genz, A.S.; Kontoes, C.P.; Marra, J.J. Annual maximum water levels from tide gauges: Contributing factors and geographic patterns. *J. Geophys. Res. Oceans* **2013**, *118*, 2535–2546.

28. De Winter, R.C.; Sterl, A.; Ruessink, B.G. Wind extremes in the North Sea Basin under climate change: An ensemble study of 12 CMIP5 GCMs. *J. Geophys. Res. Atmos.* **2013**, *118*, 1601–1612.

29. Press, W.; Teukolsky, S.A.; Vetterling, W.T.; Flannery, B.P. *Numerical Recipes 3rd Edition: The Art of Scientific Computing*, 3rd ed.; Cambridge University Press: New York, NY, USA, 2007.

30. Slangen, A.B.A.; Church, J.A.; Zhang, X.; Monselesan, D. The sea-level response to external forcings in CMIP5 climate models. *J. Clim.* **2015**, *28*, 8521–8539.

31. Tamisiea, M.E.; Mitrovica, J.X. The moving boundaries of sea level change: Understanding the origins of geographic variability. *Oceanography* **2011**, *24*, 24–39.

32. Muis, S.; Verlaan, M.; Winsemius, H.C.; Aers, J.C.J.H.; Ward, P.J. A global reanalysis of storm surges and extreme sea levels. *Nat. Commun.* **2016**, *7*, 11969.

33. Wahl, T.; Haigh, I.; Nicholls, R.; Arns, A.; Dangendorf, S.; Hinkel, J.; Slangen, A. Understanding extreme sea levels for broad-scale coastal impact and adaptation analysis. *Nat. Commun.* **2017**, Accepted.

34. Haigh, I.D.; Nicholls, R.; Wells, N. A comparison of the main methods for estimating probabilities of extreme still water levels. *Coast. Eng.* **2010**, *57*, 838–849.

35. Dangendorf, S.; Arns, A.; Pinto, J.G.; Ludwig, P.; Jensen, J. The exceptional influence of storm 'Xaver' on design water levels in the German Bight. *Environ. Res. Lett.* **2016**, *11*, 045001.

36. Erkens, G.; Bucx, T.; Dam, R.; de Lange, G.; Lambert, J.L. Sinking coastal cities. *Proc. IAHS* **2015**, *372*, 189–198.

Journal of
*Marine Science
and Engineering*

MDPI

Article

Predicting Dynamic Coastal Delta Change in Response to Sea-Level Rise

Wietse I. Van De Lageweg [1],* and Aimée B. A. Slangen [2]

[1] Geography and Geology, School of Environmental Sciences, University of Hull, Hull 7RX, UK
[2] Royal Netherlands Institute for Sea Research (NIOZ), Department of Estuarine and Delta Systems (EDS), Utrecht University, Yerseke 4400 AC, The Netherlands; aimee.slangen@nioz.nl
* Correspondence: wietse.vandelageweg@gmail.com; Tel.: +31-113-577-300

Received: 9 May 2017; Accepted: 16 June 2017; Published: 20 June 2017

Abstract: The world's largest deltas are densely populated, of significant economic importance and among the most valuable coastal ecosystems. Projected twenty-first century sea-level rise (SLR) poses a threat to these low-lying coastal environments with inhabitants, resources and ecology becoming increasingly vulnerable to flooding. Large spatial differences exist in the parameters shaping the world's deltas with respect to river discharge, tides and waves, substrate and sediment cohesion, sea-level rise, and human engineering. Here, we use a numerical flow and transport model to: (1) quantify the capability of different types of deltas to dynamically respond to SLR; and (2) evaluate the resultant coastal impact by assessing delta flooding, shoreline recession and coastal habitat changes. We show three different delta forcing experiments representative of many natural deltas: (1) river flow only; (2) river flow and waves; and (3) river flow and tides. We find that delta submergence, shoreline recession and changes in habitat are not dependent on the applied combination of river flow, waves and tides but are rather controlled by SLR. This implies that regional differences in SLR determine delta coastal impacts globally, potentially mitigated by sediment composition and ecosystem buffering. This process-based approach of modelling future deltaic change provides the first set of quantitative predictions of dynamic morphologic change for inclusion in Climate and Earth System Models while also informing local management of deltaic areas across the globe.

Keywords: sea level; delta; numerical modelling; tides; waves; coastal erosion; coastal flooding; coastal habitat; coastal management

1. Introduction

Future impacts from sea-level rise (SLR) are expected to be widespread in low-lying deltaic coastal areas [1–3]. River-dominated deltas such as the Mississippi delta have a different morphology compared to wave-influenced deltas such as the Nile delta and tide-influenced deltas such as the Ganges delta, reflecting primarily the interplay between the sediment input by rivers [4] and the marine reworking of these sediments by tides and wave action [5,6]. In addition to these global differences in river input, tides and waves, regional patterns in SLR exist [7] that can deviate substantially from the global mean [8]. A key question is whether the differences in these parameters controlling delta dynamics and evolution affect the ability of deltas to respond to SLR, and hence the impact SLR has on the different delta coasts. Such an assessment taking into account global differences in river input, tides and waves and regional patterns in SLR would allow us to develop coastal impact estimates tailored to local conditions rather than global mean values.

Coastal environments provide a range of ecosystem services for many endangered species [9]. Despite observations that coastal wetlands in the intertidal zone can persist for thousands of years [10], more recent studies indicate increased submergence of marshes due to SLR [1,11], resulting in a conversion from diverse marsh ecosystems to unvegetated subtidal habitats. Understanding and predicting such coastal response developments, their rates of change and the identification of vulnerable and resilient coastal environments is vitally important to informing future flood management and the sustainable planning of coastal habitats.

Deltaic coasts can either respond dynamically to SLR, potentially mitigating the effect of SLR, or they can be inundated [12]. Inundation assessments assume a static coastal topography which is flooded to the highest level to which the water rises (i.e., "bathtub" approach) [13]. Such assessments fail to include the dynamic response due to hydrodynamic [5], ecological [11] and morphological [14] processes such as erosion and deposition that shape deltaic coasts. The dynamic response potential for many coastal environments is quantitatively uncertain, yet crucially important to understanding how the projected SLR will be translated into coastal impact. Flooding risk and coastal erosion will increase due to the higher water levels resulting from SLR [14,15], particularly along developed coasts with fixed coastal engineering solutions [16] and low-elevation infrastructure and housing with limited capacity to respond dynamically.

Here, we conduct morphodynamic simulations using the morphodynamic model Delft3D [4,17] to systematically evaluate the dynamic response of deltas to SLR. Specifically, we aim to: (1) quantify the capability of different types of deltas (i.e., river deltas, river- and wave-influenced deltas, and river- and tide-influenced deltas) to dynamically respond to SLR; and to (2) evaluate the resultant coastal impact by assessing delta flooding, shoreline recession and coastal habitat changes. The modelling approach is generic, with boundary conditions representative for many natural deltaic systems yielding quantitative predictions for the landward translation of dynamic deltaic shorelines.

2. Materials and Methods

The numerical model Delft3D (v. 5.00.10.1983) simulates flow, sediment transport and morphologic change across a range of spatial and temporal scales in coastal and fluvial environments [4,17,18]. We conduct morphodynamic simulations by varying the type of deltaic coastal forcing (three types) and the rate of SLR (four scenarios) while holding all other factors constant. These twelve simulations of delta morphodynamics are conducted with a river discharge of 1000 m^3/s, carrying equilibrium sandy sediment concentrations into an ocean basin [4] (Table 1).

The ocean basin is forced linearly with twenty-first century projected sea-level rise scenarios from the Intergovernmental Panel on Climate Change Fifth Assessment Report, [3] (IPCC AR5): 0 cm (control run), 26 cm (lower bound of the Representative Concentration Pathway (RCP)2.6 scenario [19]), 47 cm (mean of the RCP4.5 scenario) and 82 cm (upper bound of the RCP8.5 scenario). By using a wide range of values we implicitly study the effect of spatially-variable SLR, with higher SLR projected in equatorial regions and lower SLR towards the poles [7,8].

Table 1. Model parameters and descriptive statistics of delta coasts for modeling experiments used in this study. All scenarios use a constant river discharge of 1000 m^3/s. Observations ("obs") of shoreline retreat are computed from 19 shore profiles across the modelled deltas, predictions ("pred") are based on the Bruun (1988) [20] rule applied to the same 19 shore profiles for each modelled delta. Comparisons show a smaller ability to dynamically adjust when using a different sediment transport formulation (R_82_EH), but a greater ability when using finer sediment (R_82_FI). SLR: sea-level rise. TR: tidal range. H$_{sig}$: significant wave height. St. Dev.: standard deviation. R_82_EH: sensitivity analysis in which the Engelund-Hansen sediment transport formulation is used rather than the Van Rijn formulation. R_82_FI: sensitivity analysis in which a finer sediment of 100 μm was used rather than the 225 μm sand in the main runs.

Run ID	Delta Forcing			Delta Submerged Area (d) = Dynamic Response, (s) = Static Response				Average ± St. Dev. Shoreline Retreat	
	SLR (cm)	TR (m)	H$_{sig}$ (m)	% <0 m MSL	% <+0.26 m MSL	% <+0.47 m MSL	% <+0.82 m MSL	Obs. (m)	Pred. (m)
River Only									
R_0	0	0	0	48 (d)	71 (s)	84 (s)	96 (s)	-	-
R_26	26	0	0	-	55 (d)	73 (s)	92 (s)	314 ± 454	135 ± 31
R_47	47	0	0	-	-	64 (d)	90 (s)	445 ± 262	243 ± 56
R_82	82	0	0	-	-	-	70 (d)	666 ± 320	425 ± 97
River and Tide									
RT_0	0	1.5	0	59 (d)	74 (s)	85 (s)	98 (s)	-	-
RT_26	26	1.5	0	-	67 (d)	78 (s)	93 (s)	234 ± 479	153 ± 45
RT_47	47	1.5	0	-	-	74 (d)	89 9s)	491 ± 417	276 ± 81
RT_82	82	1.5	0	-	-	-	81 (d)	843 ± 375	481 ± 140
River and Wave									
RW_0	0	0	0.5	59 (d)	79 (s)	89 (s)	98 (s)	-	-
RW_26	26	0	0.5	-	68 (d)	82 (s)	97 (s)	213 ± 254	160 ± 30
RW_47	47	0	0.5	-	-	72 (d)	94 (s)	371 ± 349	290 ± 54
RW_82	82	0	0.5	-	-	-	83 (d)	821 ± 280	505 ± 96
Sensitivity tests									
R_82_EH	82	0	0	-	-	-	76 (d)	-	-
R_82_FI	82	0	0	-	-	-	67 (d)	-	-

To explore the coastal impact for a range of deltaic environments, the SLR projections are combined with three types of deltaic coastal forcing: (1) a river-only experiment; (2) an experiment where tides with an amplitude of 0.75 m are included; and (3) an experiment where waves with a significant height of 0.5 m are applied.

The control run consists of a river continually discharging 1000 m^3/s of water with equilibrium sediment concentrations into a standing body of water, thus forming a river delta. In the tidally-influenced runs, micro-tidal forcing with a tidal range of 1.5 m is applied at the ocean boundary, and the delta is exposed to simplified tides with a sinusoidal shape. To evaluate wave-influenced deltas, the morphologic module of Delft3D is coupled to a standalone wave module (SWAN, Simulating Waves Nearshore) [21]. The wave module simulates waves nearshore with the significant wave height set to 0.5 m, a set peak period of 5 s and otherwise default parameters.

We use an ocean basin of 300 by 225 grid cells, every 25 m^2, which is identical to the model setup in Edmonds & Slingerland (2010) [4]. In agreement with their setup, initial basin depths range from 1 m to 3.5 m and white noise with a mean of 0 m and a standard deviation of 0.05 m is added to the initial bathymetry to mimic natural variations. Bed roughness is assumed constant in space and time and set to a Chezy value of 45 m$^{0.5}$/s. A uniform sand-sized sediment of 225 μm, representative of many deltaic coastal environments, is used and the Van Rijn [22,23] parameterization is selected to calculate sediment transport. A sensitivity analysis evaluating the effects of the applied sediment transport formulation (i.e., Van Rijn and the Engelund–Hansen [24] formulations) and sediment size is included and will be discussed in Section 3.2. In all experiments an initial 2.5-m-thick layer of sediment is available for erosion of the bed.

The coastal deltas are forced linearly with IPCC projected global mean sea-level rise of 0 cm (i.e., no SLR), 26 cm, 47 cm and 82 cm over a century [3], thus simulating delta dynamics as a function of projected sea level from today until approximately 100 years ahead. A modelling time step of six seconds is selected to satisfy the Courant criterion. As successfully applied in other model studies [18], one model day is assumed to represent the integrated morphologic effects of a year of river flow, a year of tidal action, or a year of wave action in nature. Therefore, 100 days of delta morphological evolution are simulated while applying a morphologic factor of 25 to scale up the rate of morphologic change and the delta volume and land area [25].

We quantify the dynamic delta response and resultant coastal impact by evaluating delta flooding, shoreline recession, and bed level changes, resulting in coastal habitat alteration for the three delta types forced by the four SLR scenarios. Therefore, delta statistics are calculated to quantitatively summarize and compare the different scenarios. Characterization of the delta morphology and shoreline is done using mathematical morphological operators [26] using default MATLAB commands. Delta land area is calculated by selecting all cells above mean sea level multiplied by the cell area. The number of bifurcations in the delta is determined by applying the MATLAB bwmorph morphologic process "thin" on channels actively transporting sediment to end up with a skeleton from which the number of nodes is observed. The number of channels at the shoreline ("endpoints") is calculated following a similar morphologic procedure selecting the channel skeleton endpoints. Shoreline sinuosity is calculated as a measure of shoreline rugosity with a higher sinuosity corresponding to a more rugose shoreline.

3. Results

3.1. Delta Morphodynamics

In each simulation, a self-formed delta with a distributary channel network is generated (Figure 1). Formation of the modelled deltas proceeds from the same processes as observed in the field [27] and physical [28] and numerical [4] modelling of low-cohesion deltas. Initially, a turbulent plume aggrades and develops subaqueous levees and a subaqueous bar seaward of the river. The subaqueous bar then stagnates, causing an unstable channel bifurcation which generally results in the closure of one arm. Continued progradation and aggradation eventually leads to subaerial bifurcations, where water and sediment are distributed across large parts of the delta topset by frequent crevasse splays breaking through the thin and easily erodible levees (see Supplementary Movies detailing the morphodynamic evolution for all twelve runs).

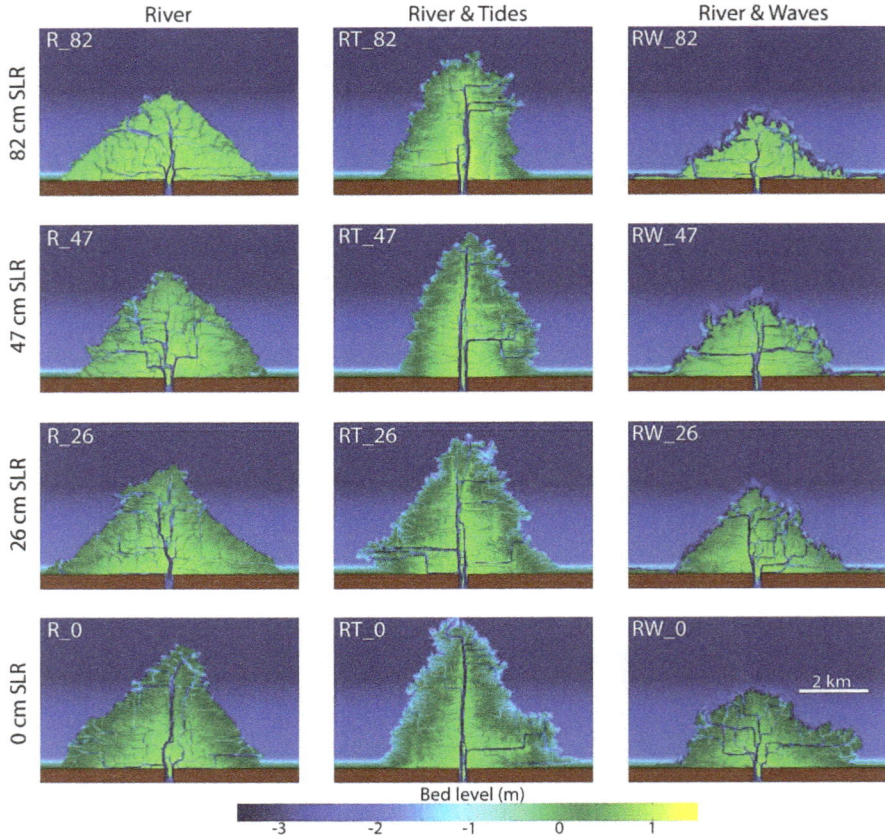

Figure 1. Deltaic bed levels (m) after 100 model-days, approximately corresponding to 100 years of natural evolution. Delta coast morphology as a function of coastal forcing (**left**: river discharge, **middle**: tides, **right**: waves) and sea-level scenario (from 82 cm at the top row to 0 cm at the bottom row). Animated videos of all scenarios showing delta dynamics over time are available as Supplementary Files.

The modelled deltas are compared after allowing sediment transport, deposition and hydrodynamic reworking for a century, which is the time scale of the SLR projections. Differences in the final sediment volume between the river-dominated, tidally-influenced and wave-influenced deltas, even in the absence of SLR (Figure 1, bottom row), are explained by the higher sediment-carrying competence of the river for higher seaward flow velocities in the river-only delta and during the ebb stages in the tidally-influenced delta.

The modelled deltas share many similarities with observations of natural deltas such as: (1) a wetted delta area of approximately 50% in the absence of SLR (Table 1), comparable to wet/dry ratios measured in the field [29]; (2) an increase in shoreline rugosity for tidally-influenced deltas and a decrease in shoreline rugosity for wave-influenced deltas, in agreement with natural deltas [5,26] (Table 2); and (3) the number of mouth bar channels at the shoreline decreases as a function of SLR (Table 2), consistent with observations from Holocene deltas [30].

Table 2. Auxiliary descriptive statistics of delta coasts for modeling experiments used in this study.

Run ID	Subaerial Area (km²)	Number of Bifurcations	Number of Channel Endpoints	Shoreline Sinuosity (-)	Minimum Elevation (m)	Median Elevation (m)	Maximum Elevation (m)
R_0	14.6	48	55	1.91	−5.92	0.02	1.25
R_26	13.5	57	39	1.54	−5.26	0.22	1.54
R_47	12.7	65	36	1.77	−5.53	0.35	1.36
R_82	11.1	33	36	1.55	−5.05	0.66	1.51
RT_0	17.2	37	50	1.97	−6.02	−0.17	1.25
RT_26	16.0	55	53	1.96	−5.15	−0.06	1.35
RT_47	14.4	28	35	1.96	−5.76	0.05	1.41
RT_82	13.0	36	32	1.84	−5.44	0.26	1.66
RW_0	12.5	69	21	1.76	−5.36	−0.08	1.29
RW_26	11.7	56	32	1.68	−5.50	0.01	1.35
RW_47	11.9	37	26	1.60	−5.45	0.01	1.31
RW_82	9.6	45	28	1.60	−5.30	−0.82	1.46
R_82_EH	-	-	-	-	−5.05	−0.03	2.40
R_82_FI	-	-	-	-	−5.26	0.71	1.54

3.2. Dynamic Delta Reponse to SLR

A key outcome of the numerical simulations is that the dynamic response to SLR is similar for the different types of modelled deltas, independent of the delta forcing (Table 1). For the highest SLR scenario of 82 cm, the modelled deltas show a land loss of 22% for river-dominated, tidally-influenced and wave-influenced deltas. Such dramatic flooding and loss of delta area for 82 cm of SLR highlight the low gradients and the lack of substantial topography on the modelled deltas, which is in agreement with many natural deltas [31]. For the lower SLR scenarios of 26 cm and 47 cm and allowing for a dynamic coastal response, the loss of delta area is smaller and amounts to 8% and 15%, respectively (Table 1).

The implication of these findings is that the regional patterns in SLR will largely determine the coastal impact in deltaic environments, irrespective of whether the delta is shaped primarily by river input, tides or waves. This means that the projected spatial patterns in SLR [7,8] can be used as a proxy for delta coastal impact globally. For example, the Amazon delta in Brazil is located in a region with a higher-than-global-mean SLR and coastal impacts are therefore expected to be more severe, while the Ganges delta in India and Bangladesh is located in a region with a lower-than-global-mean SLR and coastal impacts of climate-change driven SLR are therefore expected to be less severe. It is, however, important to note that the aforementioned coastal impacts are only related to SLR and not to land subsidence [32,33], which is substantial in the Ganges delta region and will be discussed in more detail in Section 4. Despite being shaped by different degrees of fluvial input and marine reworking, our results indicate a similar ability to dynamically respond to SLR for the aforementioned deltas thus rendering them unable to mitigate the regional differences in SLR.

Simulations performed with a finer sediment of 100 μm indicate a higher capacity of deltaic coasts to dynamically respond to SLR (Table 1, RH_82_FI). The suspended sediment fraction increases for finer-grained sediment allowing for more efficient delta aggradation to dynamically respond to SLR. This, in turn, indicates an increased resilience to coastal flooding due to SLR for finer-grained suspension-dominated delta systems compared to coarser-grained bedload-dominated deltas. Additional sensitivity tests in which we changed the sediment transport calculation from Van Rijn [22,23] to the Engelund–Hansen formulation [24] show changes in the depth of channels and the elevation of bars but the overall delta morphology and the ability to dynamically respond to SLR are similar (Table 1, RH_82_EH).

By allowing dynamic adjustment, delta coasts maintain an additional 11%, 19% and 20% of surface area emerged compared to static coasts with SLRs of 26 cm, 47 cm and 82 cm, respectively, due to morphological processes and adaptation (Table 1). The difference between a SLR of 47 cm and a SLR of 82 cm is small and indicative of the limits of delta coasts to dynamically adjust by morphological processes to these higher SLR rates. The over-prediction of land submergence by static inundation models is consistent with natural coastal systems [12]. These findings highlight the inability of static model approaches to correctly predict the flooding impact of future deltas.

In addition to the flooding of land due to SLR, the modelled deltas show a shift in the bed level distribution, affecting coastal habitat distribution (Figure 2). Particularly relevant are marshes situated on coastal platforms in the intertidal range for which minor changes in topography and hydrodynamics can lead to vegetation disturbance and thus cause rapid marsh degradation [34]. The general deepening of the delta relative to mean sea level and the gentler slope of the coastal platform in the modelled deltas for higher SLR along with a change in flow velocity range (Figure 3) may well have significant implications for coastal marsh survival. Our simulations indicate an increase of 17% (Figure 3B) and 23% (Figure 3C) of the coastal platform submerging to subtidal levels and below the wave base, respectively, representing an important threshold below which the coastal platform cannot sustain vegetation growth [11,35]. Essentially, SLR causes parts of the intertidal area to be replaced by subtidal area, limiting the diversity of coastal habitats and suitability for many coastal species.

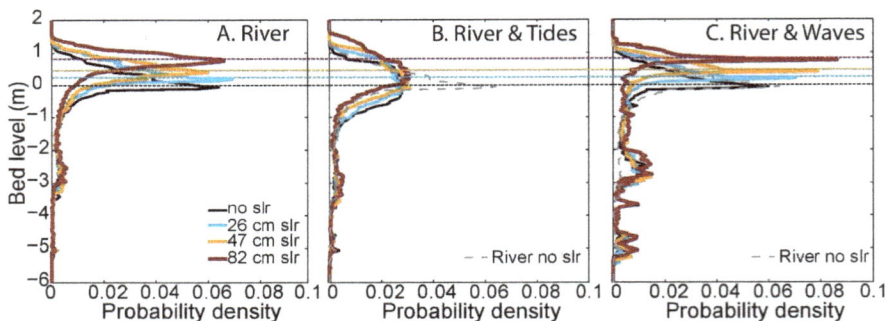

Figure 2. Delta bed level (m) as a function of sea-level rise scenario and coastal forcing. (**A**) River-dominated coasts; (**B**) River- and tidally-influenced coasts; (**C**) River- and wave-influenced coasts. Projected sea levels in 2100 indicated by horizontal lines.

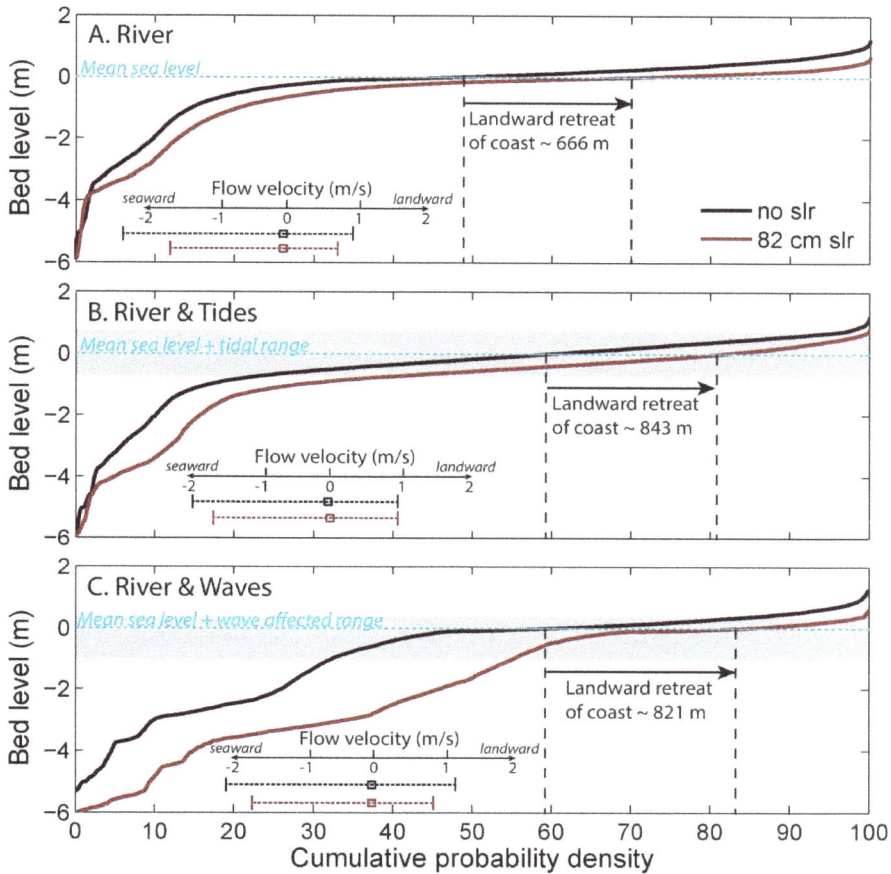

Figure 3. Integrated shoreline profiles relative to the mean sea level for deltas with SLRs of 0 cm (black) and 82 cm (red). (**A**) River-dominated delta coasts; (**B**) River- and tidally-influenced delta coasts. The simulated tidal range of 1.5 m is indicated in grey; (**C**) River- and wave-influenced delta coasts. The simulated wave range is indicated in grey with an upper range of 0.5 m, corresponding to the simulated significant wave height. The lower bound corresponds to the wave base, which is calculated as 0.5 times the wave length (in turn, calculated as wave velocity (0.5 m/s) times wave period (5 s)). The transition from subaqueous to subaerial delta is indicated by the horizontal blue dash line (see also Table 1 for descriptive statistics). The difference between the vertical black dash lines corresponds to the average shoreline retreat between 0 cm and 82 cm of SLR. Flow velocity statistics show a decreased velocity range for deltas with 82 cm SLR compared to no SLR.

3.3. Coastal Recession

SLR also drives coastal recession (Figure 3), potentially threatening natural and built environments of high economic and societal value [9,12,14]. The modelled deltas show a coastal recession of up to about 1 km over a 100-year period, equating to 10 meters per year, for an 82-cm SLR (Table 1). It is important to note that coastal recession is highly variable along the modelled deltaic coasts with some parts of the delta coast remaining fixed in position or even advancing slightly, while other parts suffer from hundreds of meters of coastal retreat. Such variability is in line with probabilistic estimates of coastal recession of natural shorelines [14] attempting to include local factors and ultimately providing coastal managers with quantitative information on the risk of retreat.

Yet, the most commonly used method to estimate coastal recession due to SLR remains the deterministic Bruun Rule [20] for its ease of application and lack of simple alternatives, despite having been a controversial tool for decades [14]. Comparing the coastal recession as observed and as predicted from the Bruun Rule for our modelled deltas, we find that the Bruun rule significantly underestimates the observed recession (Table 1), and the underestimation of coastal recession with the Bruun rule would be even larger when statically inundating the deltaic coasts. This finding reiterates the limitations of the Bruun Rule and signals the need to move to more advanced methods for reliable estimates of coastal recession to better inform coastal management.

4. Discussion

Our model simulations show a limited difference between river-dominated, wave-influenced and tide-influenced deltas in their dynamic response to SLR. The physics-based numerical modelling approach is generic, with boundary conditions representative for many natural deltaic systems. Despite many morphological similarities with natural deltas (Table 1, percentage wetted in absence of SLR; Table 2, shoreline rugosity and number of channels draining into ocean), these are idealized models that lack many of the complexities generally seen in nature. For example, salinity gradients, graded sediment, engineering structures and ecological feedbacks [36] are, amongst others, not included but are likely to affect the results. For simplicity, a yearly river flood is assumed to coincide with a representative tidal cycle or a coastal storm event in our simulations. Future studies may seek to explore time-varying marine and fluvial forcing because the timing of their morphologically relevant events (e.g., river flood, spring tide, storm wave event) determines the coastal response and the long-term (100 years) delta evolution. The applied marine delta forcing in this study is relatively small (Table 1) suggesting that the simulations are most informative for river-dominated deltas with limited marine forcing.

These findings provide the first quantitative predictions for the landward translation of deltaic shorelines while allowing for a dynamic response. Such predictions of dynamic morphologic change can be included in Climate and Earth System Models while also informing local management of deltaic areas. In essence, they confirm the over-prediction of land submergence and coastal retreat by static inundation models also observed for natural coastal systems [12]. This corroborates the inadequacy of approaches employing static models to predict flooding and coastal recession potential of future deltas. Interestingly, our observations also indicate that the over-prediction of static compared to dynamic models halts between 47 cm and 82 cm SLR due to the inability of morphologic processes to dynamically keep up with higher SLR rates for the given conditions. This would imply that deltas experiencing an SLR larger than 47 cm in 100 years become, in effect, statically inundated even when dynamical response is included. With an observational SLR rate of 3.2–3.4 mm per year for the period between 1993 and 2014 [37], this threshold between dynamically-responding deltas and deltas that become statically inundated due to the inability of morphologic processes to keep up with higher SLR rates is not too far off. Compared to the observational SLR rate, the applied SLR of 26 cm over a century will require a decrease in the current rate of SLR while the SLR scenario of 82 cm over a century requires a significant increase in the current rate of SLR.

Human engineering activities are contributing to the loss of delta land and ecosystems. In addition to a rising sea level, many coastal and delta cities suffer from land subsidence mainly due to reduced sediment loads [38] and groundwater extraction [33]. In major cities like New Orleans, Jakarta and Bangkok the combined effects of SLR and land subsidence significantly increase flood vulnerability and economic costs due to severe infrastructural damage [32]. As subsidence and SLR in effect both lead to increasingly submerged deltas, the results of the modelled deltas can also be used to understand the ability of an area to dynamically respond to subsidence. With subsidence rates exceeding present-day SLR rates up to a factor of ten in some locations [32], the ability to respond dynamically using morphological processes is likely limited and impacts are therefore expected to be significant, particularly in delta cities with fixed infrastructure.

Our results also have implications for ecosystem-based coastal defense [16] and delta restoration [4] initiatives. Delta coastal ecosystems are known to be highly dynamic environments with significant capacity to adapt to SLR due to non-linear feedbacks between the hydrodynamics, morphodynamics and ecological processes [11]. The buffer capacity that ecosystems provide in addition to the dynamic adjustment of delta morphology may well be able to overcome the loss of intertidal area as observed in this study (Figure 3), particularly in environments with suspended sediment and vegetation-enhancing sediment trapping and vertical accretion [11]. Furthermore, the initiatives in the Mississippi River delta to build new land [39] in a region with a projected SLR close to the global mean [7,8] will have to consider the dynamically adapted bed level distribution due to SLR. In the absence of ecological feedbacks and engineering structures, our simulations show an overall loss of suitable coastal habitats, particularly in the intertidal zone. Ecological processes may mitigate the impacts of SLR on the physical delta environment but the sustainability and long-term safety of such ecosystem-based coastal defenses and delta restoration initiatives can only be comprehensively explored by integrating ecosystem [11,36] and morphological dynamics in response to SLR.

5. Conclusions

This study provides the first quantitative comparison of dynamic and static delta response to SLR. We performed 12 morphodynamic simulations by varying the type of deltaic coastal forcing (three types) and the rate of SLR (four scenarios). The reported quantitative information on coastal flooding, shoreline retreat and ecological habitat loss improves our understanding of how twenty-first century projected sea-level rise is transferred to coastal impacts, provides input into Climate and Earth System models, and informs local coastal management. The idealized numerical model simulations show that the ability to dynamically respond to SLR is similar for river-dominated, wave-influenced and tide-influence deltas. Therefore, the potential for flooding and coastal retreat as a result of SLR is equally large for these delta types, and primarily governed by the regional differences in SLR. We find that static models overestimate the coastal impact. Sediment composition and ecological feedbacks may provide important mitigation mechanisms in shaping future deltas and will require further research for a comprehensive understanding.

Supplementary Materials: The following videos are available online at www.mdpi.com/2077-1312/5/2/24/s1. Video S1: R_0 (Control Run: river flow and no SLR); Video S2: R_26 (River flow and 26 cm SLR); Video S3: R_47 (River flow and 47 cm SLR); Video S4: R_82 (River flow and 82 cm SLR); Video S5: RT_0 (River flow and Tides but no SLR); Video S6: RT_26 (River flow and Tides and 26 cm SLR); Video S7: RT_47 (River flow and Tides and 47 cm SLR); Video S8: RT_82 (River flow and Tides and 82 cm SLR); Video S9: RW_0 (River flow & Waves but no SLR); Video S10: RW_26 (River flow and Waves and 26 cm SLR); Video S11: RW_47 (River flow and Waves and 47 cm SLR); Video S12: RW_82 (River flow and Waves and 82 cm SLR).

Acknowledgments: W.L. was supported by the European Community's Horizon 2020 Programme through a grant to the budget of the Integrated Infrastructure Initiative HYDRALAB+, Contract No. 654110. A.S. received funding from the NWO Netherlands Polar Programme.

Author Contributions: W.L. designed and conducted the numerical modelling and analyses. W.L. and A.S. wrote the results and conclusions and prepared the manuscript.

References

1. Nicholls, R.J.; Wong, P.P.; Burkett, V.R.; Codignotto, J.O.; Hay, J.E.; McLean, R.F.; Ragoonaden, S.; Woodroffe, C.D. Coastal systems and low-lying areas. In *Climate Change 2007: Impacts, Adaptation and Vulnerability. Contribution of Working Group II to the Fourth Assessment Report of the Intergovernmental Panel on Climate Change*; Parry, M.L., Canziani, O.F., Palutikof, J.P., van der Linden, P.J., Hanson, C.E., Eds.; Cambridge University Press: Cambridge, UK, 2007; pp. 315–356.
2. Wong, P.P.; Losada, I.J.; Gattuso, J.-P.; Hinkel, J.; Khattabi, A.; McInnes, K.; Saito, Y.; Sallenger, A. Chapter 5: Coastal Systems and Low-Lying Areas. In *Climate Change 2014: Impacts, Adaptation, and Vulnerability. Part A: Global and Sectoral Aspects. Contribution of Working Group II to the Fifth Assessment Report of the Intergovernmental Panel on Climate Change*; Field, C.B., Barros, V.R., Dokken, D.J., Mach, K.J., Mastrandrea, M.D., Bilir, T.E., Chatterjee, M., Ebi, K.L., Estrada, Y.O., Genova, R.C., et al., Eds.; Cambridge University Press: Cambridge, UK and New York, NY, USA, 2014; pp. 361–409.
3. Church, J.A.; Clark, P.U.; Cazenave, A.; Gregory, J.M.; Jevrejeva, S.; Levermann, A.; Merrifield, M.A.; Milne, G.A.; Nerem, R.S.; Nunn, P.D.; et al. Sea Level Change. In *Climate Change 2013: The Physical Science Basis. Contribution of Working Group I to the Fifth Assessment Report of the Intergovernmental Panel on Climate Change*; University Press: Cambridge, UK, 2013.
4. Edmonds, D.A.; Slingerland, R.L. Significant effect of sediment cohesion on delta morphology. *Nat. Geosci.* **2010**, *3*, 105–109. [CrossRef]
5. Galloway, W.D. Process Framework for describing the morphologic and stratigraphic evolution of deltaic depositional systems. In *Deltas, Models for Exploration*; Broussard, M.E., Ed.; Houston Geological Society: Houston, TX, USA, 1975; pp. 86–98.
6. Nienhuis, J.H.; Ashton, A.D.; Giosan, L. What makes a delta wave-dominated? *Geology* **2015**, *43*, 511–514. [CrossRef]
7. Slangen, A.B.A.; Katsman, C.A.; van de Wal, R.S.W.; Vermeersen, L.L.A.; Riva, R.E.M. Towards regional projections of twenty-first century sea-level change based on IPCC SRES scenarios. *Clim. Dyn.* **2011**, *38*, 1191–1209. [CrossRef]
8. Carson, M.; Köhl, A.; Stammer, D.; Slangen, A.B.A.; Katsman, C.A.; van de Wal, R.S.W.; Church, J.; White, N. Coastal sea level changes, observed and projected during the 20th and 21st century. *Clim. Change* **2016**, *134*, 269–281. [CrossRef]
9. Barbier, E.B.; Hacker, S.D.; Kennedy, C.; Koch, E.W.; Stier, A.C.; Silliman, B.R. The value of estuarine and coastal ecosystem services. *Ecol. Monogr.* **2011**, *81*, 169–193. [CrossRef]
10. Redfield, A.C. Development of a New England Salt Marsh. *Ecol. Monogr.* **1972**, *42*, 201–237. [CrossRef]
11. Kirwan, M.L.; Guntenspergen, G.R.; D'Alpaos, A.; Morris, J.T.; Mudd, S.M.; Temmerman, S. Limits on the adaptability of coastal marshes to rising sea level. *Geophys. Res. Lett.* **2010**, *37*, L23401. [CrossRef]
12. Lentz, E.E.; Thieler, E.R.; Plant, N.G.; Stippa, S.R.; Horton, R.M.; Gesch, D.B. Evaluation of dynamic coastal response to sea-level rise modifies inundation likelihood. *Nat. Clim. Chang.* **2016**, *6*, 1–6. [CrossRef]
13. Strauss, B.H.; Ziemlinski, R.; Weiss, J.L.; Overpeck, J.T. Tidally adjusted estimates of topographic vulnerability to sea level rise and flooding for the contiguous United States. *Environ. Res. Lett.* **2012**, *7*, 14033. [CrossRef]
14. Ranasinghe, R.; Callaghan, D.; Stive, M.J.F. Estimating coastal recession due to sea level rise: Beyond the Bruun rule. *Clim. Change* **2012**, *110*, 561–574. [CrossRef]
15. Sweet, W.V.; Park, J. From the extreme to the mean: Acceleration and tipping points of coastal inundation from sea level rise. *Earth's Future* **2014**, *2*, 579–600. [CrossRef]
16. Temmerman, S.; Meire, P.; Bouma, T.J.; Herman, P.M.J.; Ysebaert, T.; de Vriend, H.J. Ecosystem-based coastal defence in the face of global change. *Nature* **2013**, *504*, 79–83. [CrossRef] [PubMed]
17. Lesser, G.; Roelvink, J.; van Kester, J.; Stelling, G. Development and validation of a three-dimensional morphological model. *Coast. Eng.* **2004**, *51*, 883–915. [CrossRef]
18. Schuurman, F.; Kleinhans, M.G.; Marra, W.A. Physics-based modeling of large braided sand-bed rivers: Bar pattern formation, dynamics, and sensitivity. *J. Geophys. Res. Earth Surf.* **2013**, *118*, 2509–2527. [CrossRef]
19. Moss, R.H.; Edmonds, J.A.; Hibbard, K.A.; Manning, M.R.; Rose, S.K.; van Vuuren, D.P.; Carter, T.R.; Emori, S.; Kainuma, M.; Kram, T.; et al. The next generation of scenarios for climate change research and assessment. *Nature* **2010**, *463*, 747–756. [CrossRef] [PubMed]

20. Bruun, P. The Bruun rule of erosion by sea-level rise: A discussion on large-scale two-and three-dimensional usages. *J. Coast. Res.* **1988**, *4*, 627–648.

21. Booij, N.; Ris, R.C.; Holthuijsen, L.H. A third-generation wave model for coastal regions, 1, Model description and validation. *J. Geophys. Res.* **1999**, *104*, C4. [CrossRef]

22. Van Rijn, L.C. Sediment transport, part I: Bed load transport. *J. Hydraul. Eng.* **1984**, *110*, 1431–1456. [CrossRef]

23. Van Rijn, L.C. Sediment transport, part II: Suspended load transport. *J. Hydraul. Eng.* **1984**, *110*, 1613–1641. [CrossRef]

24. Engelund, F.; Hansen, E. *A Monograph on Sediment Transport in Alluvial Streams*; Teknisk Forlag: Kobenhavn, Denmark, 1967.

25. Roelvink, J.A. Coastal morphodynamic evolution techniques. *Coast. Eng.* **2006**, *53*, 277–287. [CrossRef]

26. Geleynse, N.; Voller, V.R.; Paola, C.; Ganti, V. Characterization of river delta shorelines. *Geophys. Res. Lett.* **2012**, *39*, L17402. [CrossRef]

27. Wellner, R.; Beaubouef, R.; van Wagoner, J.; Roberts, H.; Sun, T. Jet-Plume Depositional Bodies—The Primary Building Blocks of Wax Lake Delta. *Gulf Coast Assoc. Geol. Soc. Trans.* **2005**, *55*, 867–909.

28. Kleinhans, M.G.; van Dijk, W.M.; van de Lageweg, W.I.; Hoyal, D.C.J.D.; Markies, H.; van Maarseveen, M.; Roosendaal, C.; van Weesep, W.; van Breemen, D.; Hoendervoogt, R.; et al. Quantifiable effectiveness of experimental scaling of river- and delta morphodynamics and stratigraphy. *Earth Sci. Rev.* **2014**, *133*, 43–61. [CrossRef]

29. Wolinsky, M.A.; Edmonds, D.A.; Martin, J.; Paola, C. Delta allometry: Growth laws for river deltas. *Geophys. Res. Lett.* **2010**, *37*, L21403. [CrossRef]

30. Jerolmack, D.J. Conceptual framework for assessing the response of delta channel networks to Holocene sea level rise. *Quat. Sci. Rev.* **2009**, *28*, 1786–1800. [CrossRef]

31. Syvitski, J.P.M.; Saito, Y. Morphodynamics of deltas under the influence of humans. *Glob. Planet. Change* **2007**, *57*, 261–282. [CrossRef]

32. Erkens, G.; Bucx, T.; Dam, R.; de Lange, G.; Lambert, J. Sinking coastal cities. *Proc. Int. Assoc. Hydrol. Sci.* **2015**, *372*, 189–198. [CrossRef]

33. Wada, Y.; van Beek, L.P.H.; van Kempen, C.M.; Reckman, J.W.T.M.; Vasak, S.; Bierkens, M.F.P. Global depletion of groundwater resources. *Geophys. Res. Lett.* **2010**, *37*. [CrossRef]

34. Kirwan, M.L.; Murray, A.B.; Boyd, W.S. Temporary vegetation disturbance as an explanation for permanent loss of tidal wetlands. *Geophys. Res. Lett.* **2008**, *35*, L05403. [CrossRef]

35. Kirwan, M.L.; Murray, A.B. A coupled geomorphic and ecological model of tidal marsh evolution. *Proc. Natl. Acad. Sci. USA* **2007**, *104*, 6118–6122. [CrossRef] [PubMed]

36. Temmerman, S.; Kirwan, M.L. Building land with a rising sea. *Science* **2015**, *349*, 588–589. [CrossRef] [PubMed]

37. Ablain, M.; Legeais, J.F.; Prandi, P.; Marcos, M.; Fenoglio-Marc, L.; Dieng, H.B.; Benveniste, J.; Cazenave, A. Satellite Altimetry-Based Sea Level at Global and Regional Scales. *Surv. Geophys.* **2017**, *38*, 7–31. [CrossRef]

38. Syvitski, J.P.M.; Kettner, A.J.; Overeem, I.; Hutton, E.W.H.; Hannon, M.T.; Brakenridge, G.R.; Day, J.; Vörösmarty, C.; Saito, Y.; Giosan, L.; et al. Sinking deltas due to human activities. *Nat. Geosci.* **2009**, *2*, 681–686. [CrossRef]

39. Kim, W.; Mohrig, D.; Twilley, R.; Paola, C.; Parker, G. Is It Feasible to Build New Land in the Mississippi River Delta? *EOS Trans. Am. Geophys. Union* **2009**, *90*, 373–374. [CrossRef]

Journal of
*Marine Science
and Engineering*

MDPI

Article

Spatial and Temporal Clustering Analysis of Extreme Wave Events around the UK Coastline

Victor Malagon Santos [1,2,*], Ivan D. Haigh [1] and Thomas Wahl [2]

[1] Ocean and Earth Science, National Oceanography Centre, University of Southampton, European Way, Southampton SO14 3ZH, UK; I.D.Haigh@soton.ac.uk

[2] Department of Civil, Environmental, and Construction Engineering and Sustainable Coastal Systems Cluster, University of Central Florida, 12800 Pegasus Drive, Suite 211, Orlando, FL 32816-2450, USA; t.wahl@ucf.edu

* Correspondence: vmalagon@knights.ucf.edu; Tel.: +34-630-250-687

Received: 8 May 2017; Accepted: 10 July 2017; Published: 14 July 2017

Abstract: Densely populated coastal regions are vulnerable to extreme wave events, which can cause loss of life and considerable damage to coastal infrastructure and ecological assets. Here, an event-based analysis approach, across multiple sites, has been used to assess the spatial footprint and temporal clustering of extreme storm-wave events around the coast of the United Kingdom (UK). The correlated spatial and temporal characteristics of wave events are often ignored even though they amplify flood consequences. Waves that exceeded the 1 in 1-year return level were analysed from 18 different buoy records and declustered into distinct storm events. In total, 92 extreme wave events are identified for the period from 2002 (when buoys began to record) to mid-2016. The tracks of the storms of these events were also captured. Six main spatial footprints were identified in terms of extreme wave events occurrence along stretches of coastline. The majority of events were observed between November and March, with large inter-annual differences in the number of events per season associated with the West Europe Pressure Anomaly (WEPA). The 2013/14 storm season was an outlier regarding the number of wave events, their temporal clustering and return levels. The presented spatial and temporal analysis framework for extreme wave events can be applied to any coastal region with sufficient observational data and highlights the importance of developing statistical tools to accurately predict such processes.

Keywords: waves; temporal clustering; spatial footprints; coastal vulnerability and impacts; extreme value analysis; risk management

1. Introduction

Waves are a key process influencing coastal flooding, morphology, and nearshore ecology [1]. Quantifying the variability of wave parameters is essential for coastal engineering and management [2]. Similarly, understanding wave climate and probabilities of extreme events is also crucial for offshore operators. This is emphasised, for example, by a recent increase in the planned deployment of offshore wind farms in the United Kingdom (UK) shelf seas [1]. Furthermore, the electricity generation industry is regarding wave energy as a sustainable source with considerable potential and is expanding its efforts to research wave characteristics and climate [3]. Ocean waves are also crucial for coastal flood risk management. Flooding and coastal erosion due to extreme waves and storm surges are a major hazard for coastal populations [4], as they are capable of causing extensive economic, cultural, and environmental damage, and can also be associated with high mortality [5]. A combination of high water levels (caused by spring tides and storm surges) together with waves is the primary factor leading to coastal flooding. Those combined processes can cause loss of life, and damage to property and the environment by overtopping coastal defenses and inundating low-lying areas [6]. The amount of damage rendered by coastal flooding also depends on the presence of population and infrastructure along the coast [7]. Furthermore, flood risk variability along the coast depends on the spatial characteristics of the governing processes. Many studies have demonstrated that return periods of storm tides can vary considerably along the affected coasts, e.g., [8–11], and this also applies to other nearshore processes such as waves. Hence, spatial characterization of wave processes becomes essential to manage coastal hazards effectively.

In northern Europe and the UK in particular, a remarkable series of storms occurred over the winter of 2013/14, with large waves and high sea level events affecting large stretches of the UK coast. The most significant features of this storm season were the length of coastline affected by flooding (i.e., 'spatial footprints') [12] and the short inter-arrival times between extreme events (i.e., 'temporal clustering') [13]. These issues have important implications in terms of flood risk management. Economical, societal, and environmental impacts derived from extreme events are often correlated with their spatial extent [14,15]. For instance, the action of extreme waves against ports located in nearby locations can hinder trading activities between different countries [16]. On the other hand, temporal clustering of extreme wave events may have important consequences for coastal structures due to attritional effects and insufficient time to properly repair structures between storms [17]. Similarly, natural systems such as beaches may not have enough time to recover if they are affected by many consecutive extreme events [17,18], minimizing the protection of coastal areas against flooding and leading to a wide range of economic and societal consequences [19]. These two processes are particularly relevant to the UK due to its long and irregular coastline (>1800 km) and its relatively high risk of coastal flooding: about 2.5 million people and vast economic coastal assets are considered to be exposed to coastal flooding [20]. Some studies have been undertaken to understand the extreme spatial behavior and temporal clustering of storm surges and extreme sea levels around the country, e.g., [13,15,21]. However, little attention has been paid to storm waves, which also had a large contribution to the devastating consequences observed during the winter of 2013/14. Haigh et al. [15] recently developed an 'Event based' analysis approach to assess the spatial footprints and temporal clustering of extreme sea levels and storm surge events around the UK. Here we build on their approach and extend it to waves.

The main aim of this paper is to assess and better understand the spatial footprints and the temporal clustering of extreme wave events to facilitate the inclusion of such information into statistical models and coastal management. We address the following four objectives: (1) compare the dates and return periods of the extreme wave events recorded around the UK; (2) assess the spatial characteristics of the wave events around the coast, distinguishing different types of footprints of events and examining the categories and tracks of the weather systems responsible for their generation; (3) investigate the temporal variation in events over different time-scales and in particular the clustering nature of extreme storm-wave events; and (4) examine these characteristics for the exceptional 2013/14 storm season.

2. Data

The primary dataset we use comprises wave data from Datawell's Directional Waverider buoys. Surface-following moored wave buoys are the most abundant around the UK coastline and for consistency we only use data from this buoy type. Wave data is extracted from two different sources. We first use data from the Centre for Environment, Fisheries and Aquaculture Science (CEFAS) WaveNet Array [22], funded by the Environment agency [22,23]. The second main source is the Channel Coastal Observatory (CCO) wave array, which is composed of several wave boys located around the English Channel [24]. Wave buoys from both sources provide half-hourly data. In total, we used data from 18 buoys. The buoy data adopted in this study has been selected to comprise the buoys with the longest records and fewest gaps, whilst ensuring that the UK coastline is adequately covered (Figure 1a). The data from both sources undergo rigorous quality controls by the data operators (refer to the data sources for more information [22,24]). Values which the CCO had flagged suspicious were removed and our own extensive checks were applied to all datasets, removing spurious values. The mean data length for all considered records is nine years (Figure 1b, Table 1).

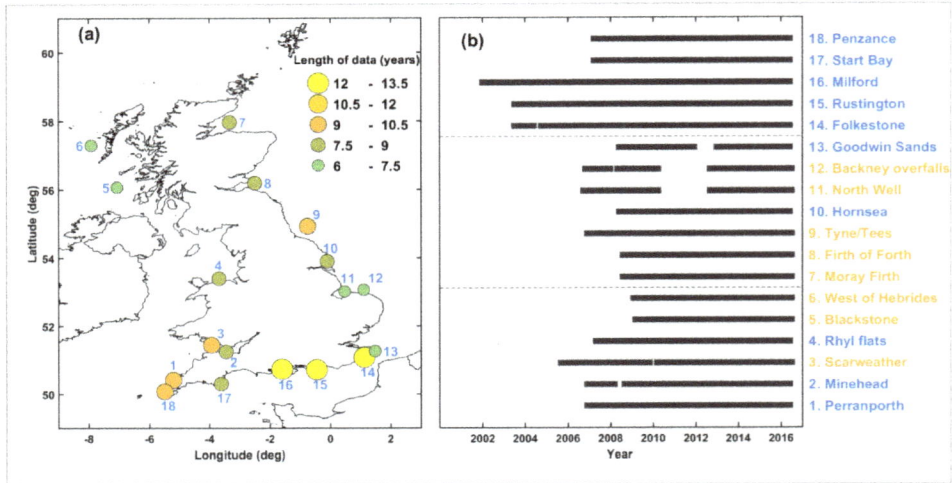

Figure 1. (**a**) Location of the selected buoys and their assigned number. (**b**) Location names and duration of the buoy records. Yellow: Centre for Environment, Fisheries and Aquaculture Science (CEFAS) WaveNet network; Blue: Channel Coastal Observatory (CCO) wave array.

Table 1. Name, location, data length, depth, and distance from the coast for each study site.

Site Number	Site Name	Longitude (deg.)	Latitude (deg.)	Range	Number of Years (Data Range)	Depth (m CD)	Distance to Coast (km)
1	Perranporth	−5.188	50.403	2006–2016	9.1 (10.5)	10	2.1
2	Minehead	−3.449	51.235	2006–2016	8.7 (10.5)	10	3.1
3	Scarweather	−3.933	51.433	2005–2016	10.2 (11.9)	35	14.7
4	Rhyl Flats	−3.686	53.393	2007–2016	8.4 (10.2)	10	9.5
5	Blackstone	−7.062	56.062	2009–2016	7.3 (8.4)	97	43.6
6	West of Hebrides	−7.914	57.292	2009–2016	7.2 (8.4)	100	29.6
7	Moray Firth	−3.333	57.966	2008–2016	7.7 (8.9)	54	26.9
8	Firth of Forth	−2.504	56.188	2008–2016	7.8 (8.9)	65	2.7
9	Tyne/Tees	−0.749	54.919	2006–2016	9.5 (10.6)	65	39.2
10	Hornsea	−0.102	53.893	2008–2016	7.8 (9.1)	12	2.95
11	NorthWell	0.475	53.008	2006–2016	6.8 (10.8)	31	6.3
12	Backney Overfalls	1.104	53.058	2006–2016	6.9 (10.7)	23	10.9
13	Goodwin Sands	1.483	51.252	2008–2016	6.8 (9)	10	4.3
14	Folkestone	1.128	51.063	2003–2016	12.3 (14)	13	0.79
15	Rustington	−0.445	50.704	2003–2016	12.4 (14)	9.9	8.8
16	Milford	−1.604	50.721	2002–2016	13.1 (15.5)	10	0.5
17	Start Bay	−3.616	50.292	2007–2016	9 (10.2)	10	1.5
18	Penzance	−5.478	50.066	2007–2016	9 (10.2)	10	5.4

The second dataset we use is gridded mean sea level pressure and near-surface wind fields to digitise storm tracks. This global meteorological dataset is taken from the 20th Century Reanalysis, Version 2 [25], which has a spatial resolution of 2° and temporal resolution of six hours from 1871–2014 (at the time of analysis). Data for 2015 and 2016 were derived from the US National Center for Environmental Predictions/National Center for Atmospheric Research's (NCEP/NCAR) comparable Reanalysis, Version 2 [26]. These fields have a horizontal resolution of 2.5° and were interpolated to 2° for consistency.

We use a third dataset comprising different atmospheric oscillation indices known to exert influence over the North Atlantic wave climate to assess their possible correlation with extreme storm-wave event occurrences [27]. This dataset contains: (1) the winter-averaged (December to March; DJFM) North Atlantic Oscillation (NAO) index described by Jones et al. [28] and obtained from [29]; (2) the DJFM East Atlantic Pattern (EA) index [30]; (3) the DJFM Scandinavian Pattern (SCAND) index [30]; and (4) the West Europe Pressure Anomaly (WEPA) as described by Castelle et al. [27].

3. Methodology

We followed the methodological approach developed by Haigh et al. [15], but adapted it for waves. The first stage of the analysis consisted of extracting significant wave heights that exceeded the 1 in 1-year return levels at all 18 wave buoy sites. This limit was chosen as a trade-off between identifying a large enough sample of events and still analyzing only waves that are considered 'extreme'. We applied a range of different extreme value analysis techniques to estimate wave return levels at each site. Given the relatively short length of records (the longest one has been recording for 13 years; Figure 1b and Table 1), we found that the Peak Over Threshold (POT) method was the most appropriate to estimate return levels. Hence, we calculated the 99.75th percentile threshold based on all observations and fitted a Generalized Pareto Distribution (GPD) to the threshold exceedances. To ensure independency, we applied a declustering algorithm. Following the meteorological independence criterion [31], we applied a 3-day storm window to separate events. Comparing the GPD fits at all sites with Gringorten's plotting positions revealed satisfactory results, i.e., the GPD performs well in modelling extreme waves (Figure 2). The significant wave heights associated with ten return periods at the 18 study sites are listed in Table 2. We then extracted all wave events, across each of the 18 sites that exceeded the 1 in 1-year return level threshold. For each wave event that reached or exceeded the threshold, we recorded the: (1) date-time of the wave event; (2) return period; (3) significant wave height; and (4) the site number.

Table 2. Hs return level estimates (meters) for each of the 18 study sites for different return periods.

Site Number	Site Name	Return Period (Years)									
		0.5	1	2	5	10	20	25	50	75	100
1	Perranporth	5.94	6.40	6.81	7.27	7.57	7.84	7.92	8.15	8.27	8.35
2	Minehead	2.44	2.59	2.70	2.81	2.88	2.93	2.95	2.98	3.00	3.01
3	Scarweather	4.93	5.37	5.77	6.25	6.59	6.91	7.01	7.30	7.46	7.57
4	Rhyl Flats	3.37	3.65	3.89	4.15	4.32	4.45	4.49	4.60	4.66	4.69
5	Blackstone	10.61	11.57	12.41	13.39	14.02	14.59	14.76	15.24	15.49	15.66
6	West of Hebrides	12.24	13.49	14.63	15.97	16.87	17.68	17.93	18.64	19.02	19.28
7	Moray Firth	4.99	5.57	6.12	6.78	7.25	7.69	7.82	8.22	8.44	8.59
8	Firth ofForth	5.37	6.09	6.78	7.65	8.27	8.87	9.06	9.62	9.94	10.16
9	Tyne/Tees	5.78	6.29	6.73	7.22	7.53	7.79	7.87	8.09	8.20	8.27
10	Hornsea	3.61	3.96	4.28	4.64	4.87	5.08	5.15	5.33	5.42	5.48
11	NorthWell	2.42	2.60	2.77	2.96	3.08	3.19	3.23	3.32	3.37	3.40
12	Backney Overfalls	3.67	4.01	4.32	4.67	4.89	5.09	5.15	5.32	5.41	5.47
13	Goodwin Sands	2.74	2.94	3.12	3.32	3.46	3.57	3.61	3.71	3.76	3.80
14	Folkestone	2.83	3.05	3.24	3.44	3.57	3.68	3.71	3.80	3.85	3.88
15	Rustington	3.87	4.20	4.51	4.87	5.12	5.34	5.40	5.60	5.71	5.78
16	Milford	3.16	3.46	3.73	4.07	4.30	4.52	4.58	4.78	4.88	4.96
17	Start Bay	3.40	3.74	4.05	4.44	4.71	4.97	5.05	5.28	5.42	5.51
18	Penzance	3.70	4.21	4.71	5.35	5.82	6.28	6.43	6.87	7.13	7.31

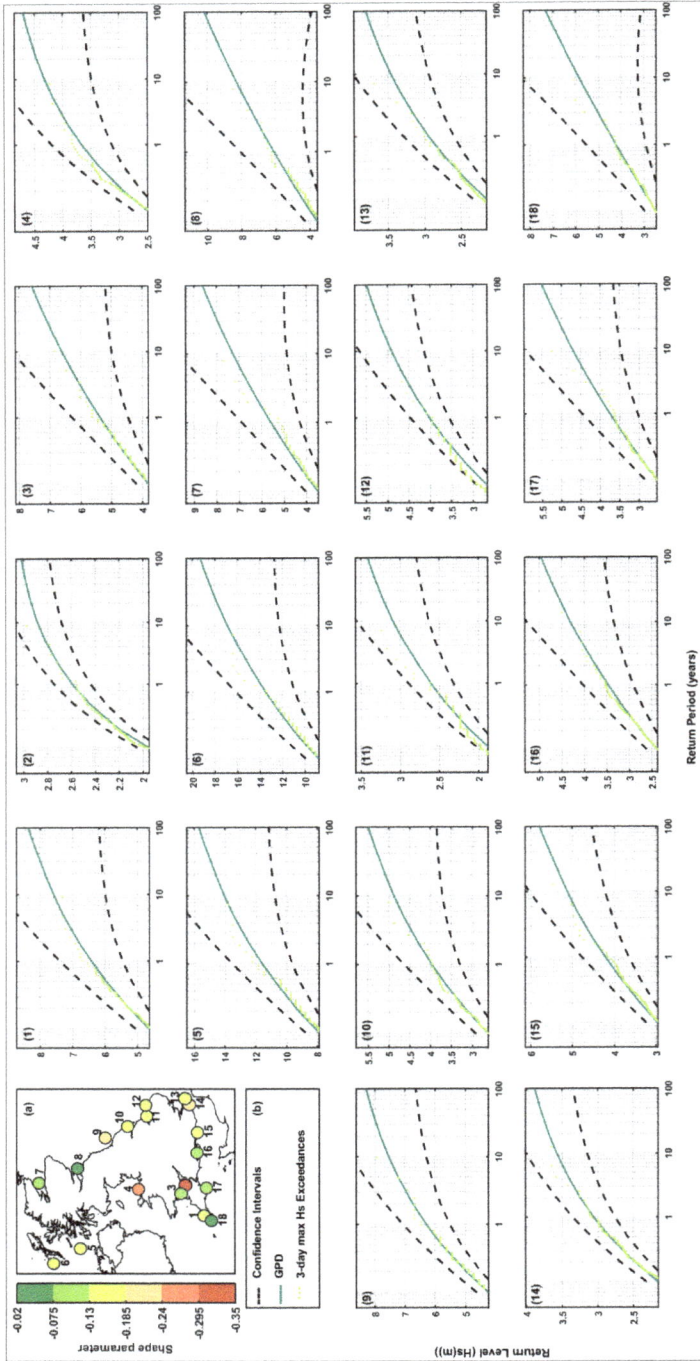

Figure 2. Generalized Pareto Distribution (GPD) fits are shown for each of the sites along with their confidence intervals (95%). The number in brackets indicates the site identity number. Both colours and dots represent the shape parameter for each location on a UK map (**a**), and there is a legend for the GPD plots (**b**).

The second stage of the analysis was to identify the distinct, extra-tropical storms that produced the extreme waves that were identified in stage one, and then to capture the meteorological information about those storms. Following the approach of Haigh et al. [15], this involved a two-step procedure. First, we used a simple 'storm window' approach, based on the fact that most storms that caused high waves in the UK typically lasted up to about three days. We started with the wave height associated to the highest return period, and found all of the other high waves that occurred within a window of one and a half days before or after at the other sites. We then assigned to these the event number 1. We set all high waves associated with event 1 aside and moved on to the event with the next highest return period, and so on. Second, we manually used the meteorological data to determine if our above-described procedure had correctly linked high waves to distinct storms. To achieve this, an interactive interface in MATLAB was used. This interface displayed the six-hourly progression of mean sea level pressure and wind vectors over the North Atlantic Ocean and Northern Europe around the time of maximum return period for each of the previously identified extreme events that exceeded a 1 in 1-year threshold. On all occasions, our simple procedure correctly identified distinct storms and associated wave events across the study sites. Using the same interactive MATLAB interface, the storm tracks associated with the extreme wave events were digitised, from when the low-pressure systems developed until they dissipated or moved beyond latitude 20° E. Storm tracks were captured by selecting the grid point of the lowest atmospheric pressure at each 6-h time step. From the start to the end of the storm the following 6-hourly information was stored: (1) time; (2) latitude and (3) longitude of the minimum pressure cell; and (4) the minimum mean sea level pressure.

The third stage of the methodology involved examining the results obtained in the previous stages to assess the spatial and temporal clustering features of events. For each storm event, we examined the spatial extent of extreme high waves by comparing the sites in which the 1 in 1-year return level was exceeded during the defined storm window (i.e., three days). Then, the spatial extent was compared for all events to find regional patterns regarding extreme wave occurrences and define distinct regions. We also assessed the wave spectral characteristics (provided by CCO and CEFAS) for the most significant events to determine whether events had sea, swell, or bi-modal characteristics. The temporal analysis of extreme wave events consisted of two steps. First, temporal trends of events were analysed by looking at both monthly and inter-annual changes. We compared the inter-annual changes of events with different atmospheric oscillations (NAO, EA, SCAND and WEPA), as done in previous wave climate studies on the west coast of Europe, e.g., [27,32–34]. Second, the temporal clustering was evaluated by comparing the time between consecutive events. The 2013/14 storm season was assessed in detail by evaluating how unusual it was in the historic (as far as data availability allows) context. The number of extreme events exceeding the 1 in 1-year return level, their actual return periods, spatial extent, and inter-arrival times were compared. A site-by-site comparison was performed for the same season to identify areas that were most affected. For correlation analyses, we use the Kendall rank correlation coefficient throughout the paper to account for integer variables.

4. Results

4.1. Wave and Storm Event Identification

First, we compare the dates and return periods of extreme waves. In total, 165 significant wave heights were found that exceeded the 1 in 1-year return level across the 18 sites. These high wave occurrences were generated by 92 distinct storms. Comparing all storm events in terms of their highest return period across all sites, we found that 43 of the distinct 92 storm events led to wave heights around the UK with maximum return periods ranging from 1 to 2 years, 22 from 2 to 5 years, and 27 had a return period that exceeded 5 years (Figure 3). Analyzing the GPD shape parameters we find pronounced spatial variability across the 18 study sites (Figure 2a). The shape parameter is also correlated with the level of the exposure. The curve of the GPD is more pronounced at sheltered sites and vice versa, with exception of the study sites located in the northwest (NW) region. These wave buoys are located further away from the coast than others with similar level of exposure. This may lead to different behavior of the GPD shape and scale parameters. Galanis et al. [35] found that the GPD was close to a straight line (in the Gumbel space) when offshore wave data was analysed.

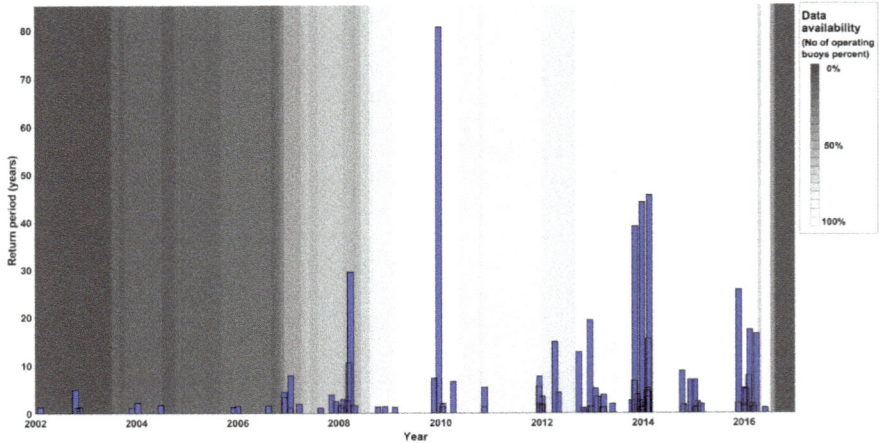

Figure 3. Maximum return period per event plotted against occurrence date. The grey color of the background represents the percentage of number of sites recording: the lighter the grey, the more data are available.

4.2. Spatial Footprint and Storm Track Analysis

Second, we examine the spatial characteristics of waves for the 92 storm events. Initially, the number of sites where the 1 in 1-year return period was reached or exceeded during an event was analysed as well as their maximum return level. More than half of the storm-wave events (51 out of 92) led to significant wave heights with a return period greater than 1 year at only one site, of which 94% (48 out of 51) had a return period between 1 and 5 years, two events had a maximum return period between 5 and 10 years, and just one had a return period between 10 and 25 years (Figure 4a). Generally, the number of sites affected by a specific event is proportional to the maximum return period. The correlation coefficient for that relationship is 0.54 and is statistically significant at the 95% confidence level (Figure 4b).

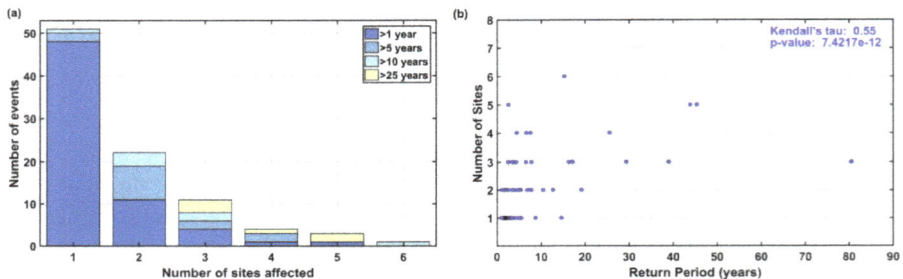

Figure 4. (**a**) Histograms showing the number of sites where the 1 in 1-year return level was exceeded. Events are also sorted by different return period thresholds. (**b**) The return period of the highest significant wave height for the 92 extreme wave events plotted against the number of sites where the 1 in 1-year return level was exceeded.

To examine the spatial patterns of extreme events along the coast we identified the most significant events, selecting those that affected several study sites (three or more sites for southern and midland study sites; two or more for northern sites due to poorer spatial resolution). Overall, our results suggest there are six categories of spatial footprints, with some overlap (Figure 5). The first type of footprint covers the southwest (SW) coast, approximately from Start Bay to Minehead (Figure 5, events xiii–xvii). The second type of footprint covers the west (W) coast, including Scarweather and Rhyl Flats in the Irish Sea (Figure 5, events xviii–xix). The third type of footprint is located in the NW of the UK (Figure 5, events xxiii–xxviii), whereas the fourth one is located in the northeast (NE) of the country (Figure 5, events xx–xxii). The fifth footprint covers the east (E) coast (Figure 5, events viii–xii), mainly including sites between Newcastle and Folkestone. The sixth footprint is located along the southeast (SE) of England, approximately between Folkestone and Milford (Figure 5, events i–vii).

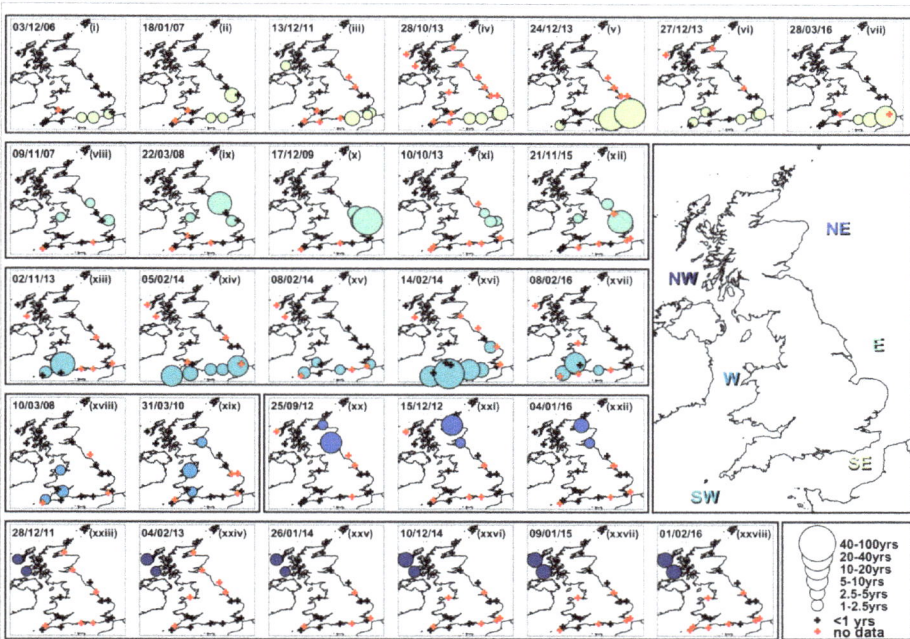

Figure 5. The spatial footprint of extreme events that affected three buoy sites (or two sites for the north of the UK). The events have been organized according to the regions they affected and subfigures have been sorted by date of occurrence within regions. The first number in the top left corner indicates the date on which they occurred and the Roman number in brackets is a label for each subfigure.

We also assessed the storm tracks that led to the extreme events considered here. This was done to identify different storm types which typically affect certain coastline stretches. As outlined below, this analysis step provides valuable insights but we also note to be careful in generalizing the results because of the limited number of events that we assessed. For the SW region, storm centers at time of the highest significant wave heights were mostly located around the north of the UK, and the storms followed similar tracks (Figure 6a). We generally observe that extreme swell events are created by storm tracks coming from the SW and heading toward Scandinavia, passing through the UK diagonally. For the W region (Figure 6b), the fact that the highest return periods of both events occurred when they were located at a very similar position suggests that wind patterns have an important role in creating extreme events. Moreover, the site protected by Ireland (Rhyl Flats, Site 4) is characterized by a spectrum (not shown) which always behaved in the same manner during the considered extreme conditions, showing that extreme wave events are usually the result of winds in the Irish Sea. Storm tracks for the NW region present very similar behavior (Figure 6c), passing between Iceland and the UK with storm centers located to the NW of the UK when the highest wave heights were observed.

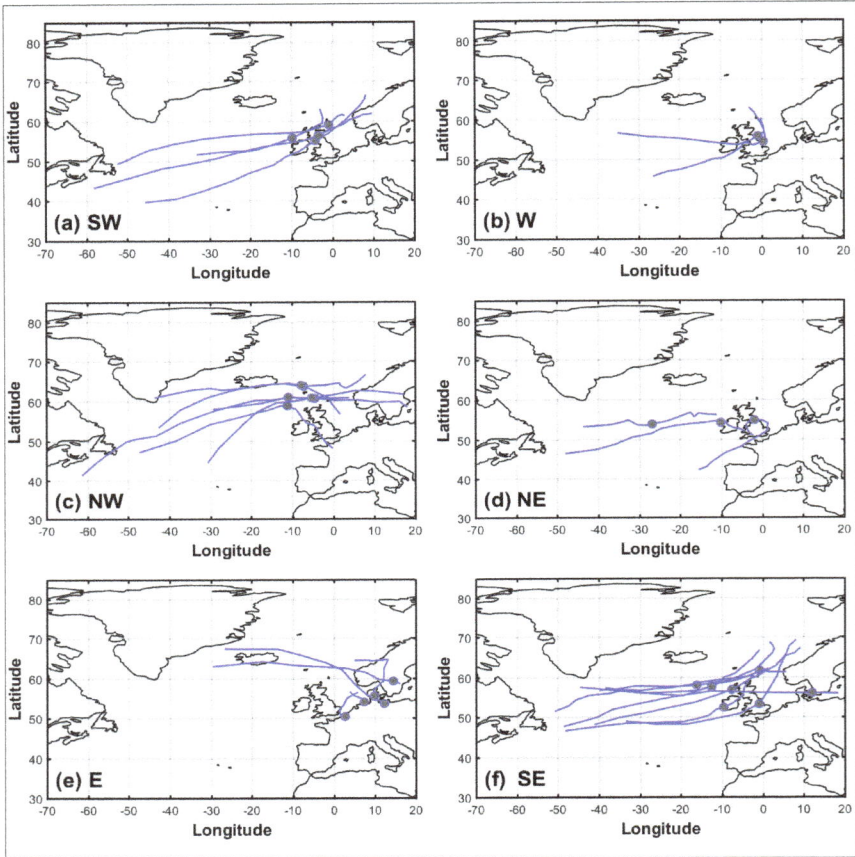

Figure 6. Storm tracks that generated the extreme wave events shown in Figure 5 for the distinct six regions identified: (**a**) SW region; (**b**) W region; (**c**) NW region; (**d**) NE region; (**e**) E region; and (**f**) SE region. The grey dots indicate where the storm centers were when the maximum significant wave height occurred.

Storm tracks leading to extreme events in the NE region show different patterns (Figure 6d), and so does the spectral density for different periods during distinct events. It is also noteworthy that extreme events in that particular area lasted twice as long as for any other region, though the short wave time-series does not allow us to examine this characteristic properly. Regarding the E region, the most significant events had swell characteristics, and all of them occurred when the storm center was over the N of Germany, the Benelux, the N of France, or Scandinavia (Figure 6e). In the SE region, storm centers at the time of the highest waves are spread out, although they are often in the N of the UK, Ireland, or Atlantic Ocean (Figure 6f). All the considered extreme events in the SE had pronounced swell characteristics and occurred with a positive NAO phase. Similar to those in the SW region, the storm tracks usually follow the same pattern in the SE, coming from the middle of the North Atlantic and bending towards the north while passing through the UK. The most exceptional extreme events occurred when the storm track had a steady longitudinal direction, constantly heading towards the SW of the country.

4.3. Seasonality and Temporal Clustering Analysis

Third, we assess temporal trends and clustering of events. The number of events per season (15 July to 14 July in the following year) and the inter-seasonal changes in the number of extreme events are assessed by examining the number of sites that they affected (Figure 7a), the maximum return period that was reached (Figure 7b), and the number of days between consecutive events (Figure 7c). The 2013/14 season had the largest number of extreme events, since 2002, and was the most unusual in terms of the extent of the spatial footprints of these events, their return periods and their short inter-arrival times (see Section 4.4 for details). The seasons of 2012/13 and 2015/16 were equally important in terms of the number of extreme events that exceeded the 1 in 1-year return period, although they present different characteristics: 2012/13 extreme events affected no more than two sites at the same time and they generally had a lower return period; 2015/16 events were more severe, with one of them affecting four sites, another event with a return period that exceeded 25 years and many events with a small number of days between them. The 2007/08 season also had a relatively large number of extreme events, but was overall calmer than the seasons discussed above. The rest of the considered seasons were relatively calm, but results have to be treated with caution before 2007, owing to poor data resolution.

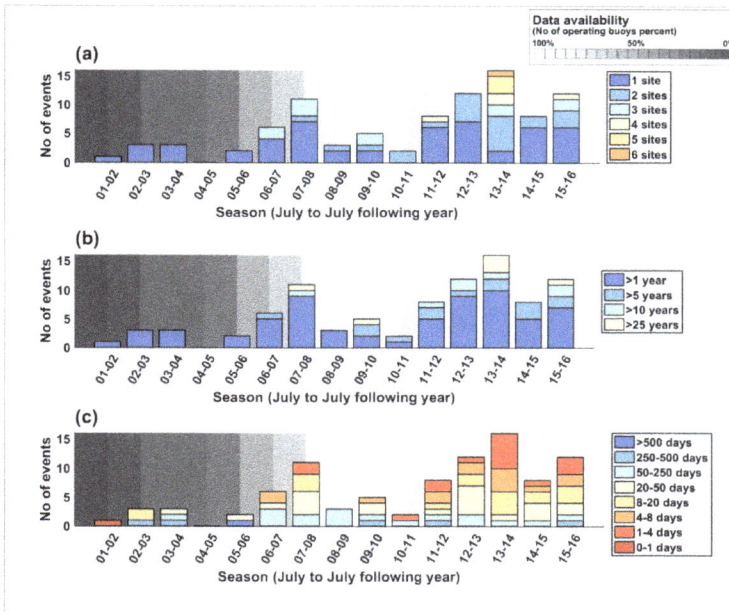

Figure 7. Number of significant wave height extreme events per season (July to July of the next year) for (**a**) distinct number of sites affected (i.e., spatial footprint), (**b**) different return level thresholds, and (**c**) ranges of days between events. The grey shading at the background represents the mean annual data availability: they lighter the grey, the more data are available.

The number of events per month over the entire time series was examined, paying attention to the spatial footprints of the events, their maximum return period and the days between events. No events were observed in July and the number grows steadily afterwards with its maximum in December/January (Figure 8). There is a sharp drop in number of events in April, and it is notable that only a small number of events occurred in September and these were even lower than those in August. Approximately 90% of all the events happened between October and March. Also, all the events with a spatial footprint extending three or more study sites occurred in those months, as well as most of the events with a return period longer than ten years and those with inter-arrival times less than eight days. November and December are the most severe months as they had more events that exceeded the 25-year return period than any other month. These two months also feature events with the shortest time between them in the entire time series. However, February is also an exceptional month, as more than half of the events affected two sites or more.

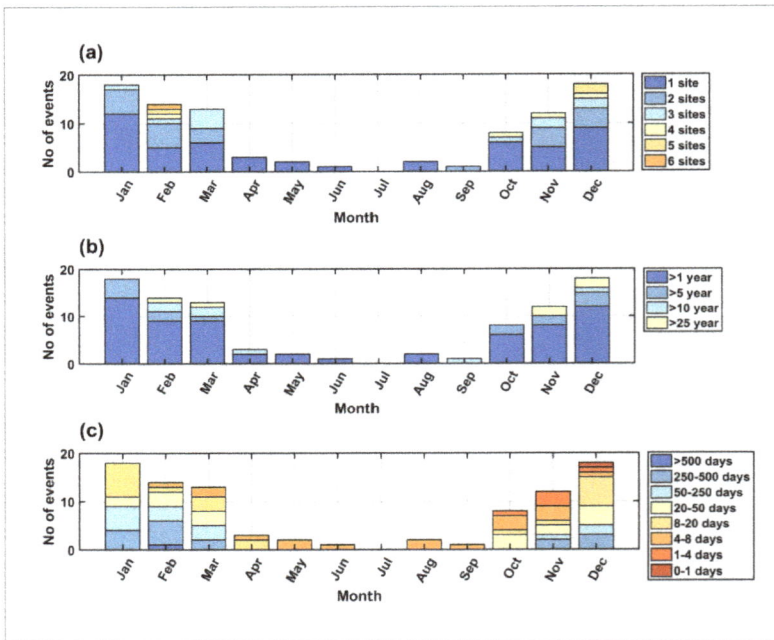

Figure 8. Number of significant wave height extreme events per month for (**a**) distinct number of sites affected (i.e., spatial footprint), (**b**) different return level threshold, and (**c**) ranges of days between events.

Finally, the clustering nature of storm-wave events was further evaluated comparing the days between events for the available records. This part of the analysis shows that about 15 events (16%) out of 92 occurred less than four days after the previous one, 29 (32%) less than eight days and 46 (50%) less than 20 days. As expected, events with the largest number of days between them occurred during periods of poor data resolution (Figure 9), and at the beginning of each season due to the low number of events during summer. The days between events have also been analysed for each site individually (not shown). Overall, there is a clear similarity between sites belonging to the same region. Furthermore, the correlation between the number of events per season and four atmospheric oscillation indices that affect the North Atlantic wave climate have been examined (Table 3); for that we use the rank correlation coefficient Kendall's tau. The highest correlation exists between the WEPA index and the total number of events with a value of 0.7 (significant at 95% confidence). No significant relationship was found for the rest of the considered atmospheric indices, but the results vary spatially. The NAO index largely explains extreme wave variability in the NW of the UK and has also strong correlation with the sites located in the SE region. The EA index is strongly correlated with events observed at the eastern wave buoy sites and one site located in the SE region. The SCAND index shows high correlation at two unconnected sites, but overall this index has the smallest influence on UK extreme wave events. Finally, strong relationships are found between the WEPA index and sites that are mainly located in the E and SW regions.

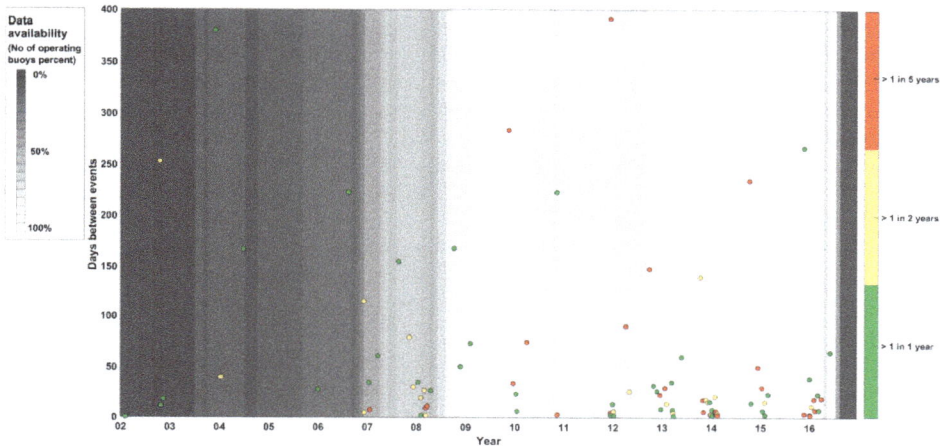

Figure 9. Number of days between consecutive significant wave height extreme events for all sites, showing different return level thresholds for each event. The grey shading at the background of the figures represents the percentage of number of sites for which data are available: the lighter the grey, the more data are available.

Table 3. Kendall rank correlation coefficient between the number of extreme wave events that exceeded the 1 in 1-year return level against different atmospheric oscillations indices known to exert influence over the wave climate in the Atlantic Ocean.

Site Number	NAO		EA		SCAND		WEPA	
	Kendall's Tau	*p*-Value	Kendall's Tau	*p*-Value	Kendall's Tau	*p*-Value	Kendall's Tau	*p*-Value
1	0.57	0.03	0.26	0.34	−0.31	0.26	0.39	0.14
2	0.48	0.08	0.34	0.22	−0.02	1.00	0.34	0.22
3	0.32	0.21	0.54	0.03	−0.47	0.06	0.51	0.04
4	0.00	1.00	0.16	0.63	−0.11	0.77	0.16	0.63
5	0.83	0.00	0.04	1.00	−0.12	0.80	0.20	0.62
6	0.56	0.10	0.22	0.60	−0.22	0.60	0.30	0.43
7	0.00	1.00	0.30	0.40	−0.15	0.72	0.30	0.40
8	−0.61	0.04	0.07	0.91	0.20	0.58	0.00	1.00
9	0.02	1.00	0.11	0.74	−0.37	0.18	0.32	0.24
10	−0.39	0.22	−0.13	0.74	0.85	0.00	−0.07	0.91
11	0.08	0.85	0.60	0.04	−0.18	0.57	0.75	0.01
12	0.03	1.00	0.55	0.06	−0.14	0.68	0.49	0.10
13	0.68	0.04	0.37	0.36	0.05	1.00	0.26	0.54
14	0.39	0.08	0.21	0.37	−0.21	0.37	0.30	0.19
15	0.55	0.02	0.33	0.17	−0.17	0.49	0.40	0.10
16	0.28	0.20	0.46	0.03	0.08	0.75	0.37	0.09
17	0.42	0.16	0.14	0.69	0.08	0.84	0.53	0.07
18	−0.08	0.85	0.39	0.17	0.23	0.44	0.54	0.05
All sites	0.36	0.07	0.36	0.07	0.05	0.84	0.70	0.00

4.4. The Unusual 2013/14 Season

Our last objective was to assess how unusual the 2013/14 season was for waves in the longer-term (back to 2002) context. The 2013/14 season was the most extreme, since 2002, in terms of both the total number of threshold exceedances (49) and the number of storm events that were responsible for the extreme waves (16). Although this season was the one with the highest number of extreme events, there were some other notable seasons such as the 2012/13 and 2015/16 seasons. However, the 2013/14 season generated the maximum-record significant wave heights and return periods at six sites (2, 13, 15, 16, 17, and 18), particularly in the SE and SW (Table 4). In addition, 5 out of those 6 maximum return periods are ranked in the top ten and it is the only season featuring three events with a return period higher than 25 years. The largest return period was found at site 17 (Penzance), being ranked 2nd with a return period of 45 years. This event occurred on the 14th of February 2014 and the driving storm also led to the highest return period on record in site 16 and had a large spatial footprint, affecting five sites in the SW and SE regions. Waves were reported to have reached 7.5 m at the Channel Light Vessel and caused coastal flooding along the English Channel [36].

Half of the events in the 2013/14 season affected at least three sites and only 11% of the events affected just one site. The season also featured the top four largest spatial footprints. The event with the largest footprint led to a maximum return period of 15 years that was reached at site 18 on the 5th of February 2014 and is ranked 10th but was not the maximum return period recorded for any site. During this extreme event, very large swell waves were reported breaking against the coast of West Wales, Cornwall and also parts of Northern Ireland and Scotland due to the joint action of waves and the storm surge [36].

Temporal clustering of extreme wave events was probably the most unusual characteristic of the season. Six out of 16 events occurred within less than four days, and 10 within eight days. The events with the shortest inter-arrival times occurred on the 3rd and the 6th of January 2014 and led to a maximum return period of 1.3 and 1.8 years, respectively. These events caused high waves in nearby locations: the former affected two sites (16 and 3) in the SE and SW respectively, whereas the latter affected just one site in the south (13). The first event led to wave heights in excess of 6 m at the Aberporth Buoy and nearly the same value in the Bristol Channel Buoy. The second one caused wave heights >4 m at the WaveNet Buoy in Pool bay, with periods of about 20 s. Both events caused damages to coastal structures and flood warnings [36].

Overall, the SE and SW were the most severely affected regions by both the number of events and their clustering nature. The most affected sites were sites 3, 13, and 16, where wave heights exceeded the 1 in 1-year threshold six times. Sites 1, 14, and 15 were also strongly affected, as they recorded between four and five extreme wave heights exceedances. However, some places did not experience any 1 in 1-year return value exceedance at all, such as those located in the NE and one of the sites located in the E region (sites 7, 8, and 9, respectively).

Table 4. Name, location, and wave variable characteristics for each of the sites.

Site Number	Site Name	Mean Hs (m)	H_0 (m)	Mean Tz (s)	Mean Tp (s)	Max Hs (m)	Max Return Period (years)	Date of Max Return Period
1	Perranporth	1.56	1.71	5.81	10.47	7.3	5.30	8 February 2016 06:30
2	Minehead	0.56	0.58	3.99	6.51	2.97	38.96	2 November 2013 20:00
3	Scarweather	1.24	1.24	4.85	8.73	6.84	17.20	8 February 2016 15:00
4	Rhyl Flats	0.69	0.69	3.21	4.29	4.22	6.53	31 March 2010 10:00
5	Blackstone	2.59	2.59	6.38	10.64	13.8	7.78	1 February 2016 21:30
6	West of Hebrides	2.94	2.94	6.73	11.00	16.39	6.85	1 February 2016 18:30
7	Moray Firth	1.10	1.10	4.21	7.21	7.66	19.24	15 December 2012 01:30
8	Firth of Forth	1.04	1.04	4.27	6.87	8.48	12.73	25 September 2012 08:00
9	Tyne/Tees	1.34	1.34	4.64	7.47	7.92	29.37	22 March 2008 02:30
10	Hornsea	0.83	0.84	4.10	7.24	4.99	14.72	4 April 2012 04:30
11	NorthWell	0.64	0.64	3.35	4.50	3.23	25.57	21 November 2015 08:30
12	Backney Overfalls	0.92	0.92	3.87	5.99	5.43	80.51	17 December 2009 20:00
13	Goodwin Sands	0.67	0.67	3.57	5.21	3.69	43.89	24 December 2013 02:30
14	Folkestone	0.58	0.58	3.55	5.19	3.65	16.35	28 March 2016 04:30
15	Rustington	0.82	0.84	3.77	6.44	5.46	30.47	24 December 2013 02:30
16	Milford	0.64	0.67	4.19	8.18	4.5	19.06	14 February 2014 22:30
17	Start Bay	0.70	0.73	4.39	8.07	5.25	45.40	14 February 2014 21:30
18	Penzance	0.64	0.67	4.43	8.62	6.06	14.43	4 February 2014 19:00

5. Discussion

Observational data has been used to investigate the spatial footprints and temporal clustering of extreme wave events around the UK. Similar to Haigh et al. [15], who analysed extreme sea levels around the UK, we find a relationship between the return period of the wave events and the number of sites they affected. The correlation coefficient is 0.5 and this is statistically significant at 95% confidence. However, the unevenly spaced distance between study sites and the differences in data length and gaps lead to an underestimation of such correlation in the present study.

Six main spatial footprint types of extreme storm-wave events were identified and classified based on the location of sites that they encompass: SW; W; NW; NE; E; and SE. However, it must be noted that it has been difficult to distinguish between the SE and SW footprints, as some extreme events affected the entire south coast and the lack of data hinders a proper assessment of possible patterns. In contrast, Haigh et al. [15] found four main regions for extreme sea levels. Nonetheless, as spatial footprint extension depends on return period for both processes, this finding could be because they used a 1 in 5-year return period threshold. Importantly, Haigh et al. [15] also found that some unconnected coastline stretches were affected by the same storm, which has also been detected for some extreme wave events. In particular, this is observed for both the W and E regions. However, the spatial resolution of buoys recording during those events is rather poor to evaluate the spatial footprints in more detail.

Generally, storm track patterns are similar within the SE, SW, W, and NW regions. The eastern regions, however, have a more irregular pattern. The storm centers at time of the highest significant wave heights typically clustered within the western regions, whereas they were more spread out in the eastern regions, which reflects the same trend as the storm track patterns. Haigh et al. [15] found a clear pattern between the location of storm centers and the occurrence of extreme sea-levels, but this trend is not as evident for extreme wave events. The fact that storm centers during the peak of the extreme sea-levels seem more synchronised could be due to a strong relationship between their generation and the storm lowest pressure, whereas waves are more related to wind velocities which do not necessarily correspond with the storm centers [37].

Most extreme wave events occur during the autumn-winter months, which agrees with the commonly observed pattern of temporal variability in wave climate in the UK [38–40]. Temporal clustering is especially observed around December, January, and February when the atmospheric oscillations in the North Atlantic possess a higher control over the weather patterns in the region [27,41]. Previous studies linked this North Atlantic wave climate variability to the NAO index, e.g., [1,32]. However, strong spatial differences in the NAO influence have been identified here and in other studies, e.g., [27], suggesting that other atmospheric oscillations are also important in terms of wave formation along the UK coast. Castelle et al. [27] developed a new atmospheric pressure anomaly index that optimally explains the European west coast wave climate between the UK and Portugal. Similar to them, we also found a high correlation between the winter-averaged NAO index and extreme wave conditions in the NW region. In the same way, we found that their newly proposed WEPA index explains wave extreme events in the SW region better than the NAO. Our results suggest that such atmospheric oscillations can also be used to describe the number of extreme events in different regions; this is crucial to help stakeholders and decision makers to improve planning and manage coastal hazards more effectively. Interestingly, we found spatial correlation differences in the southern study sites: WEPA is the index that optimally explains extreme event occurrences in the SW region, while the NAO index performs better in the SE. This highlights the existence of two different regions regarding extreme wave event generation in the south of England, but longer data records are needed to explore this characteristic further. The results regarding the EA and SCAND oscillations showed poor performances and scattered spatial patterns, suggesting that these indices are less representative of extreme waves in the UK.

We also examined how the distinct atmospheric indices explain wave extreme event occurrences by considering the entire UK coastline. We found that the WEPA index explains a large part of the total number of extreme events per season in the UK with a correlation coefficient of 0.7 (significant at 95% confidence). While the WEPA index seems promising to describe wave extreme event variability along the UK coast, caution must be taken when interpreting this relationship as this index has been developed to better represent wave activity in the SW region. The SW region has a larger number of wave-buoys and longer records than other regions in this study, which can introduce bias.

Our final objective was to evaluate how unusual the 2013/2014 storm season was in terms of extreme wave heights over the instrumental record, since 2002. Masselink et al. [42] performed a similar study for the west coast of Europe by using a combination of wave buoy and wave climate reanalysis datasets. While they analysed longer time periods, our study includes a more spatially complete wave buoy dataset and is hence very useful to assess spatial and temporal clustering around the UK. Masselink et al. [42] demonstrated that the 2013/14 season was the most unusual over the last 67 years for the west coast of Europe and indicated that the region between south Ireland and NW France was the most affected. Similarly, our results show that the 2013/14 season was the most unusual in the UK and the SW region was most affected. The season was an outlier regarding all three extreme event spatial footprints, event return periods, and event inter-arrival times. The WEPA index for 2013/14 also had a high value, which indicates that the strong relationship with extreme wave events also holds for that particular season [27].

Throughout the analysis and result interpretation, the available record length of the wave data time series must be taken into account as a limiting factor and source of uncertainty. Many studies have shown the control that storms possess over wave climate, e.g., [43–45], stating that possible changes in storminess could increase future average significant wave heights [46–48]. Also, some studies have stated a strong interdecadal variability in wave climate in the North Atlantic Ocean [46,49,50]. As the mean length of our buoy records is about nine years, this variability could not have been detected, which may introduce uncertainty in the estimated significant wave height return values. Another issue is the uneven spatial distribution of buoy locations, which does not allow a proper assessment of the limits of some of the spatial footprints of extreme wave events. Moreover, there is a clear difference between the spatial resolution in the southern regions and the northern ones, which may underestimate the extent of the events in the north of the country. Furthermore, wave buoys often fail during the most extreme events. This was a significant issue during the events of the 2013/14 season (Figure 5). For our study it could potentially lead to an underestimation of the spatial footprints as there could also be exceedances in other locations where the buoys were not recording. Some extreme events may not have been detected at all, if multiple buoys failed at the same time in a specific region. This would also have an effect on our temporal clustering analysis: missed extreme events can lead to an overestimation of the inter-arrival time between events.

There are several ways in which the results could be enhanced in future studies. Coastal flooding is mainly due to the joint action of waves, tides, surges, riverine, and rain processes. Many studies highlight the importance of considering the interaction between those properties [4,6,13,15,51,52]. Hence, coupling the present spatial and temporal analysis of extreme waves with extreme surge events and tides would provide a more holistic approach to deal with flooding issues. Wave data length can be improved by making use of more sources but wave data records in the UK before the WaveNet project were sparse and inconsistent [23]. Further, hindcast wave data could be used to lengthen wave buoy data, though the accuracy of those data relies on an appropriate validation. Past research, e.g., [3,48,53], as well as our own assessment highlight the necessity of high quality and longer wave data to better understand the wave climate. Similarly, spatial consistency in the way wave data are recorded is crucial to improve our understanding of coastal flood risk and planning [47].

6. Conclusions

The overall aim of this study was to assess the spatial behaviour and temporal features of extreme storm-wave events by developing a multi-site analysis and utilising observational wave data along the UK coast. Specifically, this was done by examining the 'spatial footprint' and 'temporal clustering' of events. There is little guidance in the present literature regarding the spatial behaviour of extreme wave events and their consequences when the inter-spacing between their occurrence is shorter than usual, e.g., [15,19], especially for wave events around the UK. Yet, these processes are significant for coastal flood risk and shoreline management as they can amplify damages [18].

Six main regions were found in terms of the spatial behaviour of extreme storm-wave events. As expected, there is a strong relationship between the spatial extent of wave events and their return periods. The most significant events were swell events, except for sheltered areas such as Liverpool Bay. Some storms had a constant direction and velocity for a long period, leading to extreme swell events mainly on the SW coast of the UK possibly due to 'trapped-fetch' processes. The most important events often occur between November and March. There are large inter-seasonal and inter-site changes in the number of extreme events and their characteristics, which are principally linked to the WEPA index. However, spatial correlation variations exist and distinct atmospheric indices perform differently along the coast of the UK [27]. The 2013/14 season has been particularly extreme in the southern regions where six study sites experienced the highest wave heights on record. For the entire time-series, the 2013/14 season was undoubtedly the outlier, considering the size of the spatial footprints of the events, their return periods and the short inter-arrival times. The incomplete and short data coverage, however, hinders a quantification of this with high statistical certainty.

The significance of this work is the novel multi-site analysis for extreme wave events. The presented approach opens the door for similar analyses of the spatial footprints of extreme wave events and their tendency to occur in clusters and is applicable to any coastal region with sufficient observational (or high quality model) data. The importance of long-term strategic monitoring of waves is shown in this paper. Advancements in wave data consistency both temporally and spatially are essential to achieve more reliable results, especially in the light of climate change and potential increase in storminess.

Acknowledgments: This work would not have been possible without the information published by CCO, which freely provides the wave data on which this study is based. We must also express our gratitude to CEFAS and especially its WaveNet project, whose wave data have also been utilised to complement those provided by CCO. Also, thanks go to Travis Manson of CCO and David Pearce of CEFAS. Their technical support has been essential to accurately interpret the wave data which is provided by their organisations. This work contributes to the objectives of the Engineering and Physical Science Research Council Flood Memory Project (grant number EP/K013513/1). T.W. has received funding from the European Union's Horizon 2020 research and innovation program under the Marie Sklodowska-Curie grant agreement No. 658025. The work for this paper was undertaken by the lead author as a research project at the University of Southampton, as part of the Engineering in the Coastal Environment MSc.

Author Contributions: I.D.H. conceived the study and co-wrote the paper. V.M.S. performed the analysis and wrote the paper. T.W. participated in technical discussions and co-wrote the paper.

Conflicts of Interest: The authors declare no conflict of interest.

References

1. Wolf, J.; Brown, J.; Howarth, M. The wave climate of Liverpool Bay—Observations and modelling. *Ocean Dyn.* **2011**, *61*, 639–655. [CrossRef]
2. Tucker, M.J.; Pitt, E.G. *Waves in Ocean Engineering*; Elsevier: Amsterdam, The Netherlands, 2001; p. 521.
3. Clément, A.; McCullen, P.; Falcão, A.; Fiorentino, A.; Gardner, F.; Hammarlund, K.; Lemonis, G.; Lewis, T.; Nielsen, K.; Petroncini, S.; et al. Wave energy in Europe: Current status and perspectives. *Renew. Sustain. Energy Rev.* **2002**, *6*, 405–431. [CrossRef]
4. Wolf, J. Coupled wave and surge modelling and implications for coastal flooding. *Adv. Geosci.* **2008**, *17*, 19–22. [CrossRef]
5. Jonkman, S.N.; Vrijling, J.K. Loss of life due to floods. *J. Flood Risk Manag.* **2008**, *1*, 43–56. [CrossRef]
6. Wolf, J. Coastal flooding: Impacts of coupled wave-surge-tide models. *Nat. Hazards* **2008**, *49*, 241–260. [CrossRef]
7. Hinton, C.; Townend, I.H.; Nicholls, R.J. *Future Flooding and Coastal Erosion Risks*; Thomas Telford: London, UK, 2007; Chapter 9; p. 514.
8. Battjes, J.A.; Gerritsen, H. Coastal modelling for flood defence. *Philos. Trans. R. Soc. Lond. A Math. Phys. Eng. Sci.* **2002**, *360*, 1461–1475. [CrossRef] [PubMed]

9. McInnes, K.L.; Hubbert, G.D.; Abbs, D.J.; Oliver, S.E. A numerical modelling study of coastal flooding. *Meteorol. Atmos. Phys.* **2002**, *80*, 217–233. [CrossRef]
10. Lewis, M.; Horsburgh, K.; Bates, P.; Smith, R. Quantifying the uncertainty in future coastal flood risk estimates for the UK. *J. Coast. Res.* **2011**, *27*, 870–881. [CrossRef]
11. Lewis, M.; Schumann, G.; Bates, P.; Horsburgh, K. Understanding the variability of an extreme storm tide along a coastline. *Estuar. Coast. Shelf Sci.* **2013**, *123*, 19–25. [CrossRef]
12. Thorne, C. Geographies of UK flooding in 2013/4. *Geogr. J.* **2014**, *180*, 297–309. [CrossRef]
13. Wadey, M.; Haigh, I.; Brown, J. A century of sea level data and the UK's 2013/14 storm surges: An assessment of extremes and clustering using the Newlyn tide gauge record. *Ocean Sci. Discuss.* **2014**, *11*, 1995–2028. [CrossRef]
14. Chailan, R.; Toulemonde, G.; Bouchette, F.; Laurent, A.; Sevault, F.; Michaud, H. Spatial Assessment of Extreme Significant Wave Heights in the Gulf of Lions. *Int. Conf. Coast. Eng.* **2014**, *1*, 17. [CrossRef]
15. Haigh, I.D.; Wadey, M.P.; Wahl, T.; Ozsoy, O.; Nicholls, R.J.; Brown, J.M.; Horsburgh, K.; Goulby, B. Spatial Footprint and Temporal Clustering of Extreme Sea Level and Storm Surge Evens around the Coastline of the UK. *Sci. Data* **2016**, *3*, 160107. [CrossRef] [PubMed]
16. Hall, J.W.; Tran, M.; Hickford, A.J.; Nicholls, R.J. (Eds.) *The Future of National Infrastructure: A System-of-Systems Approach*; Cambridge University Press: Cambridge, UK, 2016.
17. Karunarathna, H.; Pender, D.; Ranasinghe, R.; Short, A.D.; Reeve, E. The effects of storm clustering on beach profile variability. *Mar. Geol.* **2014**, *348*, 103–112. [CrossRef]
18. Ferreira, O. Storm groups versus extreme single storms: Predicted erosion and management consequences. *J. Coast. Res.* **2005**, *42*, 221–227.
19. Mailier, P. Serial Clustering of Extratropical Cyclones. Ph.D. Thesis, Department of Meteorology, University of Reading, Reading, UK, June 2007.
20. De la Vega-Leinert, A.; Nicholls, R. Potential Implications of Sea-Level Rise for Great Britain. *J. Coast. Res.* **2008**, *242*, 342–357. [CrossRef]
21. Wadey, M.P.; Haigh, I.D.; Brown, J. A temporal and spatial assessment of extreme sea level events for the UK coast. In Proceedings of the International Short Conference on Advances in Extreme Value Analysis and Application to Natural Hazards (EVAN2013), Siegen, Germany, 18–20 September 2013.
22. Cefas—WaveNet. Available online: http://www.cefas.co.uk/cefas-data-hub/wavenet/ (accessed on 14 July 2016).
23. Hawkes, P.J.; Atkins, R.; Brampton, A.H.; Fortune, D.; Garbett, R.; Gouldby, B.P. *WAVENET: Nearshore Wave Recording Network for England and Wales, Feasibility Study*; HR Wallingford: Wallingford, UK, 2001.
24. CCO Map Viewer & Data Search. Available online: http://www.channelcoast.org/ (accessed on 14 July 2016).
25. Compo, G.P.; Whitaker, J.S.; Sardeshmukh, P.D.; Matsui, N.; Allan, R.J.; Yin, X.; Gleason, B.E.; Vose, R.S.; Rutledge, G.; Bessemoulin, P.; et al. The Twentieth Century Reanalysis Project. *Q. J. R. Meteorol. Soc.* **2001**, *137*, 1–28. [CrossRef]
26. Kalnay, E.; Kanamitsu, M.; Kistler, R.; Collins, W.; Deaven, D.; Gandin, L.; Iredell, M.; Saha, S.; White, G.; Woollen, J.; et al. The NCEP/NCAR 40-Year Re analysis Project. *Bull. Am. Meteorol. Soc.* **1996**, *77*, 437–471. [CrossRef]
27. Castelle, B.; Dodet, G.; Masselink, G.; Scott, T. A new climate index controlling winter wave activity along the Atlantic coast of Europe: The West Europe Pressure Anomaly. *Geophys. Res. Lett.* **2017**, *44*, 1384–1392. [CrossRef]
28. Jones, P.D.; Jonsson, T.; Wheeler, D. Extension to the North Atlantic Oscillation using early instrumental pressure observations from Gibraltar and South-West Iceland. *Int. J. Climatol.* **1997**, *17*, 1433–1450. [CrossRef]
29. NAO Data. Available online: https://crudata.uea.ac.uk/cru/data/nao/ (accessed on 18 June 2017).
30. Climate Prediction Center—Monitoring & Data Index. Available online: http://www.cpc.ncep.noaa.gov/data/ (accessed on 18 June 2017).
31. Paolo, C.; Coco, G. (Eds.) *Coastal Storms: Processes and Impacts*; John Wiley & Sons: Hoboken, NJ, USA, 2017.
32. Bacon, S.; Carter, D. A connection between mean wave height and atmospheric pressure gradient in the North Atlantic. *Int. J. Climatol.* **1993**, *13*, 423–436. [CrossRef]
33. Bauer, E. Inter-annual changes of the ocean wave variability in the North Atlantic and in the North Sea. *Clim. Res.* **2001**, *18*, 63–69. [CrossRef]

34. Masselink, G.; Austin, M.; Scott, T.; Poate, T.; Russell, P. Role of wave forcing, storms and NAO in outer bar dynamics on a high-energy, macro-tidal beach. *Geomorphology* **2014**, *226*, 76–93. [CrossRef]

35. Galanis, G.; Chu, P.; Kallos, G.; Kuo, Y.; Dodson, C. Wave height characteristics in the north Atlantic Ocean: A new approach based on statistical and geometrical techniques. *Stoch. Environ. Res. Risk Assess.* **2011**, *26*, 83–103. [CrossRef]

36. Sibley, A.; Cox, D.; Titley, H. Coastal flooding in England and Wales from Atlantic and North Sea storms during the 2013/2014 winter. *Weather* **2015**, *70*, 62–70. [CrossRef]

37. Nissen, K.; Leckebusch, G.; Pinto, J.; Renggli, D.; Ulbrich, S.; Ulbrich, U. Cyclones causing wind storms in the Mediterranean: Characteristics, trends and links to large-scale patterns. *Nat. Hazards Earth Syst. Sci.* **2010**, *10*, 1379–1391. [CrossRef]

38. Woolf, D.K.; Challenor, P.G.; Cotton, P.D. Variability and predictability of the North Atlantic wave climate. *J. Geophys. Res.* **2002**, *107*, 3145. [CrossRef]

39. Van Nieuwkoop, J.; Smith, H.; Smith, G.; Johanning, L. Wave resource assessment along the Cornish coast (UK) from a 23-year hindcast dataset validated against buoy measurements. *Renew. Energy* **2013**, *58*, 1–14. [CrossRef]

40. Venugopal, V.; Nemalidinne, R. Wave resource assessment for Scottish waters using a large scale North Atlantic spectral wave model. *Renew. Energy* **2015**, *76*, 503–525. [CrossRef]

41. Wolf, J.; Woolf, D.K. *Storms and Waves. MCCIP Annual Report Card 2010–11*; MCCIP Science Review: Suffolk, UK, 2010; p. 15.

42. Masselink, G.; Castelle, B.; Scott, T.; Dodet, G.; Suanez, S.; Jackson, D.; Floc'h, F. Extreme wave activity during 2013/2014 winter and morphological impacts along the Atlantic coast of Europe. *Geophys. Res. Lett.* **2016**, *43*, 2135–2143. [CrossRef]

43. Wiegel, R.L. *Oceanographical Engineering*; Courier Corporation: North Chelmsford, MA, USA, 2013.

44. Young, I.R. *Wind Generated Ocean Waves*; Elsevier: Amsterdam, The Netherlands, 1999; Volume 2.

45. Dodet, G.; Bertin, X.; Taborda, R. Wave climate variability in the North-East Atlantic Ocean over the last six decades. *Ocean Model.* **2010**, *31*, 120–131. [CrossRef]

46. Wolf, J.; Woolf, D. Waves and climate change in the north-east Atlantic. *Geophys. Res. Lett.* **2006**, *33*, L06604. [CrossRef]

47. Lowe, J.A.; Howard, T.; Pardaens, A.; Tinker, J.; Holt, J.; Wakelin, S.; Milne, G.; Leak, J.; Wolf, J.; Horsburgh, K.; et al. *UK Climate Projections Science Report: Marine and Coastal Projections*; Met Office Hadley Centre: Exeter, UK, 2009.

48. Mori, N.; Yasuda, T.; Mase, H.; Tom, T.; Oku, Y. Projection of Extreme Wave Climate Change under Global Warming. *Hydrol. Res. Lett.* **2010**, *4*, 15–19. [CrossRef]

49. Gulev, S.; Hasse, L. Changes of wind waves in the North Atlantic over the last 30 years. *Int. J. Climatol.* **1999**, *19*, 1091–1117. [CrossRef]

50. Sterl, A.; Caires, S. Climatology, variability and extrema of ocean waves: The Web-based KNMI/ERA-40 wave atlas. *Int. J. Climatol.* **2005**, *25*, 963–977. [CrossRef]

51. Ozer, J.; Padilla-Hernández, R.; Monbaliu, J.; Alvarez Fanjul, E.; Carretero Albiach, J.; Osuna, P.; Yu, J.; Wolf, J. A coupling module for tides, surges and waves. *Coast. Eng.* **2000**, *41*, 95–124. [CrossRef]

52. Bunya, S.; Dietrich, J.; Westerink, J.; Ebersole, B.; Smith, J.; Atkinson, J.; Jensen, R.; Resio, D.; Luettich, R.; Dawson, C.; et al. A High-Resolution Coupled Riverine Flow, Tide, Wind, Wind Wave, and Storm Surge Model for Southern Louisiana and Mississippi. Part I: Model Development and Validation. *Mon. Weather Rev.* **2010**, *138*, 345–377. [CrossRef]

53. Jane, R.; Dalla Valle, L.; Simmonds, D.; Raby, A. A copula-based approach for the estimation of wave height records through spatial correlation. *Coast. Eng.* **2016**, *117*, 1–18. [CrossRef]

Journal of
Marine Science and Engineering

MDPI

Article

Observed Sea-Level Changes along the Norwegian Coast

Kristian Breili [1,2,*], Matthew J. R. Simpson [1] and Jan Even Øie Nilsen [3]

[1] Geodetic Institute, Norwegian Mapping Authority, NO-3507 Hønefoss, Norway; matthew.simpson@kartverket.no

[2] Faculty of Science and Technology, Norwegian University of Life Sciences (NMBU), NO-1432 Ås, Norway

[3] Nansen Environmental and Remote Sensing Center and Bjerknes Centre for Climate Research, 5006 Bergen, Norway; jan.even.nilsen@nersc.no

* Correspondence: kristian.breili@kartverket.no; Tel.: +47-3211-8211

Received: 31 May 2017; Accepted: 12 July 2017; Published: 17 July 2017

Abstract: Norway's national sea level observing system consists of an extensive array of tide gauges, permanent GNSS stations, and lines of repeated levelling. Here, we make use of this observation system to calculate relative sea-level rates and rates corrected for glacial isostatic adjustment (GIA) along the Norwegian coast for three different periods, i.e., 1960 to 2010, 1984 to 2014, and 1993 to 2016. For all periods, the relative sea-level rates show considerable spatial variations that are largely due to differences in vertical land motion due to GIA. The variation is reduced by applying corrections for vertical land motion and associated gravitational effects on sea level. For 1960 to 2010 and 1984 to 2014, the coastal average GIA-corrected rates for Norway are 2.0 ± 0.6 mm/year and 2.2 ± 0.6 mm/year, respectively. This is close to the rate of global sea-level rise for the same periods. For the most recent period, 1993 to 2016, the GIA-corrected coastal average is 3.5 ± 0.6 mm/year and 3.2 ± 0.6 mm/year with and without inverse barometer (IB) corrections, respectively, which is significantly higher than for the two earlier periods. For 1993 to 2016, the coastal average IB-corrected rates show broad agreement with two independent sets of altimetry. This suggests that there is no systematic error in the vertical land motion corrections applied to the tide-gauge data. At the same time, altimetry does not capture the spatial variation identified in the tide-gauge records. This could be an effect of using altimetry observations off the coast instead of directly at each tide gauge. Finally, we note that, owing to natural variability in the climate system, our estimates are highly sensitive to the selected study period. For example, using a 30-year moving window, we find that the estimated rates may change by up to 1 mm/year when shifting the start epoch by only one year.

Keywords: sea-level change; glacial isostatic adjustment; tide gauges; altimetry

1. Introduction

The Norwegian coast is located in the periphery of the Fennoscandian land-uplift area, experiencing uplift rates of 0.2 to 5 mm/year and associated gravitational effects on sea level of up to 0.5 mm/year [1]. The coastline is complex with fjords, islands, reefs, ocean currents, and significant variations in the tidal regime. Relative sea-level (RSL) rates vary along the coast and differ considerably from the rate of global mean sea-level rise (see, e.g., [2] for an overview of rates estimated from global mean sea level (GMSL) reconstructions). The Norwegian Mapping Authority (NMA) operates an array of 23 tide gauges over the region that continuously record water level heights (e.g., the tides, extreme sea levels, and changes to mean sea level) (see Figure 1 and Table 1). They form a north–south transect covering a latitudinal band of approximately 13 degrees. Together with an extensive network of permanently installed Global Navigation Satellite System (GNSS) receivers and lines of repeated levelling, the tide gauges contribute to Norway's national sea level observing system. In combination, the data from the observation system allow sea-level rates corrected for Glacial Isostatic Adjustment (GIA) to be computed, i.e., sea-level rates corrected for vertical land motion (VLM) and associated gravitational effects.

Figure 1. Locations of Norwegian tide gauges.

Understanding spatial variability of coastal sea-level changes is of critical importance for flood risk assessment, climate adaption in the coastal zone, and for understanding the processes that drive sea-level changes (see, e.g., [3–6]). The records from the Norwegian tide gauges have been subject to extensive previous research. A comprehensive study of the Norwegian tide gauges is [7]. Using empirical orthogonal function–analysis, the authors find the leading mode of sea-level variability for the Norwegian tide gauges. It shows a trend of 2.9 ± 0.3 mm/year for 1960 to 2010 (after correcting for VLM). There are also other investigations that have included data from the Norwegian tide gauges in wider regional analyses (e.g., [8–14]). We briefly summarize the findings from some of these studies here. [9] finds averaged regional sea-level change over Fennoscandia of 1.32 mm/year (corrected for VLM) for 1891 to 1990. [11] analyze sea-level trends from the Norwegian tide gauges for 1950 to 2009, and find markedly lower rates than reported by [7]. This could be due to slightly different periods analyzed ([11] opt to include data from the 1950s, which generally had higher sea levels than the following decades) and/or their use of a different VLM correction. In a study of tide-gauge records surrounding the North Sea, a trend of 1.6 ± 0.9 mm/year (corrected for GIA) was found for the period 1960 to 2000 [10]. [12] also address the North Sea region, analyzing data from both tide gauges and satellite altimetry. Their study presents three index time series that are the arithmetic means for each year across three subsets of GIA corrected tide-gauge records. The index time series were analyzed for a possible acceleration in regional sea-level rise over the past few decades by applying Singular Spectrum Analysis (see, e.g., [15]). Despite a linear long-term trend of roughly 1.6 ± 0.9 mm/year since 1900, no evidence was found for a significant acceleration in the North Sea region.

Table 1. Overview of the Norwegian tide gauges, ordered along the coast from north to south. The table lists longitude/latitude [degrees] of each tide gauge (from the archive of the Norwegian Mapping Authority), id-numbers in the Permanent Service for Mean Sea Level–archive, the start month of each record, the percentage of observations available for the periods 1960–2010, 1984–2014, and 1993–2016, and significant gaps (>= 1 year). (* : The tide gauge in Trondheim was relocated in 1990.)

Tide Gauge Name	Longitude (° E) Latitude (° N)	PSMSL-ID	Start yyyy.m	1960–2010 (%)	1984–2014 (%)	1993–2016 (%)	Gap
Vardø	31.104015 70.374978	524	1947.7	60	95	95	1966.2–1984.0
Honningsvåg	25.972697 70.980318	1267	1970.5	75	94	100	
Hammerfest	23.683227 70.664641	758	1957.0	88	99	99	1970.0–1971.0 1982.0–1983.0
Tromsø	18.961323 69.647424	680	1952.4	98	98	100	
Andenes	16.134848 69.326067	425	1938.0	52	78	100	1955.8–1974.0 1978.9–1982.0
Harstad	16.548236 68.801261	681	1952.2	94	97	100	
Narvik	17.425759 68.428286	312	1928.1	98	97	100	1940.3–1947.3
Kabelvåg	14.482149 68.212639	45	1948.0	97	97	100	
Bodø	14.390813 67.288290	562	1949.7	89	95	99	1953.5–1954.5 1971.0–1972.0 1972.5–1973.6
Rørvik	11.230107 64.859456	1241	1969.7	80	99	100	
Mausund	8.665230 63.869330		1988.0	39	78	89	2005.9–2008.0
Trondheim-1 *	10.391669 63.436484	34	1945.5	60	20	-	1946.5–1949.0
Trondheim-2	10.391669 63.436484	1748	1990.0	40	80	100	
Heimsjø	9.101504 63.425224	313	1928.0	99	99	100	
Kristiansund	7.734352 63.113859	682	1952.4	99	98	100	
Ålesund	6.151946 62.469414	509	1945.1	98	99	100	1946.1–1951.0
Måløy	5.113310 61.933776	486	1943.5	95	99	100	1959.0–1961.0
Bergen	5.320487 60.398046	58	1915.0	98	99	100	1941.9–1944.0
Stavanger	5.730121 58.974339	47	1919.0	96	100	100	1940.0–1946.0 1970.0–1971.3
Tregde	7.554759 58.006377	302	1927.8	99	99	99	
Helgeroa	9.856379 58.995212	1113	1965.4	64	99	100	1970.0–1981.0
Oscarsborg	10.604861 59.678073	33	1872.1	90	99	100	1883.0–1953.5
Oslo	10.734510 59.908559	62	1885.5	96	97	100	1891.0–1914.0
Viker	10.949769 59.036046	1759	1990.9	38	76	100	

In addition to ground based measurements of sea level and land uplift, the open seas and offshore waters of Norway are sampled by altimetry satellites. The records from the altimetry satellites supplement the tide gauges, and allow changes in the sea surface height (SSH) to be calculated directly in a global geodetic reference frame. To the best of our knowledge, there is no single study that focuses on sea-level trends estimated from altimetry for the Norwegian coast. However, the Norwegian coast is included in several investigations that have addressed sea-level change in the Arctic Ocean. In [16], gridded multi-mission data were analyzed using observations from the period 1992 to 2012. The gridded data were bilinearly interpolated to the positions of the Norwegian tide gauges at Kristiansund, Rørvik, Andenes, Hammerfest, Honningsvåg, and Vardø. At these locations, the SSH rates were estimated as 3.5, 4.4, 4.3, 4.0, 4.0, and 4.1 mm/year, respectively. Similar results are also reported in [11] and [17]. Both studies use the same multi-mission data originally compiled for studying the Arctic Ocean. In [11], the average rate around 11 Norwegian tide gauges (from Måløy to Hammerfest) was estimated to be 4.23 mm/year for 1993 to 2009. The authors also found that SSH changes were somewhat higher north of Sognefjorden 61° N (4 to 6 mm/year) compared to south of the fjord (2 to 4 mm/year). [17] examined sea-level rates north of 55° N and found similar results to those reported by [11]. They also estimated the sea-level rate for the Arctic region north of 66° N to be 3.6 ± 1.3 mm/year for 1993 to 2009.

In this study, we revisit sea-level changes along the Norwegian coast. Our main motivation is to understand how sea level has changed over the instrumental record and the more recent decades. Hence, we present updated sea-level rates at each tide gauge for three different time periods, i.e., 1960 to 2010, 1984 to 2014, and 1993 to 2016. We focus on the first period (1960 to 2010) because we have an understanding of the different components of sea-level change over this time and different studies have looked at the same period [6,7]. Choosing this period, we obtain for several Norwegian tide gauges the longest interval available without major voids of data (see Table 1). The second set uses data from the past 30 years (1984 to 2014) and represents present sea-level change along the Norwegian coast. The third set (1993 to 2016) is included for comparison to satellite altimetry. To quantify different contributors to sea-level change, we pay special attention to the effects of GIA on sea level and apply GIA-corrections calculated from the Norwegian GNSS network and lines of repeated levelling. Realistic error estimates taking into account time-correlated noise and systematic errors in the GIA-corrections are also presented. The fully GIA-corrected rates allow us to check for regional variation and to compare the tide-gauge rates to rates derived from SSH observed with altimetry in the vicinity of the tide gauges. Our hypothesis is that altimetry and tide gauges should observe the same sea-level rates and capture the same spatial variation in the sea-level signal. Otherwise, systematic errors may be present in the observations and/or the GIA-corrections applied.

2. Data and Methods

2.1. Analysis of the Norwegian Tide Gauges

We use data from the Permanent Service for Mean Sea Level (PSMSL, [18]) for all stations except Mausund, and follow their recommendation of only using the revised local reference datasets. These datasets are reduced to a common datum by making use of the tide-gauge datum history provided by the supplying authority; this means that any shifts in the records are removed. The PSMSL provides both monthly and annual data records. The monthly records appear to be more complete when compared to the annual records. The reason is that annual averages are computed only when 11 or 12 months of data are available. In this study, we chose to use the monthly datasets because they have shorter voids of data. For Mausund, we used data found in the archives of the NMA because this tide gauge is presently not included in the PSMSL database.

To determine long-term trends from the observed RSL changes, we conduct a least squares adjustment for each tide gauge, i.e., the model in Equation (1) was fitted to the records:

$$z(t) = a + b\,t + A_1 \sin(2\pi\,t - \phi_1) + A_2 \sin(4\pi\,t - \phi_2) + \epsilon(t). \tag{1}$$

In Equation (1), $z(t)$ is the observation at the epoch t, a is the intersect of the model, b is the rate of sea-level change, A_1, φ_1 and A_2, φ_2 are the amplitudes and phases of the annual and semiannual periodic variation in the time series, and ϵ is the error. The annual periodic term was included because visual inspection of the monthly datasets revealed significant annual variation. Annual variation arises due to variation in the steric sea level and atmospheric pressure [7]. The amplitude and phase of the annual signal may vary from year to year but were here estimated as constant parameters. If not captured by the model, the annual variation increases the standard error of the estimated rate of sea-level change. We also include the semiannual component. On the other hand, we did not include the 18.6-year periodic term representing the nodal tide. Neither was a priori corrections applied, as suggested by [19]. The nodal term is somewhat debated. For instance, [20] advocate to include it when estimating sea-level rise and acceleration. In addition, [6] include the nodal term in order to reconstruct changes in RSL along the Norwegian coast. On the other hand, [19] concludes that the majority of studies that report identified nodal signals in tide gauge records are a consequence of misidentified ocean decadal variations. Before we made our decision, we transformed the tide-gauge records into periodograms by applying the Lomb–Scargle transform [21] and searched for peaks corresponding to the nodal term. In order to evaluate the statistical significance of the peaks, detection thresholds were computed with the false alarm probability set to 5%. No significant peaks were identified at period 18.6 years. As a control experiment, we investigated the effect of including the 18.6 year period on the estimated rates. The effect was up to 0.3 mm/year, and all study periods were considered. We also computed Akaike's Information Criterion (AIC) and the Bayesian Information Criterion (BIC) for model selection. For the longest study period, AIC indicated improved model quality at 11 of 15 tide gauges when the nodal term was included. On the contrary, the BIC indicated aggregated model quality at all stations. A similar result was obtained also for the shorter and more recent periods. Hence, by not including the nodal term, we stick to one model for all tide gauges and all study periods. Using this model in Equation (1), the sea-level rates are assumed to be constant within the study period. The monthly PSMSL-records are supplied with quality flags indicating the number of days missing observations. We did not filter the data with respect to this flag. Instead, observations deviating more than three times the standard deviation from the model were flagged as outliers before final adjustment.

The regression of the tide-gauge observations is complicated by time-correlated noise. If not taken into account, the standard errors of the rates may be underestimated [22,23]. Hence, we did a preliminary fit of the model in Equation (1) and investigated the sample autocorrelation function (ACF) and the sample partial autocorrelation function (PACF) of the residuals. For all tide gauges, the sample PACFs were significantly different from zero (at the 95% level) for lag one (corresponding to one month) with values of the order of 0.25 to 0.42. For greater lags, the PACF were within the significance-threshold. The sample ACFs had values of similar magnitude as the PACFs. They were significantly different from zero for lag one at all stations and close to or within the significance threshold for lags beyond one. We therefore characterize both the PACF and the ACF functions as fast decaying functions and conclude that the residuals are moderately time-correlated. This suggests that the series of errors may be described by a first order autoregressive process (AR1), i.e., the errors can be written

$$\epsilon(t_i) = \Phi \cdot \epsilon(t_{i-1}) + w(t_i), \tag{2}$$

where $w(t_i) \sim N(\sigma)$ (white noise) and Φ is the parameter of the AR1 model. This stochastic model implies that each observation is only affected by the previous observation and by white noise. Using the AR1 noise model gave rates with standard errors 20–40% larger than the standard errors of the preliminary fit, which did not take into account time-correlated noise.

We note that other stochastic models can be used to take into account time-correlated noise in tide-gauge records. [22] are critical of only using the AR1-model. They find that the choice of model depends on the sampling rate (monthly/annual data), the length of the record, and the location. For most tide gauges along the Norwegian coast, the Generalized Gauss Markov stochastic model performs best when applying the BIC. However, the AR1 model is also found to perform satisfactorily. Therefore, we opt to stick with the AR1 model in our analysis because it is easy to implement. However, we are aware that the AR1 model may underestimate the rate uncertainties by a factor of 1.3 to 1.5 [22].

For Norway, vertical land motion due to GIA is an important component of contemporary relative sea-level change. GIA is a result of ongoing relaxation of the Earth in response to past ice mass loss and causes crustal deformations and associated changes in the Earth's gravity field. Using a simple exponential model, [24] show that the acceleration of vertical land motion due to GIA in Fennoscandia is proportional to the present GIA rates by a factor of -0.0002 year^{-1}. Hence, along the Norwegian coast, the effects of GIA on sea-level can be assumed to be constant and do not introduce significant accelerations on decadal or centennial timescales. The land motion signal can be separated from the tide-gauge records using GIA modeling and/or observations from permanent GNSS stations and levelling. In addition, it is worth remembering that GIA also affects Earth's gravity field and, therefore, acts to perturb the ocean surface. This effect needs to be taken into account if the tide-gauge data are to be "fully GIA corrected" and to help us understand the separate contributions to sea-level change (see [25]). As well as observed relative sea-level rates from the tide gauges ($\dot{S}_{tg}$), we also present rates that are what we call GIA-corrected ($\dot{S}_{giacorr}$). That is, the RSL rates are adjusted for both vertical land motion ($\dot{S}_{gnss,lev}$) and geoid changes associated with GIA ($\dot{S}_{gravgia}$). The correction for VLM is based upon observations from GNSS and levelling, whereas the geoid change is generated from a GIA model [1]. The relation between these processes is defined in Equation (3):

$$\dot{S}(\lambda, \varphi)_{tg} = -\dot{S}(\lambda, \varphi)_{gnss,lev} + \dot{S}(\lambda, \varphi)_{gravgia} + \dot{S}(\lambda, \varphi)_{giacorr}, \tag{3}$$

where λ and φ are the longitude and latitude of the tide-gauge location. The GIA-corrected rates are in principle equivalent to the SSH change caused by changes in ocean mass, density, circulation, atmospheric pressure, etc.

For each period examined, we only include tide gauges where more than 80% of the data are available (see Table 1), i.e., records with too short duration or with significant data gaps are excluded from the analysis. As the length of the tide-gauge records varies, the three data sets include different sets of tide gauges. For the period 1960 to 2010, the tide gauges at Viker, Helgeroa, Mausund, Trondheim, Rørvik, Andenes, Honningsvåg, and Vardø are omitted from our analysis. The time series from Helgeroa, Andenes, and Vardø suffer from significant data gaps, whereas the records from Viker, Mausund, Rørvik, and Honningsvåg are too short. Trondheim is omitted as the tide gauge was relocated in 1990. For 1984 to 2014, all tide gauges except Viker, Mausund, Andenes, and Vardø are used and for the same reasons as reported above. Trends are computed for 1993 to 2016 for all tide gauges, where rates calculated from altimetry are also available. Because the altimetry records are corrected for effects due to varying atmospheric pressure, we apply inverse barometer (IB) corrections to the tide gauge records from the most recent period (1993 to 2016) for consistency. The IB-corrections represent the hydrostatic response of the sea surface due to low frequency changes in atmospheric pressure. As a rule of thumb, a one hecto-Pascal increase in atmospheric pressure results in a sea surface depressed by 1 cm. The effect can be computed from Equation (4)

$$\Delta h = -0.99484(P_0 - P_{ref}), \tag{4}$$

where P_0 is observed atmospheric pressure, P_{ref} is the reference pressure of 1011 hPa and -9.99484 mm/hPa is the admittance between sea level and atmospheric pressure [26]. For the Norwegian tide gauges, we used atmospheric pressure observed at nearby meteorological stations. The observations were downloaded from the web pages of the Norwegian Meteorological Institute.

The uncertainty on the GIA-corrected rates ($\sigma^2_{giacorr}$) is calculated as the sum of the error on the tide-gauge regression (σ^2_{tg}), the observed VLM error ($\sigma^2_{gnss,lev}$), the uncertainty on the reference frame's z-drift ($\sigma^2_{z\text{-}drift}$) and scale error ($\sigma^2_{scale}$), and the geoid change error ($\sigma^2_{gravgia}$), as follows:

$$\sigma^2_{giacorr} = \sigma^2_{tg} + \sigma^2_{gnss,lev} + \sigma^2_{z\text{-}drift} + \sigma^2_{scale} + \sigma^2_{gravgia}. \tag{5}$$

Note that the observed VLM error is in the range 0.2 to 0.3 mm/year. Reference frame errors are adopted from a recent review, which concluded that the International Terrestrial Reference Frame (ITRF) is stable along each axis to better than 0.5 mm/year (z-drift) and has a scale error of less than 0.3 mm/year [27]. The uncertainty on the geoid change associated with GIA is very small and has a value of typically 0.03 mm/year. In general, the uncertainties vary according to the length of each time series and increase considerably when the uncertainties of VLM, the reference frame, and geoid changes are taken into account.

2.2. Analysis of Altimetry Data from the Norwegian Coast

Over the past 25 years, satellite altimetry has been a major technique for mapping sea surface topography and measuring sea-level changes. Compared to computing global sea level, it is more challenging to measure regional sea level within a smaller area like the Norwegian coast. This is because ocean variability is generally larger on regional scales due to redistribution effects like wind. In addition, regional altimetry is more sensitive to errors that are often negligible when calculating the global average. This could be errors in the ocean tide model, sea state corrections, and orbital errors [28].

Applications of satellite altimetry in coastal areas (closer than about 30 km to the land) are especially demanding [29]. In the coastal zones, the quality of the range measurements is degraded because the radar pulses are reflected partly from land and partly from the sea. It is also more difficult to compute accurate range corrections and the tidal patterns are more complex to model. As a consequence, altimetry observations closer than ∼30 km to the coast are normally not used. Hence, the estimates reported below for the Norwegian coast do not strictly represent sea-level change at the coast.

Two sets of altimetry data are compiled. The first combines observations from the three satellites TOPEX/POSEIDON, Jason-1, and Jason-2 and samples near-coastal waters (within approximately 20 km from the coastline) south of 66° N (see left panel of Figure 5). The second dataset combines data from ERS-1, ERS-2, ENVISAT, and Saral/AltiKa and samples the entire Norwegian coast (see right panel of Figure 5). Both datasets cover the period 1993 to 2016. The altimetry data were provided by several data centers. For TOPEX/POSEIDON, we used the generation B merged geophysical data records (GDR) distributed by NASA/JPL/PODAAC. Data from Jason-1 (GDR-C), Jason-2 (GDR-D), and SARAL/AltiKa (GDR-T) were downloaded from the AVISO portal of the Centre National d'Etudes Spatiales [30]. The data from the European satellites ENVISAT (GDR version 2) and ERS-1/2 (REAPER products) were provided by the European Space Agency and downloaded from their Earth Online portal [31].

When combining data from several missions, it is crucial to estimate possible intermission measurement biases. Biases of up to several decimeters may arise because the measurements suffer from residual errors in the observation system, the algorithms and parameters used in ground processing, and the applied range- and geophysical corrections. Often, the relative bias for a pair of missions is assumed to be time-invariant and can be assessed if data from a common period exist. Here, global cycle-averages were first computed for each satellite, and then differentiated in order to compute the bias. This procedure is straightforward for the first dataset because data from TOPEX/POSEIDON partly overlap in time with data from Jason-1, and data from Jason-1 overlap partly with data from Jason-2. The second dataset is more complex because the data from ENVISAT and Saral/AltiKa do not overlap in time. We therefore first compute the relative bias between ENVISAT and Jason-2, and then the bias between Saral/AltiKa and Jason-2. Finally, the bias between ENVISAT and Saral/AltiKa was computed by combining their relative biases to Jason-2.

Sea-level rates were computed around tide gauges and in grid points along the Norwegian coast. For each point, time series of altimetry observations were generated by computing cycle-averages for all observations within a spherical distance of 1°. Following this, a least squares adjustment was used to fit the model defined in Equation (1) to the time series. We also apply a correction for geoid changes associated with GIA using modelling results from [1].

The uncertainties of the regional estimates are difficult to assess. They cannot be estimated from the data alone because systematic effects dominate. We therefore combine the uncertainty from the regression (σ_{reg}^2) with the uncertainties of known systematic effects, i.e., the error of geoid correction ($\sigma_{\text{gravgia}}^2$), the z-drift of the reference frame ($\sigma_{\text{z-drift}}^2$), the scale rate of the reference frame (σ_{scale}^2), and the models (σ_{models}^2) used to compute the sea surface heights (e.g., the ocean tide model):

$$\sigma_{\text{altimetry}}^2 = \sigma_{\text{reg}}^2 + \sigma_{\text{gravgia}}^2 + \sigma_{\text{z-drift}}^2 + \sigma_{\text{scale}}^2 + \sigma_{\text{models}}^2. \tag{6}$$

The reference frame uncertainties are difficult to assess because the altimetry time series are defined in both ITRF2005 and CSR95 and the combined uncertainty is poorly constrained. The reference frame uncertainties applied in our tide-gauge analysis are 0.5 mm/year and 0.3 mm/year for the z-drift and scale rate, respectively [27]. These values are computed for ITRF2008, but we assume that the combined uncertainty for ITRF2005 and CSR95 is of the same order. The uncertainty of the models is poorly constrained for regional estimates. We therefore use the upper limit (0.44 mm/year) of the range of the global uncertainty reported in [32], but caution that this value may be too optimistic for regional estimates. In total, the standard errors of the regional altimetry estimates are approximately 1 mm/year.

3. Sea-Level Rates along the Norwegian Coast

Our estimated relative and GIA-corrected rates are shown in Figures 2 and 3, and are listed in Table 2. For the period 1960 to 2010, about half of the RSL rates computed are negative, i.e., RSL has fallen during this period as the rate of land uplift is greater than the rate of sea surface rise (see upper panel of Figure 2 and left panel of Figure 3). Lowest rates are found at Oslo, Oscarsborg, and in the middle part of Norway. The highest rates are found along the south and west coast of Norway and at Bodø, Tromsø, and Hammerfest.

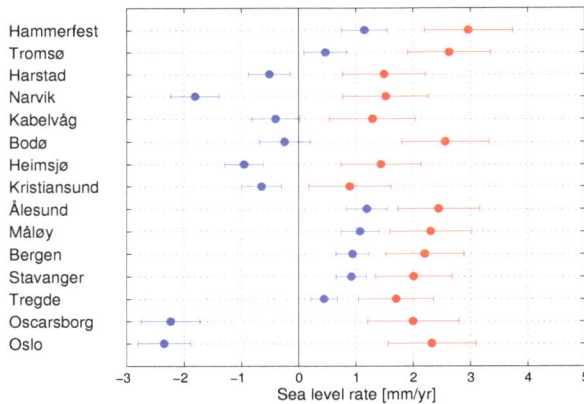

(a)

Figure 2. *Cont.*

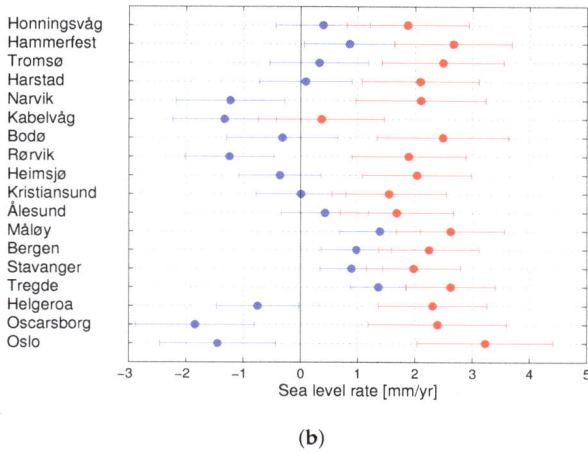

(b)

Figure 2. Relative (blue) and corrected (i.e., adjusted for vertical land motion as well as the gravity effect of glacial isostatic adjustment; red) sea-level rates with standard errors estimated from tide-gauge observations along the Norwegian coast. Rates are shown for (**a**) the period 1960 to 2010 and (**b**) the period 1984 to 2014. The stations are ordered along the coast from north to south.

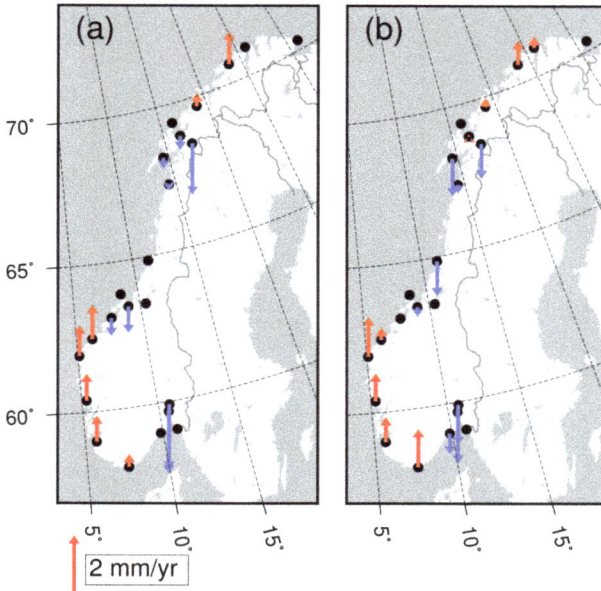

Figure 3. Relative sea-level rates at the Norwegian tide gauges for the periods 1960–2010 (**a**) and 1984–2014 (**b**). The standard error of the rates varies from 0.2 to 0.5 mm/year for the period 1960–2010 and from 0.5 to 1.0 mm/year for the period 1984–2014.

After correcting for GIA, all rates are positive but vary considerably (upper panel of Figure 2). The GIA corrected rates range from 0.9 to 3.0 mm/year and the spread between the tide gauges is calculated as ±0.6 mm/year (one standard deviation). We note that the rates at Kristiansund, Heimsjø, and Kabelvåg stand out as low. That is, they are, 1.5 to 1.0 mm/year below the rates observed at nearby stations. For 1960 to 2010, we calculate the weighted average sea-level rise along the Norwegian coast as 2.0 ± 0.6 mm/year. This rate of rise is similar to the rate of 20th century GMSL rise given in the fifth assessment report of the Intergovernmental Panel on Climate Change [33], but almost two times recent global estimates reported in [34] and [2].

For 1984–2014, the RSL rates show a similar pattern as for 1960–2010 (Figures 2 and 3). That is, for the past 30 years, the pattern of RSL change is dominated by VLM. After correcting for GIA, the rates vary between 0.4 and 3.2 mm/year. While the GIA-corrected rates observed at Kristiansund and Heimsjø appeared low for the period 1960 to 2010, we find that they are in line with the surrounding stations for the period 1984 to 2014. The rate observed at Kabelvåg, however, remains very low when compared to the other tide gauges. The cause of this apparent outlier is not known and we opt to omit this station from further analysis. With Kabelvåg excluded, the GIA-corrected rates are more uniform and have a spread of ±0.4 mm/year (one standard deviation). For 1984 to 2014, we calculate the weighted average sea-level rise along the Norwegian coast as 2.2 ± 0.6 mm/year.

Table 2. Observed relative sea-level rates and rates corrected for glacial isostatic adjustment (GIA) for selected Norwegian tide gauges, ordered along the coast from north to south. To determine the GIA-corrected sea-level rates, we adjusted the tide-gauge observations using (1) vertical land motions estimated from a combined analysis of levelling and Global Navigation Satellite System data, and (2) geoid changes generated using a GIA model [1]. Weighted averages of the rates are given for each period. The standard errors of the GIA-corrected rates and the weighted averages include uncertainties introduced by vertical land motion, geoid corrections and reference frame errors. Records with too short duration or with significant data gaps are excluded from the analysis.

Tide Gauge	Relative Rate (mm/Year) 1960–2010	GIA-Corrected (mm/Year) 1960–2010	Relative Rate (mm/Year) 1984–2014	GIA-Corrected (mm/year) 1984–2014
Honningsvåg			0.4 ± 0.8	1.9 ± 1.1
Hammerfest	1.2 ± 0.4	3.0 ± 0.8	0.8 ± 0.8	2.7 ± 1.0
Tromsø	0.5 ± 0.4	2.6 ± 0.7	0.3 ± 0.9	2.5 ± 1.1
Harstad	−0.5 ± 0.4	1.5 ± 0.7	0.1 ± 0.8	2.1 ± 1.0
Narvik	−1.8 ± 0.4	1.5 ± 0.8	−1.2 ± 1.0	2.1 ± 1.1
Kabelvåg	−0.4± 0.4	1.3 ± 0.8	−1.3 ± 0.9	0.4 ± 1.1
Bodø	−0.2 ± 0.4	2.6 ± 0.8	−0.3 ± 1.0	2.5 ± 1.2
Rørvik			−1.2 ± 0.8	1.9 ± 1.0
Heimsjø	−1.0 ± 0.3	1.4 ± 0.7	−0.4 ± 0.7	2.0 ± 0.9
Kristiansund	−0.6 ± 0.4	0.9 ± 0.7	0.0 ± 0.8	1.5 ± 1.0
Ålesund	1.2 ± 0.4	2.4 ± 0.7	0.4 ± 0.8	1.7 ± 1.0
Måløy	1.1 ± 0.3	2.3 ± 0.7	1.4 ± 0.7	2.6 ± 0.9
Bergen	0.9 ± 0.3	2.2 ± 0.7	1.0 ± 0.6	2.2 ± 0.9
Stavanger	0.9 ± 0.3	2.0 ± 0.7	0.9 ± 0.5	2.0 ± 0.8
Tregde	0.4 ± 0.2	1.7 ± 0.7	1.4 ± 0.5	2.6 ± 0.8
Helgeroa			−0.8 ± 0.7	2.3 ± 0.9
Oscarsborg	−2.2 ± 0.5	2.0 ± 0.8	−1.8 ± 1.0	2.4 ± 1.2
Oslo	−2.3 ± 0.5	2.3 ± 0.8	−1.5 ± 1.0	3.2 ± 1.2
Weighted average sea-level rise		2.0 ± 0.6		2.2 ± 0.6

Using Welch's unequal variances *t*-test [35], we determine that, in comparison to the period 1960 to 2010, the average rate of sea-level rise along the Norwegian coast for 1984 to 2014 is not significantly higher at the 95% level. Note that, before applying this test, we recalculated the coastal averages excluding Kabelvåg and using the same set of tide gauges but found this made no difference to our earlier results.

We list sea-level rates determined from nearby altimetry for the tide-gauge locations and covering the period 1993 to 2016 in Table 3. For comparison, the GIA-corrected tide-gauge records are also included. SSH rates computed from the dataset covering the region south of 66° N (TOPEX/POSEIDON, Jason-1, and Jason-2) range from 3.4 to 4.2 mm/year and have a standard deviation of 0.3 mm/year. The weighted average of the rates is found to be 3.9 ± 0.7 mm/year for 1993 to 2016. However, SSH rates computed from the dataset covering the entire Norwegian coast (ERS-1, ERS-2, ENVISAT, and SARAL/AltiKa) range from 2.4 to 5.3 mm/year and have a standard deviation of 0.7 mm/year. The corresponding weighted average is calculated as 4.0 ± 0.7 mm/year. Hence, we find that the weighted averages of the altimetry datasets agree within the errors, but the spread (one standard deviation) of the second set of rates is substantially larger than that of the first set (also when only stations south of 66° N are considered).

Table 3. Sea-level rates observed by altimetry and rates from tide-gauge records for the period 1993 to 2016. The tide gauges are ordered along the coast from north to south. The weighted average of the rates is given for all locations and for those south of 66° N. Note that the altimetry data are corrected for geoid changes associated with GIA and that the tide-gauge records are corrected for the inverse barometer effect. Records with too short duration or with significant data gaps are excluded from the analysis. The altimetry data were collected by the satellite missions TOPEX/POSEIDON (TP), Jason-1 (J1), Jason-2 (J2), ERS-1 (E1), ERS-2 (E2), ENVISAT (EN), and SARAL/AltiKa.

Tide Gauge	GIA-Corrected Rate from Tide Gauge (mm/Year) 1993–2016	Altimetry TP, J1, J2 (mm/Year) 1993–2016	Altimetry E1, E2, EN, SARAL (mm/Year) 1993–2016
Vardø	3.0 ± 0.9		2.4 ± 0.9
Honningsvåg	2.8 ± 0.9		3.1 ± 0.9
Hammerfest	3.5 ± 0.9		3.2 ± 0.9
Tromsø	3.1 ± 0.9		4.1 ± 0.9
Andenes	3.1 ± 0.9		4.0 ± 0.8
Harstad	2.9 ± 0.9		4.8 ± 0.9
Kabelvåg	3.0 ± 0.9		5.3 ± 0.9
Bodø	2.2 ± 1.0		5.2 ± 1.0
Rørvik	3.1 ± 1.0	3.7 ± 0.8	3.5 ± 0.9
Mausund	3.1 ± 1.4	3.8 ± 0.8	4.1 ± 0.8
Heimsjø	3.3 ± 0.9	4.2 ± 0.8	4.8 ± 0.9
Kristiansund	3.4 ± 0.9	4.2 ± 0.8	4.6 ± 0.8
Ålesund	2.3 ± 0.9	3.4 ± 0.8	4.5 ± 0.8
Måløy	3.8 ± 0.9	3.5 ± 0.8	4.1 ± 0.8
Bergen	2.9 ± 0.8	4.3 ± 0.8	3.5 ± 0.8
Stavanger	3.3 ± 0.8	3.9 ± 0.8	4.1 ± 0.9
Tregde	2.3 ± 0.8	4.2 ± 0.8	4.1 ± 0.9
Helgeroa	3.4 ± 1.0	3.8 ± 0.8	3.6 ± 0.9
Viker	3.9 ± 1.1	3.9 ± 0.8	3.5 ± 0.9
Weighted average sea level rise	3.2 ± 0.6		4.0 ± 0.7
Weighted average sea level rise south of 66° N	3.1 ± 0.6	3.9 ± 0.7	4.0 ± 0.7

Including the IB-corrected tide-gauge rates for comparison, we find that the coastal averages of the altimetry rates are 0.8 to 0.9 mm/year higher than the averages of the tide gauges (Table 3). Still, all averages are within one standard error. This is an encouraging result, indicating no large systematic errors in the tide-gauge records, the VLM corrections applied, the inter-mission measurement biases, or the altimetry data. In addition, the rates of sea surface rise along the Norwegian coast is significantly higher for the period 1993–2016 than for the period 1960–2010 and 1984–2014. The difference increases by approximately 0.5 mm/year when IB-corrections are not applied to the tide gauge records. It is unclear, however, to what extent this higher rate represents natural variability rather than a sustained increase owing to global warming. A detailed study of possible accelerating sea level along the Norwegian coast is not made here, but we note that the climate change signal from increasing greenhouse gases needs time to emerge from natural climate variability, and the sea-level rise observed over recent periods is not significantly larger than rates observed at other times within the past century [36–39]. This is demonstrated in Figure 4, which shows how the estimated rates vary for a 30-year moving window shifted in steps of 1 year from 1960 to 1986.

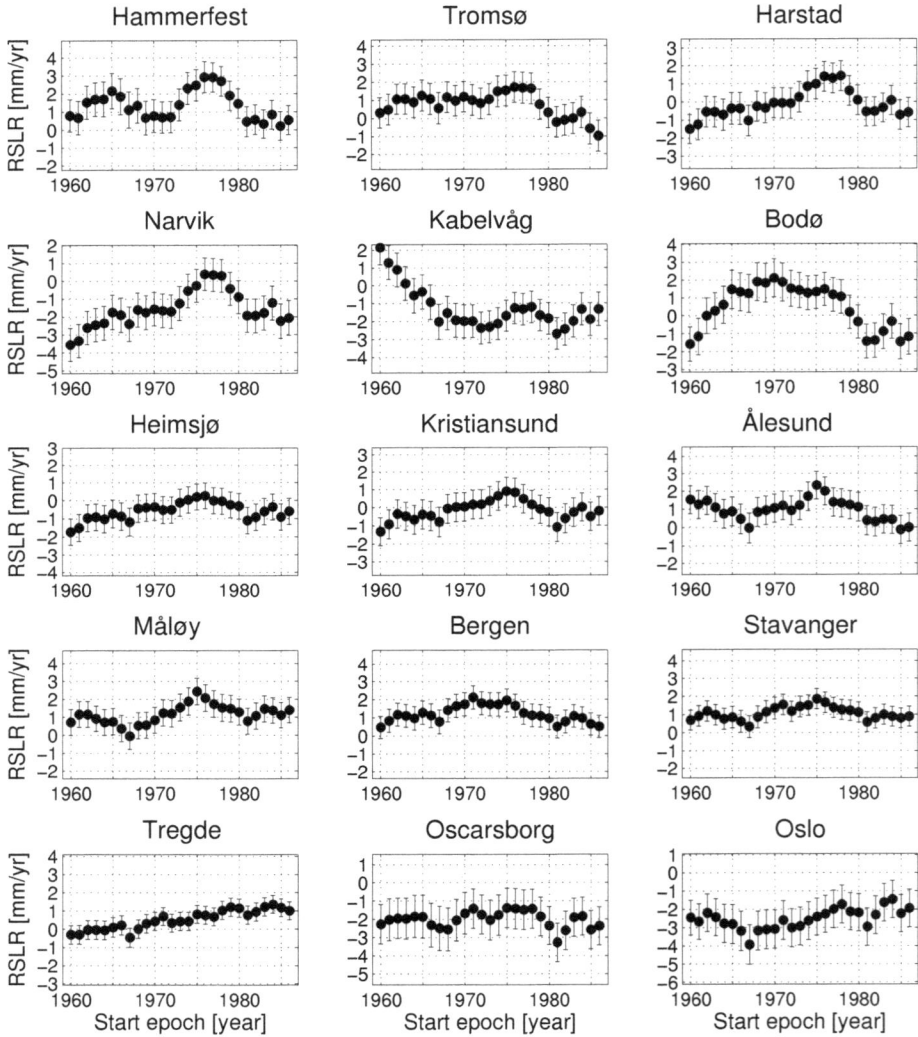

Figure 4. Relative sea-level rates from some tide-gauge records, computed for 30-year moving windows shifted in steps of 1 year as a function of the starting year of the 30-year period. The error bars indicate one standard error.

Results shown in Table 3 and Figure 5 indicate substantial spatial variations in SSH changes in all datasets. The altimetry dataset covering areas south of 66° N has highest rates offshore Bergen and west of the slope of the continental shelf at around 66° latitude, whereas the dataset covering the entire Norwegian coast shows largest rates offshore Bodø, Kabelvåg, Harstad, and Andenes and lowest rates in the northernmost part of the study area. In general, the spatial patterns of the altimetry datasets show poor agreement (Figure 5). We compute the coefficient of spatial correlation between tide gauges and altimetry, and between the two sets of altimetry rates. For the tide-gauge rates and SSH rates from TOPEX/POSEIDON, Jason-1, and Jason-2, we find r = −0.04—while, for the tide-gauge rates and rates from ERS-1, ERS-2, ENVISAT, and SARAL/AltiKa, it is r = −0.28. Thus, the altimetry measurements cannot explain the spatial variations seen in the GIA-corrected tide-gauge records and vice versa. For areas south of 66° N, the correlation between the rates estimated from our two altimetry datasets is r = 0.07.

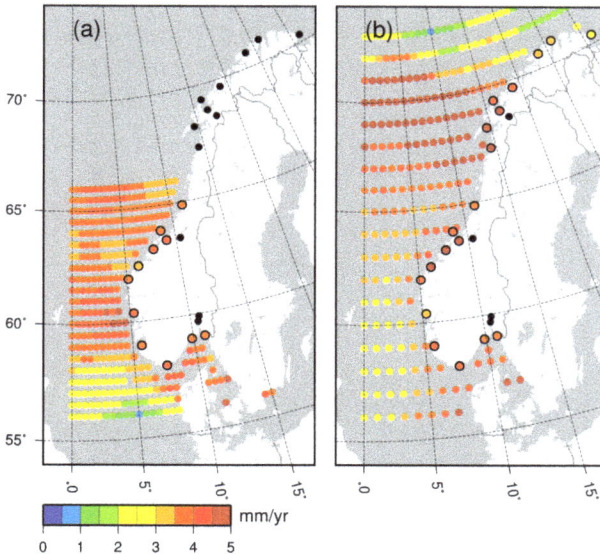

Figure 5. Sea surface height (SSH) changes measured using satellite altimetry over the period 1993 to 2016 from (**a**) the TOPEX/POSEIDON, Jason-1 and Jason-2 dataset and (**b**) the ERS-1, ERS-2, ENVISAT, and SARAL/AltiKa dataset. For the first dataset, observations are not available above 66° N owing to the orbital inclination of the satellites. Changes in SSH were computed for individual tide-gauge locations by averaging the altimetry observations within a spherical distance of 1°. The standard errors of the rates are typically 1 mm/year. Note that the data are corrected for geoid changes associated with GIA.

Compared to the rates presented in [16], our estimated sea-level rates differ by −1.7 to 1.1 mm/year depending on location. The largest differences are found at Vardø (−1.7 mm/year) and in Kristiansund (0.9 mm/year). However, this may be due to the slightly different periods analyzed. We note that [16] also present a map illustrating the pattern of SSH change including areas north of 66° N. The spatial variability shown there partly agrees with our results in Figure 5, where highest rates are found west of Bodø, Kabelvåg, Harstad, and Andenes. In addition, none of the above studies includes a correction for geoid changes associated with GIA (0.1 to 0.5 mm/year along the Norwegian coast). This is not a criticism of these investigations, but, as we opt to take this effect into account, is one reason why our altimetry results are different to these other findings.

4. Discussion

The reliability of the sea-level rates determined from the tide-gauge records depends on the quality of the measurements, the corrections made (e.g., VLM), and the appropriateness of the trend analysis applied. If we first consider the quality of the tide-gauge data, we note that the recording systems and tide-gauge technology have significantly improved over the lifespan of most of the tide gauges. The data quality is thus higher towards the end of the time series.

Secondly, concerning the GIA-corrected rates, it is important to ask how well vertical land motion is constrained at the tide-gauge sites. VLM is not directly observed using GNSS at the majority of Norwegian tide gauges. Only the tide gauges at Tregde, Andenes, and Vardø are collocated with a GNSS station. At the other tide gauges, the distance to the closest reliable GNSS station ranges from a few hundred meters to almost 100 km. For this reason, precise levelling data are also included in our VLM solution which helps better constrain land motion close to some tide gauges. If GIA is the dominant contributor to VLM, then large distances (>10 km) between the tide gauge and GNSS station and/or sparse levelling lines are not necessarily problematic. If local processes (e.g., subsidence) are at play, however, then this can cause localized motions at the tide gauge, GNSS station or along the lines of levelling. Thus, it is somewhat unclear how well the VLM solution applied here represents actual motion at tide gauges where we lack observations.

An additional challenge is that most tide gauges in Norway are not fixed to the bedrock, local processes are especially relevant at these sites. Regular control levelling is therefore conducted between the tide gauge and a nearby benchmark located on bedrock (this control levelling is made over short distances and is separate to the levelling measurements used in our VLM solution). If the control levelling detects a change in height between the tide gauge and the benchmark, the zero-point of the tide gauge is adjusted or a correction is applied to the tide-gauge record. Our analysis shows that the GIA-corrected rates show considerable spatial variability and, therefore, this might be in part due to errors in our VLM solution.

Regarding the interpretation of our tide-gauge trend analysis, it is important to be aware that the rates are extremely sensitive to the selected study period (see Figure 4). For the tide gauges south of Trondheimsfjorden and in Tromsø, the rates vary by approximately 2 mm/year. For the other tide gauges, the variation is even larger, especially at Bodø, Kabelvåg, and Narvik. At Kabelvåg, it seems like local effects strongly influence the earliest years of the record, the rate decreases by 4 mm/year when changing the start year from 1960 to 1967. For all tide gauges, we find that the estimated rates can vary by more than 1 mm/year by moving the 30-year window by just one year. These results are indicative of strong interannual to multi-decadal variability in the tide-gauge records, which is consistent with previous studies [7,13,14]. We also note that Figure 4 reveals patterns common to many tide gauges. For example, most tide gauges show a minimum in the rate series at 1967 and 1981. In addition, the majority of tide gauges indicate highest rates around 1975. This suggests that at least some of the the interannual variability in the rate estimates is due to dynamic sea-level changes that have coherent spatial pattern covering most of the Norwegian coast. As a consequence, we expect that also the coastal average may be sensitive to inter-annual to multi-decadal variation. Changing the end year of the longest study period (1960–2010) to 2000 or 2015 reduces the coastal average from 2.0 ± 0.6 to 1.8 ± 0.6 and 1.9 ± 0.6, respectively, i.e., well within one standard error. Hence, the coastal average for this particular study period appears as a robust estimate. The coastal average for the shorter and more recent period (1984–2014) is more sensitive, e.g., shifting the study period to 1975–2005 raises the coastal average to 2.9 ± 0.6 mm/year.

The poor correlations between the two sets of altimetry rates and the altimetry and GIA-corrected tide-gauge records mean we have low confidence in the ability of altimetry to monitor spatial variations in SSH changes off the Norwegian coast. The low spatial correlation may be explained by variation in rates that are of the same order as the uncertainties of the calculated rates. This suggests that the current length of the altimetry time series is too short for assessing regional sea-level changes at the millimeter per year accuracy. At the same time, we are mindful that the altimetry satellites and tide gauges do not sample at the same locations and are two completely different measurement concepts. It is possible that SSH changes offshore could be different to those at the coast where the tide gauges are located. Spatial variations may arise due to real oceanic signals and/or spurious spatial signals related to the altimetry measurement system. Errors in the reference frame or in the intermission measurement biases have long spatial wavelengths or are constant. We expect, therefore, that such errors are only small contributors to the variation in the observed rates. On the other hand, errors in the corrections applied to the altimetry measurements (e.g., sea state bias corrections, tidal corrections, and corrections for atmospheric delay) may have wavelengths of shorter spatial scales.

As mentioned above, accurate monitoring of sea-level changes within an enclosed region by altimetry is challenging. [40] emphasize this problem by arguing that the current version of the ITRF does not allow regional sea level to be monitored with millimeter per year accuracy. The challenge becomes even more complicated in the coastal zone where backscattered energy from land areas contaminates the radar pulses and sea surface conditions may have short correlation lengths. However, recent studies have demonstrated that it is possible to extract information from radar pulses in the coastal zones (see, e.g., [29,41,42]). This requires so called retracking of the radar waveforms, i.e., algorithms optimized for radar pulses backscattered from a mix of land and sea are applied. In addition, the wet tropospheric corrections must be calculated from meteorological models instead of measurements from microwave radiometers on board the altimetry satellites.

Coastal altimetry products do exist for Norway, e.g., PISTACH data for Jason-2 [43], ENVISAT data from CTOH [44,45], and PEACHI data for SARAL/AltiKa [46]. [47] computed the coastal mean dynamic topography from the PISTACH and CTOH data sets and found that they were not superior to the standard geophysical data records. However, the next generation altimeters on board the satellites Cryosat-2 and Sentinel-3 raise great expectations. These altimeters have the ability to work in Synthetic Aperture Radar (SAR) mode in the coastal zone. Among others, this will increase the spatial resolution in the along-track direction considerably, i.e., down to 250 m. The finer resolution implies that the radar pulses are less disturbed by land-features in the coastal zone and offer a better ability to resolve features with correlation lengths of less than 1 km [48]. Consequently, we envisage that SAR-mode altimetry will generate more data right at the coast. The benefits of altimetry in SAR mode along the Norwegian coast were investigated by [49]. They demonstrated that a preliminary release of Cryosat-2 SIRAL (Synthetic Aperture Interferometric Radar Altimeter) data provides reliable and comprehensive sets of data in 45×45 km boxes around most Norwegian tide gauges, at least at tide gauges close to the open ocean. Compared to tide-gauge records, Cryosat-2 SIRAL data have standard deviations down to 6 cm, i.e., at the same level as conventional pulse-limited altimetry at the open ocean. Unfortunately, none coastal altimetry records are presently sufficiently long to provide information on sea-level changes.

5. Conclusions

The relative sea-level rates estimated from the Norwegian tide-gauge network reflect the pattern of land uplift. A fall in relative sea level is observed around Oslo and in the middle of Norway, while a rise is observed along the southern and western coast of Norway and for the northernmost tide gauges. After correcting the rates for GIA, the resulting SSH rates are positive at all tide gauges.

Over the more recent period 1993 to 2016, we also estimated change in SSH from two satellite altimetry datasets. The coastal averages derived from altimetry are slightly larger than that obtained from the tide-gauge network, but within standard errors. At the same time, altimetry and tide gauges do not capture the same spatial variation. These results are important because they indicate that no systematic errors are present in the observations and corrections applied.

We also found that the rate of sea-surface rise along the Norwegian coast is significantly higher for the period 1993 to 2016 than for the period 1960 to 2010. It is unclear, however, to what extent this higher rate represents natural variability rather than a sustained increase owing to global warming. Other studies suggest that thermal expansion and melting land ice are the most prominent contributors to the observed sea-level trends for the Norwegian coastline [6], as they are for the global mean. In addition, we expect the sea-level rates to increase in the future with increased global warming [1].

Future sea-level studies for the Norwegian coast should address accelerating sea level. We also suggest to use a more advanced regression model than the one used in the present study (see Equation (1)). In order to explain a larger percentage of the variation in the tide gauge records, observed atmospheric pressure and wind speed could be included as regressors in the model. This may reduce the rates sensitivity to the chosen study period. In addition, we envisage that improved constraints on vertical land motion can be obtained by deploying GNSS-receivers directly on or close to the tide gauges. Alternatively, possible local anomalies in vertical land motion may be mapped and quantified by Interferometric Synthetic Aperture Radar (InSAR) as suggested by [50].

Acknowledgments: The authors acknowledge the open access policies of the Permanent Service for Mean Sea Level (PSMSL), and Archiving, Validation, and Interpretation of Satellite Oceanographic data (AVISO), and the eKlima service of the Norwegian Meteorological Institute. We are thankful to the European Space Agency for providing data from ERS-1, ERS-2, and ENVISAT. Support was provided by the Centre for Climate Dynamics at the Bjerknes Centre, through the project iNcREASE. The rates were estimated with the Python package Statsmodels. Finally, we thank the two anonymous reviewers for most useful comments that helped improve the manuscript.

Author Contributions: K.B. conceived the experiments, prepared the altimetry data, accomplished the time-series analysis, and made Figures 1, 2, and 4. M.J.R.S. provided the vertical land motion corrections and the gravity effect of GIA, and made Figures 3 and 5. All authors contributed to the writing of the paper and in the interpretation of results and findings.

Conflicts of Interest: The authors declare no conflict of interest. The founding sponsors had no role in the design of the study; in the collection, analyses, or interpretation of data; in the writing of the manuscript, and in the decision to publish the results.

Abbreviations

The following abbreviations are used in this manuscript:

ACF	Autocorrelation Function
AIC	Akaike's Information Criterion
AR	Autoregressive
BIC	Bayesian Information Criterion
GIA	Glacial Isostatic Adjustment
GDR	Geophysical Data Records
GMSL	Global Mean Sea Level
GNSS	Global Navigation Satellite System
IB	Inverse Barometer
InSAR	Interferometric Synthetic Aperture Radar
ITRF	International Terrestrial Reference Frame
NMA	The Norwegian Mapping Authority
PACF	Partial Autocorrelation Function
PSMSL	Permanent Service for Mean Sea Level
RSL	Relative Sea Level
SAR	Synthetic Aperture Radar
SIRAL	Synthetic Aperture Interferometric Radar Altimeter
SSH	Sea Surface Height
VLM	Vertical Land Motion

References

1. Simpson, M.J.R.; Ravndal, O.R.; Sande, H.; Nilsen, J.E.Ø.; Kierulf, H.P.; Vestøl, O.; Steffen, H. Projected 21st century sea-level changes, extreme sea levels, and sea level allowances for Norway. *J. Mar. Sci. Eng.* **2017**, submitted.

2.   Dangendorf, S.; Marcos, M.; Wöppelmann, G.; Conrad, C.P.; Frederikse, T.; Riva, R. Reassessment of 20th century global mean sea level rise. *Proc. Natl. Acad. Sci. USA* **2017**, *114*, 5946–5951, doi:10.1073/pnas.1616007114.

3.   Lewis, M.; Horsburgh, K.; Bates, P.; Smith, R. Quantifying the uncertainty in future coastal flood risk estimates for the UK. *J. Coast. Res.* **2011**, *27*, 870–881, doi:10.2112/JCOASTRES-D-10-00147.1.

4.   Lewis, M.; Schumann, G.; Bates, P.; Horsburgh, K. Understanding the variability of an extreme storm tide along a coastline. *Estuar. Coast. Shelf Sci.* **2013**, *123*, 19–25, doi:10.1016/j.ecss.2013.02.009.

5.   Hunter, J. A simple technique for estimating an allowance for uncertain sea-level rise. *Clim. Chang.* **2012**, *113*, 239–252, doi:10.1007/s10584-011-0332-1.

6.   Frederikse, T.; Riva, R.; Kleinherenbrink, M.; Wada, Y.; van den Broeke, M.; Marzeion, B. Closing the sea level budget on a regional scale: Trends and variability on the Northwestern European continental shelf. *Geophys. Res. Lett.* **2016**, *43*, 10864–10872, doi:10.1002/2016GL070750.

7.   Richter, K.; Nilsen, J.E.O.; Drange, H. Contributions to sea level variability along the Norwegian coast for 1960–2010. *J. Geophys. Res.* **2012**, *117*, doi:10.1029/2011JC007826.

8.   Douglas, B.C. Global Sea Level Rise. *J. Geophys. Res.* **1991**, *96*, 6981–6992, doi:10.1029/91JC00064.

9.   Vestøl, O. Determination of postglacial land uplift in Fennoscandia from leveling, tide-gauges and continuous GPS stations using least squares collocation. *J. Geod.* **2006**, *80*, 248–258, doi:10.1007/s00190-006-0063-7.

10.   Marcos, M.; Tsimplis, M.N. Forcing of coastal sea level rise patterns in the North Atlantic and the Mediterranean Sea. *Geophys. Res. Lett.* **2007**, *34*, doi:10.1029/2007GL030641.

11.   Henry, O.; Prandi, P.; Llovel, W.; Cazenave, A.; Jevrejeva, S.; Stammer, D.; Meyssignac, B.; Koldunov, N. Tide gauge-based sea level variations since 1950 along the Norwegian and Russian coasts of the Arctic Ocean: Contribution of the steric and mass components. *J. Geophys. Res. Oceans* **2012**, *117*, doi:10.1029/2011JC007706.

12.   Wahl, T.; Haigh, I.D.; Dangendorf, S.; Jensen, J. Inter-annual and long-term mean sea level changes along the North Sea coastline. *J. Coast. Res.* **2013**, *65*, 1987–1992, doi:10.1029/2011JC007557.

13.   Calafat, F.M.; Chambers, D.P.; Tsimplis, M.N. Inter-annual to decadal sea-level variability in the coastal zones of the Norwegian and Siberian Seas: The role of atmospheric forcing. *J. Geophys. Res. Oceans* **2013**, *118*, 1287–1301, doi:10.1002/jgrc.20106.

14.   Dangendorf, S.; Calafat, F.M.; Arns, A.; Wahl, T.; Haigh, I.D.; Jensen, J. Mean sea level variability in the North Sea: processes and implications. *J. Geophys. Res. Oceans* **2014**, *119*, 6820–6841, doi:10.1002/2014JC009901.

15.   Ghil, M.; Allen, M.R.; Dettinger, M.D.; Ide, K.; Kondrashov, D.; Mann, M.E.; Robertson, A.W.; Saunders, A.; Tian, Y.; Varadi, F.; et al. Advanced spectral methods for climatic time series. *Rev. Geophys.* **2002**, *40*, doi:10.1029/2000RG000092.

16.   Volkov, D.L.; Pujol, M.I. Quality assessment of a satellite altimetry product in the Nordic, Barents, and Kara seas. *J. Geophys. Res.* **2012**, *117*, doi:10.1029/2011JC007557.

17.   Prandi, P.; Ablain, M.; Cazenave, A.; Picot, N. A New Estimation of Mean Sea Level in the Arctic Ocean from Satellite Altimetry. *Mar. Geod.* **2012**, *35*, 61–81, doi:10.1080/01490419.2012.718222.

18.   Holgate, S.J.; Matthews, A.; Woodworth, P.L.; Rickards, L.J.; Tamisiea, M.E.; Bradshaw, E.; Foden, P.R.; Gordon, K.M.; Jevrejeva, S.; Pugh, J. New Data Systems and Products at the Permanent Service for Mean Sea Level. *J. Coast. Res.* **2013**, *29*, 493–504, doi:10.2112/JCOASTRES-D-12-00175.1.

19.   Woodworth, P.L. A Note on the Nodal Tide in Sea Level Records. *J. Coast. Res.* **2012**, *28*, 316–323, doi:10.2112/JCOASTRES-D-11A-00023.1.

20.   Baart, F.; van Gelder, P.H.A.J.M.; de Ronde, J.; van Koningsveld, M.; Wouters, B. The effect of the 18.6-Year Lunar Nodal Cycle on Regional Sea-Level Rise Estimates. *J. Coast. Res.* **2012**, *28*, 511–516, doi:10.2112/JCOASTRES-D-11-00169.1.

21.   Scargle, D.J. Studies in astronomical time series analysis. II. Statistical aspects of spectral analysis of unevenly spaced data. *Astrophys. J.* **1982**, *263*, 835–853, doi:10.1086/160554.

22.   Bos, M.S.; Williams, S.D.P.; Araújo, I.B.; Bastos, L. The effect of temporal correlated noise on the sea level rate and acceleration uncertainty. *Geophys. J. Int.* **2014**, *196*, 1423–1430, doi:10.1093/gji/ggt48.

23.   Burgette, R.J.; Watson, C.S.; Church, J.A.; White, N.J.; Tregoning, P.; Coleman, R. Characterizing and minimizing the effects of noise in tide gauge time series: Relative and geocentric sea level rise around Australia. *Geophys. J. Int.* **2013**, *194*, 719–736, doi:10.1093/gji/ggt131.

24.   Hünicke, B.; Zorita, E. Statistical Analysis of the Acceleration of Baltic Mean Sea-Level Rise, 1900–2012. *Front. Mar. Sci.* **2016**, *3*, 125, doi:10.3389/fmars.2016.00125.

25. Tamisiea, M.E.; Mitrovica, J.X. The moving boundaries of sea level change: Understanding the origins of geographic variability. *Oceanography* **2011**, *24*, 24–39, doi:10.5670/oceanog.2011.25.

26. Andersen, O.B.; Scharroo, R. Range and Geophysical Corrections in Coastal Regions: And Implications for Mean Sea Surface Determination. In *Coastal Altimetry*; Springer: Berlin, Germany, 2011; pp. 103–146.

27. Collilieux, X.; Altamimi, Z.; Argus, D.F.; Boucher, C.; Dermanis, A.; Haines, B.J.; Herring, T.A.; Kreemer, C.W.; Lemoine, F.G.; Ma, C.; et al. External evaluation of the Terrestrial Reference Frame: Report of the task force of the IAG sub-commission 1.2. In *Earth on the Edge: Science for a Sustainable Planet*; Rizos, C., Willis, P., Eds.; Springer: Berlin, Germany, 2014; pp. 197–202, doi:10.1007/978-3-642-37222-3_25.

28. Beckley, B.D.; Lemoine, F.G.; Luthcke, S.B.; Ray, R.D.; Zelensky, N.P. A reassessment of global and regional mean sea level trends from TOPEX and Jason-1 altimetry based on revised reference frame and orbits. *Geophys. Res. Lett.* **2007**, *34*, L14608, doi:10.1029/2007GL030002.

29. Cipollini, P.; Calafat, F.M.; Jevrejeva, S.; Melet, A.; Prandi, P. Monitoring Sea Level in the Coastal Zone with Satellite Altimetry and Tide Gauges. *Surv. Geophys.* **2016**, 1–25, doi:10.1007/s10712-016-9392-0.

30. The Archive, Validation, and Interpretation of Satellite Oceanographic Data (AVISO) Portal. Available online: ftp://avisoftp.cnes.fr/AVISO/pub/ (accessed on 15 July 2017).

31. ESA Earth Online. Available online: http://earth.esa.int (accessed on 15 July 2017).

32. Ablain, M.; Phillips, S.; Picot, N.; Bronner, E. Jason-2 Global Statistical Assessment and Cross-Calibration with Jason-1. *Mar. Geod.* **2010**, *33*, 162–185, doi:10.1080/01490419.2010.487805.

33. Rhein, M.; Rintoul, S.R.; Aoki, S.; Campos, E.; Chambers, D.; Feely, R.; Gulev, S.; Johnson, G.C.; Josey, S.A.; Kostianoy, A.; et al. Observations: Ocean. In *Climate Change 2013: The Physical Science Basis. Contribution of Working Group I to the Fifth Assessment Report of the Intergovernmental Panel on Climate Change*; Stocker, T., Qin, D., Plattner, G.K., Tignor, M., Allen, S.K., Boschung, J., Nauels, A., Xia, Y., Bex, V., Midgley, P.M., Eds.; Cambridge University Press: Cambridge, UK; New York, NY, USA, 2013; ISBN 978-1-107-05799-1.

34. Hay, C.C.; Morrow, E.; Kopp, R.E.; Mitrovica, J.X. Probabilistic reanalysis of twentieth-century sea-level rise. *Nature* **2015**, *517*, 481–484, doi:10.1038/nature14093.

35. Welch, B.L. The Generalization of 'Student's' Problem when Several Different Population Variances are Involved. *Biometrika* **1947**, *34*, 28–35, doi:10.2307/2332510.

36. Lyu, K.; Zhang, X.; Church, J.A.; Slangen, A.B.A.; Hu, J. Time of emergence for regional sea-level change. *Nat. Clim. Chang.* **2014**, *4*, 1006–1010, doi:10.1038/NCLIMATE2397.

37. Haigh, I.D.; Wahl, T.; Rohling, E.J.; Price, R.M.; Pattiaratchi, C.B.; Calafat, F.M.; Dangendorf, S. Timescales for detecting a significant acceleration in sea level rise. *Nat. Commun.* **2014**, *5*, doi:10.1038/ncomms4635.

38. Jordá, G. Detection time for global and regional sea level trends and accelerations. *J. Geophys. Res. Oceans* **2014**, *119*, 7164–7174, doi:10.1002/2014JC010005.

39. Dangendorf, S.; Rybski, D.; Mudersbach, C.; Müller, A.; Kaufmann, E.; Zorita, E.; Jensen, J. Evidence for long-term memory in sea level. *Geophys. Res. Lett.* **2014**, *41*, 5530–5537, doi:10.1002/2014GL060538.

40. Minster, J.B.; Altamimi, Z.; Blewitt, G.; Carter, W.E.; Cazenave, A.; Dragert, H.; Herring, T.A.; Larson, K.M.; Ries, J.C.; Sandwell, D.T.; et al. *Precise Geodetic Infrastructure: National Requirements for a Shared Resource*; The National Academies Press: Washington, DC, USA, 2010; ISBN 978-0-309-15811-4.

41. Vignudelli, S.; Cipollini, P.; Gommenginger, C.; Gleason, S.; Snaith, H.M.; Coelho, H.; Fernandes, M.J.; Lazaro, C.; Nunes, A.L.; Gomez-Enri, J.; et al. Satellite altimetry: Sailing closer to the coast. In *Remote Sensing of the Changing Oceans*; Springer: Berlin, Germany, 2011; pp. 217–238, doi:10.1007/978-3-642-16541-2_11.

42. Vignudelli, S.; Kostianoy, A.G.; Cipollini, P.; Benveniste, J.E. *Coastal Altimetry*, 1st ed.; Springer: Berlin, Germany, 2011; ISBN 978-3-642-12795-3.

43. Mercier, F.; Rosmorduc, V.; Carrere, L.; Thibaut, P. *Coastal and Hydrology Altimetry Product (PISTACH) Handbook*; Centre National D'études Spatiales: Paris, France, 2010; p. 64.

44. Roblou, L.; Lyard, F.; Le Henaff, M.; Maraldi, C. X-TRACK, a new processing tool for altimetry in coastal oceans. In Proceedings of the Envisat Symposium 2007, Montreux, Switzerland, 23–27 April 2007; pp. 23–27, doi:10.1109/IGARSS.2007.4424016.

45. Roblou, L.; Lamouroux, J.; Bouffard, J.; Lyard, F.; Le Hénaff, M.; Lombard, A.; Marsaleix, P.; De Mey, P.; Birol, F. Post-processing altimeter data towards coastal applications and integration into coastal models. In *Coastal Altimetry*; Springer: Berlin, Germany, 2011; pp. 217–246, ISBN 978-3-642-12795-3.

46. Valladeau, G.; Thibaut, P.; Picard, B.; Poisson, J.C.; Tran, N.; Picot, N.; Guillot, A. Using SARAL/AltiKa to improve Ka-band altimeter measurements for coastal zones, hydrology and ice: The PEACHI prototype. *Mar. Geod.* **2015**, *38*, 124–142, doi:10.1080/01490419.2015.1020176.

47. Ophaug, V.; Breili, K.; Gerlach, C. A comparative assessment of coastal mean dynamic topography in Norway by geodetic and ocean approaches. *J. Geophys. Res. Oceans* **2015**, *120*, 7807–7826, doi:10.1002/2015JC011145.

48. Raney, R.K.; Phalippou, L. The Future of Coastal Altimetry. In *Coastal Altimetry*; Springer: Berlin, Germany, 2011; pp. 535–560, ISBN 978-3-642-12795-3.

49. Idžanović, M.; Ophaug, V.; Andersen, O.B. Coastal Sea Level from CryoSat-2 SARIn Altimetry in Norway. *Adv. Space Res.* **2016**, Submitted.

50. Brooks, B.A.; Merrifield, M.A.; Foster, J.; Werner, C.L.; Gomez, F.; Bevis, M.; Gill, S. Space geodetic determination of spatial variability in relative sea level change, Los Angeles basin. *Geophys. Res. Lett.* **2007**, *34*, 1, doi:10.1029/2006GL028171.

Journal of
Marine Science and Engineering

MDPI

Article

South Florida's Encroachment of the Sea and Environmental Transformation over the 21st Century

Joseph Park [1,*], Erik Stabenau [1], Jed Redwine [2] and Kevin Kotun [1]

1 Physical Resources, South Florida Natural Resources Center, National Park Service,
  Homestead, FL 33030, USA; Erik_Stabenau@nps.gov (E.S.); Kevin_Kotun@nps.gov (K.K.)
2 Biological Resources, South Florida Natural Resources Center, National Park Service,
  Homestead, FL 33030, USA; Jed_Redwine@nps.gov
* Correspondence: Joseph_Park@nps.gov; Tel.: +1-305-224-4250

Received: 12 April 2017; Accepted: 18 July 2017; Published: 28 July 2017

Abstract: South Florida encompasses a dynamic confluence of urban and natural ecosystems strongly connected to ocean and freshwater hydrologic forcings. Low land elevation, flat topography and highly transmissive aquifers place both communities at the nexus of environmental and ecological transformation driven by rising sea level. Based on a local sea level rise projection, we examine regional inundation impacts and employ hydrographic records in Florida Bay and the southern Everglades to assess water level exceedance dynamics and landscape-relevant tipping points. Intrinsic mode functions of water levels across the coastal interface are used to gauge the relative influence and time-varying transformation potential of estuarine and freshwater marshes into a marine-dominated environment with the introduction of a Marsh-to-Ocean transformation index (MOI).

Keywords: South Florida; sea level rise; inundation; coastal impacts; water level exceedance

1. Introduction

Sea level rise is not evenly distributed around the globe, and the response of a regional coastline is highly dependent on local natural and human settings [1]. This is particularly evident at the southern end of the Florida peninsula where low elevations and exceedingly flat topography provide an ideal setting for encroachment of the sea. Coastal South Florida is fringed by national parks including Biscayne and Everglades National Parks, Big Cypress National Preserve and the islandic Dry Tortugas National Park. This rich natural setting and subtropical climate appeal to human interests with over six million inhabitants residing along narrow coastal strips along the Atlantic and Gulf coasts. The sustenance of these natural and human ecosystems is predicated on adequate freshwater supply, and while South Florida receives an average of 140 cm of rainfall annually, losses to evaporation are nearly as great as the rainfall itself, and water storage is limited to shallow, permeable reservoirs and thin surficial aquifers that are experiencing diminishing capacity as rising sea level drives saltwater infiltration.

Attempts to control the hydrologic resources have resulted in the construction of one of the most complex and expansive water control projects on the planet with both beneficial and detrimental impacts on human and natural populations [2,3]. Regional governments recognize the need to assess and plan for sea level rise implementing a Regional Climate Action Plan [4] with a task force specifically addressing sea level rise [5]. However, these efforts focus on urban and suburban areas with concern for property values, transportation, housing, water supply and sewer infrastructure based on global sea level rise projections that do not reflect local processes and that are not associated with specific probabilities of occurrence.

Here, we focus on the low-elevation natural areas at the southern end of the peninsula as shown in Figure 1, as these areas will experience inundation impacts prior to the urban areas, thereby serving as sensitive indicators of sea level rise. We evaluate sea level rise inundation impacts under two scenarios, a low projection and a high projection, based on a synthesis of coupled atmosphere-ocean general circulation models and tide gauge information reflecting local processes. The high projection represents an upper percentile (99%) of expected sea level rise given current models and observations, while the low projection corresponds to a median (50%) sea level rise scenario. Since models, observations and current scientific understanding are incomplete, these projections are necessarily incomplete and do not account for a rapid collapse of the Antarctic ice-sheets, a development that is currently unfolding with potential to render these projections as lower bounds [6,7].

We also examine coastal water level exceedance data, quantifying an exponential increase in low-elevation exceedances over the last decade. Application of the sea level rise projections allows us to project these exceedance curves into the future, where one can identify tipping points and time horizons for the transformation of coastal regions into marine ecosystems. Finally, we introduce a metric to characterize the transformation of a coastal wetland from a freshwater marsh into a saltwater marsh based on intrinsic mode functions of water level time series extending landward from the sea.

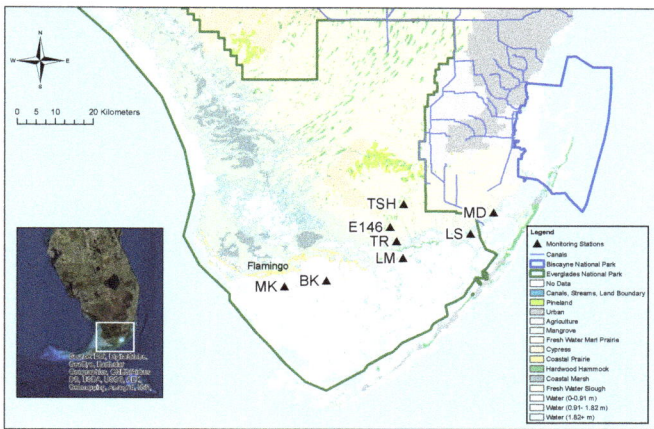

Figure 1. Physiographic map of South Florida representing different ecological domains dictated by topography, hydrology and climate. Hydrographic stations are denoted with abbreviations; for example, LM for Little Madeira Bay (Table 1). Everglades National Park (green border) covers the majority of the region with coastal hydrographic stations in Florida Bay (MK, BK, LM, LS, MD) and extending upstream to Taylor Slough (TR, E146, TSH). Urban, suburban and agricultural lands featuring water management canal infrastructure can be seen between Everglades and Biscayne National Parks (blue border).

Table 1. Hydrographic stations.

Station	Location	Latitude	Longitude	Water Level	Salinity
BK	Buoy Key	25.12111	−80.83356	WaterLog H-331	YSI 600R
E146	Taylor Slough	25.25252	−80.66626	WaterLog H-331	
LM	Little Madeira Bay	25.17580	−80.63269	WaterLog H-331	YSI 600R
LS	Long Sound	25.23516	−80.45680	WaterLog H-331	YSI 600R
MD	South Dade	25.28932	−80.39642	WaterLog H-331	YSI 600R
MK	Murray Key	25.10613	−80.94232	WaterLog H-331	YSI 600R
TR	Taylor River	25.21712	−80.64957	WaterLog H-331	YSI 600R
TSH	Taylor Slough Hilton	25.31073	−80.63100	WaterLog H-331	

2. Materials and Methods

2.1. Sea Level Rise Projection

The Intergovernmental Panel on Climate Change's (IPCC) most recent evaluation is the Fifth Assessment Report (AR5) [8] including projections of global sea level rise based on different Representative Concentration Pathway (RCP) scenarios reflecting possible future concentrations of greenhouse gases[1]. RCP 8.5, also known as the business-as-usual scenario, is the highest emission and warming scenario under which greenhouse gas concentrations continue to rise throughout the 21st Century, while RCP 6.0 and RCP 4.5 expect substantial emission declines to begin near 2080 and 2040, respectively.

The IPCC sea level rise scenarios are comprehensive, but do not include contributions from a rapid collapse of Antarctic ice sheets. However, recent evidence suggests that such a collapse may be underway [6,7]. In addition, the IPCC projections do not account for local processes such as land uplift/subsidence and ocean circulation and do not provide precise estimates of the probabilities associated with specific sea level rise scenarios.

A contemporary study that does estimate local effects and comprehensive probabilities for the RCP scenarios is provided by Kopp et al. [9] based on a synthesis of tide gauge data, global climate models and expert elicitation, including contributions from the Greenland ice sheet, West Antarctic ice sheet, East Antarctic ice sheet, glaciers, thermal expansion, regional ocean dynamics, land water storage and long-term, local, non-climatic factors, such as glacial isostatic adjustment, sediment compaction and tectonics. Even though this model includes contributions from the Antarctic ice sheets, these contributions are from dynamic equilibrium models and do not yet account for an incipient rapid collapse as noted above. Nonetheless, we find the Kopp et al. [9] projections to be among the most mature and useful sea level rise paradigms and base our South Florida projections on their results at Vaca Key, Florida.

South Florida Sea Level Rise Projection

Examination of local sea level rise projections around South Florida finds small differences between Naples, Virginia Key, Vaca Key and Key West. We chose the Vaca Key station sea level data as representative of South Florida since they best reflect local oceanographic processes that influence coastal sea levels [10].

Next, we select the RCP scenario that best fits our understanding for future greenhouse gas emissions. Although significant effort is aimed at global emission reduction, atmospheric CO_2 and emissions continue to escalate [11], and there is presently no clear socio-economic driver to depart from a carbon-based energy infrastructure. Further, recent assessments of global energy production and population conclude that the the achievement of emission scenarios corresponding to a desired 2 °C limit in global mean temperature increase require the global fraction of Renewable Energy Sources (RES) to reach 50% by 2028 [12].

We note that the International Energy Agency (IEA) reports that global RES could reach 28% by 2021 [13]. This is consistent with a 2015 estimate of 24% RES by the United Nations [14] and, if accurate, would leave seven years to achieve a near doubling to 50% to meet the Jones and Warner [12] constraint. Currently, RES is dominated by hydropower, a resource that is not easily scalable or quick to bring online. In the absence of a technological breakthrough, we conclude it is unlikely that global RES will reach 50% by 2028. This leads us to expect that the RCP 4.5 emission scenario is unobtainable and that there is significant uncertainty as to whether the RCP 6.0 scenario can be realized. We therefore restrict our projection to the RCP 8.5 scenario.

---

[1]   The number following RCP quantifies the expected thermodynamic radiative forcing relative to pre-industrial values. For example, RCP 8.5 denotes an additional 8.5 W/m^2 thermal forcing from greenhouse gases.

Finally, we select conservative projection probabilities appropriate for informing authorities of anticipated sea level rise for adaptation and planning purposes. In light of the significant uncertainties inherent in the generation of the projections and future climate dynamics, it is prudent to consider the upper percentile range of projections leading us to select the RCP 8.5 median (50th percentile) as the lower boundary and the 99th percentile as the upper boundary. Although the high projection is deemed to have a 1% chance of occurrence under current climate conditions and models, in the event of Antarctic ice sheet collapse, this high projection is consistent with estimates of the Antarctic ice melt contribution [15].

The resultant sea level rise projection for South Florida referenced to the North American Vertical Datum of 1988 (NAVD88, Appendix A) is shown in Figure 2 and tabulated in Appendix B. Projection starting points have been offset to coincide with observed mean sea level in Florida Bay over the period 2008–2015 (Appendix C). The projection does not incorporate local processes such as tides, storm surges, waves or their non-linear interactions with inundation impacts, issues that are discussed in Appendix D.

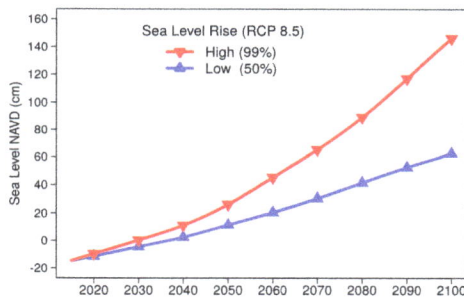

Figure 2. South Florida sea level rise projection with respect to 2015 mean sea level in Florida Bay for the RCP 8.5 greenhouse gas emission scenario. Units are cm NAVD88. Low projection is the median (50th percentile); high projection the 99th percentile. Tides and storm surges are not included in this projection. Values are tabulated in Appendix B to year 2120.

2.2. Inundation Coverage

Geospatial inundation coverages for mean sea level are created in ArcMap by application of the sea level rise projections for the years 2025, 2050, 2075 and 2100 across southern Florida. Topographical elevations are based on a synthesis of the best available high-resolution digital elevation data [16] with variable spatial resolution, but a nominal horizontal grid cell size of 50 m. The resulting inundation coverages represent a static land-masking of mean sea level at the four time horizons and do not represent influences from tides, seasonal oceanographic cycles, teleconnections, weather, such as storms, or inverse barometric adjustments, as discussed in Appendix D, or for changing morphological structure in submerged and inundated sediments or hydraulic connectivity [17]. A review of these issues and how the dynamic effects of sea level rise interact with low-gradient coastal landscapes can be found in Passeri et al. [18].

2.3. Water Level and Salinity

Water levels are obtained from eight hydrographic stations operated by Everglades National Park over the period 1 June 1994–31 December 2016 with station locations and names shown in Table 1. Water levels are collected at 6-, 15- or 60-min intervals by WaterLog shaft-encoded float gauges recorded by a Sutron SatLink2 data recorder. Water levels are then aggregated into daily mean values as shown in Figure 3.

Salinity is estimated from specific conductivity measured at 30- or 60-min intervals by a YSI 600R Water Quality Sonde and application of the International Equation of State of Seawater 1980 and Practical Salinity Scale 1978 as recommended by the United Nations Educational, Scientific and Cultural Organization (UNESCO) Joint Panel on Oceanographic Standards and Tables [19]. Daily mean salinities are shown in Figure 4, and summary statistics of the water level and salinity time series are presented in Table 2.

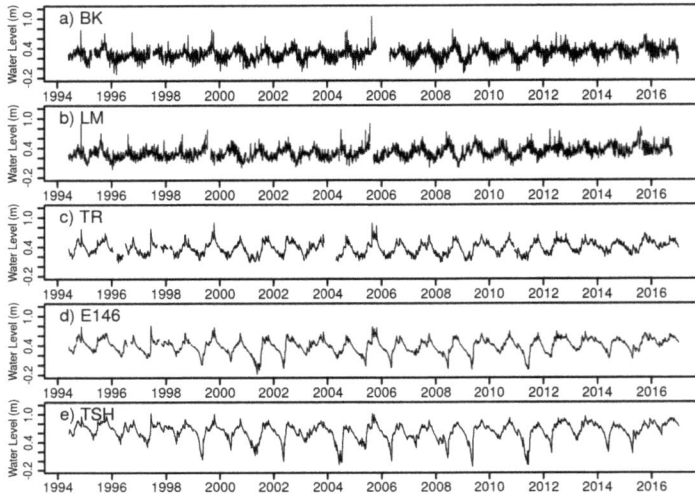

Figure 3. Daily mean water level with respect to the National Geodetic Vertical Datum of 1929 (NGVD29) at 5 stations in Florida Bay and the southern Everglades. Stations BK (**a**) and LM (**b**) are in Florida Bay; stations TR (**c**), E146 (**d**) and TSH (**e**) are within Taylor Slough.

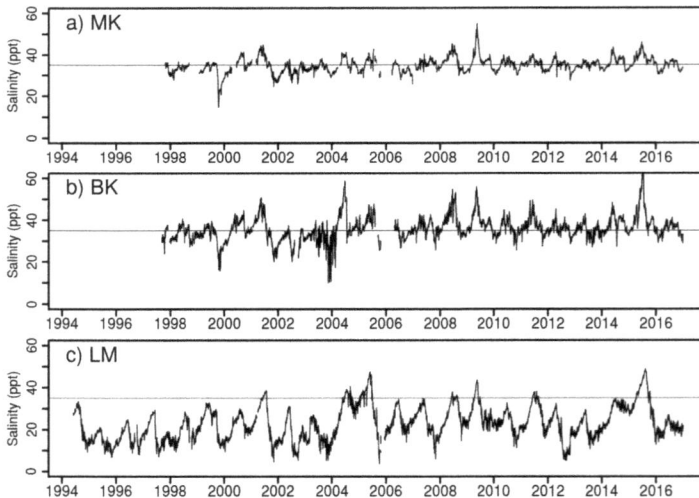

Figure 4. Daily mean salinity at 3 stations in Florida Bay. The horizontal line at 35 ppt represents nominal seawater salinity. (**a**) MK; (**b**) BK; (**c**) LM.

Table 2. Station time series statistics.

Station	Location	Water Level (m) NGVD				Salinity (ppt)			
		min	mean	max	σ	min	mean	max	σ
BK	Buoy Key	−0.12	0.29	1.03	0.109	9.94	35.91	66.07	5.70
LM	Little Madeira Bay	−0.03	0.31	0.89	0.110	3.70	22.92	48.76	8.02
TR	Taylor River	0.08	0.37	0.89	0.125				
MK	Murray Key	−0.51	0.22	0.89	0.127	14.67	34.84	54.79	3.60
E146	Taylor Slough	−0.18	0.39	0.80	0.143				
TSH	Taylor Slough Hilton	−0.12	0.63	1.02	0.176				

2.4. Empirical Mode Decomposition

Water level and salinity data are decomposed into Intrinsic Mode Functions (IMFs) and nonlinear trends through Empirical Mode Decomposition (EMD) using the Hilbert–Huang transform [20,21] as implemented in the R package hht. Application of the EMD requires uniformly-sampled data without gaps. We reconstruct missing data in our time series by using random samples drawn from distributions of all available data for a specific year day. For example, if 1 January 2000 is missing, a Gaussian kernel is fit to all available data for 1 January. A random sample is then drawn from this distribution and used as the reconstructed value. This preserves the overall distribution of the data for a year day capturing seasonal trends, while realistically allowing for variance away from the mean on the daily timescale.

2.5. Water Level Exceedance

Water level exceedances are computed from daily mean water levels by summing the number of exceedance events above an elevation threshold for each year. The probability of exceedance at a specific threshold as a function of time follows a logistic function exhibiting exponential growth followed by a linear increase, terminating in nonlinear saturation as water levels continuously exceed the threshold [22]. The logistic function suggests a growth model for water level exceedances as they enter the initial growth phase:

$$E(t) = E_0 + \alpha(t - T_L) + (1 + r)^{\frac{t - T_G}{\tau}} \tag{1}$$

where E_0 is the number of exceedances at year $t = 0$; α the linear rate of exceedance; r the growth rate; T_L and T_G the zero-crossing time of linear and exponential growth, respectively; and τ the growth time constant. This model is fit to yearly exceedance data with maximum likelihood estimation over a wide parameter space of initial conditions (Table 3), and the best-fit model from the parameter search is selected based on the minimum Akaike information criteria [23].

Table 3. Initial values and phase space search increments for the exceedance model parameters of Equation (1).

Parameter	Values	Increment
E_0	1	0
α	1	0
T_L	1990–2010	5
T_G	1995–2010	5
r	0–200	20
τ	0–60	20

To forecast the evolution of water level exceedance, we select an elevation threshold with landscape-specific relevance. For example, at the Little Madeira Bay (LM) station, inspection of coastal ridge elevations from the United States Geological Survey (USGS) mapping [24] finds a mean elevation of 70 cm NGVD29. Daily mean water levels are then extracted from the station data for the most recent three-year period, and yearly values of sea level rise from the low and high sea level rise projections are added to the dataset. Each set of yearly data is then processed to sum the total number of yearly threshold exceedances per year.

2.6. Marsh to Ocean Transformation Index

As sea levels rise, we expect a gradual transformation of freshwater coastal marshes into saltwater marshes and eventually into submarine basins. Florida Bay is largely open to the Gulf of Mexico to the west and relatively isolated from the Atlantic Ocean to the east by the island chain of the Florida Keys; as such, marine conditions can be found in western Florida Bay as shown by the tidally-dominated water levels at Buoy Key (BK) (Figure 3) and marine-like salinities at Murray Key (MK) and Buoy Key (Figure 4). As one moves eastward, the tidal signal diminishes (LM in Figure 3) with a transition to a terrestrial hydrologic cycle dominated by seasonal rainfall moving up Taylor Slough (Taylor River (TR), E146 and Taylor Slough Hilton (TSH)).

To assess this change, we decompose the water level signals shown in Figure 3 using IMFs retaining only modes with intra-annual and longer oscillatory cycles, as shown in Figure 5. These low pass versions of water levels allow one to recognize lower amplitude ocean-dominated locations such as Buoy Key (BK) and the higher amplitude, more variable marsh-dominated water levels exemplified at TSH.

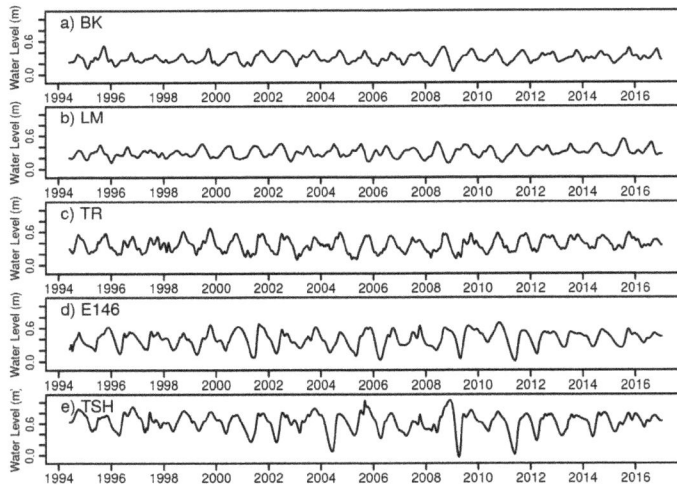

Figure 5. Low frequency cumulative IMFs of water level data in Florida Bay and Taylor Slough shown in Figure 3. (**a**) BK; (**b**) LM; (**c**) TR; (**d**) E146; (**e**) TSH.

We next identify IMFs representing ocean-dominated and freshwater marsh-dominated locales at BK and TSH, respectively, as shown in Figure 6, and use these IMFs as empirical basis functions to reconstruct the low pass water level signals at the intermediate stations LM, TR and E146. The reconstruction is based on linear combinations of weighted ocean and marsh basis functions with the goal of comparing the relative magnitudes of the ocean and marsh basis function fit coefficients as a metric describing the relative hydrologic influence of the marsh or ocean at a particular station.

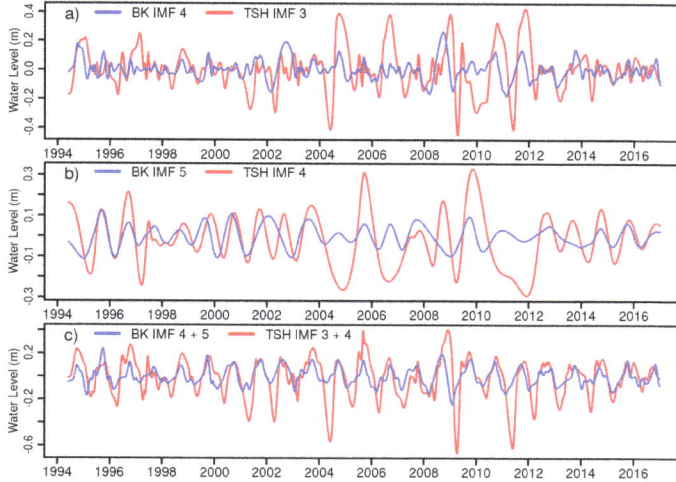

Figure 6. Low frequency IMFs at the BK and TSH stations to represent ocean-dominated and marsh-dominated hydrologic dynamics respectively. (**a**) Intra-annual modes; (**b**) annual modes; (**c**) comparison of low pass water level signals at BK and TSH constructed from the addition of the IMFs in (**a**) and (**b**).

The model is thus:

$$W(t) = \sum_{i=L}^{i=H} \omega_i \, IMF_{\Omega_i} + \mu_i \, IMF_{M_i} \tag{2}$$

where IMF_Ω represent ocean-dominated empirical basis functions, IMF_M marsh-dominated basis functions, L the IMF mode number of the lowest frequency mode or residual, H the mode number of the highest frequency mode and ω_i and μ_i fit coefficients determined by a nonlinear quasi-Newton minimization of the variance of the difference between the weighted sum of the empirical basis functions, $W(t)$, and the target time series (low pass signal of station LM, TR or E146 shown in Figure 5) [25].

The resultant coefficient vectors ω and μ are summed to produce an overall metric $\Omega = \sum \omega_i$, $M = \sum \mu_i$ representing the ocean or marsh influence. For example, with $N = 3$ empirical basis functions and using the Buoy Key (BK) time series as the target, all ω_i equal 1 with the result $\Omega = 3$, $M = 0$, while if TSH is the target then $\Omega = 0$, $M = 3$. To construct a relative metric denoted as the Marsh-to-Ocean Index (MOI), we normalize the difference of the two influence metrics by the number of basis functions N:

$$MOI = \frac{M - \Omega}{N} \tag{3}$$

so that a water level signal identical with that of Buoy Key (BK) would express MOI = −1, while a station with a signal equivalent to the upper reach of Taylor Slough (TSH) would produce MOI = 1.

The MOI discriminates between 'oceanic' and 'marsh' water level variations based on the assumption that variations in the designated ocean signal represent ocean forcing, and likewise for the marsh signal. Implicitly, a storm surge elevating coastal water levels at the ocean station is characterized as an ocean influence, while a runoff event from storm rainfall at the marsh station is attributed as a marsh water level forcing. Here, we are interested in assessing long-term transformations in hydrologic responses, basing MOI low-pass signals on intra-annual and longer cycles. The MOI methodology is general such that inclusion of higher-frequency IMFs that resolve temporally-compact events should be properly accounted for as originating from either the oceanic or marsh reference signals. The time period over which the ocean and marsh basis functions are fit to the intermediate station can also be varied to emphasize shorter-term events or longer-term processes.

3. Results

3.1. Inundation Maps for Mean Sea Level

Figures 7 and 8 present mean sea level inundation maps for the southern Florida peninsula and Dry Tortugas. Blue shadings represent the extent of projected mean sea level inundation at the four time horizons of 2025, 2050, 2075 and 2100. Grey areas indicate elevations higher than the expected mean sea level at 2100. Note that the low and high projections do not share a common legend such that the shade of blue corresponding to a specific land elevation is not shared between the low and high projections; however, the time horizon at which mean sea level reaches an elevation does correspond to the same shade of blue in both projections. Digital versions of the inundation maps are available in the Supplementary Materials.

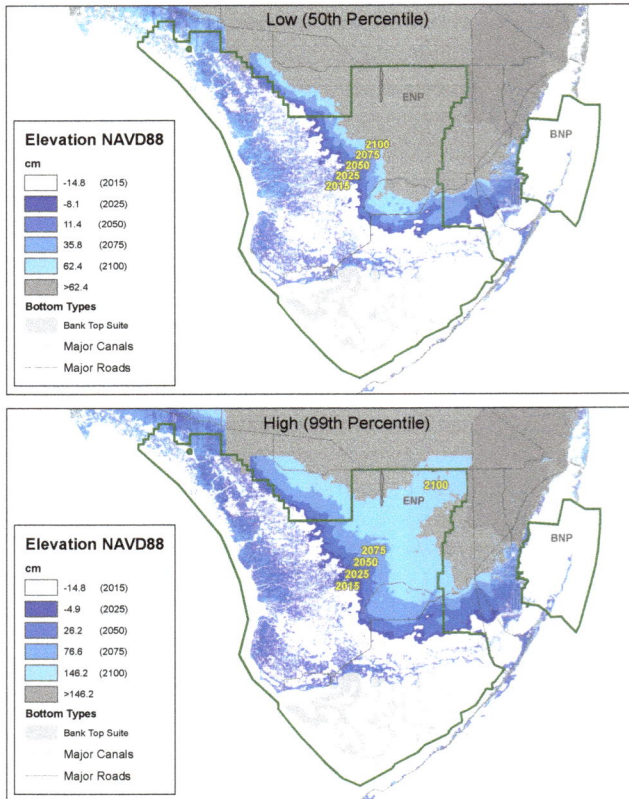

Figure 7. Mean sea level elevation maps for South Florida including Everglades and Biscayne National parks for the median (50th) and high (99th percentile) RCP 8.5 projections using current topography and the NAVD88 datum. Tides and storm surges are not included in this projection.

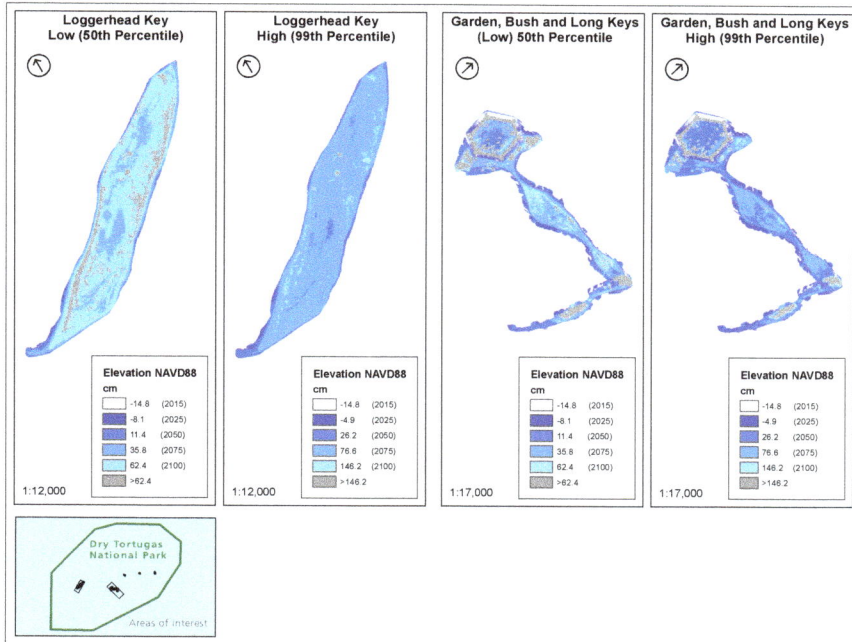

Figure 8. Mean sea level elevation maps for Dry Tortugas National Park at Loggerhead, Garden, Bush and Long Keys for the median (50th) and high (99th percentile) RCP 8.5 projections. Tides and storm surges are not included in this projection.

3.1.1. General Influence of Sea Level Rise

Over the next ten years, represented by the 2025 estimates, dramatic change in sea level is not anticipated with an expected sea level rise of 7 cm for the low scenario and 10 cm for the high projection. These changes will result in more frequent tidal inundation along coastal regions although the buttonwood ridge located along the southern peninsula and north shore of Florida Bay should remain above mean sea level. This modest increase is not likely to impact the terrestrial portions of South Florida or the Dry Tortugas.

By 2050, local sea levels are expected to increase between 26 and 41 cm. Overall, inundation at mean sea level will produce similar impacts for both scenarios with a wider fringe of saltwater inundation around the periphery of the peninsula under the high scenario. Expansion of the white zone, a low productivity area influenced by the periodic flooding of saltwater [26], is expected to continue. Under both projections, mean sea level is expected to reach the elevation of the land surrounding the Florida Power and Light Turkey Point nuclear power plant cooling canal system (vertical lines along the southwest corner of Biscayne National Park (BNP) in Figure 7).

By 2075, sea levels are anticipated to increase by 51 and 91 cm for the low and high projections, respectively. At these elevations, significant portions of the buttonwood ridge separating Florida Bay from the peninsula will be exceeded by mean sea level, and marine conditions can be expected to expand into current-day areas of the Everglades that maintain fresh and brackish-water marshes. This could signal an important tipping-point in the ecological response of freshwater marshes since freshwater basins delineated by the ridge will no longer be viable. Low-lying suburban areas along the southeastern peninsula will also be at mean sea level elevation resulting in perennial tidal flooding with significantly reduced ability to discharge rain floodwaters by gravity from the urban areas into the sea. Below ground, saltwater intrusion can be expected to reduce aquifer productivity along coastal well fields [27].

In 2100, the projected sea level rise is 77 cm for the low projection and 161 cm for the high scenario. Strikingly, in the high scenario, mean sea level can be expected to extend from the southwest peninsula to the northeast corner of Everglades National Park along the topographical depression of Shark River Slough. It is likely that widespread ecological changes will be evident around South Florida as Florida Bay expands. Many of the low-lying islands of Biscayne and Dry Tortugas national parks can be expected to become tidally submerged or dynamically redefined. Islands with coral substrate are likely to submerge, while sand- or sediment-based islands will become increasingly mobile as tidal influences trigger localized erosive and depositional dynamics.

3.1.2. Infrastructure Inundation

Figure 9 presents a comparison of projected mean sea level with land elevation surrounding infrastructure at Flamingo in Everglades National Park and Fort Jefferson in Dry Tortugas National Park where red indicates a building or infrastructure footprint. Conditions at Flamingo are mixed, with the low projection forecasting the housing and visitors center to remain above mean sea level to 2100, but with mean sea levels reaching the boat basin, maintenance yard and water plant by 2100. Under the high projection, the housing area is at mean sea level by 2100; the visitor center will be partially inundated by 2050; and the maintenance yard and water plant by 2075.

At Fort Jefferson, the projections indicate that the north coal dock and campground remain above mean sea level to 2100, while areas around the ferry dock and the isthmus to Bush and Long Keys are expected to be at mean sea level by 2075 under the low sea level rise projection. Under the high projection, much of the north coal dock and campground will be at mean sea level by 2075, as will much of the land between the ferry dock and moat, although a portion of this will be at sea level by 2050. The isthmus to Bush Key is expected to be at mean sea level by 2050.

It is important to note that mean sea level in Florida Bay fluctuates by approximately 30–40 cm (12–16 in) in a yearly oceanographic cycle, as well as up to 70 cm (2.3 ft) in daily and monthly tidal cycles so that effects of tidal inundation will be observed during high tides several years before the projected dates when mean sea level reaches a specific land elevation.

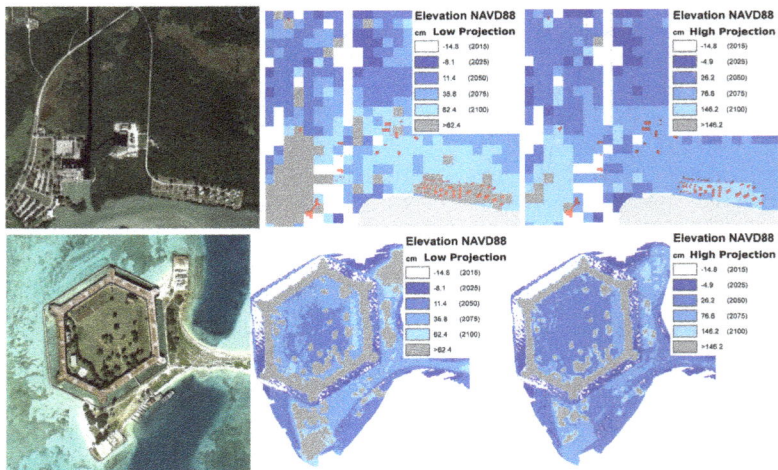

Figure 9. Sea level rise inundation maps at Flamingo in Everglades National Park (**top row**) and Fort Jefferson in Dry Tortugas National Park (**bottom row**). Building and infrastructure footprints are indicated in red.

3.2. Exceedances

As sea levels rise against a fixed elevation threshold near the mean high water tidal elevation, exceedance rate changes will follow Equation (1) experiencing nonlinear growth regardless of whether water levels are increasing at a steady or accelerated rate [22]. This is exemplified in Figure 10 at four coastal stations across the southern peninsula suggesting a transition from linear exceedance growth to exponential growth. Model parameters fit to the middle elevation threshold are shown in Table 4.

Exceedance model fits suggest a progression of exponential growth initiation times (T_G) from the eastern end of Florida Bay where freshwater marsh interaction with the coastal region is high, to the western side of the Bay where marine conditions prevail with substantial water mass exchange with the Gulf of Mexico. Generally, the transition of exceedances near the mean high water tidal threshold suggest that the late 20th to early 21st century represents a change of coastal dynamics where water level exceedances enter a growth phase. Interestingly, the model fits indicate that the doubling times (τ) increase from one decade in the marine areas to two and half decades in the eastern coastal region of Florida Bay, suggesting that environmental impacts from increased exceedances may be more acute over the next few decades along the southwestern coastal region.

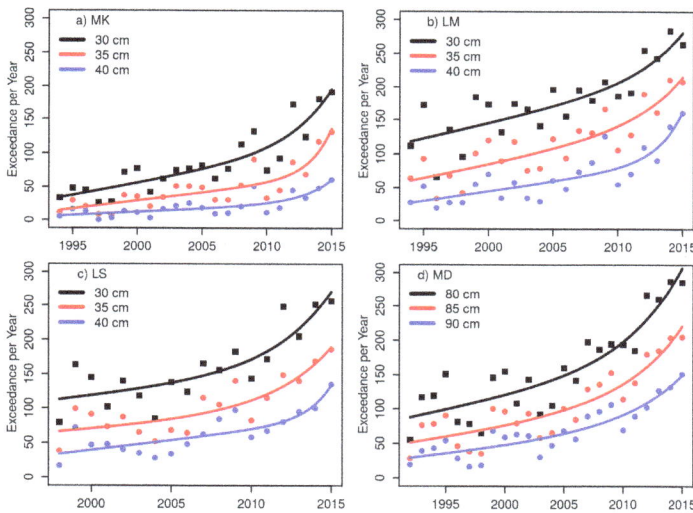

Figure 10. Yearly water level elevation exceedance data and fits to the model of Equation (1). Elevation thresholds are with respect to the NGVD29 datum. Note that the MD station is located on a higher land elevation than the other three stations. (**a**) MK; (**b**) LM; (**c**) LS; (**d**) MD.

Table 4. Exceedance model parameters at an elevation threshold of 35 cm NGVD29 at Murray Key (MK), Little Madeira Bay (LM) and Long Sound (LS). Note that the South Dade (MD) station is located on a higher land elevation and uses a threshold of 85 cm NGVD29.

Station	Threshold (cm)	E_0	α	T_L	T_G	r	τ
Murray Key (MK)	35	42.38	2.46	2005.51	2007.63	381.40	10.32
Little Madeira (LM)	35	71.49	3.99	1996.77	2000.39	134.36	16.82
Long Sound (LS)	35	83.80	1.80	2008.45	1998.01	229.10	20.36
South Dade (MD)	85	66.55	2.59	1998.34	1992.00	208.87	26.12

Exceedance Projections

Application of the sea level rise projections to exceedance data has potential to provide a meaningful environmental-change metric. For example, projection of exceedances at Little Madeira Bay based on a mean local coastal ridge elevation threshold of 70 cm NGVD29 is shown in Figure 11. Several illuminating inferences can be made from this projection; for example, it suggests that regardless of whether sea levels rise along the low or high trajectory, that between the years 2035 and 2045, mean sea level exceedances will enter a phase of exponential growth. Under the high projection, it indicates that circa 2050, mean sea level will be continuously above portions of the coastal ridge wherein one would expect marine conditions to have displaced freshwater influences. If the low projection is realized, then this transition is expected near 2070.

Such exceedance projections may therefore find utility in the identification of tipping points where the transition to an exponential increase in saltwater inundation can be expected, as well as demarcation of a time horizon upon which a fundamental transformation of the coastal environment to submarine conditions will prevail.

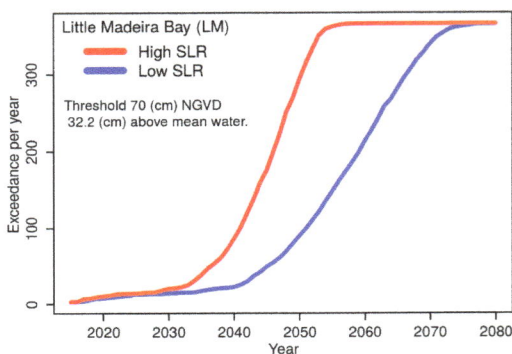

Figure 11. Projected evolution of water level exceedances at Little Madeira Bay under the low and high sea level rise scenarios. Mean water is the daily mean water level over the three-year period from January 2014–December 2016.

3.3. Trends

Empirical mode decomposition of the water level and salinity data shown in Figures 3 and 4 provides a residual signal representing the time-varying trend after all oscillating modes are removed as shown in Figure 12. Regarding salinity, Murray Key and Buoy Key in western Florida Bay exchange waters with the Gulf of Mexico, as well as fresh water runoff from the coastal Everglades, but are predominantly marine ecosystems. Seawater has a nominal salinity of 35 ppt, and we find that mean Murray Key salinity has risen by 2.8 ppt from 32.7 in 1994 to 35.5 ppt in 2016, with values at Buoy Key rising by 6.2 ppt from 30.6–36.8 ppt over the same period indicating that both stations are currently experiencing higher mean salinity and lower freshwater mixing than was common 20 years ago. Little Madeira Bay in the eastern coastal region is more influenced by freshwater runoff from the Everglades and agricultural lands with its mean salinity increasing by 3.2 ppt from 22.4 ppt in 1994 to 25.7 ppt in 2016.

Mean water levels are found to have increased in Florida Bay and the southern reaches of Taylor Slough as shown in Figure 12. In Florida Bay, water levels at Buoy Key and Little Madeira

Bay have risen from 27.7 cm NGVD29 in 1994 to 32.7 cm in 2016, an increase of 5.0 cm, with similar increases over the period observed at the Taylor Slough stations TR and E146[2].

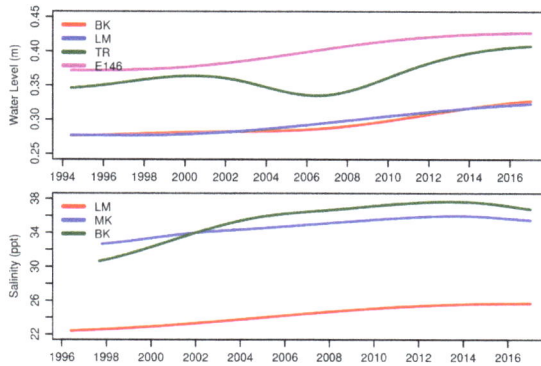

Figure 12. Nonlinear trends of water level and salinity at hydrographic stations in Florida Bay and Taylor Slough.

3.4. Marsh to Ocean Transformation

As described above, the MOI is designed to represent the relative similarity of a time series spatially intermediate with respect to two reference time series representing oceanic and marsh hydrology. Specifically, the relative similitude of water levels over the last three years at Little Madeira Bay (LM) and within Taylor Slough at stations TR and E146, with respect to the oceanic reference station of Buoy Key (BK) and marsh reference at station Taylor Slough Hilton (TSH), are shown in Table 5. MOI values of -0.21 at Little Madeira Bay (LM), 0.02 at Taylor River (TR) and 0.34 in Taylor Slough (E146) suggest that recent water levels at Little Madeira Bay follow the dynamics observed at Buoy Key more closely than those of Taylor Slough (TSH), while levels at E146 are more similar to TSH dynamics.

Table 5. Marsh-to-Ocean Index (MOI) values over the period 1 January 2014–31 December 2016 at stations LM, TR and E146 relative to the ocean-dominated station at Buoy Key (BK) and the marsh-dominated station at TSH.

Station	ω_1	ω_2	ω_3	Ω	μ_1	μ_2	μ_3	M	MOI
LM	0.00	0.62	0.00	0.62	0.00	0.00	0.00	0.00	-0.21
TR	0.46	0.38	0.00	0.84	0.32	0.58	0.00	0.90	0.02
E146	0.26	0.00	0.00	0.26	0.41	0.85	0.00	1.26	0.34

To assess changes in behavior over time, we apply the MOI to the three intermediate stations LM, TR and E146 over a one-year moving window advanced in 10-day increments, as shown in Figure 13. Also shown in Figure 13 is the monthly accumulation of streamflow measured at the confluence of the Taylor River and Little Madeira Bay. This flow represents a fraction of total flow as freshwater into Little Madeira Bay, but is representative of the annual hydrologic cycle driving marsh hydrology.

---

[2] Ignoring the nonlinear nature of these trends, one might consider a linear mean water level rise of 5 cm/21 years = 2.4 mm/year, a result coincident with linear estimates of mean sea level rise over the 20th Century; however, examination of the BK data found an increase over the last decade of 4.0 cm, suggesting a recent rate of 4 mm/year, a value somewhat larger than current global estimates of 3.4 mm/year (https://sealevel.nasa.gov/) and an illustration of difficulties in applying linear metrics to nonlinear processes.

For example, the large flow in 2005 associated with hurricanes Katrina and Wilma resulted in an increase in MOI at Little Madeira Bay from negative to positive values. Overall, it is difficult to discern patterns in the MOI dynamics, although it does appear that since 2012, MOI has remained largely positive at the Taylor River (TR) and Taylor Slough (E146) stations.

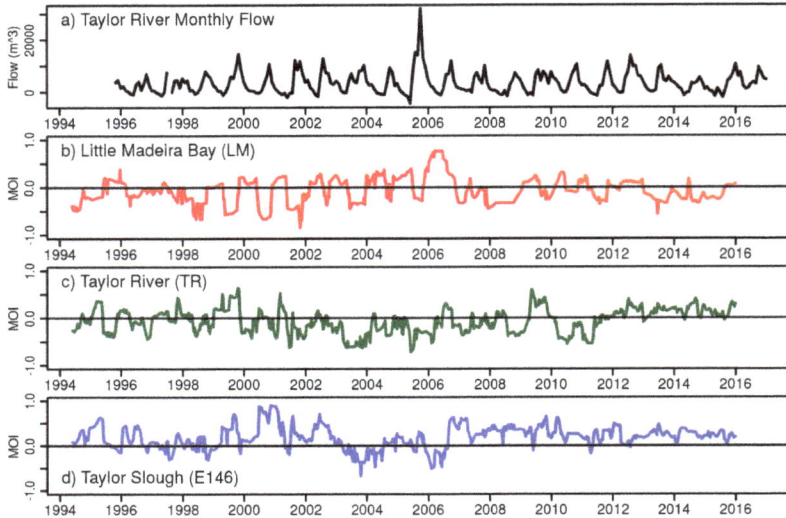

Figure 13. (**a**) Monthly stream flow at the terminus of Taylor River; (**b**) Marsh-to-Ocean Index (MOI) in Little Madeira Bay (LM) over a one year-long moving window advanced in 10-day increments; (**c**) MOI at the Taylor River (TR) station; (**d**) MOI at the E146 station in Taylor Slough.

Since the effects of environmental and hydrologic perturbations on the biomes, landscapes and ecosystems are cumulative, it makes sense to view the cumulative behavior of the time-varying MOI. In Figure 14, we plot the time integral of the MOI shown in Figure 13, that is: $\int_0^T \text{MOI}(t)\, \text{d}t$, where T is a specific day past the data origin of 1 June 1994. This integrated view of the dynamics suggests varying responses at the three stations. At Little Madeira Bay along the coastal interface between Florida Bay and the confluence of Taylor Slough, the MOI exhibits an increasing oceanic influence from 1994–2004 followed by a stable, but still negative MOI over the 2004–2013 period punctuated by the 2005 freshwater event. At the Taylor River station a stable, near-zero MOI from 1994–2000 transitioned to a steady decline from 2000–2012, followed by an increasing trend. At station E146 in Taylor Slough, the cumulative MOI has been steadily increasing since 1994.

Our interpretation of these results is that northern Florida Bay has been transitioning away from a freshwater marsh estuarine environment towards a marine environment over the last two decades. While this transition appears to be essentially continuous in Little Madeira Bay since 1994, at the lower reach of Taylor Slough (TR), hydrologic cycles appear to be transforming from marsh-dominated dynamics to ocean-dominated since 2000. These shifts in hydrologic cycles are consistent with increasing mean sea level (Figure 12) and the onset of accelerated water level exceedances as sea level rises (Figure 10). In the middle-reaches of Taylor Slough represented by station E146, there were periods of negative MOI during 2003 and 2006 (Figure 13) reflected in a stasis of the integrated MOI over this period (Figure 14), but the overall assessment is that there is no evidence of emerging ocean-dominated hydrologic cycles.

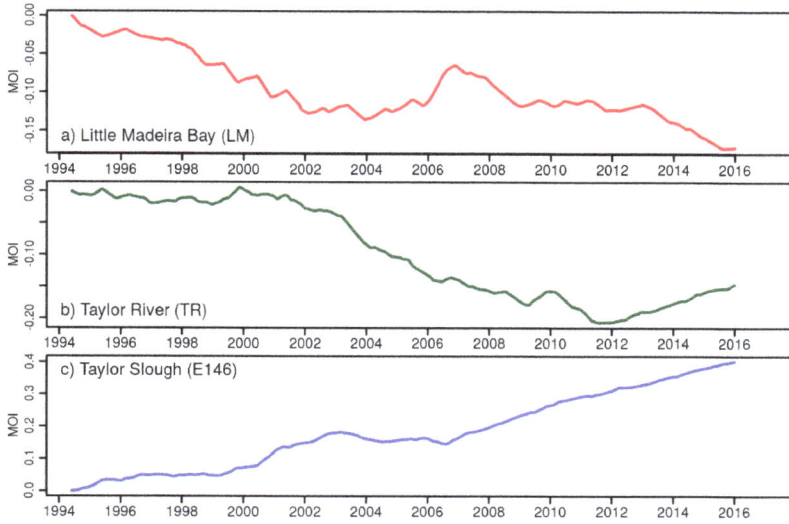

Figure 14. Integral over time of the Marsh-to-Ocean Index (MOI) at: (**a**) Little Madeira Bay (LM); (**b**) Taylor River (TR); and (**c**) Taylor Slough (E146).

4. Discussion

South Florida is ranked ninth globally among urban areas with human populations exposed to sea level rise impacts and first in terms of exposed assets [28]. South Florida is equally rich in natural assets with national parks circumscribing the southern peninsula protecting vital freshwater ecosystems that sustain both natural and human biomes. The exceedingly flat topography and low elevations provide ideal conditions for the expansion of marine bays and estuaries into existent freshwater marshes, civil infrastructure and human habitats, issues recognized by regional governments planning for future sea levels. Such planning efforts rely on global projections without a probabilistic estimate of sea level likelihood, and focus on urban areas. Here, we examine projected impacts based on a local sea level rise projection with explicit probabilities corresponding to the median and 99th percentiles focusing on the estuarine coastal fringe along the southern end of the peninsula. This fringe is generally lower in elevation than the urbanized east coast, and its natural condition provides an optimal setting to monitor sea level rise and landscape transformation.

Inundation projections indicate dramatic changes in landscape along the southern peninsula over the 21st century with the submergence of low-lying urban and suburban areas, as well as land surrounding cooling canals at the Turkey Point nuclear power plant. In the Everglades, it appears that substantial portions of existent freshwater marshes will be converted to estuarine and shallow marine zones with the potential for mean sea level to span the interior of the peninsula along Shark River Slough.

A fundamental landscape feature of the southern peninsula is a low, narrow ridge separating the marine and estuarine waters of Florida Bay from freshwater marshes of the southern Everglades. When sea levels rise above this ridge, a pronounced environmental transformation into a marine-dominated landscape is expected. We can anticipate this change by applying sea level rise projections to recent exceedance statistics at the ridge elevation to identify a tipping point horizon where water level exceedances above this elevation will grow, as well as when the elevation is forecast to become submarine, signaling a complete transformation to marine or estuarine conditions. Doing so, we find that circa 2040, the coastal region of Little Madeira Bay will enters the transition of accelerating water level exceedances above the coastal ridge, and between 2050 and 2070, the area is expected to be transformed into a marine-dominated landscape.

Transformations along the southern peninsula are inexorably coupled to ecological changes and feedbacks with the coastal landscape consisting of a dynamic surficial layer of wetland soil overlying a karstic surficial aquifer. Freshwater soils of mostly organic-rich peat support the ridge-and-slough landscape and tree islands, while coastal wetlands such as salt marshes and mangroves contain substantial organic matter along with varying amounts of inorganic sediment washed in by tides, waves or storm surge. The conversion of coastal marshes to open water from saltwater intrusion and sea-level rise is often accompanied by peat collapse and deterioration, releasing large amounts of sequestered carbon. It is estimated that mangrove forests in the Everglades store 145 tonnes per hectare [29].

These coastal wetlands possess a limited capacity to stabilize and maintain existing coastal barriers through accretion of organic matter and storm-derived sediment with an average accretion rate of 1–3 mm/year with more rapid accretion rates possible for short periods of time [30]. Global rates of sea level rise are currently estimated at 3.4 mm/year [31], and our data find local rates of 4 mm/year over the last decade. In view of the established and expected acceleration of sea level rise, the landscape may have reached a tipping point unable to sustain spatially-static coastal mangrove forests. Indeed, vegetation loss along the coast is expressed in a "white zone" of low productivity that has been shifting inland over the past 70 years [3,32], and models of expected mangrove proliferation suggest that a 66-cm sea level rise corresponding to year 2070 under the high sea rise projection and 2100 under the low scenario will transform 2000 square kilometers of freshwater marsh into mangrove forests [33].

While inundation projections point to expected changes, examination of water levels allow us to detect and quantify an acceleration of water level exceedances over the last decade. Such an acceleration is a natural product of rising sea levels against a fixed elevation whether the change in sea level itself is linear or accelerating, and we find that exponential doubling times for these exceedances are on the order of one to two decades. Ecological transformation from freshwater to saltwater biomes is driven by the spatial and temporal extent of these saltwater inundations, and in Table 6, we list the change in water level exceedances per year at elevation thresholds above the 90th percentile mean water levels in 1995. Here, we see that in the marine portion of Florida Bay at Murray Key (MK), high-water level exceedances have increased from 2–17% of the days per year over the last two decades, while along the coastal margins, these exceedances have changed from a twice-monthly occurrence, likely at the spring tides, to a nearly every-other-day occurrence.

Table 6. Number of days per year of water level exceedance in 1995 and 2015.

Station	MK		LM		LS		MD	
Threshold	40 cm		40 cm		40 cm		90 cm	
Year	1995	2015	1995	2015	1995	2015	1995	2015
Days	6	61	27	161	34	133	29	149
Percent Days	2	17	7	44	9	36	8	41

Questions regarding the spatial progression of sea level rise impacts have been addressed with inundation and exceedance projections; however, the presence of hydrographic records spanning the marine-to-freshwater interface provides an opportunity to identify spatially-explicit time series revealing a dynamic transition of water levels from marsh to ocean-dominated. This motivates us to introduce the Marsh-to-Ocean transformation Index (MOI) as a metric to quantify these changes. We find that since 1994, there is a cumulative increase in ocean-dominated hydrographic signals in Little Madeira Bay (LM), as well in the lower reach of Taylor Slough (TR). Farther upstream at station E146, there is no cumulative evidence of ocean influence.

5. Conclusions

Collectively, the data and analysis present a cohesive picture that South Florida landscapes and ecosystems are experiencing a transformation of coastal environments into marine-dominated conditions. Such a transformation will accrue benefits for marine biomes, while decreasing the productivity of coastal freshwater aquifers and presenting challenges for existent freshwater habitats, as well as human infrastructure and habitation.

It is important to note that these projections are for mean sea level and do not consider inundation due to tides or storms. Impacts from tidal inundation will first be noticed at spring tides and then from daily high tides several years or even a decade prior to mean sea level effects. These events will provide opportunities to study the impacts and responses of increasingly frequent inundation events prior to the transformation of existent coastal fringes and freshwater ecosystems into marine-dominated areas. Further, the projections do not incorporate contributions in the event of Antarctic ice-sheet collapse or from changes in ocean circulation and the Florida Current, which have the potential to increase sea levels and compress the forecast time horizons. Regardless of the specific sea level rise trajectory, restoration of the Everglades with increased freshwater flow and water levels will serve to mitigate the impacts of sea level rise over the next century and protect freshwater resources for both the natural and human inhabitants.

Supplementary Materials: GIS coverages of the sea level rise projections are available online at *Park and Stabenau* [34].

Acknowledgments: The authors are indebted to Caryl Alarcón for expert GIS analysis and support.

Author Contributions: Joseph Park and Erik Stabenau created the sea level rise projection and edited the manuscript. Joseph Park performed the exceedance analysis and suggested and computed the MOI. Jed Redwine edited the manuscript. Kevin Kotun managed the data network and edited the manuscript.

Conflicts of Interest: The authors declare no conflict of interest.

Abbreviations

The following abbreviations are used in this manuscript:

EMD	Empirical Mode Decomposition
IEA	International Energy Agency
IMF	Intrinsic Mode Function
IPCC	Intergovernmental Panel on Climate Change
MHHW	Mean High-Higher Water
MLLW	Mean Low-Lower Water
MOI	Marsh-to-Ocean Index
MSL	Mean Sea Level
NAVD88	North American Vertical Datum of 1988
NGVD29	National Geodetic Vertical Datum of 1929
NTDE	National Tidal Datum Epoch
RES	Renewable Energy Source
RCP	Representative Concentration Pathway
UNESCO	United Nations Educational, Scientific and Cultural Organization
USGS	United States Geological Survey

Appendix A. Datum and Water Level Conversions

A tidal datum [35] provides a geodetic link between ocean water level and a land-based elevation reference such as the North American Vertical Datum of 1988 (NAVD88). The National Tidal Datum Epoch (NTDE) in the United States is a 19-year period over which tidal datums specific to each tide gauge are determined. The current NTDE for the United States is 1983–2001, and sea level rise projections are referenced to the midpoint of this period (1992) consistent with design procedures determined by the U.S. Army Corps of Engineers and NOAA's National Climate Assessment [36]. Common tidal datums include Mean Sea Level (MSL), Mean High-Higher Water

(MHHW) and Mean Low-Lower Water (MLLW), as defined by the National Oceanic and Atmospheric Administration [35]. As sea level rises, tidal datum elevations also, rise and a new tidal datum is established every 20–25 years to account for sea level change and vertical adjustment of the local landmass [37].

Kopp et al. [9] assumed a water level reference of mean sea level starting at year 2000; however, the mean sea level datum at the Vaca Key tide gauge, which can be referenced to NAVD88, is with respect to the National Tidal Datum Epoch (NTDE) centered on 1992. To reference the Kopp projection to NAVD88, we must first account for sea level rise at the tide gauge from 1992–2000. We quantified this sea level rise at Vaca Key with an empirical mode decomposition resulting in value of 1.4 cm. This sea level rise offset is added to the Kopp projections to account for the fact that their projections start in the year 2000, but that the NTDE mean sea level at Vaca Key is referenced to 1992.

Next, we convert the projections with respect to the NTDE mean sea level datum to the NAVD88 geodetic datum of the topographic elevations. Table A1 lists the NTDE and NAVD88 elevations at the Vaca Key tide gauge [38], where we find that the NAVD88 datum is 25.1 cm above the NTDE MSL datum. In other words: MSL referenced to NAVD88 is equal to the NAVD88 datum elevation minus 25.1 cm. We therefore subtract 25.1 cm from all projected water levels with respect to mean sea level in order to reference them to the NAVD88 datum.

Table A1. Elevations on station datum in meters at Vaca Key, FL (NOAA station: 8723970). Tidal datum epoch: 1983–2001.

Datum	Value	Description
NAVD88	1.182	North American Vertical Datum of 1988
MHHW	1.072	Mean Higher-High Water
MHW	1.040	Mean High Water
MSL	0.931	Mean Sea Level
MLW	0.822	Mean Low Water
MLLW	0.775	Mean Lower-Low Water
STND	0.000	Station Datum

Finally, we apply the sea level rise projections with respect to NAVD88 to current mean sea level, which is not at zero elevation NAVD88. As above, we note that the NTDE (1992) mean sea level at Vaca Key is −25.1 cm NAVD88, while the current sea level in Florida Bay averaged over 2008–2015 is −14.8 cm NAVD88 (Appendix C). The difference of 10.3 cm reflects sea level rise from 1992–2015 and any local influences of dynamic height between Vaca Key and the three stations where mean sea level was estimated.

Putting this all together, the elevation of −14.8 cm NAVD88 is the starting elevation of the sea level projections, as shown in Figure 2 and Appendix B. The projections from Kopp et al. [9], which have been converted to NAVD88, are then added to this base sea level elevation to predict future mean sea level in Florida Bay.

The cautious reader might consider that there has been a double accounting of sea level rise, 1.4 cm representing the change from 1992–2000 and 10.3 cm for sea level rise from 1992–2015. However, these are two independent adjustments. The 1.4-cm adjustment was solely for the purpose of referencing the Kopp projections to the mean sea level datum (NTDE), which was then referenced to NAVD88, a datum conversion independent of the projection starting time. The 10.3-cm accounts for the fact that we choose 2015 as the starting point of the projections. Had we selected year 2000 as the starting point, then the 1.4 cm datum conversion would still apply, while the adjustment to a starting sea level of 2000 would be less than the 10.3 cm determined for a 2015 start time.

Appendix B. Tabulated Sea Level Rise Projection

Sea level rise in cm NAVD88 from Kopp et al. [9] at Vaca Key. Values between decades (2010, 2020, etc.) have been interpolated with a cubic spline. Low is the 50th percentile of the RCP 8.5

projection; high the 99th percentile. An offset of 1.4 cm has been added to account for sea level rise from 1992–2000 to convert the Kopp projections starting in 2000 to the NTDE MSL datum of 1992. The NAVD88 datum is 25.3 cm above the NTDE MSL, so that 25.3 cm has been subtracted to convert NTDE MSL to NAVD88. The projections have been offset to match observed mean sea level over the period 2008–2015 in Florida Bay of −14.8 cm NAVD88 (Appendix C).

Table A2. Sea level rise in cm NAVD88 from Kopp et al. [9].

Year	Low	High	Year	Low	High	Year	Low	High	Year	Low	High
2015	−14.8	−14.8	2045	6.8	18	2075	35.8	76.6	2105	68.3	159.9
2016	−14.2	−13.8	2046	7.7	19.6	2076	36.9	79	2106	69.5	162.7
2017	−13.6	−12.8	2047	8.6	21.1	2077	38	81.5	2107	70.8	165.4
2018	−12.9	−11.8	2048	9.6	22.8	2078	39.2	84	2108	72	168.3
2019	−12.3	−10.8	2049	10.5	24.4	2079	40.3	86.5	2109	73.2	171.2
2020	−11.6	−9.8	2050	11.4	26.2	2080	41.4	89.2	2110	74.4	174.2
2021	−10.9	−8.9	2051	12.3	27.9	2081	42.6	91.8	2111	75.6	177.2
2022	−10.2	−7.9	2052	13.2	29.7	2082	43.7	94.5	2112	76.7	180.3
2023	−9.5	−6.9	2053	14.1	31.6	2083	44.8	97.2	2113	77.9	183.5
2024	−8.8	−5.9	2054	15	33.5	2084	45.9	100	2114	79	186.8
2025	−8.1	−4.9	2055	15.9	35.4	2085	47.1	102.8	2115	80.1	190.1
2026	−7.4	−3.9	2056	16.8	37.3	2086	48.2	105.6	2116	81.2	193.4
2027	−6.7	−2.9	2057	17.7	39.3	2087	49.3	108.5	2117	82.2	196.8
2028	−6	−1.9	2058	18.6	41.2	2088	50.3	111.3	2118	83.3	200.2
2029	−5.3	−0.9	2059	19.5	43.2	2089	51.4	114.2	2119	84.4	203.7
2030	−4.6	0.2	2060	20.4	45.2	2090	52.4	117.2	2120	85.4	207.2
2031	−3.9	1.2	2061	21.4	47.1	2091	53.4	120.1			
2032	−3.2	2.2	2062	22.3	49	2092	54.4	123			
2033	−2.6	3.2	2063	23.3	51	2093	55.4	125.9			
2034	−1.9	4.3	2064	24.3	52.9	2094	56.3	128.9			
2035	−1.2	5.3	2065	25.3	54.9	2095	57.3	131.8			
2036	−0.5	6.4	2066	26.3	56.9	2096	58.3	134.7			
2037	0.2	7.6	2067	27.3	58.9	2097	59.3	137.6			
2038	0.9	8.7	2068	28.3	60.9	2098	60.3	140.5			
2039	1.6	9.9	2069	29.4	63	2099	61.3	143.3			
2040	2.4	11.2	2070	30.4	65.2	2100	62.4	146.2			
2041	3.2	12.4	2071	31.5	67.3	2101	63.5	148.9			
2042	4.1	13.8	2072	32.6	69.6	2102	64.7	151.7			
2043	5	15.1	2073	33.6	71.8	2103	65.9	154.4			
2044	5.9	16.6	2074	34.7	74.2	2104	67.1	157.1			

Appendix C. Mean Sea Level in Florida Bay

MSL was determined by averaging data over the last seven years at three sea level stations across Florida Bay. Sea levels were first aggregated into daily averages, followed by a 30-day moving average at each station. The MSL estimate consists of an average of these three stations from 1 July 2008–1 July 2015, as shown in Figure A1, and this MSL value of 0.97 ft NGVD29 or −14.8 cm NAVD88 (−0.49 feet NAVD88) is used as the starting point of the projections in 2015.

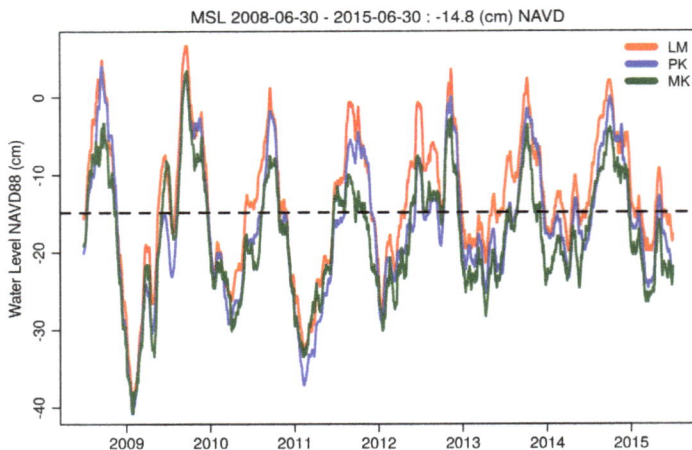

Figure A1. Thirty-day moving averages of daily mean sea level at Murray Key (MK), Peterson Key (PK) and Little Madeira Bay (LM) in Florida Bay. The dashed line is the mean of all three datasets.

Appendix D. Processes Not Included in the Projections

The mean sea level projections presented in this paper represent the contemporary state-of-the-art in local sea level rise forecasts. However, knowledge of all processes and feedbacks driving sea levels is limited, and the models on which these projections are based are necessarily incomplete. The models do not have the spatial resolution and physical process representation required to resolve fine-scale oceanographic processes such as tides and changes in the Florida Current. This means that inundation will be observed during high tides and peaks of seasonal sea level cycles several years before the projected dates when mean sea level reaches a specific land elevation.

Appendix D.1. Tides and Seasonal Cycles

Tides represent the most regular and familiar sea level changes at a coast, but are highly variable in height and timing depending on regional and local bathymetric features. Along the Cape Sable region, tides produce a water level change of up to 70 cm (2.3 ft) in daily and monthly cycles. There is also a regional yearly cycle of water level from atmospheric and oceanographic forcings producing water level changes of 30–40 cm (Appendix C).

Appendix D.2. Florida Current

The Florida Current is one of the strongest and most climatically-important ocean currents forming the headwater of the Gulf Stream [39]. As the Florida Current fluctuates in intensity, sea levels along the Atlantic coast of Florida respond to a geostrophic balance by falling when the current increases and rising when the current decreases [40].

The Gulf Stream and Florida Current are components of the Atlantic Meridional Overturning Circulation (AMOC), a component of the global ocean conveyor belt. Climate models agree that as the ocean warms and fresh meltwater is added, there will be a decline in the strength of the AMOC [41]. A weakening AMOC is expected to result in a weakening of the Florida Current and a subsequent increase in sea levels. The extent of this change is difficult to forecast, but recent evidence suggests that a 10% decline in transport has contributed 60% of the roughly 7-cm increase in sea level at Vaca Key over the last decade [10]. Continued reduction of the AMOC and Florida Current could be expected to contribute an additional 10–15 cm of sea level rise to South Florida over this century. This potential is not reflected in the sea level rise projections, but should be considered by authorities and planners that use them.

Appendix D.3. Storm Surge

Although sea level rise and increases in coastal flooding are important physical stresses on South Florida's natural areas, it is the infrequent, high-impact storm surge events that drastically change the landscape over the course of a few hours. For example, Hurricane Wilma in 2005 had a profound impact on the ecology of the Cape Sable region of Everglades National Park [42,43] producing extensive damage at the Flamingo Visitor Center of Everglades National Park, permanently closing the Flamingo Lodge and Buttonwood Cafe.

Storm surge is highly dependent on the severity and path of the storm, as well as the local bathymetric and topographic features of the coast, and since they occur infrequently, it is difficult to develop robust predictions of these rare events. A popular approach is to fit an extreme-value probability distribution to the highest water levels observed at a water level monitoring gauge. However, gauges have short periods of record, typically a few decades at most, and they fail or are destroyed during extreme storms such that peak water levels are not recorded. A predictive storm surge database, SurgeDat, was developed in part to address this shortcoming by providing a statistical combination of data from multiple events within an area of interest [44]. SurgeDat records storm surge water levels from all available sources, often from post-event high-water marks where gauge data are not available. SurgeDat then applies a statistical regression to estimate storm surge recurrence intervals. A recurrence interval is the length of time over which one can expect a storm surge to meet or exceed a specific inundation level. A familiar example is the 100-year flood level, which is really a 100-year recurrence interval at the specified flood level. In other words, in any one year, there is a 1/100, or 1% chance that the flood level will be matched or exceeded. An excellent discussion of this can be found at the United States Geological Survey web page water.usgs.gov/edu/100yearflood.html.

Relevant to South Florida, a subset of SurgeDat storm surge events was selected within a 25-mile radius of 25.2° N, 80.7° W to represent Florida Bay impacts and is tabulated in Table A3. Based on these events, the SurgeDat projection for storm surge recurrence intervals are shown in Figure A2 and tabulated in Table A4, suggesting that a 180-cm (6 ft) surge event can be expected every 20 years. This same level of sea level rise is not anticipated to occur until at least 2100.

Table A3. SurgeDat database entries for a 25-mile radius centered on 25.2° N, 80.7° W in Florida Bay.

Storm Name	Year	Longitude	Latitude	Surge (m)	Datum	Location
Katrina	2005	−81.0369	25.1294	1.22	Extreme	SW Florida
Rita	2005	−80.7200	24.8605	1.22	NGVD29	Middle and Upper Keys
Wilma	2005	−81.0352	25.3523	2.50		Shark River 3
Gordon	1994	−80.5139	25.0108	1.22	Above Sea Level	Upper Florida Keys
Andrew	1992	−80.9120	25.1431	1.50		Flamingo
David	1979	−80.6263	24.9231	0.61	Above Normal	Islamorada
Gladys	1968	−80.5135	25.0110	0.15	Above Normal	Tavernier
Inez	1966	−80.5297	24.9976	1.10	Above Normal	Plantation Key
Alma	1966	−80.5135	25.0110	0.30	Above Normal	Tavernier
Betsy	1965	−80.5148	25.0096	2.35	Mean Low Water	Tavernier
Donna	1960	−80.6353	24.9133	4.11		Upper Matecumbe Key
Labor Day	1935	−80.7375	24.8516	5.49		Lower Matecumbe
Unnamed	1929	−80.3885	25.1848	2.68	Mean Sea Level	Key Largo

The recurrence interval projection is by necessity based on a sparse dataset, and caution should be used in its interpretation. As projection intervals become longer, it is more likely that the observed data are inadequate to robustly represent all possibilities. Furthermore, these projections do not incorporate changes from sea level rise or from a changing climate, which can alter the strength and frequency of storms. An important aspect of sea level rise is that it significantly shortens the expected recurrence intervals of storm surge. For example, under a median sea level rise projection at Key West,

Park et al. [45] find that a one-in-50-year storm surge based on historic data in 2010 can be expected to occur once every five years by 2060.

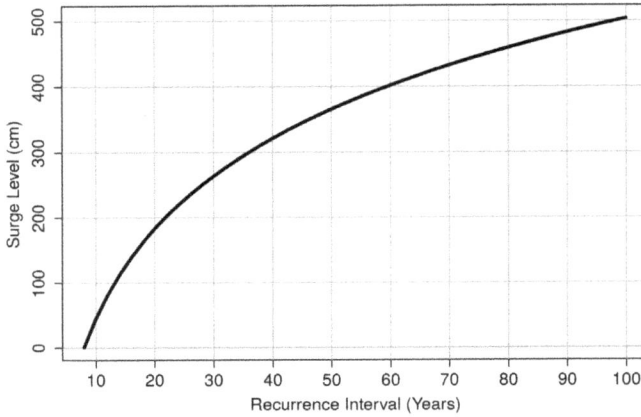

Figure A2. Storm surge recurrence intervals from the SurgeDat database and return period predictor for a 25-mile radius centered on 25.2° N, 80.7° W.

Table A4. Recurrence interval projection in years from the Florida Bay SurgeDat data. Note that this projection does not take into account future sea level rise.

Interval (Year)	Surge (m)	Interval (Year)	Surge (m)
10	0.45	56	3.88
12	0.82	58	3.95
14	1.12	60	4.02
16	1.39	62	4.08
18	1.62	64	4.15
20	1.83	66	4.21
22	2.02	68	4.27
24	2.19	70	4.33
26	2.35	72	4.38
28	2.50	74	4.44
30	2.64	76	4.49
32	2.77	78	4.54
34	2.89	80	4.59
36	3.00	82	4.64
38	3.11	84	4.69
40	3.21	86	4.73
42	3.31	88	4.78
44	3.40	90	4.82
46	3.49	92	4.87
48	3.57	94	4.91
50	3.65	96	4.95
52	3.73	98	4.99
54	3.81	100	5.03

References

1. Cazenave, A.; Le Cozannet, G. Sea level rise and its coastal impacts. *Earth's Future* **2013**, *2*, 15–34, doi:10.1002/2013EF000188.
2. Light, S.S.; Dineen, J.W. Water control in the Everglades: A historical perspective. In *Everglades: The Ecosystem and its Restoration*; Davis, S.M., Ogden, J.C., Park, W.A., Eds.; St. Lucie Press: Delray Beach, FL, USA, 1994; p. 826, ISBN 0963403028.

3. National Research Council (NRC). *Progress toward Restoring the Everglades: The Fifth Biennial Review—2014*; Committee on Independent Scientific Review of Everglades Restoration Progress, Water Science and Technology Board, Board on Environmental Studies and Toxicology; Division on Earth and Life Studies; National Research Council: Washington, DC, USA, 2014; p. 320, doi:10.17226/18809.2014. Available online: http://www.nap.edu/catalog.php?record_id=18809 (accessed on 25 July 2017).

4. Southeast Florida Regional Climate Change Compact (RCCC). *A Region Responds to a Changing Climate Southeast Florida Regional Climate Change Compact Counties, Regional Climate Action Plan*; October 2012; p. 84. Available online: https://southeastfloridaclimatecompact.files.wordpress.com/2014/05/regional-climate-action-plan-final-ada-compliant.pdf (accessed on 25 July 2017).

5. Southeast Florida Regional Climate Change Compact (RCCC). *The Sea Level Rise Task Force*; 2013. Available online: http://www.miamidade.gov/planning/boards-sea-level-rise.asp (accessed on 25 July 2017).

6. Holland, P.R.; Brisbourne, A.; Corr, H.F.J.; McGrath, D.; Purdon, K.; Paden, J.; Fricker, H.A.; Paolo, F.S.; Fleming, A.H. Oceanic and atmospheric forcing of Larsen C Ice-Shelf thinning. *Cryosphere* **2015**, *9*, 1005–1024, doi:10.5194/tc-9-1005-2015.

7. Wouters, B.; Martin-Español, A.; Helm, V.; Flament, T.; van Wessem, J.M.; Ligtenberg, S.R.M.; van den Broeke, M.R.; Bamber, J.L. Dynamic thinning of glaciers on the Southern Antarctic Peninsula. *Science* **2015**, *348*, 899–903, doi:10.1126/science.aaa5727.

8. The Intergovernmental Panel on Climate Change (IPCC). *Climate Change 2013: The Physical Science Basis*; Contribution of Working Group I to the Fifth Assessment Report of the Intergovernmental Panel on Climate Change; 5th Assessment report; Stocker, T.F., Qin, D., Plattner, G.-K., Tignor, M., Allen, S.K., Boschung, J., Nauels, A., Xia, Y., Bex, V., Midgley, P.M., Eds.; Cambridge University Press: Cambridge, UK; New York, NY, USA, 2013.

9. Kopp, R.W.; Horton, R.M.; Little, C.M.; Mitrovica, J.X.; Oppenheimer, M.; Rasmussen, D.J.; Strauss, B.H.; Tebaldi, C. Probabilistic 21st and 22nd century sea-level projections at a global network of tide gauge sites. *Earth's Future* **2014**, *2*, 383–406, doi:10.1111/eft2.2014EF000239.

10. Park, J.; Sweet, W. Accelerated sea level rise and Florida Current transport. *Ocean Sci.* **2015**, *11*, 607–615, doi:10.5194/os-11-607-2015.

11. National Oceanic and Atmospheric Administration (NOAA). *Earth Systems Research Laboratory, CO_2 Annual Mean Growth Rate for Mauna Loa, Hawaii*; 2017. Available online: https://www.esrl.noaa.gov/gmd/ccgg/trends/gr.html (accessed on 25 July 2017).

12. Jones, G.A.; Warner, K.J. The 21st century population-energy-climate nexus. *Energy Policy* **2016**, *93*, 206–212, doi:10.1016/j.enpol.2016.02.044.

13. International Energy Agency. *Medium-Term Renewable Energy Market Report 2016*; Available online: https://www.iea.org/newsroom/news/2016/october/medium-term-renewable-energy-market-report-2016.html (accessed on 25 July 2017).

14. United Nations REN21. *Renewables 2016 Global Status Report (Paris: REN21 Secretariat)*; ISBN 978-3-9818107-0-7. 2016. Available online: www.ren21.net/status-of-renewables/global-status-report/ (accessed on 25 July 2017).

15. DeConto, R.M.; Pollard, D. Contribution of Antarctica to past and future sea-level rise. *Nature* **2016**, *531*, 591–597, doi: 10.1038/nature17145.

16. Fennema, R.; James, F.; Bhatt, T.; Mullins, T.; Alarcon, C. *EVER Elevation (Version 1): A Multi-Sourced Digital Elevation Model for Everglades National Park*; National Park Service: Homestead, FL, USA, 2015.

17. Lentz, E.; Thieler, E.; Plant, N.; Stippa, S.; Horton, R.; Gesch, D. Evaluation of dynamic coastal response to sea-level rise modifies inundation likelihood. *Nat. Clim. Chang.* **2016**, *6*, 696–700, doi:10.1038/nclimate2957.

18. Passeri, D.L.; Hagen, S.C.; Medeiros, S.C.; Bilskie, M.V.; Alizad, K.; Wang, D. The dynamic effects of sea level rise on low-gradient coastal landscapes: A review. *Earth's Future* **2015**, *3*, 159–181.

19. United Nations Educational, Scientific, and Cultural Organization (UNESCO). *The Practical Salinity Scale 1978 and the International Equation of State of Seawater 1980*; Unesco Technical Papers in Marine Science 36; 1981. Available online: http://unesdoc.unesco.org/images/0004/000461/046148eb.pdf (accessed on 25 July 2017).

20. Huang, N.E.; Wu, Z. A review on Hilbert-Huang transform: Method and its applications to geophysical studies. *Rev. Geophys.* **2008**, *46*, RG2006, doi:10.1029/2007RG000228.

21. Chambers, D.P. Evaluation of empirical mode decomposition for quantifying multi-decadal variations and acceleration in sea level records. *Nonlin. Process. Geophys.* **2015**, *22*, 157–166, doi:10.5194/npg-22-157-2015.

22. Sweet, W.; Park, J. From the extreme to the mean: Acceleration and tipping points of coastal inundation from sea level rise. *Earth's Future* **2014**, *2*, 579–600, doi:10.1002/2014EF000272.

23. Akaike, H. A new look at the statistical model identification. *IEEE Trans. Autom. Control* **1974**, *19*, 716–723, doi:10.1109/TAC.1974.1100705.

24. USGS. *Measuring and Mapping the Topography of the Florida Everglades for Ecosystem Restoration, FS-021-03*; March 2003. Available online: https://sofia.usgs.gov/publications/fs/021-03/factsheet02103-Desmond.pdf (accessed on 25 July 2017).

25. Byrd, R.H.; Lu, P.; Nocedal, J.; Zhu, C. A limited memory algorithm or bound constrained optimization. *SIAM J. Sci. Comput.* **1995**, *16*, 1190–1208.

26. Ross, M.S.; Gaiser, E.E.; Meeder, J.F.; Lewin, M.T. Multi–Taxon Analysis of the "White Zone", a Common Ecotonal Feature of South Florida Coastal Wetlands. In *Linkages Between Ecosystems in the South Florida Hydroscape: The River of Grass Continues*; CRC Press: Boca Raton, FL, USA, 2001; pp. 205–238.

27. Dausman, A.; Langevin, C.D. *Movement of the Saltwater Interface in the Surficial Aquifer System in Response to Hydrologic Stresses and Water-Management Practices, Broward County, Florida*; USGS Scientific Investigations Report 2004-5256; 2005. Available online: http://pubs.usgs.gov/sir/2004/5256/ (accessed on 25 July 2017).

28. Hanson, S.; Nicholls, R.; Ranger, N.; Hallegatte, S.; Corfee-Morlot, J.; Herweijer, C.; Chateau, J. A global ranking of port cities with high exposure to climate extremes. *Clim. Chang.* **2011**, *104*, 89–111, doi:10.1007/s10584-010-9977-4.

29. Hutchison, J.; Manica, A.; Swetnam, R.; Balmford, A.; Spalding, M. Predicting Global Patterns in Mangrove Forest Biomass. *Conserv. Lett.* **2014**, *7*, 233–240, doi:10.1111/conl.12060.

30. Wanless, H.; Parkinson, R.; Tedesco, L. Sea level control on stability of Everglades wetlands. In *Everglades: The Ecosystem and Its Restoration*; Davis, S.M., Ogden, J.C., Eds.; St. Lucie Press: Delray Beach, FL, USA, 1997; pp. 199–223.

31. National Aeronautic and Space Administration (NASA). *Sea Level Change, Observations from Space*; 2017. Available online: https://sealevel.nasa.gov/ (accessed on 10 July 2017).

32. Fuller, D.O.; Wang, Y. Recent Trends in Satellite Vegetation Index Observations Indicate Decreasing Vegetation Biomass in the Southeastern Saline Everglades Wetlands. *Wetlands* **2014**, *34*, 67–77, doi:10.1007/s13157-013-0483-0.

33. Doyle, T.W. *Predicting Future Mangrove Forest Migration in the Everglades under Rising Sea Level*; USGS Fact Sheet FS-030-03, U.S. Department of the Interior; U.S. Geological Survey, March 2003. Available online: https://www.nwrc.usgs.gov/factshts/030-03.pdf (accessed on 25 July 2017).

34. Park, J.; Stabenau, E. South Florida sea level rise projections, U.S. Department of Interior, National Park Service, South Florida Natural Resources Center: Homestead, Florida, USA. High projection: http://nps.maps.arcgis.com/home/webmap/viewer.html?layers=b61db3e154104ea486528c031390066c. Low projection: http://nps.maps.arcgis.com/home/webmap/viewer.html?layers=87e87e094680431eab085a18adb36836 (accessed on 27 July 2017).

35. National Oceanic and Atmospheric Administration (NOAA). *Tidal Datums*; 2016. Available online: http://tidesandcurrents.noaa.gov/datum_options.html (accessed on 25 July 2017).

36. US Army Corps of Engineers (USACE). *Procedures to Evaluate Sea Level Change: Impacts, Responses and Adaptation*; U.S. Army Corps of Engineers: Washington, DC, USA. Technical Letter No. 1100-2-1, 30 June 2014. Available online: http://www.publications.usace.army.mil/Portals/76/Publications/EngineerTechnicalLetters/ETL_1100-2-1.pdf (accessed on 25 July 2017).

37. National Oceanic and Atmospheric Administration (NOAA). *Tidal Datums and Their Applications, Special Publication NOS CO-OPS 1*; National Oceanic and Atmospheric Administration, National Ocean Service Center for Operational Oceanographic Products and Services, 2001; p. 111. Available online: http://tidesandcurrents.noaa.gov/publications/tidal_datums_and_their_applications.pdf (accessed on 25 July 2017).

38. National Oceanic and Atmospheric Administration (NOAA). *Vaca Key Tidal Datums*; 2016. Available online: http://tidesandcurrents.noaa.gov/datums.html?id=8723970 (accessed on 25 July 2017).

39. Gyory, J.; Rowe, E.; Mariano, A.; Ryan, E. *The Florida Current*; Ocean Surface Currents. The Rosenstiel School of Marine and Atmospheric Science, University of Miami, 1992. Available online: http://oceancurrents.rsmas.miami.edu/atlantic/florida.html (accessed on 25 July 2017).

40. Montgomery, R.B. Fluctuations in Monthly Sea Level on Eastern U. S. Coast as Related to Dynamics of Western North Atlantic Ocean. *J. Mar. Res.* **1938**, *1*, 165–185.

41. Rahmstorf, S.; Box, J.; Feulner, G.; Mann, M.; Robinson, A.; Rutherford, S.; Schaffernicht, E. Exceptional twentieth-century slowdown in Atlantic Ocean overturning circulation. *Nat. Clim. Chang.* **2015**, *5*, 475–480, doi:10.1038/nclimate2554.

42. Smith, T.J., III; Anderson, G.H.; Balentine, K.; Tiling, G.; Ward, G.A.; Whelan, K.R.T. Cumulative Impacts of Hurricanes on Florida Mangrove Ecosystems: Sediment Deposition, Storm Surges and Vegetation. *Wetlands* **2009**, *29*, 24–34, doi:10.1672/08-40.1.

43. Whelan, K.R.T.; Smith, T.J.; Anderson, G.H.; Ouellette, M.L. Hurricane Wilma's impact on overall soil elevation and zones within the soil profile in a mangrove forest. *Wetlands* **2009**, *29*, 16–23.

44. Needham, H.F.; Keim, B.D.; Sataraj, D.; Shafer, M. A Global Database of Tropical Storm Surges. *EOS Trans.* **2013**, *94*, 213–214.

45. Park, J.; Obeysekera, J.; Irizarry, M.; Trimble, P. Storm Surge Projections and Implications for Water Management in South Florida. *Clim. Chang.* **2011**, *107*, 109–128, doi:10.1007/s10584-011-0079-8.

Journal of
Marine Science and Engineering

MDPI

Article

Sea Level Forecasts Aggregated from Established Operational Systems

Andy Taylor [1,*] and Gary B. Brassington [2]

1   Bureau of Meteorology, 700 Collins St, Docklands 3008, Australia
2   Bureau of Meteorology, 300 Elizabeth St, Darlinghurst 1300, Australia; Gary.Brassington@bom.gov.au
*   Correspondence: Andy.Taylor@bom.gov.au; Tel.: +61-3-9669-4650

Received: 30 May 2017; Accepted: 25 July 2017; Published: 1 August 2017

Abstract: A system for providing routine seven-day forecasts of sea level observable at tide gauge locations is described and evaluated. Forecast time series are aggregated from well-established operational systems of the Australian Bureau of Meteorology; although following some adjustments these systems are only quasi-complimentary. Target applications are routine coastal decision processes under non-extreme conditions. The configuration aims to be relatively robust to operational realities such as version upgrades, data gaps and metadata ambiguities. Forecast skill is evaluated against hourly tide gauge observations. Characteristics of the bias correction term are demonstrated to be primarily static in time, with time varying signals showing regional coherence. This simple approach to exploiting existing complex systems can offer valuable levels of skill at a range of Australian locations. The prospect of interpolation between observation sites and exploitation of lagged-ensemble uncertainty estimates could be meaningfully pursued. Skill characteristics define a benchmark against which new operational sea level forecasting systems can be measured. More generally, an aggregation approach may prove to be optimal for routine sea level forecast services given the physically inhomogeneous processes involved and ability to incorporate ongoing improvements and extensions of source systems.

Keywords: forecasting; sea level; tides; Australia; operational oceanography

1. Introduction

1.1. Routine Sea Level and Operations

Of the activities that now constitute 'operational oceanography' [1], sea level forecasting possibly has the most historical baggage as well as the most widespread application. Day-to-day routine decisions are based on quantitative expectations of still water level [2] at the coast. For example in marina and managed estuary operations, maritime under keel clearance systems and coastal works scheduling. Such routine decisions do not involve sea level extremes such as during tropical cyclones and tsunamis—and rare extremes are not addressed here. The focus of this paper is routine sea level forecasting that includes the superposition of relatively moderate phenomena. Figure 1 is an illustrative example of how such forecast guidance can be presented.

All models are wrong, but some are useful [3]; and some are operational. Forecast systems that enjoy ongoing and reliable operational support are of particular relevance to users. Existing operational systems also set a relevant benchmark for the justification of the development of new systems.

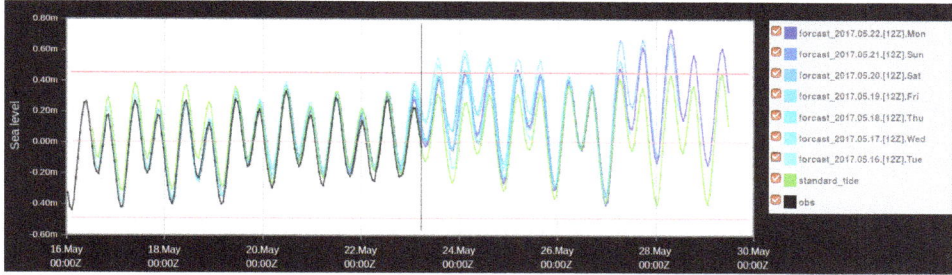

Figure 1. Illustrative sea level forecast for St Kilda at Melbourne, Australia. Sequential 7-day forecasts are shown overlaid in shades of blue. Observed sea level is shown in black and standard tide prediction in green. Red horizontal lines indicate conventional tidal planes LAT, MSL and HAT. Some spread is apparent between forecast updates. Sea level is not especially extreme, but is forecast to cross the reference HAT in coming days.

Observed sea level is a manifestation of diverse physical processes and scales; some local in nature, but many involving signal propagation [4]. The balance of contributions varies across the vast geographic range of the Australian coastline [5–10].

Specialised forecasting approaches have evolved to target different scales and subcomponents of sea level [11,12]. A variety of systems relevant to sea level are now in side-by-side, but isolated, operation in organisations such as the Australian Bureau of Meteorology (BoM). These capabilities include in situ and remote observations, conventional tide predictions, wave spectra forecasts, river run-off routing and tsunami models. More recently, data assimilating primitive equation ocean models [13] have come into operational centres via a path broadly analogous to the development of numerical weather prediction [14].

Inevitably the foundations of operational global forecasts are being leveraged for localised downscaled dynamic operational ocean models: e.g., [15–18].

However, spatial and temporal coverage is non universal; and relevance to everyday decision makers is not necessarily a direct function of increased model resolution.

1.2. Data-Driven Conventional Tides

Conventional tide predictions are a remarkably successful data-driven forecast product that provides an omni-present reference for coastal activities. By design these make no account for aperiodic phenomena. Production involves time series statistics based on historical records of observations at each forecast site. Importantly, the observations need not be recently collected; let alone delivered in real time.

Tidal techniques exploit the significance of periodic sea level variations observed to be coherent with the 'astronomical tide producing forces (ATGF)' [19]. Many variations exist for implementing this general approach (e.g., [20–23]). Useful sea level predictions can thereby be produced many years into the future. Tidal methods based on a harmonic decomposition of the ATGF have long been typical of bodies promulgating 'official' tidal products; including BoM. Official tidal products have come to have a special status; for instance, statistical properties of tidal predictions define elevation references for mapping and legal applications [24].

The ongoing value of standard tidal predictions reflects the fact that periodic signals generally dominate routine coastal still water levels. However, the physical drivers of sea level represented in tidal predictions need not be gravitational at all. Treatment of non-gravitational signals is an practical consideration in tidal analysis proceedures; notably for the long-period constituents Sa and Ssa ([25] Sect. 3.7) but also at shorter timescales such as constituent S1 [26]. The extent to which conventional harmonic tidal predictions are 'physics free' actually allows for a pragmatic flexibility to represent almost all of the everyday rise and fall of coastal sea level at a place.

Even when relatively high resolution dynamic tidal models are available, the standard tide predictions at a place are commonly considered a superior estimate of routine sea level [27,28].

1.3. Real Time Observations

Conventional tide methods have had such a long time to become deeply embedded [11] thanks to the ability to produce useful forecasts via access to observations only in a much lagged batch mode. In contrast, tide gauge observations are increasingly communicated in near real time to operational centres and general users. BoM operates its own network of tide gauges [29] but the majority of available observations are shared by partner or '3rd party' organisations. Gathering observations from diverse organisations is valuable but can raise issues with data quality and metadata management.

Despite the nominal co-location of real-time tide gauge observations and various forecasting systems, presentation of useful verifiable guidance has been found to be surprisingly elusive.

While more real time observations in principle opens opportunities for assimilation into dynamic models or various 'trained' forecast systems [30], such use can place much weight on the reliability and quality control of live data streams; a non-trivial concern over large regions and multiple agencies [31].

2. Forecast System Description

2.1. Motivation

This work is founded on the operational maturity of a global ocean forecast system (OceanMAPS) within a agency that also provides weather, tide and river forecasts. The demonstration of limited non-tidal sea level forecast skill in earlier versions of OceanMAPS [32] motivated investigation of potential practical applications. Liaison with forecastors and forecast-users lead to the current configuration; routine 7-day forecasts that can be directly evaluated against tide gauge observations. A secondary motivation was to establish a performance benchmark against which new sea level forecast capabilities can be evaluated.

2.2. Superposition

The configuration is a linear superposition of time series derived from heterogeneous operational systems, schematically illustrated in Figure 2 and Equation (1). Although the superposition itself is linear, subsets of non-linear hydrodynamics are internally represented within the ocean circulation model and the tidal harmonic fit respectively.

Component systems *included* are listed here and described further below: (1) Global ocean circulation forecasts—Section 2.3; (2) Global numerical weather prediction—Section 2.3; (3) adjusted harmonic tide predictions—Section 2.4 and (4) observations-based bias correction—Section 2.5.

There are some notable *exclusions* from the current configuration. Spectral wave forecasts are available but not included on the basis that observation sites are located outside of surf zones (similar to [33]). River levels or outflow forecasts are not available in suitable form or coverage. In contrast, sea level forecasts are an input into hydrological models [34].

As the native time disctritisation of the input differ, each timeseries is projected onto a common 1-hourly format using an integral spline method. In the case of model inputs this is interpolation from 3-hourly averaged model outputs to 1-hourly. For observations the same method provides a down-sampling from from 1-min or 15-min samples to 1-hourly averages.

$$h(t) = \eta_T(t) - \eta_{HA}(t) + \eta_{SLA}(t) + \eta_{LIB}(t) + b(t0) \tag{1}$$

where $h(t)$ is the aggregated sea level value at forecast time t. Signal components η correspond to inputs in schematic Figure 2 and subscripts T and HA indicate tide and harmonic adjustment respectively. Bias correction b is a fixed value across each forecast.

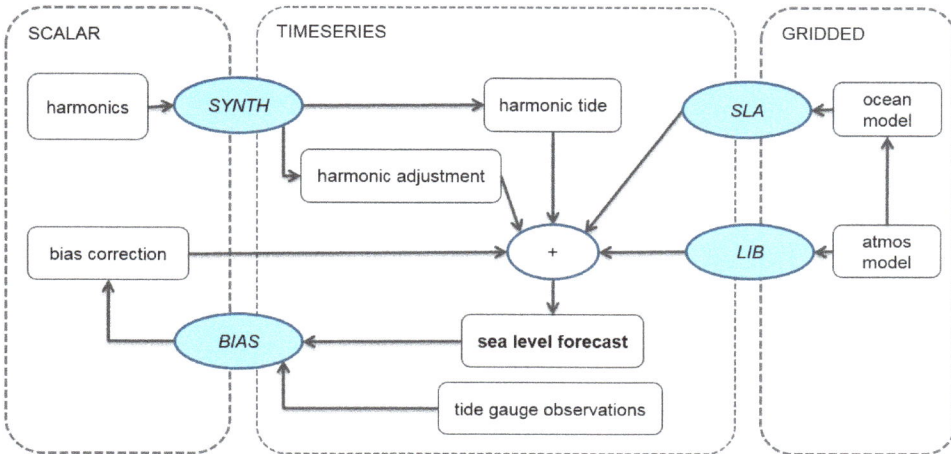

Figure 2. Schematic illustration of aggregation configuration. Source systems are heterogeneous but mapped onto time series that can be directly compared against 1-hourly observations. (**SYNTH**) indicates tidal synthesis, (**SLA**) sea level anomaly from OceanMAPS, (**LIB**) local inverse barometer approximation based on atmospheric pressure forecast, (**BIAS**) non-causal filter bias correction scheme.

2.3. Input: Data Assimilating Primitive Equation Forecasts

Near-global ocean forecasts have been in operational production at the BoM now for 10 years via several versions of the Ocean Model, Analysis and Prediction System ('OceanMAPS') [35–38]. The dynamic ocean model component of OceanMAPS is based on the Modular Ocean Moel ('MOM') [39] configured with 0.1×0.1 degree regular structured horizontal resolution, hydrostatic free surface, z-level and split-implicit scheme; where the barotropic calculation is performed at a finer time stepping.

Gravitation tidal forces are intentionally *not* included in OceanMAPS.

Land run-off fresh water fluxes are only roughly approximated with a climatological annual cycle. Australian rivers are typically dry for very long periods with intermittant flooding. The climatological river input is only included for maintaining global mass balance and has essentially no skill for Australian river outflow impacts on sea level.

Initial conditions for the ocean state are constrained using an ensemble optimal interpolation data assimilation scheme [40] which ingests a large number and range of remote and in situ ocean observations. Importantly for sea level, tide gauge observations are *not* assimilated and are independent. Satellite altimeter observations of sea level are assimilated, but not inshore of the shelf break; nominally cut off at the 200 m isobath.

Atmospheric fluxes, excluding barometric pressure, are applied directly from the global numerical weather prediction (NWP) system ACCESS-G [41]. ACCESS-G is also based on a data assimilating primitive equation model. It is not coupled with any ocean model beyond use of a persisted SST analysis boundary condition. These flux fields are generated on a N512 gaussian grid with an indicative spatial resolution of 25 km.

OceanMAPS produces a new ocean state forecast each day for the next 7-days using a multi-cycle ensemble schedule [42].

As a result, 7-day forecasts of a sea level anomaly (η_{SLA}) quantity are reliably available each day. This data is output as 3-hourly averages. The quantity η_{SLA} is not directly observable, but in the open water is observationally constrained by corrected altimeter observations. η_{SLA} is quantified relative to the model rest state, which nominally represents a geopotential surface like mean sea level.

The regular spatial discretisation of OceanMAPS does not resolve features smaller than 10 km. And some nominally larger embayments have been intentionally excluded; such as Port Phillip Bay in South Eastern Australia. The Arakawa B-grid discretization imposes a numerical requirement to exclude 1-cell bays. An minumum depth of three z-levels equivalent to 15 m is also imposed.

The representation of the coast is illustrated in Figure 3.

In order to map the gridded η_{SLA} field to a tide gauge location, a generic one-to-one 'nearest coastal neighbour' algorithm is applied. A manual exception for cell selection is applied to the Port Phillip case to ensure alignment with the single bay entrance.

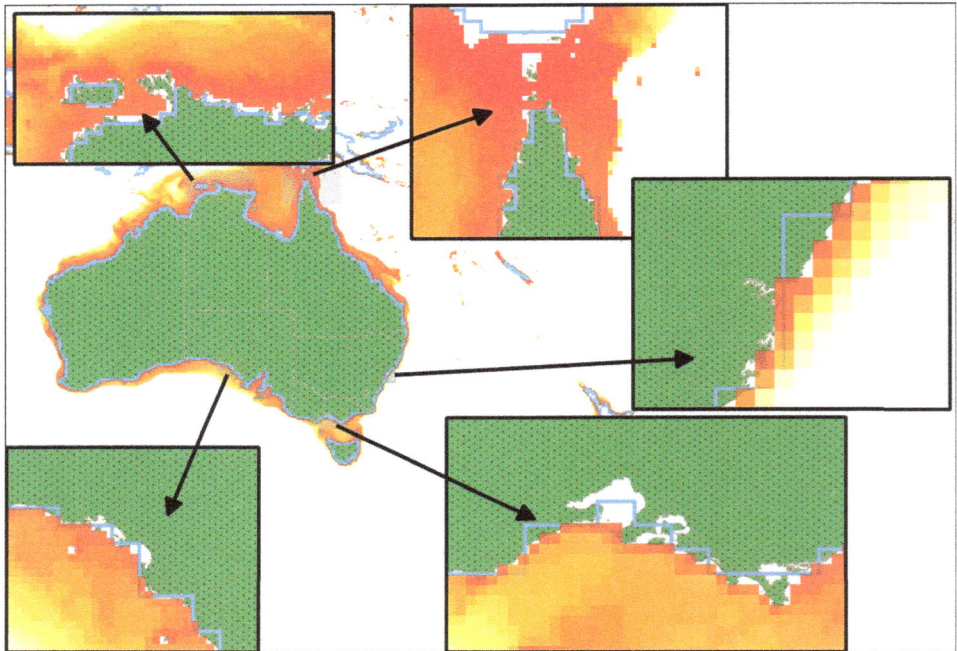

Figure 3. Illustration of coastal discretisation. Red coloured grid cells indicate ocean model bathymetry the inner extent of which is the model coastline. The thick blue line indicates the edge of the atmospheric model land/sea mask. Most small scale features and embayments at tide gauge locations are only approximated at scales above 10 km in the ocean and 80 km in the atmosphere.

OceanMAPS does not represent the effect of atmospheric pressure forces on the ocean. Subsequently, a local inverse barometer approximation is applied as per Equation (2). This formulation was chosen for being simple and robust, but is acknowledged as a compromise with regard to atmospheric representation and non-instantaneous ocean responses [43]. A fixed conventional reference pressure, rather than one derived from the NWP, was considered appropriate given the generic offset adjustment built into the bias correction scheme (Section 2.5).

$$\eta_{LIB} = \frac{p_{NWP} - p_{ref}}{\rho g} \tag{2}$$

where reference pressure is fixed at p_{ref} = 101,325 Pa, and bulk sea water density is also kept fixed at 1027 kg/m^3. Only a small subset of tide gauges are co-located with real-time barometer instruments.

2.4. Input: Tidal Harmonics

Officially promulgated tide predictions have a special relevance, as raised in Section 1.2. The aggregation configuration intentionally aims to align with the BoM's existing tide tables.

However, the heterogeneous nature of harmonic tide analysis and OceanMAPS configuration leads to a situation where the respective sea level signals are not cleanly complimentary. In particular some of the sea level variation in the ocean model may be seen as 'tidal', whereas some of the variation in the tidal harmonics may be seen as meteorological. In isolation, this spectral overlap is generally not problematic. However, linear superposition of the OceanMAPS η_{SLA} with standard tides can lead to undesirable double-counting. This is most obvious at longer time scales in Northern Australia; where relatively powerful seasonal sea level changes have projected onto tidal harmonics.

A pragmatic approach is taken that aims to address both of these motivations; namely to align with other official tidal products and mitigate effects of spectral overlap. The BoM's operationally supported tidal synthesis software is applied to two versions of tidal harmonics for each location:

- full set of tidal harmonics: typically 114 constituents.
- subset assigned apriori to be primarily non-gravitational in expression: Sa, SSa, Mm, Msf and S1.

The subset time series is designated as a harmonic adjustment signal that is then subtracted within the superposition of signals.

2.5. Input: Bias Correction

Near real-time observations are available at an increasing number of tide gauge locations, and intuitively serve as a source of guidance to users. Typical practice, though often not formalised, is to inspect recent tidal error (residual) evolution and project as a correction to tide tables into the near future.

Operational availability of this source of information is exploited in a generic and un-trained bias correction scheme. The bias term (b in Equation (1)) is a constant added to each forecast. The value of b is derived as a weighted mean of recent errors between observed sea level and previous forecasts between 0–24 lead times. The most recent value persists in the case of observation drop-outs and gaps. Observations are pre-processed with a median spike filter to mitigate the impact of communications glitches and noise. Weighting is tapered such that the influence of older error values decreases with time prior to the present and it is 'causal' in the sense that no observations after the forecast base time have any influence. The total window length of the filter spans 21 days. Identical settings are applied to all locations.

The scheme aims to cover multiple needs. Firstly, to address alignment of reference datums between sources. It is the temporal evolution of η_{SLA} and η_{LIB} that is expected to contain real information—not the absolute values. Moreover, metadata ambiguity with regard to the reference datum of real-time observational streams (from 3rd parties) relative to tide prediction datums has proven to be a real issue. Secondly, the correction serves as a crude data assimilation method to adjust the vertical offset of the forecast in light of long period error evolution.

The actual behaviour of this term is discussed further in Section 4.

3. Evaluation

Evaluation results are presented based on aggregate sea level forecasts produced in an operational mode at BoM between June 2016 to May 2017.

3.1. Geographic Overview

Evaluation is restricted to a series of point locations. These locations correspond to tide gauges from which real-time information was reported into the operational centre over the study period. An overview map of these locations is shown in Figure 4. The inconsistent coastal distance between locations is an outcome of human population distribution and inter-organisational arrangements. It is foreseeable that the number of sites could expand significantly in the future.

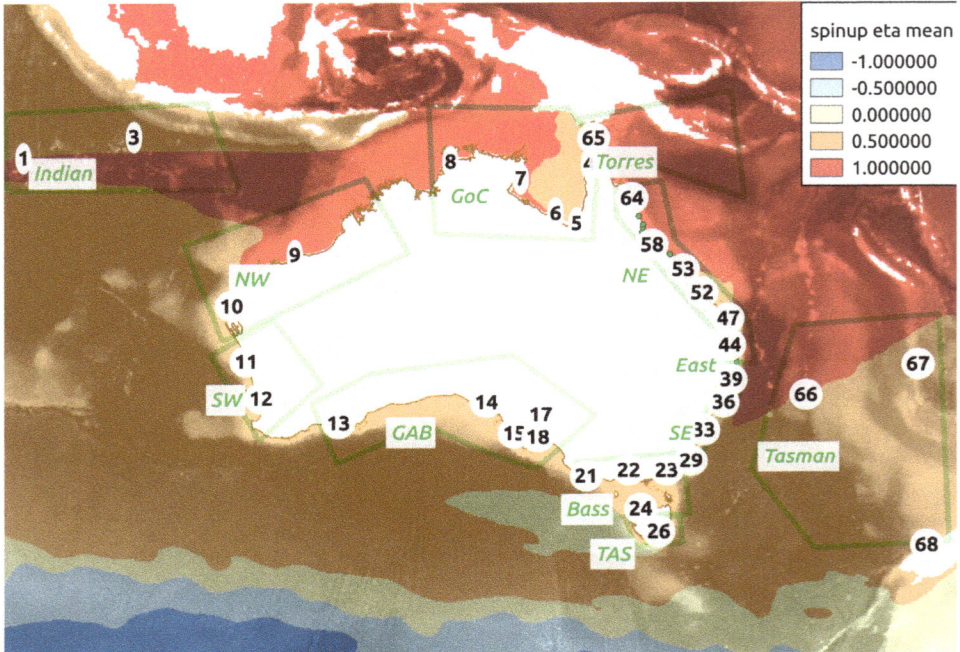

Figure 4. Geographic overview of Australia region. Ocean model bathymetry is indicated by grey background shading. Colour contours illustrate the 'spin up' model mean dynamic topography referenced below. Tide gauge locations are identified with sequential numbers with order chosen to loosely align with anti-cyclonic coastal wave propagation direction. Regional groupings are referenced in subsequent results.

The observation locations are quite diverse with regard to exposure to the open ocean, instrumentation, sample frequency and communications quality.

3.2. Forecast Goodness as Routine Guidance

In general, forecast goodness cannot be properly judged against any single measure. The evaluation below is informed by the the concepts of 'quality' and 'value' following Murphy [44]. Even if the component parts contain comparative technical deficiencies, the whole package may offer real value above available alternatives. For this routine guidance use-case, only full time series statistics are presented as an evaluation. Categorical and event-based measures are not addressed in this paper.

Harmonic tide predictions considered in light of recent residuals commonly offer remarkably good guidance for coastal sea level. Such a 'tide + persisted residual' scheme is formulated in Equation (3) and taken as an appropriate benchmark against which to evaluate the aggregated forecast. For the new scheme to offer value at a particular location it must demonstrate superior skill relative to this benchmark.

$$h_{T+r}(t) = \eta_T(t) + r(t0) \tag{3}$$

Following Equation (1) where $r(t0)$ is the difference between observations and tide prediction η_T at or just prior to forecast base time.

3.3. Highlighted Behaviour

The time series shown in Figure 5 is included as an example of a skilful dynamic model contribution. This location is subject to relatively powerful sea level variations associated with mid-latitude weather. Of interest is the fact that the embayment in which the observations are taken is not represented at all in the ocean model. This indicates the large influence of sea level outside of the bay at these timescales.

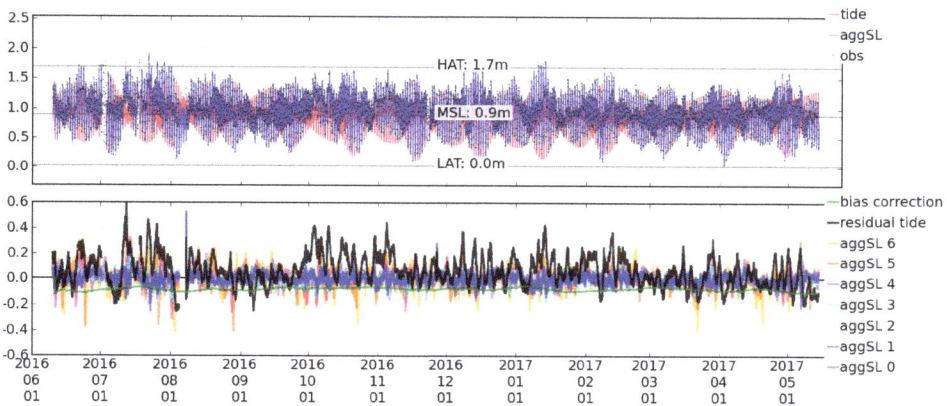

Figure 5. Example time series record for location at Melbourne (**23** in Figure 4); **Upper panel** shows total still water levels relative to conventional tidal benchmarks. **Lower panel** shows error signals and bias correction evolution. Errors ('residuals') are shown overlaid for standard tides and aggregate forecasts ('aggSL') categorised by forecast lead time; each continuous aggSL timeseries is concatenated from forecast lead times of 0–24 h (aggSL0), 24–48 h (aggSL1), ... , 144–168 h (aggSL6). An overall reduction in error variance is apparent between standard tides and aggregate forecasts; errors grow with forecast lead time. Large reduction in error variance relative to harmonic tides is evident.

Contrasting behaviour is illustrated in Figures 6 and 7 by means of error statistics. At all three locations the aggregated forecasts offer 'quality' in the sense of matching observations; but for different reasons and different degrees of potential value.

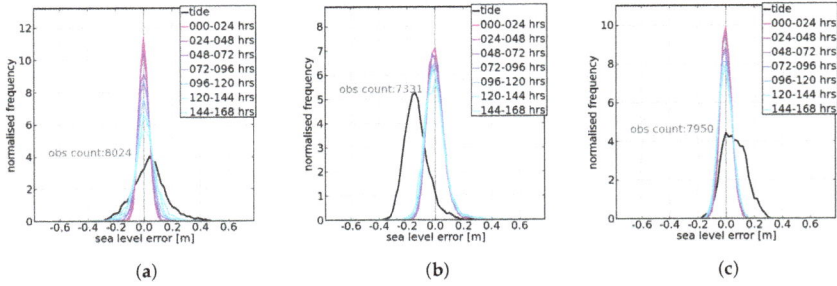

Figure 6. Forecast error distributions at selected contrasting locations. Skill improvement over standard harmonic tides is driven by different aspects of the generic aggregation process. (**a**) shows skill gain due to forecast signals associated with mid-latitudee weather (**b**) shows the practical problem of mismatched reference datums between real-time observational data and tide predictions (see Section 4), (**c**) is a location at which longer period deviations from tide predictions are relatively powerful. (**a**) Site 13, Mid-latitude weather; (**b**) Site 43, Tide datum mismatch; (**c**) Site 3, Long period anomalies.

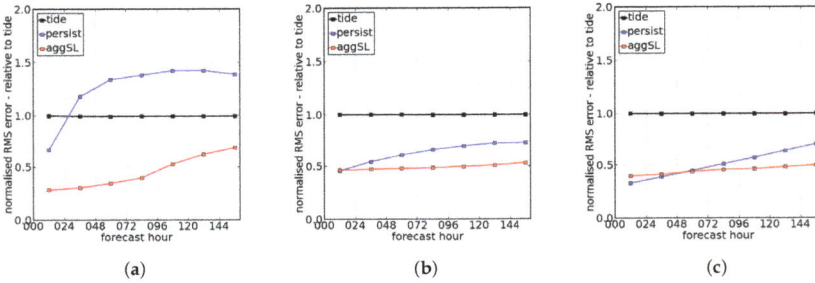

Figure 7. Relative error metrics at select locations as per Figure 6. RMSE at increasing forecast lead times is normalised and plotted relative to the standard tidal residual—such that a lower value represents greater accuracy. Error growth with forecast age is evident. (**a**) As per (a) in Figure 6; (**b**) As per (b) in Figure 6; (**c**) As per (c) in Figure 6.

3.4. Skill Summary: Non-Tidal Information Decay

In order to highlight the differentiation between dynamic forecast skill between locations, Figure 8 summarises average forecast evolution characteristics using Taylor diagrams [45] for time series that have been de-tided using a band-limited harmonic tide. Each 'comet' contains a point for each 24 h period of the 7-day forecasts. Site number is located at the 1st day forecast point. Both the reference observations and forecasts have been de-tided using only the nominally gravitational tide signal described in Section 2.4. All statistics are centered and normalised relative to the reference observation value $\hat{h}_{OBS}$ as defined in Equation (4).

$$\hat{h}_{OBS}(t) = h_{OBS}(t) - (\eta_T(t) - \eta_{HA}(t))$$
$$\hat{h}_{FC}(t) = \eta_{SLA}(t) + \eta LIB(t) + b(t0)$$

(4)

Following Equation (1). Where $\hat{h}(t)$ is sea level de-tided using a band-limited tidal time series.

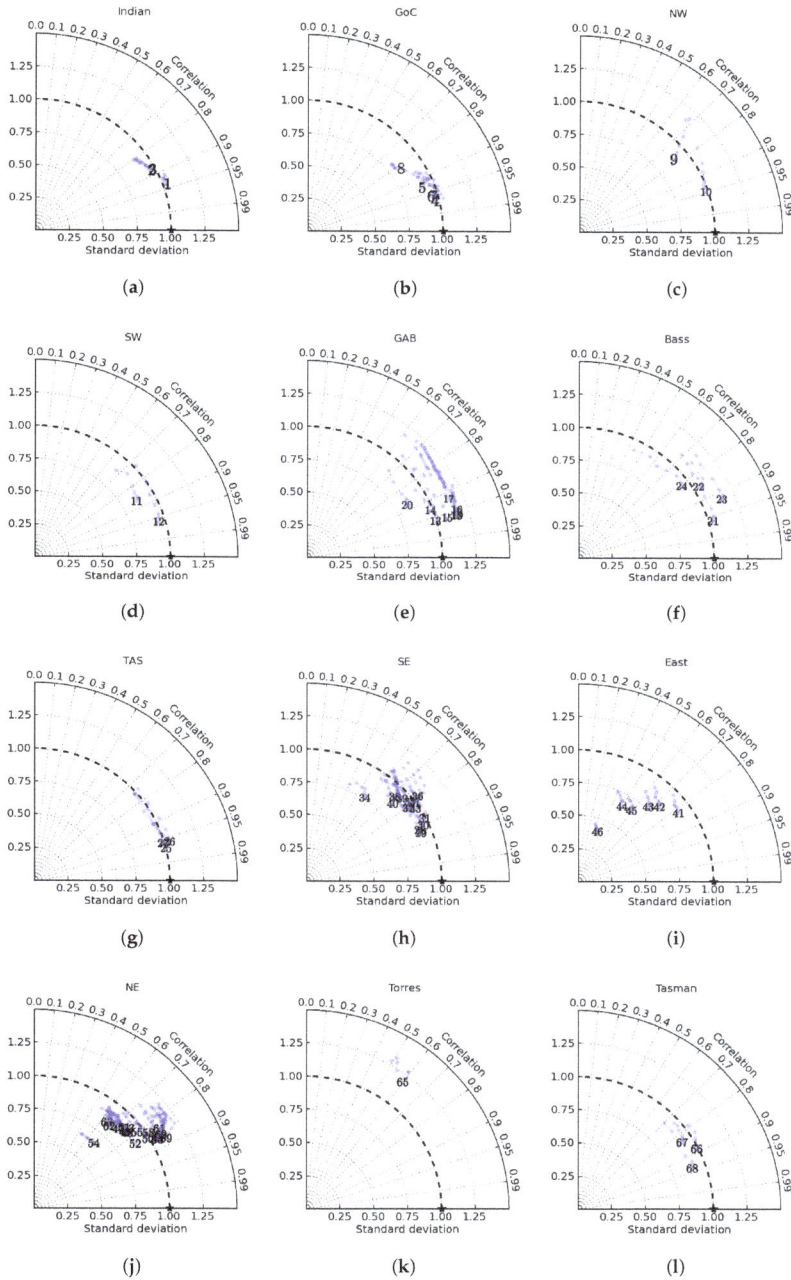

Figure 8. Taylor diagrams summarising dynamic skill evolution across 7-day forecasts. Each 'comet' describes average statistics for one location. Panels are divided according to regions shown in Figure 4. Statistics are derived from filtered time series, not total sea level, and are normalised relative to the reference observations. (**a**) Indian; (**b**) Gulf of Carpentaria; (**c**) North West; (**d**) South West; (**e**) Great Aust Bight; (**f**) Bass Strait; (**g**) Tasmania; (**h**) South East; (**i**) East; (**j**) North East; (**k**) Torres Stait; (**l**) Tasman Sea.

A notable feature of this visualisation is the de-correlation with forecast lead time. This is expected behaviour of a skillful deterministic forecast model where errors grow due to explicit numerical approximation of chaotic dynamics. Regional differences are apparent in degree of initial correlation and rate of de-correlation.

The under-prediction of variability in the 'East' region (panel i) appears to reflect the noisy observational data streams from these 3rd party sites - such that the observation reference variability is inflated by communications glitches.

Anomalous variability growth for station 9 in 'NW' region (panel c) was found to reflect the influence of a small number of over-forecast tropical cyclone events. While the resolution limitations of the atmospheric forcing rule out applicability to extreme storm surge forecasts, NWP systems do evolve TCs that subsequently drive sea level signals in the OceanMAPS. In this case, the relative size of these over-forecasts at longer lead times is reflected in the Taylor diagram 'comet'.

Torres Strait stands out for general poor performance and will be the subject of future investigation.

3.5. Skill Summary: Comparison Against 'Tide + Persisted Residual'

Based on the relative RMSE evaluation shown in Figure 7a traffic-light summary map of results is shown in Figure 9. A red symbol indicates that aggregated forecast RMSE score is lower (better) than that of the reference 'tide + persisted residual' for the specified forecast lead time, a blue symbol indicates the opposite. Where neither is better than standard tides, a black symbol is allocated.

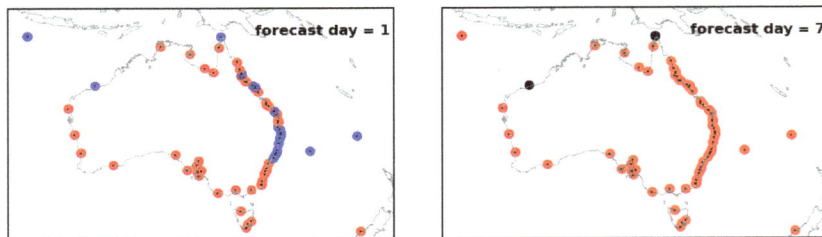

Figure 9. Summary map showing which forecast source provides best information on average with regard to RMS error reduction. Red symbols at locations where aggregated forecast is best, blue for 'tide + persisted residual' and black for standard tides. Forecast lead times of 0–24 h are shown in **left panel**, and 144–168 h on the **right panel**.

4. Role of Bias Correction Component

The practical behaviour of the bias correction term is of special interest for evaluation. It is the only data driven term allowed to evolve in operations as described in Section 2.5. An example time series is shown in the lower panel of Figure 5. The relatively static nature is typical of other sites.

To characterise behaviour each bias correction record was decomposed into a mean and temporarily varying signal.

The mean bias correction is primarily aligned with the ocean model representation of mean dynamic topography MDT (compare [46]). Such a reference surface in model space is the most common strategy used in the assimilation of altimetry observations, which are themselves constructed as anomalies from a reference surface in observation space. For OceanMAPS this surface was derived from a long free 'spin up' run of the model [47]. Figure 10 shows the correspondence between model MDT (η_{spinup}) and the *negative* of the bias correction mean. The wider spatial distribution of MDT is indicated by contours in Figure 4.

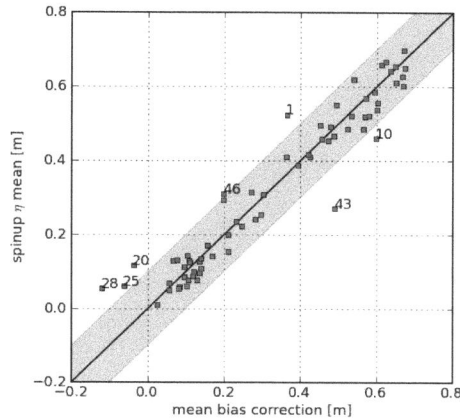

Figure 10. Magnitude of the temporal mean of bias correction matched against model MDT surface at each location. An arbitrary mismatch threshold of 10 cm is used to highlight apparent outliers.

As the model MDT is known apriori, this correspondance supports the expectation that it is a good first guess of the bias term.

Several sites deviate from this alignment by more than a fixed arbitrary threshold of 10 cm, and these are distributed across the domain. A large deviation indicates that the bias correction term systematically adjusted for information not available in the modeling systems alone.

Site 46 is an esturine location where observations are strongly effected by surrounding sand sediment. The bias term at this location is partially adjusting to the asymmtery of the choked tidal signal.

Model sources of bias are feasible and likely. However, of special note is the possibility of datum metadata mismatch between the available real-time observation data stream and the standard tide predictions. Site 43 is an example of a 3rd party datastream with such a mismatch.

This is a symptom of organisational rather than modeling factors. Australian tide gauges are owned and operated by different bodies under a variety of data sharing arrangements. Consistent metadata management from these diverse agencies remains problematic, despite being a nominally simple matter.

Tide predictions in Australia are reported relative to 'prediction datum' which *typically* aligns with the promalgated lowest astronomical tide (LAT) value. From time to time either the value for LAT or the alignment of prediction datum may be updated. While overall the real-time data is expected to be reported relative to either instrument zero, Australian Height Datum (AHD) or tidal prediction datum, operational systems avaiable at the time of writing cannot reliably manage these differences.

The temporally varying bias correction signals for each station are arranged in numbered order in Figure 11. Column order is such that adjacent stations are together, though separation distance varies greatly. In this visualisation it is apparent that the time varying signal has relatively small amplitude of <7 cm for the majority of sites. The single station located Torres Strait is a stand-out exception. The sea level signal at this location is the subject of further investigation, but for the present purpose will be disregarded as not valid.

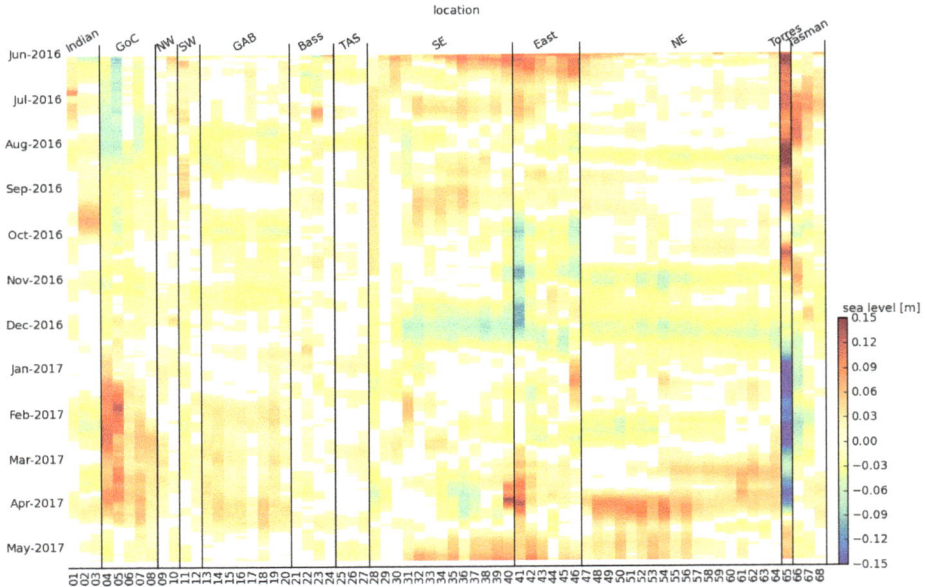

Figure 11. Temporal signal from bias correction histories after subtraction of respective mean offsets. Absolute values less than 0.01 m are shown as white. A degree of coherence is apparent between adjacent sites within the regional groupings-shown on map in Figure 4.

5. Discussion

5.1. Adequacy and Value

The aggregation method offers generally improved sea level forecasts by drawing a pragmatic balance between the strengths of existing operational systems. A common configuration for all sites was employed for robustness and to facilitate expansion to new locations. Although real-time observations are exploited when available, the simple bias correction scheme is relatively robust to data gaps and noise. The balance of contributions from component terms varies with timescale and location around the Australian coast. Regional groupings are apparent in the skill characteristics of the forecasts. By producing a quantity that can be directly compared to observations and reference tidal planes, forecasts can be presented intuitively along with recent real-time sea level and provide ongoing and immediate verification—as is Figure 1.

The skill level and value offered by these forecasts sets a benchmark against which any new sea-level forecasting capabilities can be compared and contrasted.

5.2. Spatial Interpolation

The characteristics demonstrated by the bias correction scheme indicate that meaningful forecasts may be produced at intermediate sites at which real-time observations are not available. The patterns of regional coherence indicate that spatial interpolation of bias correction values may be worthy of further investigation. Figure 11 indicates that spatial interpolation within regions 'GAB' and 'NE' is particularly promising.

Appropriate consideration of geography and other factors on validity will be required. Strongly contrasting bias correction characteristics at Torres Strait (site 65) highlight the importance of not ignoring geography.

J. Mar. Sci. Eng. **2017**, *5*, 33

While the irregular spatial sampling across the domain is undesirable, there is real prospect to obtain access to many more tide gauges that are known to exist but not report real time data to the BoM. Such additional locations will facilitate future investigation of spatial interpolation approaches.

5.3. Extensions

The aggregation concept presented is flexible. And it is foreseeable that additional or alternative operational inputs could be incorporated. The potential to seamlessly include short-range higher resolution forecast information, while maintaining the 7-day outlook, is a logical extension given the aim of exploiting existing capabilities. On the other hand, NWP forecasts currently cover out to 10-days lead time. The option of extending the length of ocean forecasts to match is worthy of consideration in light of this sea level evaluation (such as in Figure 7).

Uncertainty information can already be roughly indicated by means of presenting overlaid sequential forecasts as in Figure 1. Optimal treatment of the lagged ensemble and communicating error growth to users is an outstanding need. Developments in ensemble NWP systems within the operational center are expected to provide additional sampling of uncertainty and enable further development of this important aspect of sea level forecasts.

Acknowledgments: Thanks for input from Kevin Walsh at University of Melbourne, James Chittleborough at Bureau Tidal Unit.

Author Contributions: Andy Taylor conceived, implemented and analyzed the system and also wrote this paper. Gary B. Brassington lead the project and provided substantial critical input.

Conflicts of Interest: Both authors are employed by the Bureau of Meteorology, but declare no conflict of interest with regard to the results presented. The Bureau of Meteorology beyond the authors had no role in the design, analysis or presentation of the results.

Abbreviations

The following abbreviations are used in this manuscript:

aggSL	Aggregate Sea Level
AHD	Australian Height Datum
ATGF	Astronomical Tide Producing Forces
BoM	Australian Bureau of Meteorology
HAT	Highest Astronomical Tide—conventional tidal plane
LAT	Lowest Astronomical Tide—conventional tidal plane
LIB	Local Inverse Barometer Approximation
MDT	Mean Dynamic Topography
MSL	Mean Sea Level—conventional tidal plane
MOM	Modular Ocean Model
NWP	Numerical Weather Prediction
OceanMAPS	Ocean Model, Analysis and Prediction System
RMSE	Root Mean Square Error
SLA	Sea Level Anomaly

References

1. Bell, M.J.; Lefèbvre, M.; Le Traon, P.Y.; Smith, N.; Wilmer-Becker, K. GODAE The Global Ocean Data Assimilation Experiment. *Oceanography* **2009**, *22*, 14–21.
2. Pugh, D.; Woodworth, P. *Sea-Level Science: Understanding Tides, Surges, Tsunamis and Mean Sea-Level Changes*; Cambridge University Press: Cambridge, UK, 2014.
3. Box, G. Robustness in the strategy of scientific model building. *Robust. Stat.* **1979**, doi:10.1016/ b978-0-12-438150-6.50018-2.
4. Melet, A.; Almar, R.; Meyssignac, B. What dominates sea level at the coast: A case study for the Gulf of Guinea. *Ocean Dyn.* **2016**, *66*, 623–636.

5. Haigh, I.D.; Wijeratne, E.M.S.; MacPherson, L.R.; Pattiaratchi, C.B.; Mason, M.S.; Crompton, R.P.; George, S. Estimating present day extreme water level exceedance probabilities around the coastline of Australia: Tides, extra-tropical storm surges and mean sea level. *Clim. Dyn.* **2013**, *42*, 121–138.

6. Haigh, I.D.; MacPherson, L.R.; Mason, M.S.; Wijeratne, E.M.S.; Pattiaratchi, C.B.; Crompton, R.P.; George, S. Estimating present day extreme water level exceedance probabilities around the coastline of Australia: Tropical cyclone-induced storm surges. *Clim. Dyn.* **2013**, *42*, 139–157.

7. Woodham, R.; Brassington, G.B.; Robertson, R.; Alves, O. Propagation characteristics of coastally trapped waves on the Australian Continental Shelf. *J. Geophys. Res. Oceans* **2013**, *118*, 4461–4473.

8. Ridgway, K.R. The 5500-km-long boundary flow off western and southern Australia. *J. Geophys. Res.* **2004**, doi:10.1029/2003jc001921.

9. Church, J.A.; Freeland, H. Coastal-trapped waves on the east Australian continental shelf. I: Propagation of modes. *J. Phys. Oceanogr.* **1986**, *16*, 1929–1943.

10. Allen, S.C.R.; Greenslade, D.J.M. *A Spectral Climatology of Australian and South-West Pacific Tide Gauges*; CAWCR Technical Reports; Centre for Australian Weather and Climate Research: Melbourne, VIC, Australia, 2009.

11. Cartwright, D.E. *Tides*; Cambridge University Press: Cambridge, UK, 2000.

12. Petersen, A. *Simulating Nature*; Chapman and Hall/CRC: London, UK, 2012.

13. Schiller, A.; Brassington, G.B. *Operational Oceanography in the 21st Century*; Springer Science + Business Media: Berlin, Germany, 2011.

14. Harper, K.C. *Weather by the Numbers*; The MIT Press: Cambridge, MA, USA, 2008.

15. Paramygin, V.; Sheng, Y.; Davis, J. Towards the Development of an Operational Forecast System for the Florida Coast. *J. Mar. Sci. Eng.* **2017**, *5*, 8.

16. Yang, Z.; Richardson, P.; Chen, Y.; Kelley, J.; Myers, E.; Aikman, F.; Peng, M.; Zhang, A. Model Development and Hindcast Simulations of NOAA's Gulf of Maine Operational Forecast System. *J. Mar. Sci. Eng.* **2016**, *4*, 77.

17. Wei, E.; Zhang, A.; Yang, Z.; Chen, Y.; Kelley, J.; Aikman, F.; Cao, D. NOAA's Nested Northern Gulf of Mexico Operational Forecast Systems Development. *J. Mar. Sci. Eng.* **2014**, *2*, 1–17.

18. Peng, M.; Schmalz, R.A., Jr.; Zhang, A.; Aikman, F., III. Towards the Development of the National Ocean Service San Francisco Bay Operational Forecast System. *J. Mar. Sci. Eng.* **2014**, *2*, 247–286.

19. Hendershott, M. Long waves and ocean tides. In *Evolution of Physical Oceanography*; The MIT Press: Cambridge, MA, USA, 1981.

20. Foreman, M.G.G.; Cherniawsky, J.Y.; Ballantyne, V. Versatile Harmonic Tidal Analysis: Improvements and Applications. *J. Atmos. Ocean. Technol.* **2009**, doi:10.1175/2008JTECHO615.1.

21. Groves, G.W.; Reynolds, R.W. An Orthogonalized Convolution Method of Tide Prediction. *J. Geophys. Res.* **1975**, *80*, 4131–4138.

22. Leffler, K.; Jay, D. Enhancing tidal harmonic analysis: Robust (hybrid L1/L2L1/L2) solutions. *Cont. Shelf Res.* **2009**, *29*, 78–88.

23. Smith, A.; Ambrosius, B.; Wakker, K.F.; Woodworth, P.L.; Vassie, J.M. Comparison between the harmonic and response methods of tidal analysis using TOPEX/POSEIDON altimetry. *J. Geod.* **1997**, *71*, 695–703.

24. Mapping, I.C.o.S. *Australian Tides Manual*; Technical Report; Permanent Committee on Tides and Mean Sea Level: Wollongong, Australia, 2014.

25. Parker, B.B. *Tidal Analysis and Prediction*; Technical Report; National Oceanic and Atmospheric Administration—Center for Operational Oceanographic Products and Services: Silver Spring, MD, USA, 2007.

26. Ray, R.; Egbert, G.D. The Global S1 Tide. *J. Phys. Oceanogr.* **2004**, *34*, 1922.

27. Horsburgh, K.J.; Williams, J.A.; Flowerdew, J.; Mylne, K. Aspects of operational forecast model skill during an extreme storm surge event. *J. Flood Risk Manag.* **2008**, *1*, 213–221.

28. Egbert, G.D.; Bennett, A.F. Data assimilation methods for ocean tides. *Elsevier Oceanogr. Ser.* **1996**, doi:10.1016/s0422-9894(96)80009-2.

29. Greenslade, D.J.M.; Warne, J.O. Assessment of the Effectiveness of a Sea-Level Observing Network for Tsunami Warning. *J. Waterw. Port Coast. Ocean Eng.* **2012**, *138*, 246–255.

30. Horsburgh, K.; De Vries, H. *Guide to Storm Surge Forecasting*; Technical Report; World Meteorological Organization: Geneva, Switzerland, 2011.

31. Mourre, B.; Crosnier, L.; Provost, C.L. Real-time sea-level gauge observations and operational oceanography. *Philos. Trans. R. Soc. A* **2006**, *364*, 867–884.

32. Taylor, A.J.; Brassington, G.B.; Nader, J. *Assessment of BLUElink OceanMAPSv1.0b Against Coastal Tide Gauges*; Technical Report; Centre for Australian Weather and Climate Research: Melbourne, VIC, Australia, 2010.

33. Tilburg, C.E.; Garvine, R.W. A Simple Model for Coastal Sea Level Prediction. *Weather Forecast.* **2004**, doi:10.1175/1520-0434(2004)019<0511:ASMFCS>2.0.CO;2.

34. Taylor, A.J.; Smith, A.; Wang, W.; Robinson, J.; Brassington, G.B. Ocean meets river: Connecting Bureau of Meteorology ocean forecasts and river height predictions for improved flood warnings. In Proceedings of the 19th International Congress on Modelling and Simulation, Perth, Australia, 12–16 December 2011.

35. Brassington, G.B.; Pugh, T.; Spillman, C.; Schulz, E.; Beggs, H.; Schiller, A.; Oke, P.R. BLUElink> Development of Operational Oceanography and Servicing in Australia. *J. Res. Pract. Inf. Technol.* **2007**, *39*, 151–164.

36. Bureau of Meterology. *Implementation of OceanMAPS (BLUElink> Ocean Forecast System)*; Technical Report; Bureau of Meteorology: Melbourne, VIC, Australia, 2007.

37. Bureau of Meterology. *Operational Upgrades to OceanMAPS (BLUElink> Ocean Forecast System)*; Technical Report; Bureau of Meteorology: Melbourne, VIC, Australia, 2011.

38. Brassington, G.B.; Freeman, J.; Huang, X.; Pugh, T.; Oke, P.; Sandery, P.A.; Taylor, A.J.; Andreu-Burillo, I.; Schiller, A.; Griffin, D.; et al. *Ocean Model, Analysis and Prediction System: Version 2*; Technical Report; Centre for Australian Weather and Climate Research: Melbourne, VIC, Australia, 2012.

39. Griffies, S.M.; Harrison, M.; Pacanowski, R. *A Technical Guide to MOM4*; Technical Report; NOAA/Geophysical Fluid Dynamics Laboratory: Princeton, NJ, USA, 2008.

40. Oke, P.; Brassington, G.B.; Griffin, D. The Bluelink ocean data assimilation system (BODAS). *Ocean Model.* **2008**, *21*, 46–70.

41. Bureau of Meterology. *APS2 Upgrade to the ACCESS-G Numerical Weather Prediction System*; Technical Report; Bureau of Meteorology: Melbourne, VIC, Australia, 2016.

42. Brassington, G.B. Multicycle ensemble forecasting of sea surface temperature. *Geophys. Res. Lett.* **2013**, *40*, 6191–6195.

43. Mathers, E.L.; Woodworth, P.L. A study of departures from the inverse-barometer response of sea level to air-pressure forcing at a period of 5 days. *Q. J. R. Meteorol. Soc.* **2004**, *130*, 725–738.

44. Murphy, A.H. What Is a Good Forecast? An Essay on the Nature of Goodness in Weather Forecasting. *Weather Forecast.* **1993**, *8*, 281–293.

45. Taylor, K. *Summarizing Multiple Aspects of Model Performance in a Single Diagram*; Technical Report; Program For Climate Model Diagnosis And Intercomparison University Of California, Lawrence Livermore National Laboratory: Livermore, CA, USA, 2000.

46. Slobbe, D.C.; Verlaan, M.; Klees, R.; Gerritsen, H. Obtaining instantaneous water levels relative to a geoid with a 2D storm surge model. *Cont. Shelf Res.* **2013**, *52*, 172–189.

47. Oke, P.; Griffin, D.; Schiller, A. Evaluation of a near-global eddy-resolving ocean model. *Geosci. Model Dev.* **2013**, *6*, 591–615.

Journal of
Marine Science and Engineering

MDPI

Article

Integrating Long Tide Gauge Records with Projection Modelling Outputs. A Case Study: New York

Phil J. Watson

School of Civil and Environmental Engineering, University of New South Wales, Sydney 2052, Australia;
philwatson.slr@gmail.com

Received: 23 May 2017; Accepted: 2 August 2017; Published: 5 August 2017

Abstract: Sea level rise is one of the key artefacts of a warming climate which is predicted to have profound impacts for coastal communities over the course of the 21st century and beyond. The IPCC provide regular updates (5–7 years) on the global status of the science and projections of climate change to assist guide policy, adaptation and mitigation endeavours. Increasingly sophisticated climate modelling tools are being used to underpin these processes with demand for improved resolution of modelling output products (such as predicted sea level rise) at a more localized scale. With a decade of common coverage between observational data and CMIP5 projection model outputs (2007–2016), this analysis provides an additional method by which to test the veracity of model outputs to replicate in-situ measurements using the case study site of New York. Results indicate that the mean relative velocity of the model projection products is of the order of 2.5–2.8 mm/year higher than the tide gauge results in 2016. In the event this phenomena is more spatially represented, there is a significant role for long tide gauge records to assist in evaluating climate model products to improve scientific rigour.

Keywords: mean sea level; velocity; acceleration; CMIP5 projection modelling; New York

1. Introduction

Climate change is predicted to have far reaching physical, social, environmental and economic impacts [1–5]. The capacity for mankind to adapt will (in part) be governed by the pace at which impacts will manifest and the success of global adaptation endeavours which might offset (or delay) the inevitability of impacts from longer term commitments such as sea level rise.

Sea level rise is one of the more insidious (or irreversible) of the postulated climate change impacts, due in part to the fact that thermal expansion (as one of the key elements of the sea level rise budget) will continue for centuries after stabilization of radiative forcing owing to the thermal inertia of the ocean water mass and the long response time scale of the deep ocean [6] and ice sheets [7]. The continued trend for coastal global population migration [4] fuels the increasing projected risks associated with sea level rise.

The Intergovernmental Panel on Climate Change (IPCC) Fifth Assessment Report (AR5) [8] provides the most authoritative and up-to-date global assessment of the state of climate science, including sea level change, past, present and future [9,10]. The Coupled Model Intercomparison Project—Phase 5 (CMIP5), developed in conjunction with AR5, provides the means by which to assess the differences in future model projections of dynamical sea level changes at fine resolution scale for the benefit of climate research, policy setting and adaptation planning [11,12].

This paper provides a methodology for improved comparison and integration of long tide gauge record data with CMIP5 model outputs at a specific location using New York as a case study. The analysis uses state-of-the-art techniques for resolving the mean sea level signal and associated kinematic properties from the long tide gauge record at Battery Park, New York with improved temporal resolution [13,14]. These techniques have been extended to the ensemble model outputs for total sea level rise from CMIP5 at the nearest ocean model grid point to New York with complete model data coverage and normalized to the tide gauge record.

From the analysis undertaken, it is of particular interest to note that associated error margins from the ensemble model outputs for sea level rise for all Representative Concentration Pathway (RCP) experiments [15], are $\approx$5–10 times that of the mean sea level trend analysis from the long New York tide gauge record over the period of overlapping coverage (2007 to 2016). When considering key kinematic properties of mean sea level (such as velocity) over the projection timescale to 2100, the error margins highlight the comparatively wide spread in the model ensemble outputs.

It is also noted that at the 95% confidence level, the initial mean relative velocity of the projected mean sea level ensemble (i.e., at 2007) is estimated at 6.3 (3.6 to 9.0) mm/year for RCP2.6, 6.0 (3.0 to 9.0) mm/year for RCP4.5 and 6.0 (3.5 to 8.5) mm/year for RCP8.5 compared to 3.7 (3.5 to 3.9) mm/year estimated from the tide gauge record at New York. If such analyses in other ocean basins of the world reveal similar artefacts, then this might be something requiring attention in the model evaluation processes of the CMIP6 design [16] for improved utility in the development of associated projection modelling outputs for AR6 [17].

2. Data and Methods

The Battery Park, New York tide gauge used in this study is the longest publicly available ocean water level record along the east coast of the USA. Figure 1 also identifies seven additional quality tide gauge records from the PSMSL extending back prior to 1940 within proximity of the Battery Park record. The selection of the Battery Park record for the task at hand is based on the fact the key temporal regional characteristics evident in each of these time series records are also captured adequately by this much longer record. In addition, the longer record has improved utility for Singular Spectrum Analysis (SSA) to isolate the trend component.

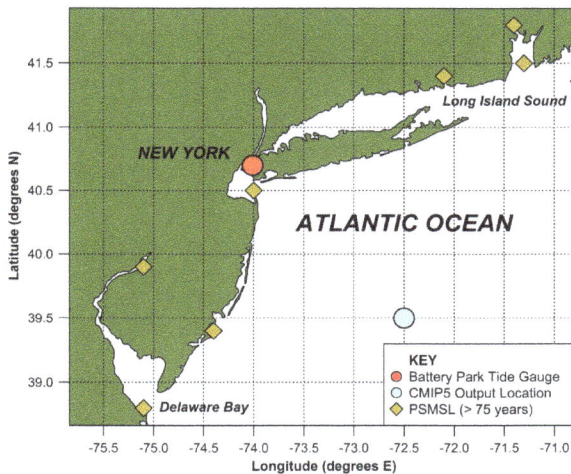

Figure 1. Location of data sources used in the study.

CMIP5 regional sea level data from IPCC AR5 [9] has been used to extract model ensemble outputs for total projected relative sea level rise for each of the respective RCP2.6, 4.5 and 8.5 experiments. The RCPs were developed for the climate modeling community as a basis for long-term and near-term modeling experiments, based on together spanning the range of year 2100 radiative forcing values found in the open literature, i.e., from 2.6 to 8.5 W/m^2 [18].

Time series data for each of the respective RCP experiments have been extracted at the nearest grid point to the Battery Park tide gauge record, for which there is complete ensemble model coverage (refer Figure 1).

It is important to note that no allowance for vertical land motion has been applied to the tide gauge analysis. The reason for this is that the CMIP5 outputs for which we are comparing the tide gauge results too, have already been corrected for glacial isostatic adjustment to project sea surface height "relative" to the land.

All analysis and graphical outputs have been developed by the author from customized scripting code within the framework of the R Project for Statistical Computing [19] and are available upon request. The applied methodology can be appropriately partitioned into analysis of the historical tide gauge record and that of the CMIP5 ensemble projection model outputs.

2.1. Historical Tide Gauge Analysis

Annual average time series data for the Battery Park, New York tide gauge have been analysed for the period spanning the timeframe 1853–2016, which are available from the public archives of the Permanent Service for Mean Sea Level (PSMSL) [20,21]. This record contains 18 missing years of data which have been filled from the extensive composite time series work by Hogarth using near neighbour tide gauge records [22].

Analysis of the observational tide gauge record at Battery Park, New York is based upon the use of the "msltrend" extension package in R [23]. This package has been specifically built as a state-of-the-art analysis tool for decomposing annual average ocean water level records to estimate mean sea level with improved temporal resolution [13,14]. The package development has been underpinned by time series analysis testing and parameter optimisation using all records in the PSMSL exceeding 100 years in length [24] that are a minimum of 85% complete. Details of the methodology underpinning "msltrend" and the analysis of the Battery Park record can be broadly summarised in the following three steps:

Step 1: Estimation of mean sea level. The time series is decomposed using a one dimensional Singular Spectrum Analysis (SSA). The method decomposes the original record into a series of components of slowly varying trend, oscillatory components with variable amplitude, and a structure-less noise [25]. The trend (or in this case mean sea level) can be isolated by reconstructing only the components that possess distinctly "trend-like" characteristics. Trend components are automatically detected and reconstructed based on the singular value having a relative contribution threshold $\geq 75\%$ contained within the low frequency bin ≤ 0.01.

Step 2: Estimation of mean sea level velocity. The approach adopted to estimating the time varying velocity is based on the first derivative of a fitted cubic smoothing spline. This approach provides a realistic representation of a smoothly varying trend, so long as the cubic smoothing spline model fit can accurately describe the reconstructed trend-like components of the SSA decomposition.

Trial and error on a wide variety of long records in PSMSL indicates that by fitting a cubic smoothing spline to the trend (determined in Step 1) with approximately 1 degree of freedom per every eight years of record length, optimises the fit whilst removing the extraneous effects of the "sawtoothing". This so-called "sawtoothing" effect can occur because the isolated trend components from the SSA decomposition are portions of linearly additive components that reconstruct the original time series and thus are not precisely smooth or curvilinear at point to point scale.

For the numerous records tested, the coefficient of determination (R^2) of the fitted spline to the estimated mean sea level (trend) exceeds 0.99 in all cases, providing a high degree of confidence in this form of model to estimate the associated time varying velocity.

Step 3: Estimation of errors. The estimation of errors in the trend and associated velocity is one of the more significant features of the "msltrend" package and is based on bootstrapping techniques. This process initially involves fitting an autoregressive time series model to remove the serial correlation in the residuals between the SSA derived trend and the original annual average time series [26].

The uncorrelated residuals are then tested to identify change points in the statistical variance along the time series.

Where a change point is detected, bootstrapping processes to randomly sample uncorrelated residuals are quarantined between identified variance change points (otherwise known as "block" bootstrapping). The randomly sampled uncorrelated residuals are then added to the SSA derived trend and the process repeated 10,000 times (Steps 1 and 2). From the extensive pool of outputted trends and associated velocities, standard deviations are readily calculated to derive robust confidence intervals.

2.2. CMIP5 Projection Model Output Analysis

The CMIP5 models used in AR5 provide projection outputs of sea surface height at each grid point for model experiments that meet requisite evaluation protocols [27]. These data are publicly available in netCDF format from the Integrated Climate Data Center (ICDC) [28] with yearly outputs spanning the period 2007 to 2100 on a spatial resolution grid of $1° \times 1°$ for RCP2.6, 4.5 and 8.5 experiments. The CMIP5 multi-model ensemble contains only 16 models for the RCP2.6 experiment however, the RCP4.5 and 8.5 experiments are based on all 21 models (Dr Mark Carson, Institute of Oceanography, ICDC, University of Hamburg, 2017, pers.comm., 19 June). These outputs are based upon modelled responses to dynamic ocean responses, atmospheric loading, land ice, terrestrial water sources and glacial isostatic adjustment in order to estimate sea surface height "relative" to the land. These time series are extracted from the respective RCP2.6, 4.5 and 8.5 netCDF format files available from the ICDC [28] using customised R scripting code at the point of interest (refer Figure 1). The following methodology has then been applied to these data products in order to both assimilate and compare them to the observational tide gauge data:

Step 1: Normalise ensemble model outputs to tide gauge datum. The ensemble model sea surface height output products are based on a 20 year moving average with the modelling start point set at 1986–2005 (i.e., centred around 1995). The annual time series output products from the ICDC for AR5 start at 2007 and have therefore been normalized to the Battery Park tide gauge record by using the estimate of mean sea level derived from the SSA decomposition in 1995.

Step 2: Estimation of projected mean sea level velocity. The approach adopted is similar to that described for the tide gauge record analysis above, whereby a cubic smoothing spline has been fitted to each of the respective ensemble sea surface height time series from 2007 to 2100 to estimate the associated time varying velocity. The only difference is that less degrees of freedom are necessary for the smoothing spline (1 degree of freedom per 15 years) in order to achieve the same quality of model fit (R^2) owing to the very different characteristics of the sea surface height time series from the model projections. The selection of the optimum spline stiffness for the model outputs is somewhat arbitrary based on trial and error.

Step 3: Estimation of means and errors. The pool of outputted sea surface height time series and associated velocities for each of the respective RCP experiments enables provision to calculate simple arithmetic means and standard deviations from which to estimate robust confidence intervals.

3. Results

The results from the decomposition of the Battery Park tide gauge record are diagrammatically summarised in Figure 2. The decomposition of the annual time series in the top panel highlights the nature of the internal climate variability influence on mean sea level at this location, with an amplitude in the range of ≈50–60 mm. The removal of such influences reveals the mean sea level signal which can be assumed to approximate the climate change signal resulting from external forcings. The higher resolution of the kinematic properties of the mean sea level signal provide more instruction on the how relative mean sea level (that is "relative" to the land) has been changing from 1853 to present at this gauge location. Points to note include:

- The relative mean sea level has been rising continuously at this location over the 163 year record (≈465 mm)
- The relative velocity has continued to increase steadily over the course of the record peaking at around 3.7 (3.3 to 4.1) mm/year (95% CI) in 2008;
- At the 95% confidence level, the relative velocity in 2016 at 3.5 (3.0 to 4.0) mm/year is higher than the velocity at the start of the record in 1853; and
- Time varying relative velocity increasing over time suggests the presence of a positive acceleration.

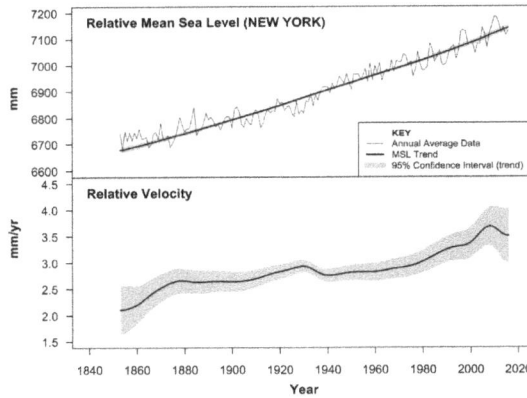

Figure 2. Estimated mean sea level and associated velocity from the Battery Park (New York) tide gauge record.

Figures 3–5 summarise the analysis of the ensemble outputs corresponding to the RCP2.6, 4.5 and 8.5 experiments, respectively, integrated with the observational tide gauge analysis (Figure 2). The top panel in each figure integrates the tide gauge time series with the ensemble projection outputs. The middle panel provides an estimate of relative velocity in mean sea level from both data sets while the bottom 2 panels provide zoomed in versions of the analysis for the period of common coverage (2007–2016) of both the observational record and CMIP5 ensemble projection modelling products at the $1° \times 1°$ grid resolution. All three figures use the same scales for ready comparison between RCP's.

It is evident that internal modes of climate variability built into the projection modelling products (to replicate ENSO, PDO, etc.) would appear visually to be of comparable scale and amplitude to that evident over the course of the observational record at this location.

As anticipated, the mean velocity of the projection model ensembles increase in line with the increased radiative forcing associated with the respective RCP experiments. Further, the mean velocity stabilises mid-century before declining slightly to 2100 under the RCP2.6 experiment, increases slightly to mid-century remaining relatively steady to 2100 under RCP4.5 and increases steadily to 2100 under the RCP8.5 experiment. These temporal characteristics mirror those of the global mean sea level projections for the respective RCP experiments advised in AR5 [9], though the scale of the initial velocities are considerably higher at this grid location.

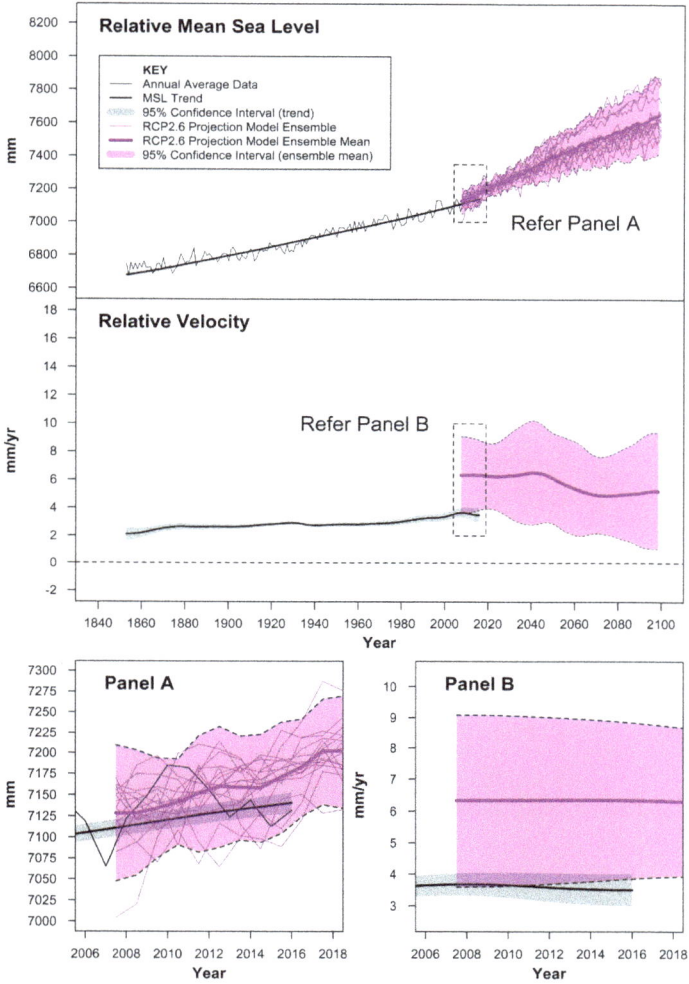

Figure 3. Integrating estimated mean sea level and associated velocity for the Battery Park (New York) tide gauge record with CMIP5 projection modelling (RCP2.6).

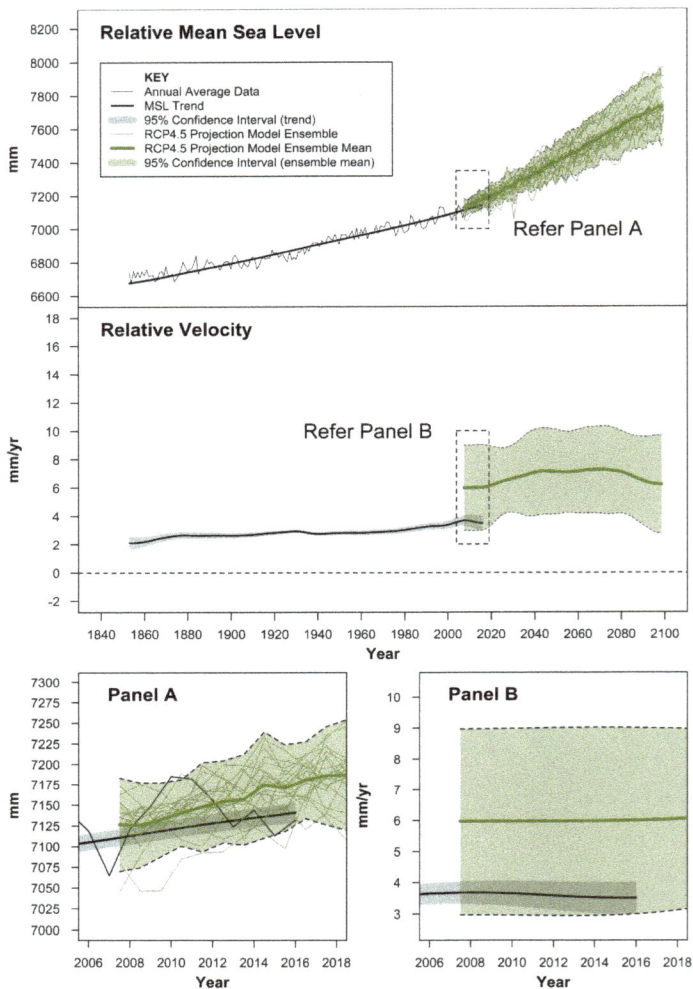

Figure 4. Integrating estimated mean sea level and associated velocity for the Battery Park (New York) tide gauge record with CMIP5 projection modelling (RCP4.5).

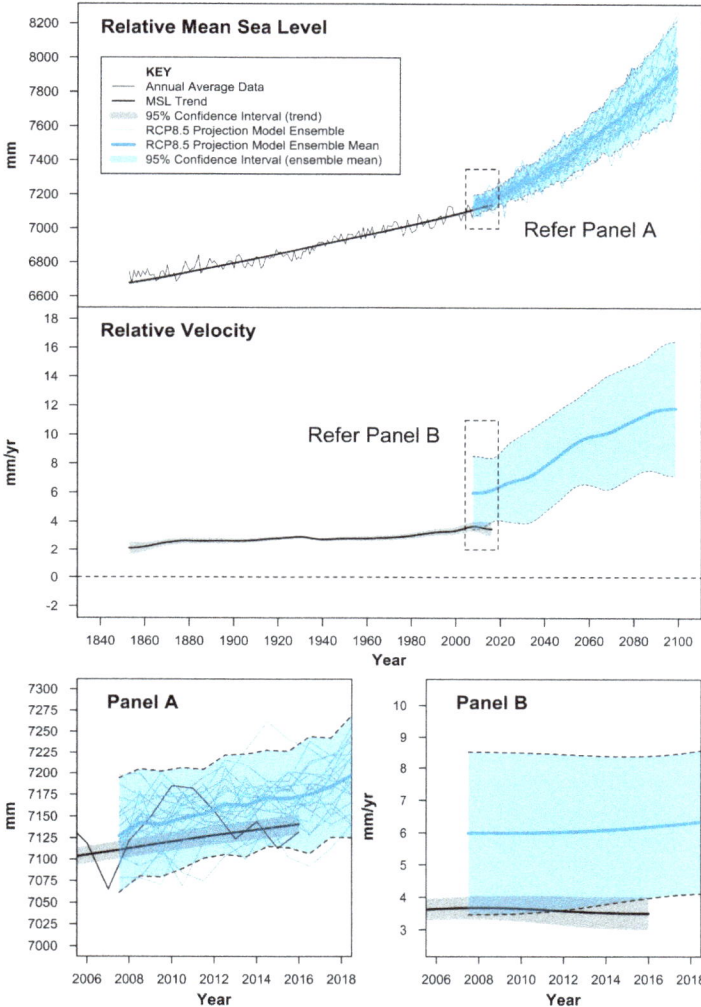

Figure 5. Integrating estimated mean sea level and associated velocity for the Battery Park (New York) tide gauge record with CMIP5 projection modelling (RCP8.5).

The mean of the ensemble for all RCP experiments at this location exhibit a higher gradient (and therefore rate of rise) over the period of common coverage than for the tide gauge record.

Specifically, at the 95% confidence level, the mean velocity of the projection model ensembles estimated in 2016 for RCP2.6, 4.5 and 8.5 are 6.3 (3.8 to 8.8), 6.0 (3.0 to 9.0) and 6.2 (4.0 to 8.4) mm/year, respectively compared to 3.5 (3.0 to 4.0) mm/year for the tide gauge record. In 2016, the mean relative velocity of the model projection products is of the order of 2.5–2.8 mm/year higher than the estimate observed from the tide gauge record.

4. Discussion

The techniques espoused in this case study provide improved means by which to test the veracity of the CMIP5 (and future CMIP6) sea level projections to replicate observational data at increasingly localized scales, using long quality tide gauge records.

The differing nature of the respective data sets present challenges in aligning analysis products to be directly comparable. For example, with the tide gauge analysis, it is a relatively straightforward task to remove the internal modes of climate variability (and other cyclical dynamic influences) from the time series record using SSA as described. The resulting mean sea level (or trend) is assumed to be principally attributable to external (or climate change) forcings. However, the CMIP5 projection model ensembles are designed to encompass a wide range of possible climate outcomes embedded with randomly phased internal modes of climate variability over the future projection horizon. The larger the pool of ensemble model outputs, the better the chance that internal climate variability can be accommodated (or averaged out) to reveal the more consistent sustained response to the external (or climate change) forcing.

However, there are only 16 CMIP5 ensemble sea surface height model projection products available for RCP2.6 and 21 for the RCP4.5 and 8.5 experiments. Thus, owing to the intrinsic nature of the projection modelling products, the ensemble mean sea level and associated velocity will be associated with much larger error margins than the analysis of a singular tide gauge time series which is clearly evident in Figures 3–5. By virtue of these large error margins, at least from a statistical perspective at the 95% confidence level, the projected mean sea level and associated velocity over the period of common coverage for all RCP experiments are comparable to that of the tide gauge record for this location. Unless the CMIP6 framework [16] results in a substantial increase in the available number of sea surface height model projections, then the spread of the ensemble products is likely to remain similar for AR6, irrespective of increased performance and improved resolution of the models.

To tease out whether the CMIP5 projection outputs at this case study site are indeed aligned with the observational record, we perhaps need to look a bit deeper with additional resources available. As advised in Section 3, the mean relative velocity of the projection modelled outputs for 2016 are of the order of 2.5–2.8 mm/year higher than that from the Battery Park tide gauge analysis. The CMIP5 ensemble outputs provided by the ICDC have the Glacial Isostatic Adjustment (GIA) correction inverted so the projection outputs provide continuity of the picture relative to the land (Dr Mark Carson, Institute of Oceanography, ICDC, University of Hamburg, 2017, pers.comm., 31 March). For the CMIP5 grid point considered (refer Figure 1), the GIA estimate is approximately -1.4 mm/year, whereas the total vertical land motion (VLM) which includes GIA, measured from a GPS station collocated at the Battery Park tide gauge site is approximately -2.12 ± 0.62 mm/year (1σ) [29,30]. Thus if the relative tide gauge record were corrected to incorporate GIA only (and not all VLM sources) in order to be more directly comparable to the CMIP5 ensemble model outputs, then the actual gap between the relative tide gauge velocity and the projection modelled outputs over the common period would be wider again. It is noted that there are other GPS stations within 20 km of the Battery Park tide gauge with measured VLM ranging from -1.02 ± 0.30 (Willets Point) to -2.65 ± 0.27 mm/year (Sandy Hook) [29,30]. Whilst it is acknowledged that estimates of GIA and VLM are sources of uncertainty, measured VLM will play an increasingly important role for augmenting mean sea level records from tide gauges, particularly as the GPS measurements lengthen.

It is also worth noting that this paper attempts to compare the characteristics of the overlapping parts of the tide gauge record and projection modelling products for a comparatively small time window (10 years) at the end and start of the respective records. Although every effort has been made to take advantage of state-of-the-art analytical techniques to improve the resolution of the mean sea level signal from the tide gauge record, in reality time series analysis techniques are inherently limited by the ubiquity of end effects. Extensive time series analysis testing and optimization for sea level research [24,31] has limited these influences, but, notwithstanding, the broadened error margins at the ends of the respective records take some account of the uncertainties of the respective mean estimates

J. Mar. Sci. Eng. **2017**, *5*, 34

(mean sea level and velocity) near the ends of such records. The utility of such analyses will therefore continue to improve as the length of the overlapping records increase into the future.

One element not examined during the course of this case study is whether there would be any particular physical or oceanographic reason why the rise in mean sea level at a point ≈200–230 km offshore (projection modelling grid point) would necessarily be occurring at a faster rate to that observed at the land/sea interface by a tide gauge. This issue might be a worth investigating on a more global scale within the context of the ensemble sea surface height model projection products available, but is beyond the scope of this study.

5. Conclusions

The analysis of the long Battery Park record enables a more temporally resolved and accurate estimate of mean sea level through the identification and removal of contaminating influences of decadal to multi-decadal timescale (e.g., [32–37]). This presents an opportunity to improve evaluation of climate models to the rate of mean sea level rise at increasingly finer resolution. Similarly, these processes present an opportunity to augment the broader, global Archiving, Validation and Interpretation of Satellite Oceanographic (AVISO) data products [38] currently used to evaluate CMIP5 dynamic sea surface heights [27], that are only 20–25 years in length. The longer decadal and multi-decadal influences are not able to be removed from these satellite data products at this point in time.

The higher mean relative velocity of the CMIP5 model projection products compared to that observed from the tide gauge record over the common time frame (2007–2016) from this case study might prove to be site specific. However, if this proves to be a more common phenomenon across other ocean basins in key locations, then this might raise the necessity to improve the evaluation of the ocean component of the climate models for CMIP6 using a key set of long PSMSL gauge records and the techniques espoused in this paper as a guide. In addition to integrating more advanced oceanographic phenomena at increasingly finer resolution into the ocean model components, e.g., [39], the techniques espoused in this paper might also be considered part of the evolutionary process by which to improve the robustness and veracity of these critical projection modelling tools at increasingly finer resolution over the course of CMIP6 and AR6 and beyond.

Acknowledgments: This research work has not benefitted from any grants or financial assistance. The author would like to thank Aimée Slangen (Royal Netherlands Institute for Sea Research, The Netherlands) for wide ranging advice and discussions on climate modelling products that improved the study; Mark Carson (Institute of Oceanography, University of Hamburg, Germany) who provided advice on specifics of the CMIP5 sea surface height projection model outputs used in AR5, made available via the Integrated Climate Data Center, University of Hamburg; and Thiago Dos Santos (Land and Atmospheric Science, University of Minnesota, USA) who provided advice on R scripting code to facilitate extraction of data from netCDF format files. The author would also like to thank the PSMSL, ICDC and SONEL for their publicly accessible data repositories.

Conflicts of Interest: The author declares no conflict of interest in the preparation of this study.

References

1. Houser, T.; Hsiang, S.; Kopp, R.; Larsen, K. *Economic Risks of Climate Change: An American Prospectus*; Columbia University Press: New York, NY, USA, 2015.

2. Intergovernmental Panel on Climate Change. *Climate Change 2014: Impacts, Adaptation, and Vulnerability. Part A: Global and Sectoral Aspects*; Contribution of Working Group II to the Fifth Assessment Report of the Intergovernmental Panel on Climate Change; Field, C.B., Barros, V.R., Dokken, D.J., Mach, K.J., Mastrandrea, M.D., Bilir, T.E., Chatterjee, M., Ebi, K.L., Estrada, Y.O., Genova, R.C., et al., Eds.; Cambridge University Press: Cambridge, UK; New York, NY, USA, 2014; p. 1132.

3. Melillo, J.M.; Richmond, T.T.; Yohe, G.W. *Climate Change Impacts in the United States: The Third National Climate Assessment*; U.S. Global Change Research Program: Washington, DC, USA, 2014; p. 841.

4. Neumann, B.; Vafeidis, A.T.; Zimmermann, J.; Nicholls, R.J. Future coastal population growth and exposure to sea-level rise and coastal flooding-a global assessment. *PLoS ONE* **2015**, *10*, 0118571. [CrossRef] [PubMed]

5. Watkiss, P. *The ClimateCost Project*; Final Report, Volume 1: Europe; Stockholm Environmental Institute: Stockholm, Sweden, 2011; ISBN 978-91-86125-35-6.
6. Zickfeld, K.; Eby, M.; Alexander, K.; Weaver, A.J.; Crespin, E.; Fichefet, T.; Goosse, H.; Philippon-Berthier, G.; Edwards, N.R.; Holden, P.B.; et al. Long-term climate change commitment and reversibility: An EMIC inter-comparison. *J. Clim.* **2013**, *26*, 5782–5809. [CrossRef]
7. Levermann, A.; Clark, P.U.; Marzeion, B.; Milne, G.A.; Pollard, D.; Radic, V.; Robinson, A. The multimillennial sea-level commitment of global warming. *Proc. Nat. Acad. Sci. USA* **2013**, *110*, 13745–13750. [CrossRef] [PubMed]
8. Intergovernmental Panel on Climate Change. *Climate Change 2013: The Physical Science Basis*; Stocker, T.F., Qin, D., Plattner, G.-K., Tignor, M., Allen, S.K., Boschung, J., Nauels, A., Xia, Y., Bex, V., Midgley, P.M., Eds.; Contribution of Working Group I to the Fifth Assessment Report of the Intergovernmental Panel on Climate Change; Cambridge University Press: Cambridge, UK; New York, NY, USA, 2013; 1535p. [CrossRef]
9. Church, J.A.; Clark, P.U.; Cazenave, A.; Gregory, J.M.; Jevrejeva, S.; Levermann, A.; Merrifield, M.A.; Milne, G.A.; Nerem, R.S.; Nunn, P.D.; et al. Sea Level Change. In *Climate Change 2013: The Physical Science Basis*; Stocker, T.F., Qin, D., Plattner, G.-K., Tignor, M., Allen, S.K., Boschung, J., Nauels, A., Xia, Y., Bex, V., Midgley, P.M., Eds.; Contribution of Working Group I to the Fifth Assessment Report of the Intergovernmental Panel on Climate Change; Cambridge University Press: Cambridge, UK; New York, NY, USA, 2013; pp. 1137–1216. [CrossRef]
10. Rhein, M.; Rintoul, S.R.; Aoki, S.; Campos, E.; Chambers, D.; Feely, R.A.; Gulev, S.; Johnson, G.C.; Josey, S.A.; Kostianoy, A.; et al. Observations: Ocean. In *Climate Change 2013: The Physical Science Basis*; Stocker, T.F., Qin, D., Plattner, G.-K., Tignor, M., Allen, S.K., Boschung, J., Nauels, A., Xia, Y., Bex, V., Midgley, P.M., Eds.; Contribution of Working Group I to the Fifth Assessment Report of the Intergovernmental Panel on Climate Change; Cambridge University Press: Cambridge, UK; New York, NY, USA, 2013; pp. 255–316.
11. World Climate Research Programme. *Coupled Model Intercomparison Project—Phase 5*; Special Issue of the CLIVAR Exchanges Newsletter; WCRP: Geneva, Switzerland, 2011; No. 56; Volume 15(2).
12. Taylor, K.E.; Stouffer, R.J.; Meehl, G.A. An overview of CMIP5 and the experiment design. *Bull. Am. Meteorol. Soc.* **2012**, *93*, 485–498. [CrossRef]
13. Watson, P.J. Acceleration in US Mean Sea Level? A New Insight using Improved Tools. *J. Coast. Res.* **2016**, *32*, 1247–1261. [CrossRef]
14. Watson, P.J. Acceleration in European Mean Sea Level? A New Insight Using Improved Tools. *J. Coast. Res.* **2017**, *33*, 23–38. [CrossRef]
15. Moss, R.H.; Edmonds, J.A.; Hibbard, K.A.; Manning, M.R.; Rose, S.K.; Van Vuuren, D.P.; Carter, T.R.; Emori, S.; Kainuma, M.; Kram, T.; et al. The next generation of scenarios for climate change research and assessment. *Nature* **2010**, *463*, 747–756. [CrossRef] [PubMed]
16. Eyring, V.; Bony, S.; Meehl, G.A.; Senior, C.A.; Stevens, B.; Stouffer, R.J.; Taylor, K.E. Overview of the Coupled Model Intercomparison Project Phase 6 (CMIP6) experimental design and organization. *Geosci. Model Dev.* **2016**, *9*, 1937–1958. [CrossRef]
17. Intergovernmental Panel on Climate Change. Press Release—IPCC Agrees Special Reports. AR6 Workplan, 2016/03/PR. 14 April 2016. Available online: http://www.ipcc.ch/news_and_events/pdf/press/160414_pr_p43.pdf (accessed on 15 April 2017).
18. Van Vuuren, D.P.; Edmonds, J.; Kainuma, M.; Riahi, K.; Thomson, A.; Hibbard, K.; Hurtt, G.C.; Kram, T.; Krey, V.; Lamarque, J.F.; et al. The representative concentration pathways: An overview. *Clim. Chang.* **2011**, *109*, 5–31. [CrossRef]
19. R Project for Statistical Computing. Available online: https://www.r-project.org/ (accessed on 10 April 2017).
20. Holgate, S.J.; Matthews, A.; Woodworth, P.L.; Rickards, L.J.; Tamisiea, M.E.; Bradshaw, E.; Foden, P.R.; Gordon, K.M.; Jevrejeva, S.; Pugh, J. New data systems and products at the permanent service for mean sea level. *J. Coast. Res.* **2012**, *29*, 493–504. [CrossRef]
21. Permanent Service for Mean Sea Level (PSMSL). Available online: http://www.psmsl.org (accessed on 10 April 2017).
22. Hogarth, P. Preliminary analysis of acceleration of sea level rise through the twentieth century using extended tide gauge data sets (August 2014). *J. Geophys. Res. Oceans* **2014**, *119*, 7645–7659. [CrossRef]

23. *Msltrend: Improved Techniques to Estimate Trend, Velocity and Acceleration from Sea Level Records*, version 1.0; 12 January 2016. Available online: https://cran.r-project.org/web/packages/msltrend/index.html (accessed on 8 April 2017).

24. Watson, P.J. Improved Techniques to Estimate Mean Sea Level, Velocity and Acceleration from Long Ocean Water Level Time Series to Augment Sea Level (and Climate Change) Research. Ph.D. Thesis, University of New South Wales, Sydney, Australia. (under review, submitted for examination).

25. Golyandina, N.; Nekrutkin, V.; Zhigljavsky, A.A. *Analysis of Time Series Structure: SSA and Related Techniques*; Chapman & Hall/CRC Monographs on Statistics & Applied Probability; Taylor & Francis: Boca Raton, FL, USA, 2001; p. 320.

26. Foster, G.; Brown, P.T. Time and tide: Analysis of sea level time series. *Clim. Dyn.* **2015**, *45*, 291–308. [CrossRef]

27. Flato, G.; Marotzke, J.; Abiodun, B.; Braconnot, P.; Chou, S.C.; Collins, W.; Cox, P.; Driouech, F.; Emori, S.; Eyring, V.; et al. Evaluation of Climate Models. In *Climate Change 2013: The Physical Science Basis*; Stocker, T.F., Qin, D., Plattner, G.-K., Tignor, M., Allen, S.K., Boschung, J., Nauels, A., Xia, Y., Bex, V., Midgley, P.M., Eds.; Contribution of Working Group I to the Fifth Assessment Report of the Intergovernmental Panel on Climate Change; Cambridge University Press: Cambridge, UK; New York, NY, USA, 2013; pp. 741–866. [CrossRef]

28. Integrated Climate Data Center (ICDC) for AR5 Sea Level Data. Available online: http://icdc.cen.uni-hamburg.de/1/daten/ocean/ar5-slr.html (accessed on 12 April 2017).

29. SONEL. Available online: http://www.sonel.org/spip.php?page=gps&idStation=2722 (accessed on 11 April 2017).

30. Santamaría-Gómez, A.; Gravelle, M.; Collilieux, X.; Guichard, M.; Míguez, B.M.; Tiphaneau, P.; Wöppelmann, G. Mitigating the effects of vertical land motion in tide gauge records using a state-of-the-art GPS velocity field. *Glob. Planet. Chang.* **2012**, *98*, 6–17. [CrossRef]

31. Watson, P.J. Identifying the Best Performing Time Series Analytics for Sea Level Research. In *Time Series Analysis and Forecasting*; Rojas, I., Pomares, H., Eds.; Contributions to Statistics; Springer: Geneva, Switzerland, 2016; pp. 261–278.

32. Calafat, F.; Chambers, D. Quantifying recent acceleration in sea level unrelated to internal climate variability. *Geophys. Res. Lett.* **2013**, *40*, 3661–3666. [CrossRef]

33. Chambers, D.P.; Merrifield, M.A.; Nerem, R.S. Is there a 60-year oscillation in global mean sea level? *Geophys. Res. Lett.* **2012**, *39*. [CrossRef]

34. Houston, J.R.; Dean, R.G. Effects of sea-level decadal variability on acceleration and trend difference. *J. Coast. Res.* **2013**, *29*, 1062–1072. [CrossRef]

35. Minobe, S. Resonance in bidecadal and pentadecadal climate oscillations over the North Pacific: Role in climatic regime shifts. *Geophys. Res. Lett.* **1999**, *26*, 855–858. [CrossRef]

36. Qiu, B.; Chen, S. Multidecadal sea level and gyre circulation variability in the northwestern tropical Pacific Ocean. *J. Phys. Oceanogr.* **2012**, *42*, 193–206. [CrossRef]

37. Sturges, W.; Douglas, B.C. Wind effects on estimates of sea level rise. *J. Geophys. Res. Oceans* **2011**, *116*. [CrossRef]

38. AVISO. Available online: http://www.aviso.altimetry.fr/en/my-aviso.html (accessed on 14 April 2017).

39. Slangen, A.B.A.; Carson, M.; Katsman, C.A.; Van de Wal, R.S.W.; Köhl, A.; Vermeersen, L.L.A.; Stammer, D. Projecting twenty-first century regional sea-level changes. *Clim. Chang.* **2014**, *124*, 317–332. [CrossRef]

Journal of
Marine Science and Engineering

MDPI

Article

Projected 21st Century Sea-Level Changes, Observed Sea Level Extremes, and Sea Level Allowances for Norway

Matthew J. R. Simpson [1,*], Oda R. Ravndal [2], Hilde Sande [2], Jan Even Ø. Nilsen [3], Halfdan P. Kierulf [1], Olav Vestøl [1] and Holger Steffen [4]

[1] Geodetic Institute, Norwegian Mapping Authority, 3507 Hønefoss, Norway; halfdan.kierulf@kartverket.no (H.P.K.); olav.vestol@kartverket.no (O.V.)
[2] Hydrographic Service, Norwegian Mapping Authority, Professor Olav Hanssens vei 10, 4021 Stavanger, Norway; oda.ravndal@kartverket.no (O.R.R.); hilde.sande.borck@kartverket.no (H.S.)
[3] Nansen Environmental and Remote Sensing Center, and Bjerknes Centre for Climate Research, Thormøhlensgt. 47, 5006 Bergen, Norway; even@nersc.no
[4] Lantmäteriet, Lantmäterigatan 2C, 80182 Gävle, Sweden; holger.steffen@lm.se
* Correspondence: matthew.simpson@kartverket.no; Tel.: +47-3211-8306

Received: 31 May 2017; Accepted: 18 July 2017; Published: 14 August 2017

Abstract: Changes to mean sea level and/or sea level extremes (e.g., storm surges) will lead to changes in coastal impacts. These changes represent a changing exposure or risk to our society. Here, we present 21st century sea-level projections for Norway largely based on the Fifth Assessment Report from the Intergovernmental Panel for Climate Change (IPCC AR5). An important component of past and present sea-level change in Norway is glacial isostatic adjustment. We therefore pay special attention to vertical land motion, which is constrained using new geodetic observations with improved spatial coverage and accuracies, and modelling work. Projected ensemble mean 21st century relative sea-level changes for Norway are, depending on location, from −0.10 to 0.30 m for emission scenario RCP2.6; 0.00 to 0.35 m for RCP 4.5; and 0.15 to 0.55 m for RCP8.5. For all RCPs, the projected ensemble mean indicates that the vast majority of the Norwegian coast will experience a rise in sea level. Norway's official return heights for extreme sea levels are estimated using the average conditional exceedance rate (ACER) method. We adapt an approach for calculating sea level allowances for use with the ACER method. All the allowances calculated give values above the projected ensemble mean Relative Sea Level (RSL) rise, i.e., to preserve the likelihood of flooding from extreme sea levels, a height increase above the most likely RSL rise should be used in planning. We also show that the likelihood of exceeding present-day return heights will dramatically increase with sea-level rise.

Keywords: Norway; sea-level change; regional sea-level projections; IPCC AR5; glacial isostatic adjustment; extreme sea levels; ACER extreme value prediction; sea level allowances; tide gauges

1. Introduction

In the global setting, Norway is commonly perceived as being at low risk from sea-level rise. The coastline is largely characterized by steep topography and exposed bedrock that is resistant to erosion. In addition to this, the land surface is experiencing on-going uplift due to the loss of the large ice sheet that once covered Fennoscandia (a process known as glacial isostatic adjustment, GIA; e.g., [1,2]). It is well recognized that this uplift process will act to mitigate future sea-level rise (e.g., [3]). These factors suggest that, in comparison to other coastal countries, Norway has a generally low physical vulnerability to increasing sea levels [4].

There are, nevertheless, several reasons for wanting to gain a better understanding of future sea-level changes and their associated risks. Firstly, while the general perception that Norway is at low risk from sea-level rise is broadly true, it is not the case in certain parts of the country and on local scales. In trying to understand this issue, one can start by briefly considering the nature of the Norwegian landscape, which has been shaped over numerous glaciations. The coastline is long, rugged and complicated, with an estimated 240,000 islands. There are large areas of steep exposed bedrock but also some low-lying areas of continuous and soft sediment deposits (mostly glacial till). The majority of Norway's cities and numerous towns and villages are situated along the coast and, in recent years, many large developments have been undertaken close to the shoreline. There is also an extensive infrastructure of bridges, tunnels and ports connecting these communities. Previous vulnerability assessments have helped identify the low-lying coastal areas of Norway that are at risk from sea-level rise owing to their important cultural and economic value [4,5]. These studies have shown that, in some specific areas where economic activities are concentrated close to the coast (e.g., oil, fishing and shipping industries), the impacts of sea-level rise could be significant.

Secondly, there is a clear need to take into account our changing climate in planning and/or adaption policy. Here it is the job of researchers to provide up-to-date scientific information on sea level to help decision makers. In our view, this information should be updated at least as often as the Intergovernmental Panel for Climate Change (IPCC) publishes its assessment reports, which is every 6 to 7 years (IPCC AR5 was released in 2013). Providing such information means setting up a framework for how to apply sea-level projections on a local scale. It also requires close cooperation of the institutions involved in different aspects of sea level science (no single institution has overall responsibility for sea level science in Norway). Once we have reviewed and updated our scientific understanding of sea level, then a decision can be made on how that information is used in policy. On that point, we note that although advice on sea-level rise is normally taken into account in planning, sea level is not specifically addressed in national legislation (i.e., Norway's building code) [4,5].

Relative Sea Level (RSL) change is defined as the change in ocean surface height with respect to the solid Earth. Paleo observations from across Fennoscandia, including Norway, show a spatial pattern of RSL change over the past ~10,000 years that largely reflects vertical land uplift due to GIA (e.g., [6]). Norway has therefore experienced an overall *fall* in RSL as the solid Earth surface has rebounded, in some places as much as several hundred meters. As well as being evident in the paleo record, analyses of the Norwegian tide-gauge network (Figure 1) also show that GIA is an important component of twentieth century RSL change (e.g., [7–10]). These tide-gauge data, which are unevenly spaced along the coast, indicate that over the past ~100 years some areas of Norway underwent a RSL fall while other areas underwent a limited RSL rise (that is, values somewhat below the global mean rise). In parts of the country where RSL is rising, climate driven sea-level rise is now greater than land uplift. This represents a reversal of the general trend of late-Holocene relative sea-level fall.

To be able to quantify the effect of sea-level change on flood risk assessments, information on future sea levels is required in a probabilistic form (i.e., for a specific future date, an assessment is made of the probability or likelihood of a certain sea level occurring). Up until relatively recently, such projections for mean sea-level changes were not achievable. The assessment made by Church et al. [11] in IPCC AR5 was that confidence in projecting sea-level changes had improved due to: (1) our improved understanding of observed and modelled 20th century global sea-level changes (e.g., [12]); and (2) progress made with quantifying ice-dynamic contributions (e.g., [13]). This has meant that probabilistic projections are now possible, but there remain significant challenges to overcome before projections of sea-level change can be considered truly robust. In IPCC AR5, future global sea-level rise is assessed as *likely* ($p > 66\%$) to be within the range of the projections [11]. Partly owing to insufficient and inconsistent evidence, however, the IPCC AR5 authors refrained from trying to quantify the probability of levels above the *likely* range (i.e., the upper tail of the probability distribution that essentially relates to the poorly constrained future behaviour of the ice sheets). The same main authors also reached the same conclusion in a more recent review [14].

Not having information on the complete probability distribution of the sea-level projections presents a problem for their use in practical applications. Understanding of low probability but potentially large impact future sea-level changes is important for coastal management [15]. Furthermore, knowledge (or an assumption) about the complete probability distribution of future sea-level rise is also required for calculating sea level "allowances" [16,17], a possible attractive option in coastal planning for Norway. Allowances give the height by which an asset needs to be raised so that the probability of flooding remains preserved for an *uncertain* future sea-level change [16,17]. As IPCC AR5 did not provide information beyond the *likely* ranges, some studies have instead made use of expert elicitations to help quantify the complete probability distribution of future sea-level rise (e.g., [18,19]). At the same time, post IPCC AR5, there have also been advances in our understanding of present and future ice sheet changes. Most notably, observations of glacier losses in Antarctica suggest the early stages of a collapse may have begun (e.g., [20,21]). Recent modelling studies of Antarctica have tried to quantify how such a marine collapse might contribute to future sea-level rise from the continent with diverging results (e.g., [22,23]). The amounts and timing of these potential future contributions from Antarctica are very uncertain. What is apparent, however, is that this area of sea level science is rapidly evolving.

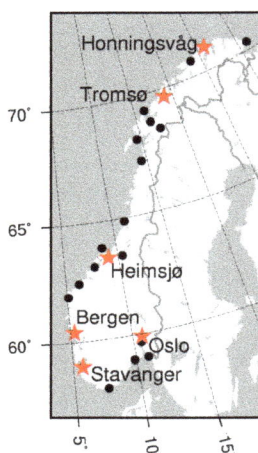

Figure 1. The 23 tide gauges that form the Norwegian tide gauge network. The red stars with names indicate the tide gauge sites chosen as key locations in this article (i.e., here we show sea-level projections and return periods for extreme sea levels).

This article is structured as follows. We begin by examining vertical land motion (VLM) in Norway, an important component of present sea-level change in Scandinavia, which we constrain using new Global Positioning System (GPS) measurements and modelling [2]. The VLM solution is a key component used in the sea-level projections presented here. Our *regional* projections of 21st century sea-level changes for Norway are largely based on findings from IPCC AR5 and Coupled Model Intercomparison Project phase 5 (CMIP5) output [24]. The main aim of the work is to show sea-level projections for Norway updated with IPCC AR5 science. We go on to look at how our sea-level projections can be combined with statistics on extreme sea levels. Here we use the Average conditional exceedance rate (ACER) method for estimating the return periods for extreme sea levels. This is the approach used for calculating the official return periods for Norway [25]. We then adapt a method for calculating sea level allowances [16,17] for use with our preferred ACER return heights. Finally, allowances are calculated for the Norwegian coast using different normal distributions of future sea-level rise.

2. Materials and Methods

2.1. Present-Day Vertical Land Motion in Norway

There is a long history of GIA research in Fennoscandia [26]. A wide variety of observations are available for the study of GIA, for example, paleo sea level, tide gauge, levelling and terrestrial gravity measurements. Now, over the past two decades, the satellite based observation systems of GPS and the Gravity Recovery and Climate Experiment (GRACE) have provided new insights into the process of GIA (see Steffen and Wu [27], for a review of all the datasets).

Such observations of GIA in Fennoscandia have traditionally been used to infer details of Earth's viscosity structure and/or the region's ice history (e.g., [1,6,28,29]). They also inform us on vertical land motion—an important component of present-day RSL change for Norway. The development of GPS, in particular, has enabled us to image crustal deformation to a high degree of precision. These observations show that present-day VLM across Fennoscandia is dominated by the on-going relaxation of the Earth in response to past ice mass loss (e.g., [1]). While the broad pattern of land motion in Norway reflects GIA, a number of other physical processes can also cause vertical movements, for example, tectonics, elastic loading effects, sediment deposition or compaction, or groundwater storage changes. Generally speaking these processes are thought to be small in Norway but they can be significant in some areas and especially on local scales (e.g., [30]).

Here we make use of new GPS observations and GIA modelling work (Kierulf et al. [2]), and updated precise levelling measurements. The new results are employed to determine vertical land motion, with corresponding uncertainties, for the Norwegian coast. We also show gravitational effects on sea level associated with GIA. These changes in the gravity field are largely driven by the movement of mantle material from the forebulge areas peripheral to Fennoscandia back towards the center as the region uplifts. The movement of mantle mass acts to increase gravitational attraction which, in turn, causes sea surface heights to increase. Such gravitational changes in Fennoscandia have been observed using both satellite gravity data from GRACE and ground-based gravity measurements (e.g., [31]). Previous work has shown, based on empirical derivation, that where the vertical rates are around 10 mm/year at the centre of uplift in Fennoscandia the corresponding geoid change is 0.6 mm/year (i.e., the sea surface height change is ~6% of the land uplift signal) [32,33].

Our findings are compared to the GIA solution applied in IPCC AR5, which is based on a combination of the ICE-5G [34] and ANU-ICE models (Lambeck et al. [6] and subsequent improvements).

2.1.1. Permanent GPS and Levelling Networks and Analysis

The establishment of permanent GPS networks in the Nordic countries (Norway, Sweden, Finland, and Denmark) began in the early 1990s. A dense network exists in the region today, which is primarily used for real-time positioning and navigation, but also geodynamic and geophysical studies (Figure 2). Many GIA-related GPS investigations have been completed under the Baseline Inferences for Fennoscandian Rebound Observations, Sea level, and Tectonics (BIFROST) project [35]. Crustal deformation rates from the BIFROST network have been published regularly [1,35–38], largely incorporating Swedish and Finnish stations and also some Norwegian and North European (e.g., Germany, Poland, and Estonia) stations.

The Norwegian GPS network is currently comprised of 160 permanent stations. As the network has been gradually built up over a number of years, some GPS sites have longer time series than others. This has implications for the reliability of the crustal velocities estimated from GPS observations in different parts of the network. In the first analysis of the entire Norwegian network, Kierulf et al. [39] examined the relationship between time series length and the accuracy of the velocity estimates. The authors suggested that only vertical velocities estimated from more than 3 years of data can be considered reliable. Using this 3-year cut-off and data up until the beginning of 2011, Kierulf et al. [39] were able to determine vertical velocities for 65 (~40%) of the 160 GPS stations in the network. In the more recent study of Kierulf et al. [2], where they use data up until the beginning of 2013 (i.e., 2 more years of data), the authors were able to determine vertical velocities estimates for 92 (~60%) of the 160 GPS stations in the network (Figure 2a). Thus, velocities presented in Kierulf et al. [2] and used in this study have a better spatial coverage than before and, owing to longer time series, improved accuracies.

In their analysis Kierulf et al. [2] employ the GAMIT/GLOBK software [40] to derive daily solutions for GPS stations across the Nordic countries. The velocities presented here are realized in the ITRF2008 reference frame [41]. Note that a recent review concluded that the ITRF is stable along each axis to better than 0.5 mm/year and has a scale error of less 0.3 mm/year [42]. For more information on the analysis strategy, see Kierulf et al. [2]. Note that 4 GPS stations, Tregde, Tjøme, Moldjord and Mysen, are removed from further analysis in this work as subsequent inspection of the data suggests they may be unstable.

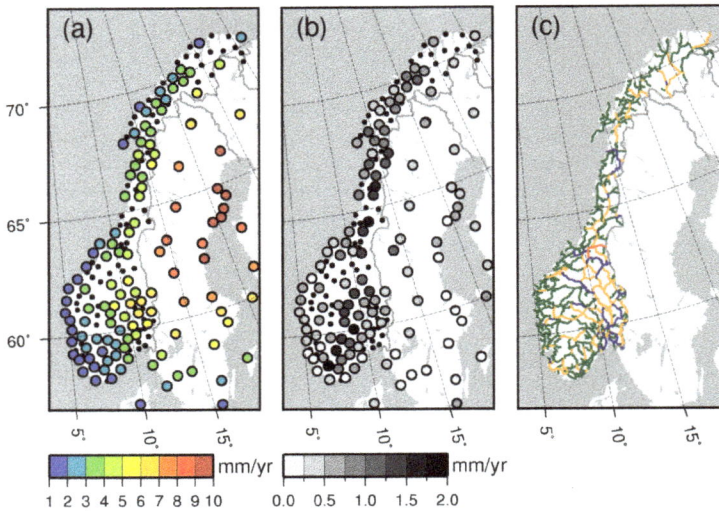

Figure 2. (**a**) Observed vertical land motion from the GPS measurements: black dots mark stations with less than three years of data, which are not included in this study as they are considered unreliable; (**b**) uncertainty on the observations (standard error); and (**c**) the levelling lines used. Red lines have been measured four times, orange lines three times, blue lines two times and green lines once. Levelling data from outside of Norway are included in our least-squares collocation solution (see below) but are not shown here.

In addition to the GPS stations which observe crustal motion in a geodetic reference frame, we also make use of repeated precise levelling data which provide a measure of relative land movements. The levelling data help us to better constrain VLM and are of particular use in areas between the GPS stations. The Norwegian levelling data have been collected over several years and date from 1916 to present. Levelling lines in Norway, including the number of times they have been measured, are shown in Figure 2c.

2.1.2. Defining a Vertical Velocity Field for Norway

The GPS observations indicate that vertical land motion over Norway varies between 1 and 7 mm/year. Coastal locations generally have uplift rates lower than 5 mm/year (Figure 2a). The average minimum distance between the 92 Norwegian GPS stations for which we have velocities is

57 km. Thus, we consider the spatial coverage of the observations to be good, but there are coastal areas in the middle and in northern Norway where we currently lack crustal velocity estimates.

Figure 2b shows the standard errors on the observations. The GPS stations in coastal areas in the middle of Norway have higher uncertainties than elsewhere. These stations have generally shorter time series and consequently larger uncertainties. The average uncertainty on the 92 velocity estimates in Norway is ±0.75 mm/year (standard error). We explore two different approaches to defining a continuous vertical crustal velocity field, with corresponding uncertainties, using (1) results from GIA modelling and (2) applying least-squares collocation to the levelling and GPS observations (see Vestøl [33] for details of the methodology). The GIA model is briefly described below.

2.1.3. Glacial Isostatic Adjustment Modelling

A GIA model is generally composed of three components: a model of grounded past ice evolution (for Fennoscandia and other ice covered areas), a sea level model to compute the redistribution of ocean mass for a given ice and Earth model, and an Earth model to compute the solid Earth deformation associated with the ice-ocean loading history. Note that the sea level model is based on the "sea level equation" [43] and includes subsequent improvements to allow for coastline migration and changes in Earth rotation [44,45]. GIA models allow the calculation of vertical and horizontal land motion, sea level, gravitational, rotation and stress changes associated with a selected ice-Earth model combination. Here we essentially use modelling results from Kierulf et al. [2] and show predictions of vertical land motion, which are compared to the GPS observations, and gravitational effects on sea level associated with GIA.

In their analysis, Kierulf et al. [2] test two different types of GIA models (finite element and normal mode) together with three different global ice models. We do not go into the details of the analysis of Kierulf et al. [2] but instead show results from the GIA model which they determine provides best fit to the GPS observations. Spherical harmonic expansions were truncated at degree and order 192.

The Earth model is one-dimensional and employs the normal mode method [46]. A Maxwell viscoelastic rheology is used and the Earth model is spherically symmetric, self-gravitating and compressible. The elastic and density structure are taken from seismic constraints [47] and depth parameterized with a resolution of 15–25 km. The radial viscosity structure is depth parameterized more crudely into three layers: an elastic lithosphere (i.e., very high viscosity values are assigned), an isoviscous upper mantle bounded by the base of the lithosphere and the 670 km deep seismic discontinuity, and an isoviscous lower mantle continuing below this depth to the core mantle boundary.

The ice model is made of two parts: The Fennoscandian and Barents Sea ice sheets are represented by the model of Lambeck et al. [6], which has been shown to provide good fit to paleo sea level data from the region. For other areas of the globe, they use the ICE-3G ice sheet reconstruction of Tushingham and Peltier [48]. This is the same model setup as used in former BIFROST studies [1,28,37].

Past GIA modelling studies have used both paleo sea level data (e.g., [6,29]) and/or GPS observations (e.g., [1,28,49]) to help constrain Earth model parameters. These investigations have shown that it is not yet possible to uniquely constrain Earth's viscosity structure for the Fennoscandian region. Such studies, however, are able to provide a range of Earth parameter values that satisfy the various GIA observables. Based on values from former GIA studies, Kierulf et al. [2] examine laterally homogeneous Earth models bracketing 60–160 km for lithospheric thickness, $(0.1–40) \times 10^{20}$ Pa s for upper mantle viscosity and $(0.1–10) \times 10^{22}$ Pa s for lower mantle viscosity.

2.2. Computing Regional Sea-Level Projections

Observations show that past sea-level changes have been spatially variable, so we expect that future changes will also be of this nature [50]. Regional sea level can be substantially different from global mean changes owing to spatial variations in: (1) ocean density, ocean mass redistribution, and dynamics; (2) ocean mass changes and associated gravitational effects on sea level; and (3) vertical land motion and associated gravitational effects on sea level. We show *regional* sea-level projections for Norway's coastal municipalities using findings largely from IPCC AR5 [11] and CMIP5 model output [24]. Sea-level projections are given for the emissions scenarios RCP2.6, RCP4.5 and RCP8.5.

The difference between our results and those shown in IPCC AR5 are that we adopt a new VLM field with corresponding gravity changes (Section 3.1) and include an estimate of sea-level changes caused by the crustal displacements and gravitational effects of ocean mass redistribution (so-called self-attraction and loading). Here we make use of projections from the earth system model NorESM, where the self-attraction and loading contribution along the Norwegian coast was found to range from 1–2 cm for RCP2.6–8.5 over the 21st century [51].

Unlike in IPCC AR5, the contribution from changes in atmospheric pressure is not considered separately here. This contribution, known as the inverse barometer effect, is instead combined with the projected ocean density, ocean mass redistribution and circulation fields from the coupled Atmosphere-Ocean General Circulation Models (AOGCMs). The inverse barometer effect is relatively small. It is projected to be positive in the Arctic regions, up to 1.5 cm for RCP4.5 and 2.5 cm for RCP8.5 [52], but likely less than this along most of the Norwegian coast.

For the sea surface height CMIP5 model data the uncertainties are computed from the multi model ensemble spread (this is the *steric/dyn* signal given below). The ice sheet, glacier and land water storage regional uncertainties are calculated by multiplying their global uncertainties by the respective normalized "sea level equation" patterns [43–45]. The methods used to compute uncertainties associated with GIA (*obsvlm* and *gravgia*) are given in Sections 2.1 and 3.1. Gravitational effects on sea level due to ocean mass redistribution (*sal*) are described above (see also Richter et al. [51]). To combine the uncertainties, contributions that correlate with global warming have correlated uncertainties, so they are added linearly. Other contributions are assumed to be uncorrelated and are thus added in quadrature. The regional uncertainty is found by (adapted from Church et al. [11]):

$$\sigma_{total}^2 = \left(\sigma_{\frac{steric}{dyn}} + \sigma_{antsmb} + \sigma_{greensmb} \right)^2 + \sigma_{glaciers}^2 + \sigma_{obsvlm}^2 + \sigma_{gravgia}^2 + \sigma_{grdwater}^2$$
$$+\sigma_{antdyn}^2 + \sigma_{greendyn}^2 + \sigma_{sal}^2 \tag{1}$$

steric/dyn = Global thermal expansion uncertainty and uncertainty associated with changes in ocean density, mass redistribution and circulation changes from the AOGCMs (includes the inverse barometer effect uncertainty).

antsmb = Antarctic ice sheet surface mass balance uncertainty.

greensmb = Greenland ice sheet surface mass balance uncertainty.

glaciers = glacier uncertainty.

obsvlm = observed vertical land motion uncertainty (here includes reference frame errors).

gravgia = modelled gravity changes owing to GIA uncertainty.

grdwater = land water storage uncertainty.

antdyn = Antarctic ice sheet rapid dynamics uncertainty.

greendyn = Greenland ice sheet rapid dynamics uncertainty.

sal = self-attraction and loading uncertainty (gravitational effects of ocean mass redistribution).

Note that for each contribution σ is the standard error except for *grdwater*, *antdyn* and *greendyn*. These have uniform probability distributions in the global projections and, therefore, the half-range of their distributions was used as σ.

2.3. Methodology for Calculating Return Heights for Extreme Sea Levels

In the above we have focussed on mean sea level. For practical purposes, however, it is often shorter-term sea-level changes that are of interest. Tides occur on a daily basis but their heights vary on a range of timescales (the different tidal constituents) and also along the coast. When quantifying coastal impacts and risk, the even higher extreme sea levels associated with storm surges need to be considered.

A return height for extreme sea levels is a height the sea level exceeds on average only once during a given return period (e.g., the 200-year return height). Return heights are typically calculated using statistical analysis of tide gauge data. Here we use the average conditional exceedance rate method for estimating the return heights [25]. Comparisons of alternative methodologies have found that the ACER method is suitable for estimating extreme sea level return periods for Norway and, furthermore, that it provides certain advantages compared to the Gumbel method and other classical methods [53,54]. Below we give an overview of the method, but for a detailed description the reader is referred to Næss and Gaidai [55] and Skjong et al. [54].

The difference between the ACER method and many of the classical methods is that the ACER method focuses on the exceedance rates of the extreme sea level instead of only the yearly maximum values. This makes the method less sensitive to gaps of missing data and outliers than some other methods. The ACER method also takes into account the dependency of the data, that is, that several subsequent peaks exceeding some level might be from the same extreme event. Working with extreme sea levels, this means that two subsequent high waters exceeding the given level are likely to be caused by the same storm surge as the storm surge often persists over more than one tidal cycle. Thus, we calculate the rate of which peaks exceed a given level given the condition that the previous peaks did not exceed; hence, we get a conditional exceeding rate.

More precisely, one defines the ACER function $\epsilon_k(\eta)$ as the rate at which the sea level crosses the threshold η given $k - 1$ previous non-exceedances. The ACER method attempts to capture the sub-asymptotic behaviour of the data by assuming that sub-asymptotically this function is given by:

$$\epsilon_k(\eta) = q_k(\eta)exp(-a_k(\eta - b_k)^{c_k}), \ \eta \geq \eta_1 \tag{2}$$

for a given level η_1 called the tail marker. In practice, the function $q_k(\eta)$ is varying slowly compared to the exponential function when η is large, and it is therefore replaced by a constant value q_k.

We assume that one year of sea level data is one realization of the process. The conditional exceedance rate is calculated for discrete levels of η and the average is taken as the estimate. Since the conditional upcrossings are assumed to be independent for a high enough value of k, the ACER function is calculated for different values of k and the value of k for which the process starts to converge is chosen. For sea level data this happens for $k = 3$. A curve with the form given by Equation (2) is then fitted to the estimated ACER functions, determining the parameters q_k, b_k, a_k and c_k. The confidence intervals are found in a similar way, by fitting curves to the lower and upper confidence bounds of the estimate of the ACER function. The different return heights can now be estimated by extrapolating the fitted curve to high values of η. The return height z_m where m is the return period is given by the following formula, where N is the average number of peaks in the data during one year:

$$z_m = b_k + \left[\frac{1}{a_k} \left(\ln(q_k N) - \ln \left[-\ln \left(1 - \frac{1}{m} \right) \right] \right) \right]^{\frac{1}{c_k}} \tag{3}$$

Return heights are available for all 23 Norwegian tide gauges except Mausund, which has not been fully integrated in the tide gauge network and has a shorter time series than the other gauges. Return heights given in this paper are calculated using data up to December 2014 and are currently the official levels for use in coastal planning [25]. Note that return heights are normally updated every 5 years; this ensures that changes in: (1) the frequency and height of extreme sea levels; and (2) the mean sea level are taken into account. The 20-, 200- and 1000-year return heights are given in Table 1 together with the first year of data used for each tide gauge. Figure 3 shows all return periods with confidence intervals for the six key locations.

Table 1. Return heights for extreme sea levels calculated for the Norwegian tide gauges, given in meters above mean sea level (1996–2014). The 5% and 95% confidence levels are given in parentheses.

Tide Gauge	Start	20-Year	200-Year	1000-Year
Vardø	1947	2.19 (2.12, 2.26)	2.37 (2.28, 2.46)	2.48 (2.37, 2.58)
Honningsvåg	1970	2.01 (1.90, 2.08)	2.21 (2.05, 2.30)	2.33 (2.14, 2.44)
Hammerfest	1957	2.01 (1.91, 2.07)	2.19 (2.03, 2.27)	2.29 (2.11, 2.39)
Tromsø	1952	2.03 (1.97, 2.07)	2.21 (2.13, 2.26)	2.32 (2.22, 2.37)
Harstad	1952	1.75 (1.68, 1.79)	1.92 (1.82, 1.98)	2.03 (1.91, 2.09)
Andenes	1991	1.84 (1.70, 1.93)	2.08 (1.88, 2.20)	2.23 (1.99, 2.38)
Kabelvåg	1988	2.45 (2.30, 2.54)	2.71 (2.49, 2.82)	2.87 (2.60, 3.00)
Narvik	1931	2.59 (2.46, 2.65)	2.85 (2.65, 2.94)	3.02 (2.77, 3.11)
Bodø	1949	2.25 (2.16, 2.31)	2.47 (2.35, 2.55)	2.61 (2.47, 2.69)
Rørvik	1969	2.08 (1.95, 2.14)	2.30 (2.12, 2.38)	2.43 (2.22, 2.53)
Trondheim	1989	2.21 (2.10, 2.27)	2.38 (2.24, 2.45)	2.49 (2.33, 2.57)
Heimsjø	1928	1.94 (1.87, 1.99)	2.10 (2.00, 2.17)	2.20 (2.08, 2.28)
Kristiansund	1952	1.80 (1.72, 1.84)	1.96 (1.85, 2.02)	2.06 (1.93, 2.13)
Ålesund	1961	1.70 (1.60, 1.76)	1.88 (1.73, 1.95)	1.98 (1.80, 2.07)
Måløy	1943	1.53 (1.48, 1.57)	1.66 (1.60, 1.71)	1.74 (1.66, 1.80)
Bergen	1915	1.29 (1.25, 1.32)	1.41 (1.35, 1.46)	1.48 (1.41, 1.54)
Stavanger	1919	1.01 (0.95, 1.04)	1.15 (1.06, 1.19)	1.23 (1.13, 1.29)
Tregde	1927	0.95 (0.89, 1.00)	1.12 (1.01, 1.19)	1.23 (1.09, 1.32)
Helgeroa	1965	1.26 (1.12, 1.34)	1.51 (1.29, 1.62)	1.67 (1.39, 1.81)
Oslo	1914	1.53 (1.39, 1.62)	1.86 (1.62, 1.99)	2.09 (1.77, 2.25)
Oscarsborg	1953	1.42 (1.29, 1.50)	1.67 (1.49, 1.76)	1.83 (1.61, 1.93)
Viker	1990	1.39 (1.18, 1.52)	1.66 (1.35, 1.84)	1.83 (1.46, 2.05)

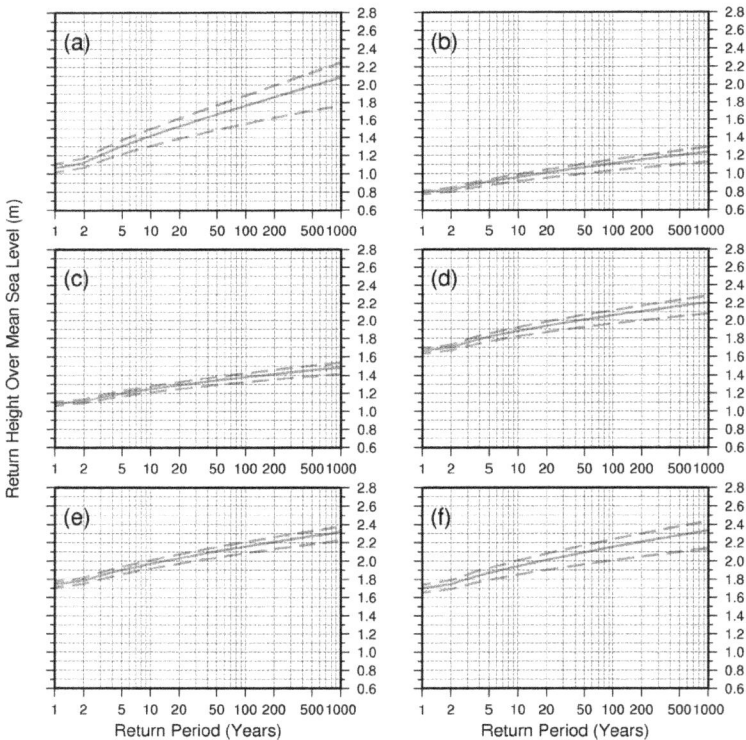

Figure 3. Return heights for extreme sea levels for the six key locations: (**a**) Oslo; (**b**) Stavanger; (**c**) Bergen; (**d**) Heimsjø; (**e**) Tromsø; and (**f**) Honningsvåg. Dashed lines are the 5% and 95% confidence intervals.

Since the amplitude and time of the tide vary along the Norwegian coast, the return heights calculated for a particular tide gauge cannot simply be applied to another point of the coastline. Hence, we need to be able to extrapolate these extreme sea levels along the coast. To assist this extrapolation procedure, there are several hundred shorter data series available from temporary tide gauges, dating from the beginning of the 20th century to present-day. These data series are analysed so that the relationship between the tidal behaviour in the area of a temporary tide gauge and that of a permanent tide gauge can be quantified. Following this procedure, the Norwegian coastline has been divided into zones of similar tidal properties. When extrapolating the water level to a point away from the permanent tide gauges, therefore, the astronomical tide is first determined using the tidal zones as described above and then added to the meteorological effect as seen at the closest permanent tide gauge(s). This approach assumes that the meteorological conditions affecting sea level are large-scale phenomena that have a similar effect on sea level over a large area and vary smoothly. The return heights are then estimated using the corrected water level series, in the same way as for the permanent tide gauges. Figure 4 shows water levels extrapolated to the 290 coastal municipalities in Norway.

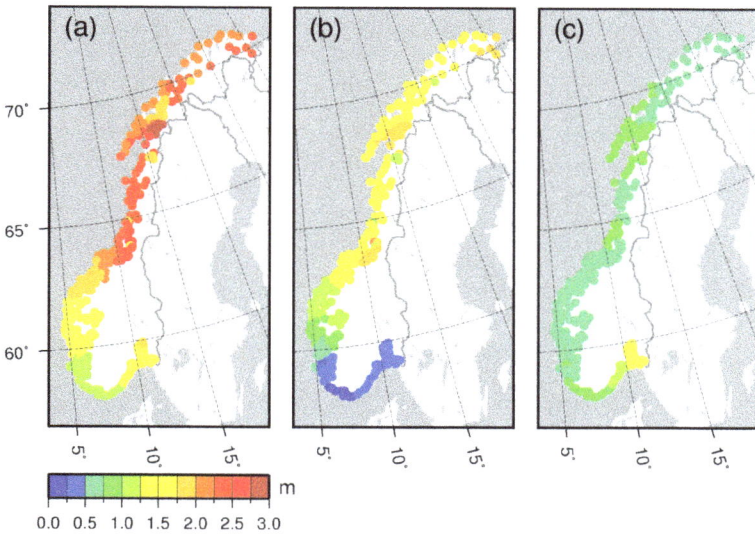

Figure 4. Water levels extrapolated to the 290 coastal municipalities in Norway: (**a**) the 200-year return height in meters above mean sea level; (**b**) highest astronomical tide (HAT) in meters above mean sea level; and (**c**) the 200-year return height in meters above HAT in order to compare the significance of the extreme levels.

2.4. Combining Storm Surge Statistics with Sea-Level Projections

Sea-level projections can be combined with the storm surge statistics to provide "allowances" (see, e.g., Hunter [16] and Hunter et al. [17]). Here, we adapt this approach for use with the ACER method for calculating the return heights. We define the expected number of exceedances of a level z over a given time period to be given by:

$$N_{AC} = Nq \, \exp\left(-a(z-b)^c\right) \tag{4}$$

where N, q, a, b and c are the ACER parameters for a given k as described above. We now assume that mean sea level is raised by $\Delta z + z'$, where Δz is the mean value of the sea-level rise and z' is a random variable with a probability distribution given by $P(z')$ and a zero mean. As our level z has now been reduced by $\Delta z + z'$, we try to find how much we have to raise this level with in order to keep the same exceedance rate as z had before this rise of the mean sea level. This amount will be the allowance A. We denote the overall number of excedances of the level $z - \Delta z - z' + A$ as $N_{ov,AC}$, that is,

$$N_{ov,AC} = \int_{-\infty}^{\infty} P(z') Nq \, \exp\left(-a(z - \Delta z - z' + A - b)^c\right)dz' \tag{5}$$

The objective is to calculate the allowance A so that $N_{ov,AC} = N_{AC}$, that is, the expected number of exceedances before and after the sea level rise remains constant. By manipulating Equation (5) we obtain

$$N_{ov,AC} = Nq \, \exp\left(-a(z - b)^c\right) \int_{-\infty}^{\infty} P(z') \, \exp\left(a(z - b)^c\right) \, \exp\left(-a(z - \Delta z - z' + A - b)^c\right)dz' \tag{6}$$

Thus, we have $N_{ov,AC} = N_{AC}$ if

$$\int_{-\infty}^{\infty} P(z') \, \exp\left(a(z - b)^c\right) \, \exp\left(-a(z - \Delta z - z' + A - b)^c\right)dz' = 1 \tag{7}$$

It is not possible to eliminate the return height z from Equation (7). Thus, by combing the approach of Hunter [16] with the ACER method, instead of the Gumbel method, it is no longer the case that the allowance remains the same for any return height z given.

In a preliminary test of the method and its application to Norway, we examine the case of a future sea level rise assuming a normal uncertainty distribution with zero mean and a standard error σ such that

$$P(z') = \frac{1}{\sigma\sqrt{2\pi}} \exp\left(-\frac{z'^2}{2\sigma^2}\right) \tag{8}$$

3. Results

3.1. Present-Day Vertical Land Motion in Norway

3.1.1. Results from GIA Modelling

The goodness-of-fit between the GPS observations and modelled vertical crustal velocities is tested for changes in Earth model parameters. In total 1089 Earth models are tested (model output from Kierulf et al. [2]) and the best fitting Earth model is determined to have 140 km lithospheric thickness, 7×10^{20} Pa s upper mantle viscosity and 4×10^{21} Pa s lower mantle viscosity. Predicted vertical velocities generated using the best-fit GIA model (Figure 5) show a familiar pattern of land motion (e.g., [1]). All of mainland Norway is predicted to be uplifting; rates along the Norwegian coast vary between 1 and 5 mm/year. Visual inspection of Figure 5 shows that the pattern of modelled uplift is, broadly speaking, in good agreement with the observations. At 44 of the 92 Norwegian GPS stations, the model fits to within one-sigma of the observations.

Selecting a subset of 10 stations where we have long time series and are confident in the velocity estimates, we find the RMS error between the model and GPS measurements is 0.9 mm/year (this test is useful for a comparison later on). Residuals between the best-fit GIA model and GPS data show that the model tends to overpredict rates of uplift in the middle of Norway, around $64°$ N (Figure 5b).

In order to incorporate the GIA results within our regional sea-level projections it is necessary to define a mean GIA field along with lower and upper 90% uncertainty bounds. For modelled GIA we opt to define the mean and uncertainty bounds as follows. Of the 1089 Earth models tested in the analysis of Kierulf et al. [2] a subset of 61 are identified as having comparably good fit to the observations and classified as the best-fit models (at the one standard deviation confidence level). Using this subset of 61 models, we calculate the mean and 90% uncertainty bounds from the spread in the vertical crustal velocity field predictions. In this manner, we obtain a model uncertainty which is tightly constrained to the GPS observations. This approach is preferable to simply using the full range of 1089 Earth models tested as the vertical velocity predictions are highly sensitive to the assumed Earth structure. The average GIA model uncertainty for the coastal municipalities is ±0.2 mm/year (standard error). It is important to note that the GIA uncertainty given here only accounts for changes in Earth model parameters (i.e., it ignores possible errors in the assumed ice loading history).

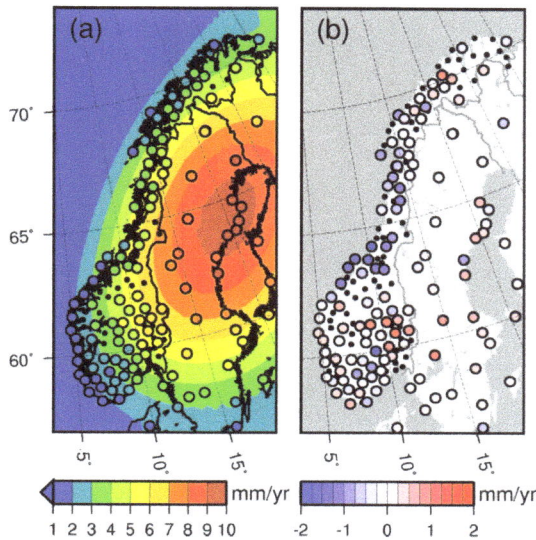

Figure 5. (**a**) Modelled vertical land motion from the best-fit GIA model. The vertical velocities from the GPS observations are shown as circles. Black dots mark stations with less than three years of data, these observations are not included in this study as they are considered unreliable. (**b**) Residuals (observed minus modelled).

As mentioned, we also take into account gravitational effects on sea level associated with GIA. Ocean surface height changes are typically between 5% and 10% of the vertical land motion signal [56] so this is a relatively small effect. We calculate the mean field and 90% uncertainty bounds for ocean surface height changes associated with GIA. This is done with same subset of 61 Earth models used to calculate the mean and 90% uncertainty for the vertical land motion rates as above. For the mean field, modelled ocean surface changes associated with GIA vary between 0.2 and 0.5 mm/year along the Norwegian coast, the average model uncertainty is ±0.03 mm/year (standard error).

3.1.2. Results from Least-Squares Collocation

In our second approach, we base our calculation purely on geodetic observations and use a method of least-squares collocation to combine the levelling and GPS observations and determine VLM in Norway (e.g., [33]). Our mean VLM solution is shown in Figure 6a and residuals with the GPS data shown in Figure 6b. The pattern of uplift based purely on the combined observations is broadly similar to our GIA model. Note that, for the least-squares collocation method, the input data from levelling and GPS act to mutually control each other, which increases the reliability of the VLM solution. In addition, the method filters the observations making the solution more robust to errors and outliers (see [33]).

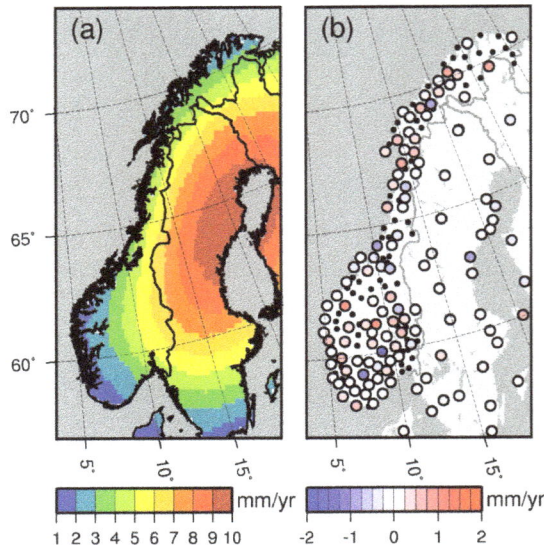

Figure 6. Vertical land motion determined from least-squares collocation of the levelling and GPS observations for: (**a**) our mean solution; and (**b**) the residuals (GPS observations—mean solution). Black dots mark stations with less than three years of data, these observations are not included in this study as they are considered unreliable.

In a test of the least-squares collocation method, we recalculate the VLM solution but omit a subset of 10 GPS stations where we have long time series and are confident in the velocity estimates (the same test as performed for the GIA model and using the same 10 GPS stations). For these 10 Norwegian stations, the RMS error between the recalculated VLM solution at the GPS observations is 0.3 mm/year. This is an encouraging result and gives us confidence in the solution where observations are few or lacking. From our least-squares collocation solution we calculate the mean field and 90% uncertainty bounds for the Norwegian coastal municipalities, the average uncertainty for these locations is ±0.2 mm/year (standard error).

3.1.3. Comparison of GIA Model and Least-Squares Collocation Approach

We have shown two different approaches to predicting VLM for the coastal municipalities: (1) a GIA model that is calibrated to the GPS observations; and (2) compute land motion by combining the levelling and GPS observations through least-squares collocation. It is of clear interest to determine which method best describes land motion for coastal Norway and, therefore, which is preferable when calculating future sea-level changes. In a simple test, we calculated the RMS error between the two different solutions and 10 Norwegian GPS stations where we have long time series and are confident in the velocity estimates (see also above). The RMS errors for the GIA model and least-squares collocation solutions are 0.9 mm/year and 0.3 mm/year, respectively. This comparison therefore indicates that the least-squares collocation approach performs best, i.e., the method has a better accuracy in predicting VLM in areas where we do not have observations.

For the 290 coastal municipalities we find that the average uncertainties on the GIA model and least-squares collocation solutions are the same (both ±0.2 mm/year). As a consequence of the method and additional constraint of the levelling data, the average uncertainty on the least-squares collocation solution is somewhat less than that we could expect to obtain from the GPS stations alone (the average uncertainty on the 92 GPS velocity estimates in Norway is ±0.75 mm/year). For the GIA model, the uncertainty is tightly constrained by the GPS observations and, as discussed, only accounts for changes in Earth model parameters. Better quantifying the GIA uncertainty is a challenging task but more rigorous methods are forthcoming (e.g., [57]). Finally, it is important to recognise that neither of these uncertainty estimates account for possible systematic errors in the reference frame. We address this point below.

We opt to use the least-squares collocation solution to define our vertical velocity field for use in the sea-level projections. There are a few caveats with the approach that should be highlighted. Firstly, by using the observed vertical land motion in our sea-level projections, we assume that the observed rates will persist unchanged over the 21st century. We argue that this is a reasonable assumption as GIA dominates present-day vertical land motion in Norway (e.g., [1,2,28,39,49]) and that the viscoelastic response time of the Earth is so long we would not expect any significant changes in the uplift rates over next ~100 years. Furthermore, the generally good fit between GIA model and GPS observations shows we have a good understanding of physical process causing uplift and gives us confidence that the observations can be extrapolated in this way.

As shown in Figure 5, however, there are some significant misfits between the GIA model and observations. (We also expect similar differences between the GIA model and least-squares collocation solution but this is not specifically examined here). Such residuals may be explained by: (1) errors in the GIA model, for example, that the ice loading is incorrect or that lateral variations in the Earth model should considered; (2) other geophysical processes contributing to vertical land motion, for example, recent work by Olesen et al. [30] identified neotectonic deformations of around 1 mm/year in the north of Norway; and (3) errors in the GPS and/or levelling observations. As noted above, four GPS stations have been removed from our final results as the data suggests they may be unstable. One advantage of using the least-squares collocation approach means that, to some extent, our solution is less sensitive to outliers. These issues should be kept in mind when interpreting the results.

3.1.4. Our GIA Solution and Comparison to IPCC AR5

Our full GIA solution, as a contribution to RSL change, is the VLM field determined from least-squares collocation added to the modelled gravitational effects on sea level associated with GIA. When computing uncertainties on our GIA solution, note that we include systematic errors on the reference frame's z-drift (0.5 mm/year) and scale error (0.3 mm/year). Taken together, the average uncertainty on our GIA solution is 0.6 mm/year.

The mean GIA field in IPCC AR5 is evaluated as the mean of the ICE-5G model [34] and the ANU-ICE model (Lambeck et al. [6] and subsequent improvements). These GIA models are based upon global ice sheet reconstructions where the loading history of the ice sheets is essentially inferred from paleo sea level observations. The one standard error of the GIA field is taken as the difference between the separate models. This approach is adequate when considering regional patterns of sea-level change on a global scale, as in IPCC AR5, but here we are focussed on changes for Norway and at a local level. It is also of interest to compare our GIA solution (based on geodetic observations) to the modelled solution applied in IPCC AR5.

In Figure 7, we show the future RSL contribution from our GIA solution and that used in IPCC AR5 at our six key locations. At most locations the solutions agree to within their respective uncertainties. Our GIA solution has an almost uniform uncertainty, but the IPCC AR5 uncertainty is small in some locations (e.g., Oslo) but much larger in others (e.g., Tromsø). This simply reflects differences between the ICE-5G and ANU models. We find that the mean IPCC AR5 GIA solution lies within our 90% uncertainty bounds for 270 of the 290 coastal municipalities (figure not shown). Our mean GIA solution gives on average 4 cm higher values than that applied in IPCC AR5; i.e., our mean GIA solution indicates on average a smaller RSL fall. We have more confidence in our GIA solution, the values and uncertainties being essentially based upon the GPS and levelling observations.

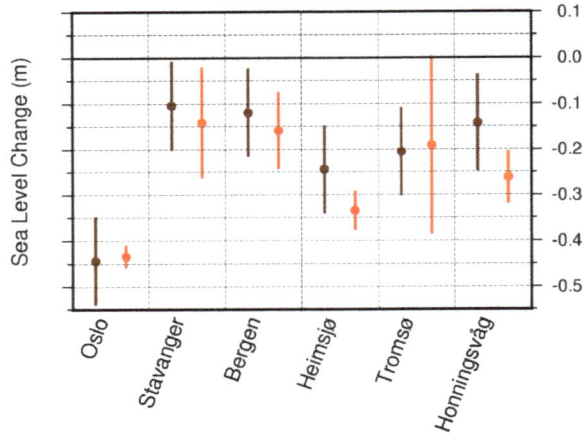

Figure 7. Regional relative sea-level change (m) due to GIA and associated gravitational effects over the period 1986–2005 to 2081–2100 for our GIA solution (brown) and IPCC AR5 (red). Dots show mean values and vertical bars the 5% to 95% uncertainty.

3.2. Regional Sea-Level Projections

Here we present regional sea-level projections for Norway using model output from the CMIP5 and for the emission scenarios RCP2.6, RCP4.5 and RCP8.5. The projections are given with corresponding 5% to 95% ensemble spread, these ranges are defined as the *likely* ranges in IPCC AR5 ($p > 66\%$).

3.2.1. Contributions to 21st Century Sea-Level Changes for Norway

The separate contributions to projected sea level are shown in Figure 8 for RCP4.5. This shows that the largest projected contributions and also largest uncertainties are from rapid ice dynamics in Antarctica, the steric/dyn signal and observed vertical land motion. The observed vertical land motion signal shows the largest variation between the six tide gauge sites examined, indicating that this signal will dominate the pattern of 21st century sea-level changes along the Norwegian coast. The steric/dyn and glacier projections contribute to the spatial variability to a lesser extent.

Changes in the distribution of mass on the Earth's surface produces a non-uniform sea level pattern due to gravitational changes [43]. This sea level response is often referred to as a "fingerprint" as it can be used to identify the source and size of ice mass variations. As Norway sits in the near field of Greenland, the projections indicate a small or even negative sea-level change along the coast. That is, owing to gravitational changes, Norway is predicted to experience a sea-level change between −40% and 10% of the global average sea-level rise due to Greenland surface mass balance loss. On the other hand, ice mass losses in Antarctica are predicted to produce an above average sea-level change for Norway.

Modelling results indicate an above average steric/dyn sea-level rise for Norway over the 21st century but with relatively large uncertainties attached to the projections (Figure 8). We note the results from the CMIP5 ensemble indicate that areas that are projected to have above average sea-level change coincide with areas that have larger uncertainties [58]. That appears to be the case for Norway. Past modelling studies have focused on identifying the contributing factors to regional differences in projected ocean density and circulation changes (e.g., [53,59–61]). These generally show that in the nearby North Atlantic positive thermosteric changes are partially compensated by a negative halosteric signal, whereas, in the Arctic Ocean, the halosteric term is positive and dominates due to ocean freshening. Related to these steric changes is a mass redistribution term that could be important for the shallow shelf seas around Norway [51].

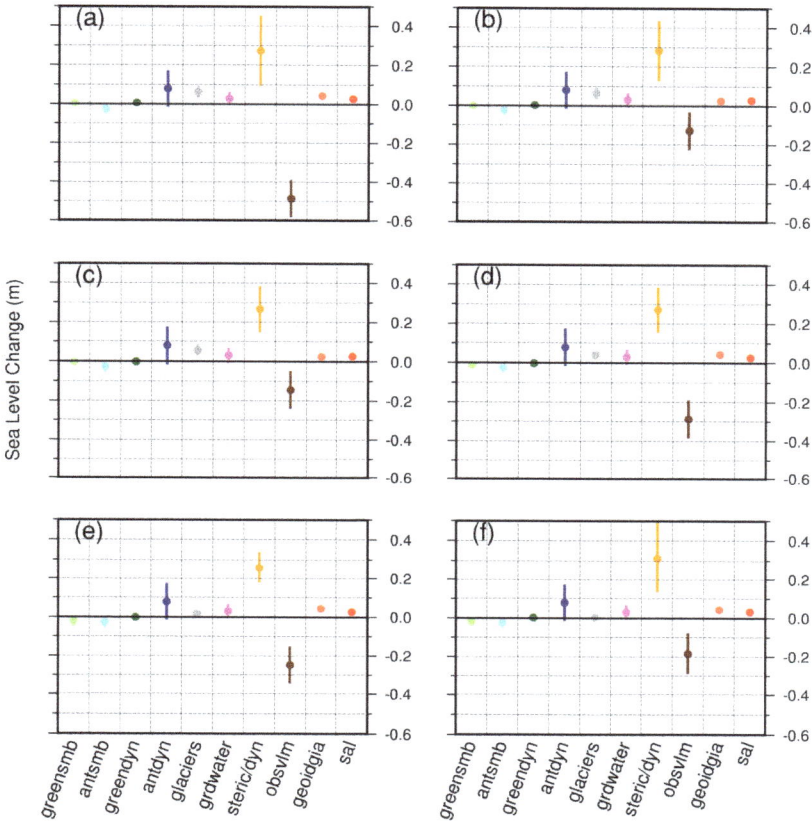

Figure 8. Contributions to projected relative sea-level change for RCP4.5 over the period 1986–2005 to 2081–2100 for the six key locations: (**a**) Oslo; (**b**) Stavanger; (**c**) Bergen; (**d**) Heimsjø; (**e**) Tromsø; and (**f**) Honningsvåg. The ensemble mean and spread (5% to 95%) are shown by the circles and vertical bars, respectively.

3.2.2. Projected 21st Century Sea-Level Changes for Norway

Here we present our regional relative sea-level projections (Figure 9 and Table 2) which take account of changes to the contributions described in Sections 2.2 and 3.2.1 The projections show, unsurprisingly, that the pattern of twenty-first century relative sea-level changes for Norway is governed by GIA. For all RCPs, projected ensemble mean changes indicate that the majority of Norway will experience a RSL rise over the period 1986–2005 to 2081–2100. Thus, climate driven sea-level rise will dominate over VLM over the next 100 years.

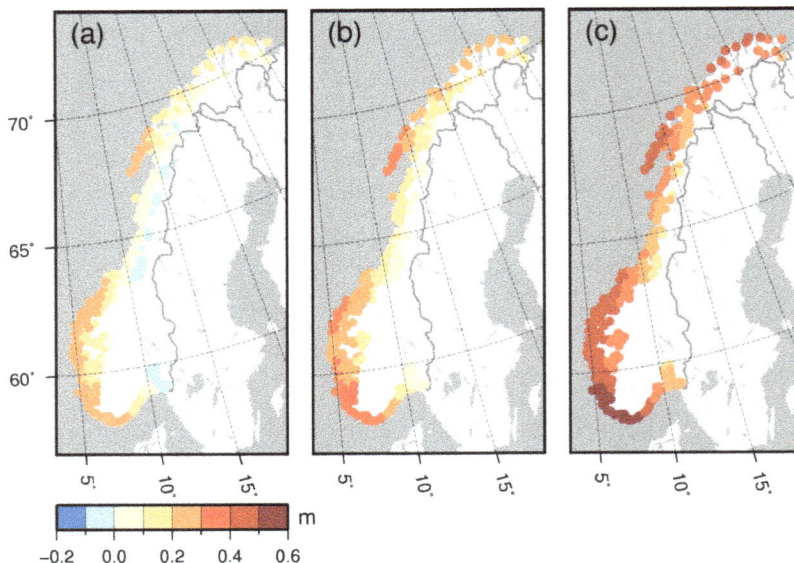

Figure 9. Projected ensemble mean relative sea-level change (m) for the 290 coastal municipalities in Norway over the period 1986–2005 to 2081–2100 for: (**a**) RCP2.6; (**b**) RCP4.5; and (**c**) RCP8.5.

Table 2. Comparison of mean relative sea-level projections at six key locations for the period 1986–2005 to 2081–2100 and the period 1986–2005 to 2100. The 5% to 95% ensemble spread is given in the parentheses. Units are in centimetres.

Location	RCP2.6 2081–2100	RCP2.6 2100	RCP4.5 2081–2100	RCP4.5 2100	RCP8.5 2081–2100	RCP8.5 2100
Oslo	−7 (−32 to 16)	−8 (−36 to 19)	0 (−25 to 24)	0 (−29 to 28)	18 (−12 to 47)	23 (−11 to 56)
Stavanger	28 (5 to 50)	30 (4 to 54)	35 (12 to 58)	38 (12 to 63)	52 (25 to 79)	59 (28 to 90)
Bergen	23 (3 to 42)	23 (2 to 45)	31 (10 to 51)	33 (11 to 55)	48 (23 to 72)	53 (26 to 80)
Heimsjø	7 (−17 to 27)	7 (−14 to 28)	16 (−5 to 37)	17 (−6 to 40)	30 (8 to 57)	36 (8 to 65)
Tromsø	6 (−10 to 23)	8 (−11 to 27)	15 (−3 to 34)	15 (−5 to 35)	29 (5 to 55)	32 (3 to 63)
Honningsvåg	18 (−7 to 44)	20 (−8 to 48)	27 (0 to 53)	29 (−2 to 59)	43 (10 to 76)	48 (11 to 86)

Projected ensemble mean RSL changes along the Norwegian coast over the period 1986–2005 to 2081–2100 are, depending on location, for RCP2.6 between −0.10 and 0.30 m; for RCP4.5 between 0.00 and 0.35 m; and for RCP 8.5 between 0.10 and 0.55 m. Expressed as a percentage of global mean change, mean regional RSL changes for Norway are for RCP2.6 from −30% to 70% (global mean change of 0.40 m); for RCP4.5 from −10% to 75% (global mean change of 0.47 m); and for RCP8.5 from 20% to 85% (global mean change of 0.63 m). For all RCPs the mean regional RSL change is therefore projected to be below the global mean.

Over the period 1986–2005 to 2081–2100 the ensemble mean RSL change averaged over the coastal municipalities is for RCP2.6 0.1 m (90% uncertainty bounds are −0.10 to 0.35 m), for RCP4.5 0.2 m (−0.05 to 0.45 m) and for RCP8.5 0.35 m (0.10 to 0.65 m). If we ignore the effects of GIA, i.e., we look at the projected sea surface height change, these numbers are for RCP2.6 0.35 m (0.15 to 0.55 m), for RCP4.5 0.4 m (0.20 to 0.65 m) and for RCP8.5 0.6 m (0.30 to 0.85). For the different RCPs the projected sea surface height changes are between 80% and 90% of the global mean change. This is because the projected above average input from steric/dyn sea-level changes and the rapid ice dynamic contribution from Antarctica are more than compensated for by the below average contributions from Greenland and glaciers.

Examining the projected RSL time series (Figure 10) shows that there are only small differences between the RCPs up until 2050. Going towards 2100 the separate projections from the RCPs begin to diverge but there are still large overlaps between their respective uncertainties. In fact, inspection of the vertical bars in Figure 10 shows that differences between the ensemble means for the different RCPs are somewhat smaller than the projections ensemble spread (5% to 95%).

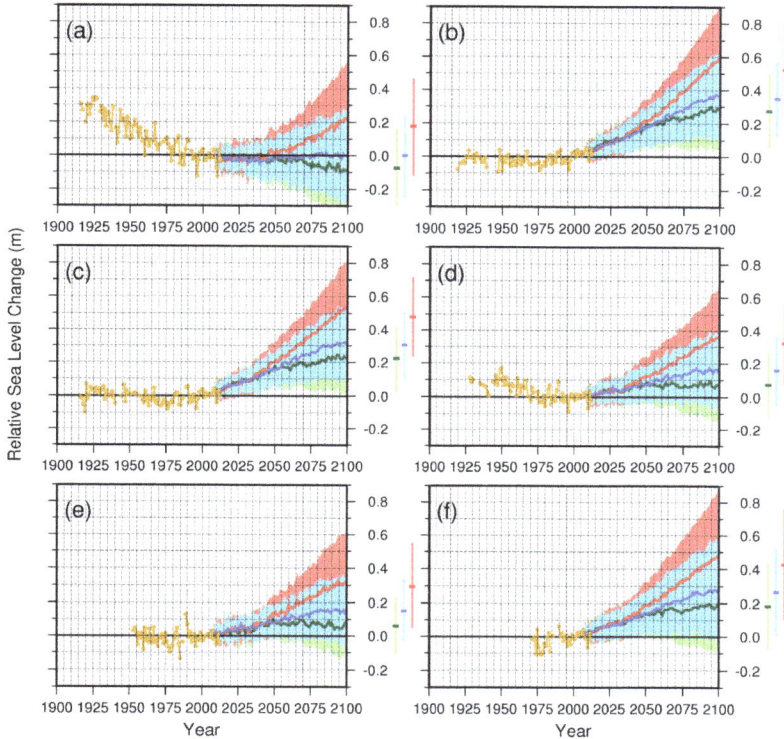

Figure 10. Relative sea-level projections for RCP2.6 (green), RCP4.5 (blue) and RCP8.5 (red) for the six key locations: (**a**) Oslo; (**b**) Stavanger; (**c**) Bergen; (**d**) Heimsjø; (**e**) Tromsø; and (**f**) Honningsvåg. The vertical bars on the right side of the panels represent the ensemble mean and ensemble spread (5% to 95%) for RSL change for 2081–2100. Annual mean tide gauge observations are shown in yellow.

The reasons for this become apparent when we examine the contributions from rapid ice dynamics in Antarctica, the steric/dyn signal and vertical land motion (Figure 8). These are the largest contributions to projected RSL change in Norway and, in general, the contributions with the largest uncertainties. Firstly, we note the rapid ice dynamic contribution from Antarctica is the same for all RCPs (i.e., it is RCP independent). IPCC AR5 assess that the scientific community is currently unable to quantify how Antarctic rapid ice dynamics relate to emission scenario, but such a dependency is expected to exist (see also Discussion). Secondly, we see that there is considerable overlap between the steric/dyn uncertainties for the different RCPs owing to the relatively large ensemble spread. For the period 1986–2005 to 2081–2100 the ensemble spread (5% to 95%) for the steric/dyn contribution is projected to be 0.05 to 0.35 m for RCP2.6; 0.13 to 0.43 m for RCP4.5; and 0.25 to 0.63 for RCP8.5.

These values are for the average projected change across the Norwegian coastal municipalities. Thirdly, the contribution from vertical land motion is based upon the extrapolation of observed rates from the permanent GPS stations and levelling lines. This contribution is, therefore, clearly not related to present-day climate change and is also RCP independent.

Figure 11 shows the projected rates of sea-level change, these show considerable variability owing to interannual to decadal variability in the steric/dyn component (due to changes in heat uptake and wind stress simulated by the AOGCMs). For 2081–2100, we find that projected mean rates for RCP2.6 and RCP4.5 are broadly similar, whereas, for RCP8.5, ensemble mean rates are higher for the same period and the ensemble spread shows that rates may exceed 10 mm/year.

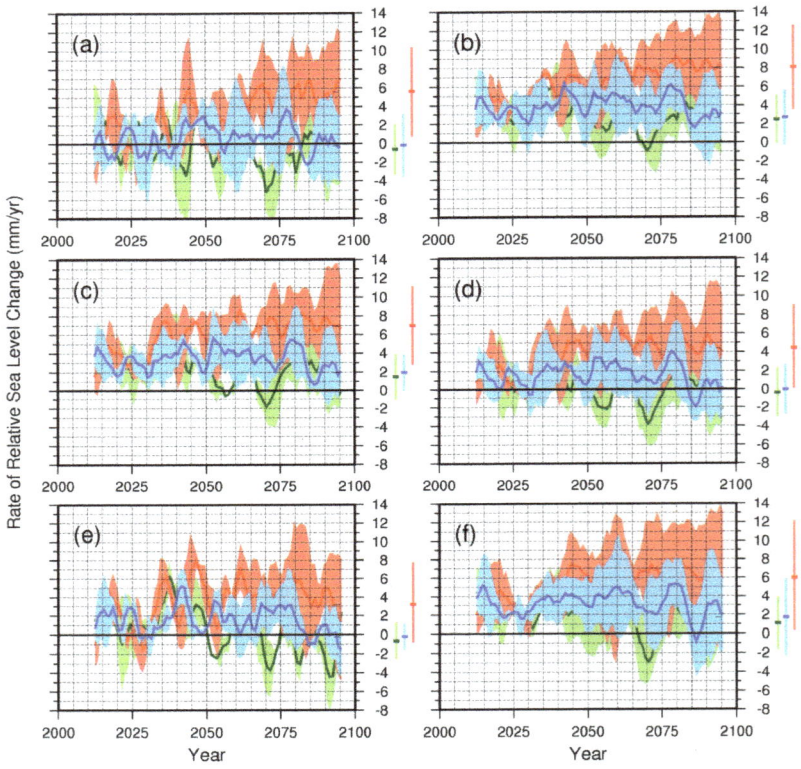

Figure 11. Projected rates of sea-level change for RCP2.6 (green), RCP4.5 (blue) and RCP8.5 (red) for the six key locations: (**a**) Oslo; (**b**) Stavanger; (**c**) Bergen; (**d**) Heimsjø; (**e**) Tromsø; and (**f**) Honningsvåg. Rates are calculated as linear trends over a 10-year moving window, the x-axis represents the mid-point of each window. The vertical bars on the right side of the panels represent the ensemble mean and ensemble spread (5% to 95%) for rates calculated for the period 2081 to 2100.

3.3. Combining Storm Surge Statistics with Sea-Level Projections

3.3.1. Changes in Return Heights and Return Periods with Sea-Level Rise

Figure 12 shows changes in return heights after adding our twenty-first century mean relative sea-level projections. Changes in the likelihood of exceeding present-day return heights are dependent on both the projected sea-level change and the statistics of the observed sea level extremes (i.e., the spread between the different return heights, which determines the gradients of the lines shown in Figure 12) (see also Kopp et al. [19]). For Oslo, which has a relatively *small* projected sea-level change but has relatively *large* differences between the return heights, we expect only small changes in the frequency of exceedance (Figure 12a). However, for Stavanger and Bergen, the reverse is true and we therefore expect a large increase in the frequency of exceedance (Figure 12b,c).

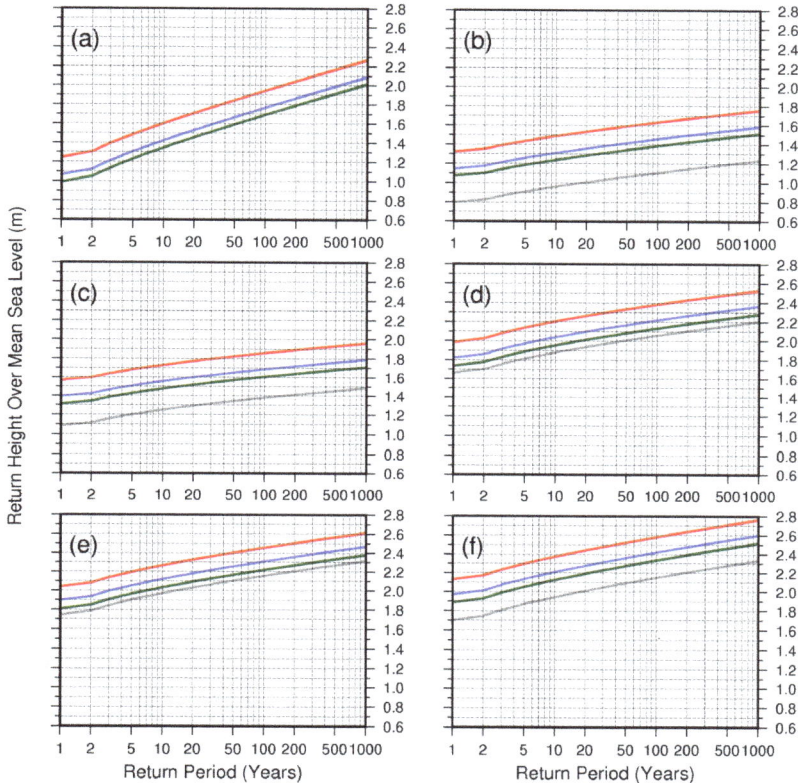

Figure 12. Return heights for stationary sea level (grey) and for ensemble mean RSL projections over the period 1986–2005 to 2081–2100 for RCP2.6 (green), RCP4.5 (blue) and RCP8.5 (red) at the six key locations: (**a**) Oslo; (**b**) Stavanger; (**c**) Bergen; (**d**) Heimsjø; (**e**) Tromsø; and (**f**) Honningsvåg. Note that the grey and blue lines coincide for Oslo.

An extreme sea level with a return period of 200 years will on average occur once every 200 years or, equivalently, has a 0.5% probability of occurring any given year. Figure 13 shows how the likelihood of exceeding the present 200-year return height can be dramatically increased with sea-level rise. By 2050, and only examining the ensemble mean projections, we see that the probability of exceeding the present 200-year return height will be ~10% (or a 1 in 10 year event) in some areas. By the end of the 21st century, the 200-year return heights could be exceeded at least every year.

This is not a surprising result. For example, in Bergen, the 200-year return height lies ~1.4 m above present mean sea level (grey line in Figure 12c). An extreme sea level of this height will cause nuisance flooding in the historic harbour area of the city. The ensemble mean projection for RCP4.5 indicates an RSL increase of ~0.3 m over 1986–2005 to 2081–2100 (blue line in Figure 12c). For this projection, therefore, we see that extreme sea levels reaching a height 1.4 m above present mean sea level will have a one-year return period by the end of the 21st century. For larger increases in RSL, we clearly expect flooding to occur more frequently, multiple times every year. In Bergen, the difference in height between the 200-year return period and the highest astronomical tide is ~0.5 m, i.e., with a RSL rise larger than this, then flooding at the level of the present 200-year return height will start to occur as part of the tidal regime.

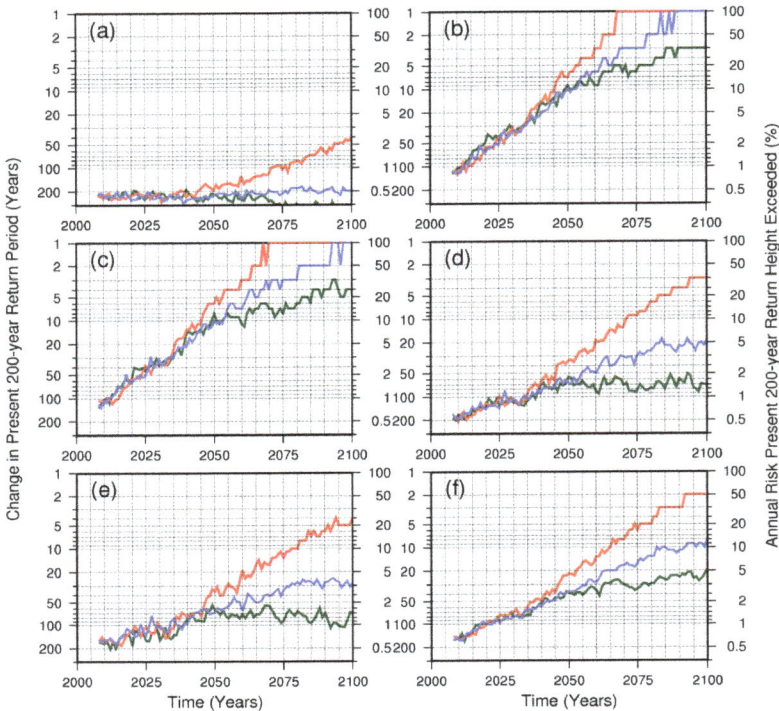

Figure 13. Changes in the 200-year return period (or risk) for the ensemble mean RSL projection for RCP2.6 (green), RCP4.5 (blue) and RCP8.5 (red) and at the locations: (**a**) Oslo; (**b**) Stavanger; (**c**) Bergen; (**d**) Heimsjø; (**e**) Tromsø; and (**f**) Honningsvåg.

3.3.2. Sea Level Allowances

Using our regional sea-level projections and ACER parameters for each tidal zone, we obtain allowances for the coastal municipalities in Norway (Tables 3–5 and Figures 14 and 15). A disadvantage of using the ACER method in this respect (as opposed to the Gumbel method; see Hunter [16] and Hunter et al. [17]) is that our allowances are dependent on the return height of interest. However, comparison of Tables 3–5 shows that differences due to changes in return height are generally no larger than 10 cm.

Table 3. Allowances (in cm) calculated for six key locations using the 20-year return heights and projected ensemble mean RSL change over the period 1986–2005 to 2081–2100 (the 5% to 95% ensemble spread is given in parentheses). Allowances are given for two different choices of the level of likelihood assigned to the ensemble spread. The parentheses next to the allowances give the value of the standard error above the ensemble mean projected RSL.

Location	RCP2.6 2081–2100	RCP2.6 Allowance 5–95%	RCP2.6 Allowance 17–83%	RCP4.5 2081–2100	RCP4.5 Allowance 5–95%	RCP4.5 Allowance 17–83%	RCP8.5 2081–2100	RCP8.5 Allowance 5–95%	RCP8.5 Allowance 17–83%
Oslo	−7 (−32 to 16)	0 (7)	12 (17)	0 (−25 to 24)	7 (15)	21 (25)	18 (−12 to 47)	28 (36)	45 (48)
Stavanger	28 (5 to 50)	40 (41)	60 (51)	35 (12 to 58)	48 (49)	69 (59)	52 (25 to 79)	70 (68)	96 (80)
Bergen	23 (3 to 42)	32 (35)	48 (43)	31 (10 to 51)	41 (43)	59 (52)	48 (23 to 72)	62 (63)	85 (73)
Heimsjø	7 (−17 to 27)	−15 (19)	29 (28)	16 (−5 to 37)	25 (29)	40 (38)	30 (8 to 57)	45 (46)	65 (58)
Tromsø	6 (−10 to 23)	12 (16)	22 (24)	15 (−3 to 34)	22 (27)	34 (35)	29 (5 to 55)	42 (45)	63 (56)
Honningsvåg	18 (−7 to 44)	30 (34)	50 (45)	27 (0 to 53)	39 (43)	61 (54)	43 (10 to 76)	62 (63)	93 (77)

Table 4. Allowances (in cm) calculated for six key locations using the 200-year return heights and projected ensemble mean RSL change over the period 1986–2005 to 2081–2100 (the 5% to 95% ensemble spread is given in parentheses). Allowances are given for two different choices of the level of likelihood assigned to the ensemble spread. The parentheses next to the allowances give the value of the standard error above the ensemble mean projected RSL.

Location	RCP2.6 2081–2100	RCP2.6 Allowance 5–95%	RCP2.6 Allowance 17–83%	RCP4.5 2081–2100	RCP4.5 Allowance 5–95%	RCP4.5 Allowance 17–83%	RCP8.5 2081–2100	RCP8.5 Allowance 5–95%	RCP8.5 Allowance 17–83%
Oslo	−7 (−32 to 16)	0 (7)	14 (17)	0 (−25 to 24)	8 (15)	22 (25)	18 (−12 to 47)	28 (36)	47 (48)
Stavanger	28 (5 to 50)	42 (41)	65 (51)	35 (12 to 58)	50 (49)	75 (59)	52 (25 to 79)	72 (68)	104 (80)
Bergen	23 (3 to 42)	35 (35)	54 (43)	31 (10 to 51)	44 (43)	65 (52)	48 (23 to 72)	66 (63)	93 (73)
Heimsjø	7 (−17 to 27)	17 (19)	34 (28)	16 (−5 to 37)	27 (29)	45 (38)	30 (8 to 57)	47 (46)	72 (58)
Tromsø	6 (−10 to 23)	13 (16)	25 (24)	15 (−3 to 34)	23 (27)	38 (35)	29 (5 to 55)	45 (45)	70 (56)
Honningsvåg	18 (−7 to 44)	32 (34)	56 (45)	27 (0 to 53)	42 (43)	67 (54)	43 (10 to 76)	66 (63)	102 (77)

Table 5. Allowances (in cm) calculated for six key locations using the 1000-year return heights and projected ensemble mean RSL change over the period 1986–2005 to 2081–2100 (the 5% to 95% ensemble spread is given in parentheses). Allowances are given for two different choices of the level of likelihood assigned to the ensemble spread. The parentheses next to the allowances give the value of the standard error above the ensemble mean projected RSL.

Location	RCP2.6 2081–2100	RCP2.6 Allowance 5–95%	RCP2.6 Allowance 17–83%	RCP4.5 2081–2100	RCP4.5 Allowance 5–95%	RCP4.5 Allowance 17–83%	RCP8.5 2081–2100	RCP8.5 Allowance 5–95%	RCP8.5 Allowance 17–83%
Oslo	−7 (−32 to 16)	0 (7)	14 (17)	0 (−25 to 24)	8 (15)	23 (25)	18 (−12 to 47)	29 (36)	48 (48)
Stavanger	28 (5 to 50)	43 (41)	69 (51)	35 (12 to 58)	51 (49)	79 (59)	52 (25 to 79)	74 (68)	109 (80)
Bergen	23 (3 to 42)	36 (35)	58 (43)	31 (10 to 51)	46 (43)	70 (52)	48 (23 to 72)	68 (63)	97 (73)
Heimsjø	7 (−17 to 27)	18 (19)	37 (28)	16 (−5 to 37)	28 (29)	48 (38)	30 (8 to 57)	49 (46)	77 (58)
Tromsø	6 (−10 to 23)	14 (16)	27 (24)	15 (−3 to 34)	24 (27)	41 (35)	29 (5 to 55)	46 (45)	75 (56)
Honningsvåg	18 (−7 to 44)	33 (34)	59 (45)	27 (0 to 53)	43 (43)	71 (54)	43 (10 to 76)	68 (63)	108 (77)

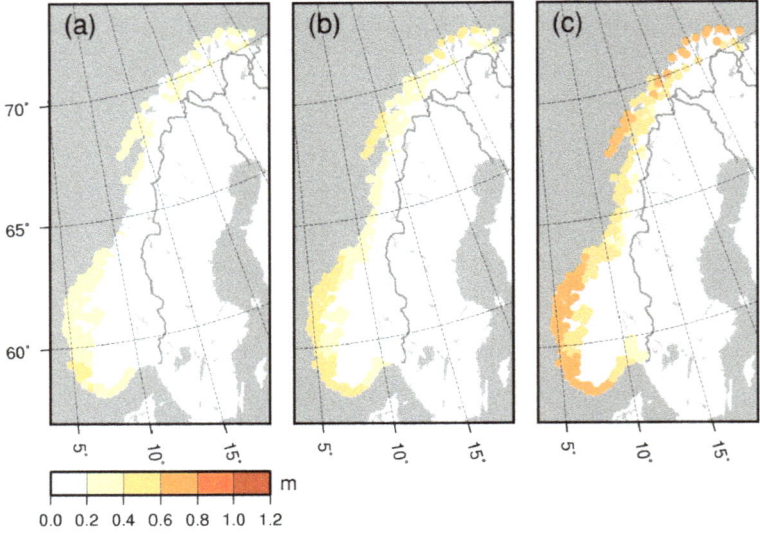

Figure 14. Allowances (m) for the 290 coastal municipalities in Norway calculated using the 200-year return heights and projected RSL change over the period 1986–2005 to 2081–2100 for: (**a**) RCP2.6; (**b**) RCP4.5; and (**c**) RCP8.5. We assume that the model range corresponds to the 5% to 95% probability bounds and fit a normal distribution.

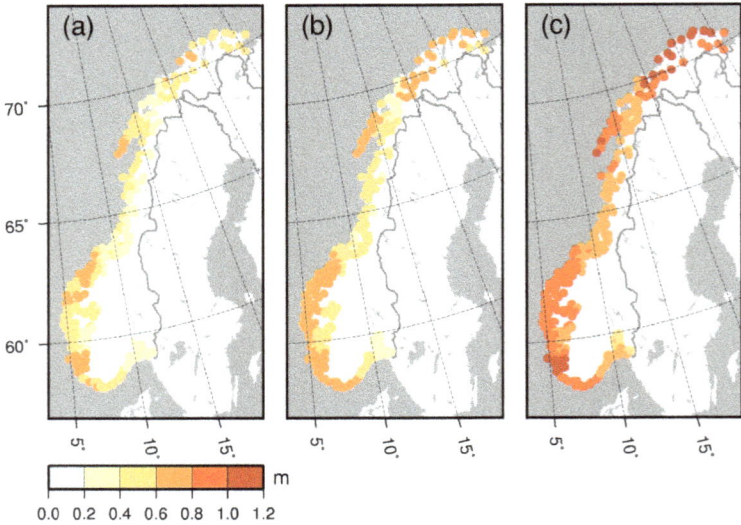

Figure 15. Allowances (m) for the 290 coastal municipalities in Norway calculated using the 200-year return heights and projected RSL change over the period 1986–2005 to 2081–2100 for: (**a**) RCP2.6; (**b**) RCP4.5; and (**c**) RCP8.5. We assume that the model range corresponds to the 17% to 83% probability bounds and fit a normal distribution.

Allowances are calculated in two alternative ways using a normal distribution: (1) assuming that the model spread does in fact correspond to the 5% and 95% probability bounds (e.g., [62,63]) and (2) assuming that the model spread being defined as the *likely* range in IPCC AR5 (*p* > 66%) corresponds to the 17% and 83% probability bounds. Note that the latter is one interpretation of the *likely* range. The use of *likely* in IPCC AR5 means that there is a probability of 33% or less that future sea level will lie outside the model spread (and that 33% is not necessarily symmetrically distributed) [14]. The difference between the two sets of allowances calculated here, therefore, is simply the difference in the spread of the sea-level projections. For allowances computed using the model spread as the 5% to 95% probability bounds, then the allowance generally lies around the mark of the standard error above the mean, but below the top of the IPCC AR5 *likely* range (see also Hunter et al. [17]), whereas, when calculating allowances using the model range as the 17% to 83% probability bounds, then the allowances generally lie above the mark of the standard error above the mean, and also above the top of the IPCC AR5 *likely* range (Tables 3–5). The allowances are therefore quite sensitive to the assumed spread of the probability distribution of future sea-level rise. Note that all allowances calculated give values above the ensemble mean, i.e., to preserve the likelihood of flooding from extreme sea levels, a height increase above the most likely RSL rise should be used in planning. The pattern of allowance heights (Figures 14 and 15) reflects spatial differences in both the RSL projections and statistics of the extreme sea levels.

4. Discussion

Projecting future sea-level change is a challenging task as it requires a sound understanding of many different aspects of the Earth-climate system. Key to improving our understanding of sea level is being able to identify the separate contributions to regional sea-level change. For Norway, part of achieving that goal means maintaining and improving the national sea level observing system (the tide gauge network and geodetic observation used to constrain land motion).

Our regional sea-level projections are based on findings from IPCC AR5 and CMIP5 model output. The projections take into account spatial variations in: (1) ocean density, ocean mass redistribution, and dynamics; (2) ocean mass changes and associated gravitational effects on sea level; and (3) vertical land motion and associated gravitational effects on sea level. We pay special attention to land motion due to GIA, which is better constrained using new GPS [2] and levelling observations. The generally good fit between the GPS data and GIA modelling results gives us confidence that the observations can be extrapolated into the future. Our full GIA solution, as a contribution to RSL change, is the VLM field determined from least-squares collocation (e.g., [33]) added to modelled gravitational effects on sea level associated with GIA.

Projected ensemble mean 21st century RSL changes in Norway are, depending on location, from −0.10 to 0.30 m for RCP2.6; 0.00 to 0.35 m for RCP 4.5; and 0.15 to 0.55 m for RCP8.5. The projections show, unsurprisingly, that the pattern of twenty-first century RSL changes for Norway is governed by GIA. For all RCPs, projected ensemble mean changes indicate that the majority of Norway will experience a RSL rise over the period 1986–2005 to 2081–2100. Thus, climate driven sea-level rise will dominate over VLM over the next 100 years. This represents a reversal of the late-Holocene trend of RSL fall (see e.g., [6]). By the end of the 21st century, the ensemble spread shows rates of RSL rise may approach or exceed ~10 mm/year for RCP8.5.

The projections presented here are given with corresponding 5% to 95% model ranges which are defined as the likely range in IPCC AR5 (*p* > 66%). Quantifying the probability of levels above the likely range (i.e., the upper tail of the probability distribution) remains difficult because information is lacking [14]. Of particular concern is the ice sheet contribution, which potentially could be quite large, especially if a collapse of the marine portions of the Antarctic ice sheet were to be triggered (see Alley et al. [64] for a review). As mentioned, recent modelling studies of Antarctica have tried to quantify how such a marine collapse might contribute to future sea-level rise from the continent with diverging results (e.g., [22,23]). We note that the modelling study by Ritz et al. [22] is broadly in line

J. Mar. Sci. Eng. **2017**, *5*, 36

with IPCC AR5, and indicates a complex and skewed distribution of future ice loss from Antarctica, whereas DeConto and Pollard [23] show by incorporating new processes—the hydrofracturing of buttressing ice shelves and the collapse of ice cliffs—future ice mass loss from Antarctica is strongly RCP dependent and that the ice sheet has the potential to contribute more than a metre of sea-level rise by 2100. This would suggest that projections based on IPCC AR5 science are systematically biased low, however, we caution that this is the result from only one study.

Regional sea-level change beyond 2100 is not dealt with in this paper. However, it is clear sea level will continue to rise after this time owing to the long response times of the oceans and ice sheets (e.g., [65]). Evidence from the paleo record shows that, with even moderate warming, that is, temperatures close to those we observe today, the Greenland and Antarctic ice sheets contributed to a multi-meter sea-level rise above present-day levels [66].

Return heights for extreme sea levels are calculated using statistical analysis (the ACER method) of observations from the Norwegian tide gauge network. Return heights for other points along the coast are found by analysing extrapolated observations where temporary tide gauges are used to quantify the relationship between the tidal behaviour of the temporary and permanent tide gauges. In this work we have assumed no future changes to the amplitude or frequency of the extremes. (However, as shown, sea level extremes and their associated risks will nevertheless increase with mean sea-level rise). Changes in extreme sea levels can arise due to changes in storminess and/or the wave climate and wave setup. Examining evidence on this from an earlier study of the tide gauge records, we find that the picture is somewhat mixed. Some Norwegian tide gauges indicate a small but statistically significant positive late 20th century trend in storm surge heights when compared to the mean sea-level change, while others indicate a negative or insignificant trend [67]. Projections of wind and wave climate, relevant indicators for changes in storm surges, show in general little expected change in our regions. Sterl et al. [68] forced a storm surge model with wind data from an ensemble of 17 climate model runs for 1950–2100. The authors found no significant change in the direction of strong winds (near gale winds or more) in southern and western Norway over this period. Debernard and Røed [69] performed a similar study based on the moderate IPCC scenarios and projected a weak (2–6%) increase in storm surges along the Norwegian coast. However, the authors also point out that storm surge events are extremely dependent on the local conditions (topography and the movement of the storms). Other model studies project little change in wave climate, however, the same studies also caution that confidence in projections of waves and storm surges is very low (e.g., [70]).

Using our regional sea-level projections and ACER parameters for each tidal zone, we have calculated allowances for the 290 coastal municipalities in Norway. We have made some tests of the method by assuming that our regional sea-level projections are normally distributed. If the distribution were skewed, however, this would presumably result in higher allowances than those presented here [71]. A disadvantage of using the ACER method (as opposed to the Gumbel method) is that our allowances are dependent on the return height of interest. All the allowances calculated give values above the projected ensemble mean RSL rise, i.e., to preserve the likelihood of flooding from extreme sea levels, a height increase above the most likely RSL rise should be used in planning. Finally, we note that changes in the likelihood of flooding are dependent on both the projected sea-level change and the statistics of the observed extreme sea levels. The likelihood of exceeding present-day return heights can be dramatically increased with sea-level rise.

Acknowledgments: The work was commissioned and partly funded by the Norwegian Environment Agency. Support was also provided by the Centre for Climate Dynamics at the Bjerknes Centre, through the project iNcREASE.

Author Contributions: M.J.R.S. and J.E.Ø.N. conceived and designed the experiments; M.J.R.S., O.R.R., H.S., O.V. and H.S. performed the experiments and analysed the data; J.E.Ø.N. and H.P.K. contributed materials; and all authors contributed to writing the paper.

Conflicts of Interest: The authors declare no conflict of interest.

References

1. Milne, G.A.; Davis, J.L.; Mitrovica, J.X.; Scherneck, H.-G.; Johansson, J.M.; Vermeer, M.; Koivula, H. Space-geodetic constraints on glacial isostatic adjustment in Fennoscandia. *Science* **2001**, *291*, 2381–2385. [CrossRef] [PubMed]

2. Kierulf, H.P.; Steffen, H.; Simpson, M.J.R.; Lidberg, M.; Wu, P.; Wang, H. A GPS velocity field for Fennoscandia and a consistent comparison to glacial isostatic adjustment models. *J. Geophys. Res. Solid Earth* **2014**, *119*. [CrossRef]

3. Simpson, M.; Breili, K.; Kierulf, H.P. Estimates of twenty-first century sea-level changes for Norway. *Clim. Dyn.* **2014**, *42*, 6613–6629. [CrossRef]

4. Aunan, K.; Romstad, B. Strong coasts and vulnerable communities: Potential implications of accelerated sea-level rise for Norway. *J. Coast. Res.* **2008**, *24*, 403–409. [CrossRef]

5. Anders, A.J.; Hygen, H.O. Impacts of sea level rise towards 2100 on buildings in Norway. *Build. Res. Inf.* **2012**, *40*, 245–259.

6. Lambeck, K.; Smither, C.; Johnston, P. Sea-level change, glacial rebound and mantle viscosity for northern Europe. *Geophys. J. Int.* **1998**, *134*, 102–144. [CrossRef]

7. Ekman, M. A consistent map of the post glacial uplift of Fennoscandia. *Terra Nova* **1996**, *8*, 158–165. [CrossRef]

8. Henry, O.; Prandi, P.; Llovel, W.; Cazenave, A.; Jevrejeva, S.; Stammer, D.; Meyssignac, B.; Koldunov, N. Tide gauge-based sea level variations since 1950 along the Norwegian and Russian coasts of the Arctic Ocean: Contribution of the steric and mass components. *J. Geophys. Res.* **2012**, *117*. [CrossRef]

9. Richter, K.; Nilsen, J.E.Ø.; Drange, H. Contributions to sea level variability along the Norwegian coast for 1960–2010. *J. Geophys. Res.* **2012**, *117*. [CrossRef]

10. Breili, K.; Simpson, M.J.R.; Nilsen, J.E.Ø. Observed sea-level changes along the Norwegian coast. *J. Mar. Sci. Eng.* **2017**, *5*, 29. [CrossRef]

11. Church, J.A.; Clark, P.U.; Cazenave, A.; Gregory, J.M.; Jevrejeva, S.; Levermann, A.; Merrifield, M.A.; Milne, G.A.; Nerem, R.S.; Nunn, P.D.; et al. Sea level change. In *Climate Change 2013: The Physical Science Basis. Contribution of Working Group I to the Fifth Assessment Report of the Intergovernmental Panel on Climate Change*; Stocker, T.F., Qin, D., Plattner, G.-K., Tignor, M., Allen, S.K., Boschung, J., Nauels, A., Xia, Y., Bex, V., Midgley, P.M., Eds.; Cambridge University Press: Cambridge, UK; New York, NY, USA, 2013.

12. Gregory, J.M.; White, N.J.; Church, J.A.; Bierkens, M.F.P.; Box, J.E.; van den Broeke, M.R.; Cogley, J.G.; Fettweis, X.; Hanna, E.; Huybrechts, P.; et al. Twentieth-century global-mean sea-level rise: Is the whole greater than the sum of the parts? *J. Clim.* **2012**, *26*. [CrossRef]

13. Nick, F.M.; Vieli, A.; Andersen, M.L.; Joughin, I.; Payne, A.; Edwards, T.L.; Pattyn, F.; van de Wal, R.S. Future sea-level rise from Greenland's major outlet glaciers in a warming climate. *Nature* **2013**, *497*, 235–238. [CrossRef] [PubMed]

14. Clark, P.U.; Church, J.A.; Gregory, J.M.; Payne, A.J. Recent progress in understanding and projecting regional and global mean sea level change. *Curr. Clim. Chang. Rep.* **2015**, *1*, 224–246. [CrossRef]

15. Hinkel, J.; Jaeger, C.; Nicholls, R.J.; Lowe, J.; Renn, O.; Peijun, S. Sea-level rise scenarios and coastal risk management. *Nat. Clim. Chang.* **2015**, *5*, 188–190. [CrossRef]

16. Hunter, J. A simple technique for estimating an allowance for uncertain sea-level rise. *Clim. Chang.* **2012**, *113*, 239–252. [CrossRef]

17. Hunter, J.R.; Church, J.A.; White, N.J.; Zhang, X. Towards a global regionally varying allowance for sea-level rise. *Ocean Eng.* **2013**, *71*, 17–27. [CrossRef]

18. Jevrejeva, S.; Grinsted, A.; Moore, J.C. Upper limit for sea level projections by 2100. *Environ. Res. Lett.* **2014**, *9*, 104008. [CrossRef]

19. Kopp, R.E.; Horton, R.M.; Little, C.M.; Mitrovica, J.X.; Oppenheimer, M.; Rasmussen, D.J.; Strauss, B.H.; Tebaldi, C. Probabilistic 21st and 22nd century sea-level projections at a global network of tide-gauge sites. *Earth's Future* **2014**, *2*, 383–406. [CrossRef]

20. Joughin, I.; Smith, B.E.; Medley, B. Marine ice sheet collapse potentially under way for the Thwaites Glacier Basin, West Antarctica. *Science* **2014**, *244*. [CrossRef] [PubMed]

21. Rignot, E.; Mouginot, J.; Morlighem, M.; Seroussi, H.; Scheuchl, B. Widespread, rapid grounding line retreat of Pine Island, Thwaites, Smith, and Kohler glaciers, West Antarctica, from 1992 to 2011. *Geophys. Res. Lett.* **2014**, *41*, 3502–3509. [CrossRef]

22. Ritz, C.; Edwards, T.L.; Durand, G.; Payne, A.J.; Peyaud, V.; Hindmarsh, R.C.A. Potential sea-level rise from Antarctic ice-sheet instability constrained by observations. *Nature* **2015**, *528*, 115–118. [CrossRef] [PubMed]

23. Deconto, R.M.; Pollard, D. Contribution of Antarctica to past and future sea-level rise. *Nature* **2016**, *531*, 591–597. [CrossRef] [PubMed]

24. Taylor, K.; Stouer, R.J.; Meehl, G.A. An overview of CMIP5 and the experiment design. *Bull. Am. Meteorol. Soc.* **2012**, *93*, 485498. [CrossRef]

25. Ravndal, O.R.; Sande, B.H. *Ekstremverdianalyse av Vannstandsdata Langs Norskekysten*; Technical Report NDDF 16-1; Norwegian Mapping Authority, Hydrographic Service: Stavanger, Norway, January 2016; p. 26.

26. Ekman, M. A concise history of post glacial land uplift research (from its beginning to 1950). *Terra Nova* **1991**, *3*, 358–365. [CrossRef]

27. Steffen, H.; Wu, P. Glacial isostatic adjustment in Fennoscandia—A review of data and modeling. *J. Geodyn.* **2011**, *52*, 160–2004. [CrossRef]

28. Milne, G.A.; Mitrovica, J.X.; Scherneck, H.G.; Davis, J.L.; Johansson, J.M. Continuous GPS measurements of postglacial adjustment in Fennoscandia: 2 modeling results. *J. Geophys. Res.* **2004**, *109*. [CrossRef]

29. Steffen, H.; Kaufmann, G. Glacial isostatic adjustment of Scandinavia and northwestern Europe and the radial viscosity structure of the Earth's mantle. *Geophys. J. Int.* **2005**, *163*, 801–812. [CrossRef]

30. Olesen, O.; Kierulf, H.P.; Brönner, M.; Dalsegg, E.; Fredin, O.; Solbakk, T. Deep weathering, neotectonics and strandflat formation in Nordland, northern Norway. *Nor. J. Geol.* **2013**, *93*, 189–213.

31. Steffen, H.; Gitlein, O.; Denker, H.; Müller, J.; Timmen, L. Present rate of uplift in Fennoscandia from GRACE and absolute gravimetry. *Tectonophysics* **2009**, *474*, 69–77. [CrossRef]

32. Ekman, M.; Mäkinen, J. Recent post glacial rebound, gravity change and mantle flow in Fennoscandia. *Geophys. J. Int.* **1996**, *126*, 229–234. [CrossRef]

33. Vestøl, O. Determination of postglacial land uplift in Fennoscandia from leveling, tide-gauges and continuous GPS stations using least squares collocation. *J. Geod.* **2006**, *80*, 248–258. [CrossRef]

34. Peltier, W.R. Global glacial isostasy and the surface of the ice-age earth: The ICE-5G (VM2) model and GRACE. *Annu. Rev. Earth Planet. Sci.* **2004**, *32*, 111–149. [CrossRef]

35. Scherneck, H.-G.; Johansson, J.M.; Mitrovica, J.X.; Davis, J.L. The BIFROST project: GPS determined 3-D displacement rates in Fennoscandia from 800 days of continuous observations in the SWEPOS network. *Tectonophysics* **1998**, *294*, 305–321. [CrossRef]

36. Johansson, J.M.; Davis, J.L.; Scherneck, H.G.; Milne, G.A.; Vermeer, M.; Mitrovica, J.X.; Bennett, R.A.; Jonsson, B.; Elgered, G.; Elósegui, P.; et al. Continous GPS measurements of postglacial adjustment in Fennoscandia 1 Geodetic result. *J. Geophys. Res.* **2002**, *107*, 2157. [CrossRef]

37. Lidberg, M.; Johansson, J.; Scherneck, H.G.; Davis, J. An improved and extended GPS-derived velocity field for the glacial isostatic adjustment in Fennoscandia. *J. Geod.* **2007**, *81*, 213–230. [CrossRef]

38. Lidberg, M.; Johansson, J.M.; Scherneck, H.G.; Milne, G.A. Recent results based on continuous GPS observations of the GIA process in Fennoscandia from BIFROST. *J. Geodyn.* **2010**, *50*, 8–18. [CrossRef]

39. Kierulf, H.P.; Ouassou, M.; Simpson, M.J.R.; Vestøl, O. A continuous velocity field for Norway. *J. Geod.* **2012**, *87*, 337–349. [CrossRef]

40. Herring, T.; King, R.; McClusky, S. *Introduction to GAMIT/GLOBK Release 10.4*; Technical Report; Massachusetts Institute of Technology: Cambridge, MA, USA, 2011.

41. Altamimi, Z.; Collilieux, X.; Metivier, L. ITRF2008: An improved solution of the international terrestrial reference frame. *J. Geod.* **2011**, *85*, 457–473. [CrossRef]

42. Collilieux, X. External evaluation of the terrestrial reference frame: Report of the taskforce of the IAG sub-commission 1.2. In *Earth on the Edge: Science for a Sustainable Planet*; Rizos, C., Willis, P., Eds.; Springer: Berlin/Heidelberg, Germany, 2014; pp. 197–202.

43. Farrell, W.E.; Clark, J.A. On postglacial sea level. *Geophys. J. R. Astron. Soc.* **1976**, *46*, 647–667. [CrossRef]

44. Milne, G.A.; Mitrovica, J.X. Postglacial sea-level change on a rotating Earth. *Geophys. J. Int.* **1998**, *133*, 1–19. [CrossRef]

45. Kendall, R.; Latychev, K.; Mitrovica, J.X.; Davis, J.E.; Tamisiea, M. Decontaminating tide gauge records for the influence of Glacial Isostatic Adjustment: The potential impact of 3-D Earth structure. *Geophys. Res. Lett.* **2006**, *33*, L24318. [CrossRef]

46. Wu, P. Response of a Maxwell Earth to Applied Surface Mass Loads: Glacial Isostatic Adjustment. Master's Thesis, Department of Physics, University of Toronto, Toronto, ON, Canada, 1978.

47. Dziewonski, A.M.; Anderson, D.L. Preliminary reference Earth model. *Phys. Earth Planet. Inter.* **1981**, *25*, 297–356. [CrossRef]

48. Tushingham, A.M.; Peltier, W.R. Ice–3G: A new global model of late Pleistocene deglaciation based upon geophysical predictions of post-glacial relative sea level change. *J. Geophys. Res.* **1991**, *96*, 4497–4523. [CrossRef]

49. Zhao, S.; Lambeck, K.; Lidberg, M. Lithosphere thickness and mantle viscosity inverted from GPS-derived deformation rates in Fennoscandia. *Geophys. J. Int.* **2012**, *190*, 278–292. [CrossRef]

50. Milne, G.A.; Gehrels, W.R.; Hughes, C.W.; Tamisiea, M.E. Identifying the causes for sea-level change. *Nat. Geosci.* **2009**, *2*, 471–478. [CrossRef]

51. Richter, K.; Riva, R.E.M.; Drange, H. Impact of self-attraction and loading effects induced by shelf mass loading on projected regional sea level rise. *Geophys. Res. Lett.* **2013**, *40*. [CrossRef]

52. Yin, J.; Griffies, S.M.; Stouffer, R.J. Spatial variability of sea-level rise in the twenty-first century projections. *J. Clim.* **2010**, *23*, 4585–4607. [CrossRef]

53. Haug, E. *Extreme Value Analysis of Sea Level Observations*; Technical Report DAF 12-1; Norwegian Mapping Authority, Hydrographic Service: Stavanger, Norway, January 2012; p. 180.

54. Skjong, M.; Næss, A.; Næss, O.E.B. Statistics of extreme sea levels for locations along the Norwegian coast. *J. Coast. Res.* **2013**, *29*, 1029–1048. [CrossRef]

55. Næss, A.; Gaidai, O. Estimation of extreme values from sampled time series. *Struct. Saf.* **2009**, *31*, 325–334. [CrossRef]

56. Tamisiea, M.E.; Mitrovica, J.X. The moving boundaries of sea level change: Understanding the origins of geographic variability. *Oceanography* **2011**, *24*, 24–39. [CrossRef]

57. Love, R.; Milne, G.A.; Tarasov, L.; Engelhart, S.E.; Hijma, M.P.; Latychev, K.; Horton, B.P.; Törnqvist, T.E. The contribution of glacial isostatic adjustment to projections of sea-level change along the Atlantic and Gulf coasts of North America. *Earth's Future* **2016**, *4*, 440–464. [CrossRef]

58. Yin, J. Century to multi-century sea level rise projections from CMIP5 models. *Geophys. Res. Lett.* **2012**, *39*, L17709. [CrossRef]

59. Landerer, F.W.; Jungclaus, J.H.; Marotzke, J. Regional dynamic and steric sea level change in response to the IPCC-A1B scenario. *J. Phys. Oceanogr.* **2007**, *37*, 296–312. [CrossRef]

60. Katsman, C.A.; Hazeleger, W.; Drijfhout, S.S.; van Oldenborgh, G.J.; Burgers, G. Climate scenarios of sea level rise for the northeast Atlantic Ocean: A study including the effects of ocean dynamics and gravity changes induced by ice melt. *Clim. Chang.* **2008**. [CrossRef]

61. Pardaens, A.K.; Gregory, J.M.; Lowe, J.A. A model study of factors influencing projected changes in regional sea level over the twenty-first century. *Clim. Dyn.* **2011**, *36*, 2015–2033. [CrossRef]

62. McInnes, K.L.; Church, J.; Monselesan, D.; Hunter, J.; O'Grady, J.; Haigh, I.; Zhang, X. Information for Australian impact and adaptation planning in response to sea-level rise. *Aust. Meteorol. Oceanogr. J.* **2015**, *65*, 127–149. [CrossRef]

63. Zhai, L.; Greenan, B.J.; Hunter, J.; James, T.S.; Han, G.; MacAulay, P.; Henton, J.A. Estimating sea-level allowances for Atlantic Canada using the Fifth assessment report of the IPCC. *Atmosphere-Ocean* **2015**, *53*, 476–490. [CrossRef]

64. Alley, R.B.; Anandakrishnan, S.; Christianson, K.; Horgan, H.J.; Muto, A.; Parizek, B.R.; Pollard, D.; Walker, R.T. Oceanic forcing of ice-sheet retreat: West Antarctica and more. *Ann. Rev. Earth Planet. Sci.* **2015**, *43*, 207–231. [CrossRef]

65. Clark, P.U.; Shakun, J.D.; Marcott, S.A.; Mix, A.C.; Eby, M.; Kulp, S.; Levermann, A.; Milne, G.A.; Pfister, P.L.; Santer, B.D.; et al. Consequences of twenty-first-century policy for multi-millennial climate and sea-level change. *Nat. Clim. Chang.* **2016**, *6*, 360–369. [CrossRef]

66. Dutton, A.; Carlson, A.E.; Long, A.J.; Milne, G.A.; Clark, P.U.; DeConto, R.; Horton, B.P.; Rahmstorf, S.; Raymo, M.E. Sea-level rise due to polar ice-sheet mass loss during past warm periods. *Science* **2015**, *349*, 6244. [CrossRef] [PubMed]

67. Menéndez, M.; Woodworth, P.L. Changes in extreme high water levels based on a quasi-global tide-gauge data set. *J. Geophys. Res.* **2010**, *115*, C10011. [CrossRef]

68. Sterl, A.; van den Brink, H.W.; de Vries, H.; Haarsma, R.; van Meijgaard, E. An ensemble study of extreme North Sea storm surges in a changing climate. *Ocean Sci.* **2009**, *5*, 369–378. [CrossRef]

69. Debernard, J.B.; Røed, L.P. Future wind, wave and storm surge climate in the Northern Seas: A revisit. *Tellus A* **2008**, *60*, 427–438. [CrossRef]

70. Hemer, M.A.; Fan, Y.; Mori, N.; Semedo, A.; Wang, X.L. Projected changes in wave climate from a multi-model ensemble. *Nat. Clim. Chang.* **2013**. [CrossRef]

71. Slangen, A.B.A.; Roderik, S.W.W.; Reerink, T.J.; de Winter, R.C.; Hunter, J.R.; Woodworth, P.L.; Edwards, T. The impact of uncertainties in ice sheet dynamics on sea-level allowances at tide gauge locations. *J. Mar. Sci. Eng.* **2017**, *5*, 21. [CrossRef]

Journal of
*Marine Science
and Engineering*

MDPI

Article

Applying Principles of Uncertainty within Coastal Hazard Assessments to Better Support Coastal Adaptation

Scott A. Stephens [1,*], Robert G. Bell [1] and Judy Lawrence [2]

[1] National Institute of Water and Atmospheric Research, Hamilton 3251, New Zealand; rob.bell@niwa.co.nz
[2] New Zealand Climate Change Research Institute, Victoria University of Wellington, Wellington 6140, New Zealand; judy.lawrence.@vuw.ac.nz
* Correspondence: scott.stephens@niwa.co.nz; Tel.: +64-7-856-7026

Received: 30 May 2017; Accepted: 24 August 2017; Published: 29 August 2017

Abstract: Coastal hazards result from erosion of the shore, or flooding of low-elevation land when storm surges combine with high tides and/or large waves. Future sea-level rise will greatly increase the frequency and depth of coastal flooding and will exacerbate erosion and raise groundwater levels, forcing vulnerable communities to adapt. Communities, local councils and infrastructure operators will need to decide when and how to adapt. The process of decision making using adaptive pathways approaches, is now being applied internationally to plan for adaptation over time by anticipating tipping points in the future when planning objectives are no longer being met. This process requires risk and uncertainty considerations to be transparent in the scenarios used in adaptive planning. We outline a framework for uncertainty identification and management within coastal hazard assessments. The framework provides a logical flow from the land use situation, to the related level of uncertainty as determined by the situation, to which hazard scenarios to model, to the complexity level of hazard modeling required, and to the possible decision type. Traditionally, coastal flood hazard maps show inundated areas only. We present enhanced maps of flooding depth and frequency which clearly show the degree of hazard exposure, where that exposure occurs, and how the exposure changes with sea-level rise, to better inform adaptive planning processes. The new uncertainty framework and mapping techniques can better inform identification of trigger points for adaptation pathways planning and their expected time range, compared to traditional coastal flooding hazard assessments.

Keywords: sea-level rise; coastal hazard assessment; uncertainty; coastal adaptation; climate change

1. Introduction

Coastal hazards are physical phenomena that expose a coastal area to risk of property damage, loss of life and environmental degradation [1]. Coastal hazards include flooding during high storm-tides, large waves or tsunami, as well as the more gradual hazards of coastal erosion, high-tide inundation and rising groundwater levels, due to sea-level rise (SLR). Coastal hazards are an increasing problem. Sea level has been relatively stable during the last 2000–3000 years [2]. Civilization has developed near the upper limits of the sea's reach on the premise of a relatively "stable" sea level [3,4]. Global sea level began to rise in the late 1800s, due mainly to anthropogenic greenhouse gas emissions [5,6]. Anthropogenic SLR over this century and beyond will cause more frequent flooding of coastal land and saltwater intrusion into groundwater, geomorphological adjustment of the coastline, rising groundwater levels and vegetation change, e.g., [7]. With a SLR of ~0.2 m since 1900, in low-lying areas of New Zealand there is an increased incidence of coastal storm flooding [8,9]. In the USA, SLR is causing deeper floods during extreme sea-level events, and more regular "nuisance" flooding during high tides, resulting in millions of dollars of insurance claims [10]. The rate of SLR is projected to accelerate over this century and beyond [3–5], which will greatly increase the frequency of flooding, e.g., [9–12], and exacerbate coastal erosion, e.g., [13] forcing communities to adapt in some way. Communities will need to decide when and how to adapt. For example, "adaptation tipping points" [14,15] might be set to when the 1 in 100-year event becomes a 1 in 5-year event, or when the 1 in 5-year event occurs several times per year, erosion reaches a pre-determined distance from houses, or to some measure of community coping capacity.

Government policies generally recognize the need to curtail rising coastal hazard risks over short to long timescales arising from SLR. The 2010 New Zealand Coastal Policy Statement (NZCPS), which has statutory power, requires the identification of areas in the coastal environment that are "potentially affected" by coastal hazards, and assessment of the associated risks over at least the next 100 years (Policy 24). The NZCPS requires a *risk*-based approach to managing coastal hazards (Policies 24–25 and 27)—which requires determination of the *likelihoods* of different magnitude events and their *consequences*, i.e., risk = likelihood × consequence. However, likelihood can be difficult to assign over the long-term due to uncertainties, yet consequences could be high. The uncertainty framework presented in this paper was motivated by a need to guide local government in New Zealand, when commissioning coastal hazard assessments (to give effect to the NZCPS policies) for input to the dynamic adaptive policy pathways (DAPP) process [16]. The framework and concepts were developed while revising the coastal hazards and climate change guidance manual for local government in New Zealand [17]. The revision is due for final release in late 2017.

The purpose of a coastal hazard assessment is to provide the exposure information for risk and vulnerability assessments necessary for decision making, including the uncertainties, in a way that is clearly understood. Such assessments must identify the spatial extent and magnitude of hazard exposure, both now and with future higher sea level, and must quantify the likelihood of occurrence of the hazards, recognizing the uncertainties in the future by distinguishing under what conditions probabilistic approaches are appropriate or where scenarios supported by expert judgement are more appropriate. The hazard and uncertainty information is required by planners, asset managers and decision makers, and for input to engagement processes with potentially affected communities (property owners and residents) and wider stakeholders.

When considering the ongoing, but increasing effects of climate change on coastal hazards, uncertainty is fundamental to how the problem is addressed. For coastal areas, it is "virtually certain" that SLR will continue beyond 2100 for many centuries [5]—but what is deeply uncertain is the rate of rise in sea level and magnitudes at junctures over long timeframes [5,6,18]. This uncertainty results in a wide future window within which further substantial exposure could occur. There is more certainty in the near-term for adaptation decisions, e.g., global SLR by 2040–2060 is projected to be in a relatively narrow likely range of 0.16–0.33 m (above 1986–2005 base) across all emission scenarios, compared with the range at 2100 and beyond [5,18,19]. This means that near-term decisions need to build in flexibility, both to reduce exposure and to enable changes to actions, or pathways that can accommodate higher sea levels over longer timeframes. Such actions should integrate the decision life-time, so as not to lock in path dependency arising from the inflexibility of the decision made now [16,19,20].

The clear identification and separation of uncertainty sources is important in any coastal hazard assessment, because confusion could lead to false representation of true uncertainty, resulting in sub-optimal adaptation planning and decision making. Walker et al. [21] introduced an uncertainty framework aimed at providing a conceptual basis for the systematic treatment of uncertainty in model-based decision support, such as the coastal hazard assessment considered here. In Section 3 we briefly review this framework and recent revisions [22,23], and apply it to coastal hazard assessment considering SLR and coastal storm flooding likelihood.

Assessment and adaptation approaches that explicitly deal with uncertainty and the changing character of risk need to be used in coastal areas to avoid inflexible and path-dependent decisions. Such approaches can assess the *consequences* component of *risk*, but *likelihood* of potentially-large future

SLR and climate change impacts is highly uncertain over longer timeframes. The DAPP process is a method for planning under conditions of uncertainty [16]; Haasnoot et al. [16] set out the basis of DAPP, review the literature leading to its development and provide examples of its application. DAPP integrates two existing adaptive planning approaches, Adaptive Policymaking [24,25] and Adaptation Pathways [26]. Adaptive Policymaking provides a stepwise approach for developing a basic plan or policy, and contingency planning to adapt the plan or policy to new information over time. Adaptation Pathways provide insight into the sequencing of actions over time, potential lock-ins (i.e., a path taken now may lock in future negative consequences), and path dependencies [16]. Adaptation Pathways uses the concept of "adaptation tipping point" [27], which is the point at which a particular action is no longer adequate for meeting the agreed objectives and a new action is therefore necessary [16]. The exact timing of a tipping point is not necessary; but bracketing the time period should provide a clear indication—for example, "on average the tipping point will be reached within 50 years, at earliest within 40 years, and at latest within 60 years" [16]. Adaptive Policymaking uses "trigger points", which specify the conditions under which a pre-specified action to change the plan, is to be taken [16]. The combination of Adaptive Policymaking and Adaptation Pathways, DAPP, results from using the strengths of both approaches. This integrated approach includes: transient scenarios representing a variety of relevant uncertainties and their development over time; different types of actions to handle vulnerabilities and opportunities; Adaptation Pathways describing sequences of promising actions; and a monitoring system with related contingency actions to keep the plan on track with the objectives [16]. The basis of the DAPP process is that, given uncertainty about the future, one needs to design *dynamic* adaptive plans that allow future decisions to be changed in the light of new information (e.g., extensive monitoring of the impacts, changes in frequency of events and drivers such as SLR), with inherent flexibility to change course once certain trigger points are signaled. The change in course (to another pathway) may be delayed if slower than anticipated SLR occurs and the trigger point takes longer to reach, and conversely, an earlier change may be implemented if SLR is more rapid than expected, or if progress on reducing global emissions is limited. The DAPP process is now being used internationally to plan for adaptation to rising risk over time to anticipate tipping points for future decisions, irrespective of how the timing of climate change impacts unfolds, e.g., [14,19,28].

In a coastal context, the DAPP process focuses on coastal hazard risks, particularly the consequences, and developing alternate pathways, and trigger points (with approximate bracketed time windows). It encompasses "testing" responses to climate change against a wide range of future (SLR/hazard) scenarios, which are used to develop dynamic adaptive policy pathways. Subsequent evaluation of these pathways can assess the accrued benefits, or otherwise, over their useful life, covering a range of possible timing for the trigger point being reached, Within the DAPP, coastal hazard assessments can be used to identify vulnerabilities and thresholds of intolerable or "nuisance" risk, to design adaptive policy pathways, and to identify triggers (decision points) for when to switch pathways before the threshold (tipping point) is reached (i.e., anticipatory rather than reactive) and objectives are no longer being met [16]. This enables adaptation to occur before thresholds are reached.

Since coastal hazard assessments take considerable time and resources, one difficulty faced by local government is knowing what hazard scenarios to model to provide an appropriate range of information to support the DAPP process. The 'uncertainty framework for coastal hazard assessment' outlined in Section 4, provides more targeted guidance on the scenarios to model that are tuned to the situation faced, the planning timeframe, and the appropriate management of the uncertainty. This meets the requirements of the DAPP process to examine the consequence, through testing scenarios, of situation-dependent SLR and storm-tide (or erosion) hazards. For example, the framework considers that a wide range of SLR scenarios is necessary when adapting to existing development (leaving aside low-risk assets for non-habitable use), and focuses on higher SLR scenarios over long timeframes if the aim is to avoid hazard risk when undertaking significant new development or change in land use.

The goals of this paper are to: (i) reveal the multiple levels of uncertainty associated with SLR within the context of a formal uncertainty framework; (ii) show how an uncertainty framework might be applied to guide local government when commissioning coastal hazard assessment studies, to ensure that uncertainty is appropriately and transparently accounted for and the assessments provide information appropriate to the decision-making process within the DAPP; and (iii) demonstrate enhancements of coastal flood exposure mapping, which are tailored for adaptive decision making compared with conventional maps showing only the horizontal flooding extent. By isolating both flooding depth and frequency, such maps can clearly show the degree of hazard exposure and likelihood, where that hazard occurs (presently or emergent), and how the hazard exposure changes with SLR. These maps enable more informed community engagement and decision making around tolerability of risks.

The paper is organized as follows. In Section 2 we present SLR scenarios developed during revision of the New Zealand coastal hazard and climate change guidance, focusing on the earliest and latest arrival times for several SLR increments, to support the DAPP and hazard assessment processes. In Section 3 we review the conceptual basis for the systematic treatment of uncertainty and we apply it to consider the levels of uncertainty present within a coastal hazard assessment. In Section 4 we outline the 'uncertainty framework for coastal hazard assessment'. In Section 5 we provide a case study to demonstrate enhancements of coastal flood hazard mapping, and discuss the utility of the maps for the DAPP process. In Section 6 we provide a hypothetical example demonstrating the integration of all the preceding steps within the DAPP. Conclusions are given in Section 7.

2. Sea-Level Rise Scenarios

Changes in the rate of SLR depend on future greenhouse gas emissions [5]. The Intergovernmental Panel on Climate Change (IPCC) presented four greenhouse gas representative concentration pathways (RCP) in their Fifth Assessment Report, based on global climate modeling [5]. For each RCP, probability distributions have been developed that describe the *statistical* uncertainty of future SLR for the unique RCP scenario, but the scenarios cannot be assigned a specific likelihood [18,29]. The SLR scenarios for different RCPs are in relatively close agreement over the next few decades, but substantially diverge beyond about the year 2080 (Table 1). Furthermore, within each RCP, the SLR uncertainty widens considerably with time. The IPCC provided detailed SLR projections out to 2100, and only indicative projections beyond 2100, because of major uncertainty in the upper plausible range of SLR, due to the unknown future dynamical response of the polar ice sheets to warming and other potentially-unknown feedback mechanisms [5]. Therefore, a challenge for coastal hazard assessments is to account for the different types of uncertainty associated with SLR, such as *statistical*, *scenario*, and *deep* uncertainty [30]. These uncertainty terms are defined in Section 3.

Table 1 provides four SLR scenarios out to the year 2150, which are based around three greenhouse gas representative concentration pathways (RCP2.6, RCP4.5 and RCP8.5). Three of the scenarios are derived from the median projections of global SLR for three of the four RCPs presented by IPCC in their Fifth Assessment Report out to 2100 [5] and extended to 2150 by applying the rate of rise from the global projections of Kopp et al. [18]. The fourth 'H^+' scenario is at the upper-end of the "likely range" (i.e., 83rd-percentile) of the large ensemble of SLR projections based on RCP8.5 [18]. In particular, this higher scenario reflects the possibility of future surprises (*deep uncertainty*, Section 3) towards the upper range in SLR projections of an RCP8.5 scenario (recognizing that higher rises cannot be ruled out as shown by higher percentiles in Kopp et al. [18]). These more rapid rates of SLR could occur in the later part of this century and beyond, primarily from emerging polar ice sheet instabilities or as-yet uncertain understanding of dynamic ice sheet processes [18,31,32]. Note: All SLR scenarios in Table 1 have had a small offset of up to 0.05 m by 2100 (pro-rated to 2150) applied to account for slightly higher SLR projections in the regional sea around New Zealand compared to the global mean [33].

The four scenarios were used to develop bracketed timeframes to reach a specific increment of SLR, from the earliest to latest time across the RCP2.6, RCP4.5, RCP8.5 and H^+ scenarios (Table 1).

These timeframes can assist with the timing of triggers (decision points) in the DAPP process. They can be used where particular SLR triggers or associated thresholds for changes in frequency of flooding events have been established, based on vulnerability and risk assessments.

Table 1. Approximate years, from possible earliest to latest, when specific sea-level rise increments (meters above 1986–2005 baseline) could be reached for various projection scenarios of sea-level rise (SLR) for the wider New Zealand region. The earliest year listed is based on the representative concentration pathway RCP8.5 (83rd percentile) or H^+ projection and the next three columns are based on the median projections of the RCP8.5, 4.5 and 2.6 scenarios.

SLR (m)	RCP8.5 H^+ (83rd Percentile)	RCP8.5 (Median)	RCP4.5 (Median)	RCP2.6 (Median)
0.3	2045	2050	2060	2070
0.4	2055	2065	2075	2090
0.5	2060	2075	2090	2110
0.6	2070	2085	2110	2130
0.7	2075	2090	2125	2155
0.8	2085	2100	2140	2175
0.9	2090	2110	2155	2200
1.0	2100	2115	2170	>2200
1.2	2110	2130	2200	>2200
1.5	2130	2160	>2200	>2200
1.8	2145	2180	>2200	>2200
1.9	2150	2195	>2200	>2200

3. How Certain are We? Uncertainty is Important

The clear identification and separation of uncertainty sources is important for coastal hazard assessment. Walker et al. [21] developed an uncertainty matrix as a framework for identifying and characterizing the uncertainty in model-based decision support. This framework has been adapted and modified in several studies, and was revised by Kwakkel et al. [22], and again by Walker et al. [23]. The framework suggests that uncertainty is a three-dimensional concept defined by: the *location* in the analysis, the *level* and the *nature* of the uncertainty. The *location* of the uncertainty could be in the conceptual model, the computer model, the input data, model implementation, or processed output data. The *level* denotes the degree or severity of the uncertainty, ranging from deterministic knowledge to total ignorance. The *nature* of the uncertainty arises from our lack of knowledge about the phenomena or to the inherent variability in the phenomena, or, to ambiguity because the same data can be interpreted differently by different persons depending on differences in frames and values. The nature of the uncertainty matters in choosing a strategy for handling uncertainty, because if the uncertainty is inherent variability, then more research will not help [22].

We focus here on how the different *levels* of uncertainty can be treated within a coastal hazard assessment, where the risk is rising with ever-widening spread of plausible SLR projections. A coastal hazard assessment can combine different sources of flooding, and because these sources have different levels of uncertainty, the uncertainty levels can be confused. For example, confusion of statistical and scenario uncertainty could give a misleading assurance of the true likelihood of outcomes and thus misinform decision making. Horton et al. [34] combine the probability distribution functions (pdf) of RCP4.5 and RCP8.5 SLR projections, assuming a 50% likelihood for each to produce a single pdf for SLR, which they then combine with storm-tide distribution. The study provides statistical confidence of extreme sea levels being reached, but the combined distribution provides a false assurance, because the true likelihood of SLR scenarios used is largely unknown (within a wide range of possible futures beyond 2100 and uncertainty around how quickly global carbon emissions can be curbed). One might argue that this approach represents an expert ranking of the RCP scenarios, but our concern is that the RCP ranking then becomes conflated in the analysis and the implications are not able to be explored within a subsequent decision-making process. Focusing on only a single RCP scenario is another example where levels of uncertainty are not fully addressed. Several recent coastal erosion and flooding studies in New Zealand have calculated the statistical probability of future flooding and erosion hazards following [35,36], but, possibly due to the computational expense, have only considered SLR projections associated with the continued high-emissions RCP8.5 median scenario, e.g., [37,38], rather than exploring sensitivity to other higher and lower scenarios. The resulting hazard maps consider only the statistics within the RCP8.5 scenario, and the possibility that the hazards could be quite different under another scenario is not available for consideration within a DAPP process. This becomes extremely contested in communities if such maps are used directly for statutory zoning within which planning controls are then exercised as has occurred in New Zealand [39]. Fortunately, it is common practice elsewhere to separately consider multiple SLR scenarios, e.g., [10,18,40].

We have therefore applied the five uncertainty levels described by Walker et al. [23], and matched them to some of the uncertainty sources within a coastal hazard assessment in Table 2. Some or all of these levels of uncertainty will typically be involved in decision making in practice [21]. Walker et al. [21] used three descriptive terms to describe the level of uncertainty: *statistical* (level 2) and *scenario* (levels 3–4) uncertainty, and *recognized ignorance* (level 5). We use these terms in Section 4 to develop an 'uncertainty framework for coastal hazard assessment', because they are readily applicable to the treatment of SLR uncertainty. *Deep uncertainty* is defined as the situation where analysts *do not know*, or the parties to a decision *cannot agree on*, the appropriate conceptual models, the probability distributions used to represent uncertainty, and/or how to value the desirability of alternative outcomes [41]. Walker et al. [23] refer to level 4 and 5 uncertainties as *deep uncertainty*, and assign the *do not know* portion of the definition to level 5, and the *cannot agree upon* portion of the definition to level 4 uncertainties. SLR can be considered *deeply uncertain* towards the end of this century since experts *cannot agree* which of the scenarios (a multiplicity of unranked plausible futures, Table 2) is more likely and cannot assign relative probabilities to each RCP (due to multiple uncertainties such as, how emission policies, landuse, technological and socio-economic factors will evolve and the degree of response of polar ice sheets [18,31,42]). However, there is some evidence that the RCP2.6 scenario for SLR is increasingly unlikely [43], and so could be ranked as less likely than the other SLR scenarios (level 3 uncertainty, Table 2). Although not possible to assign statistical probabilities to future SLR, estimates have been made of the largest plausible projections of SLR by 2100 [29,44].

Statistical probabilities can be calculated from the historical record for some hazard sources such as storm-tide, which equates to level 2 uncertainty. The frequency and magnitude of present-day storm tides (a combination of storm surge and high tide) can be modelled by fitting an extreme-value model to the historical observations of very high (e.g., annual maxima) sea-levels (Figure 1). The extreme-value model has a maximum-likelihood estimate (solid line in Figure 1), and a *statistical uncertainty* (level 2) around the maximum-likelihood estimate (dashed 95% confidence intervals in Figure 1). Climate change may also alter the frequency and magnitude of storm tides in future, but there is recognized ignorance about exactly how this will occur, other than it is likely to be a second order effect compared to SLR. One solution is to undertake sensitivity tests for various hazard scenarios (leaving aside SLR) such as exploring ±10% change, for example.

A challenge for coastal hazard assessment arises because the degree of uncertainty within individual hazard sources changes with time. Additionally, Le Cozannet et al. [45] showed that the relative importance of the various sources of uncertainties changes over the time—local coastal processes such as storm-tide and wave runup are the most important during the first part of this century, whereas uncertainties of future SLR scenarios largely dominate beyond the year 2080. In other words, level 2 *statistical* uncertainty is relatively important over short-term planning timeframes (before year 2060), but after a transition period (2060–2080), level 3–5 *scenario* and *deep* uncertainties become dominant over longer planning timeframes (after the year 2080), driven mainly by the increasing uncertainty in the rates of SLR [45]. For coastal hazard modeling, we suggest that uncertainty surrounding future rates of SLR must be dealt with by evaluating the hazard from various SLR scenarios (*scenario uncertainty*), and using higher H^+ type scenario(s), e.g., [46] as a proxy for exploring some implications of *deep uncertainty* in our understanding of the hazard from possible upper-range SLR, and evaluating their consequence within the DAPP [16]. This is the approach recommended within the 'uncertainty framework for coastal hazard assessment' in Section 4.

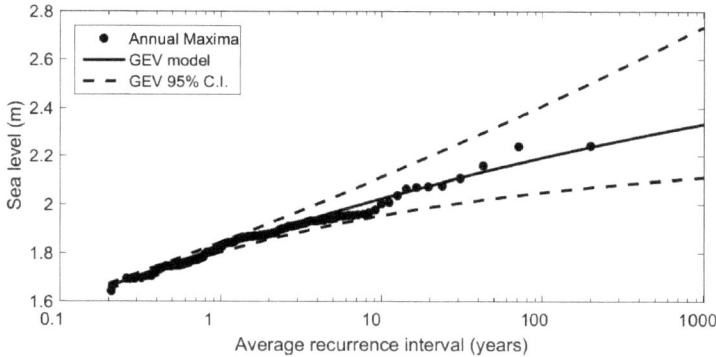

Figure 1. Generalized extreme-value model fitted to annual maxima sea level at Auckland, NZ. Dashed lines mark the 95% confidence intervals (C.I.) (statistical uncertainty). Data supplied by Auckland Council.

The New Zealand Parliamentary Commissioner for the Environment recommended "in revising [New Zealand] central government direction and guidance on SLR, to specify that "best estimates" with uncertainty ranges for all parameters be used in technical assessments of coastal hazards" [47]. While it is desirable to specify statistical uncertainty ranges for some parameters, this may not always be possible. The longer the planning timeframe, the increasing dominance of SLR on the outcome [45], and neither a best estimate, nor statistical uncertainty, can be robustly derived for SLR [18,29]. In any case the optimal *risk* will occur at a higher SLR than the best estimate of the hazard, due to the tail in the probability distribution for SLR for any of the RCPs [12]. In such situations, the *likelihood* component of risk must be handled some other way, such as using adaptive approaches like the DAPP process [16,19]. With consequences for existing development rising non-linearly with increasing SLR, use of a "most likely" SLR (i.e., hazard exposure) is not commensurate with managing risk as required by the NZCPS. The DAPP process interactively embeds the likelihood or emergence aspect, where the time to reach pre-agreed decision or trigger points can be adjusted through regular monitoring and reviews as climate change effects and the ability to cope with them unfolds [16]. This is an appropriate way of addressing future coastal vulnerability and risk management in an adaptive manner, which will enable uncertainties to be worked around, rather than adapting now to a pre-determined future by selecting a best or likely SLR estimate or a "worst-case" scenario.

Table 2. The five *levels* of uncertainty within the uncertainty framework described by Walker et al. [23], including descriptions of the uncertainty *levels* at the context and system-model *locations*. We have provided examples of how the uncertainty levels could be related to the treatment of SLR and other hazard sources within a coastal hazard assessment.

Location	Level 1	Level 2	Level 3	Level 4	Level 5
Context [23]	A clear enough future (with sensitivity)	Alternate futures (with probabilities)	Alternate futures (with ranking)	A multiplicity of plausible futures (unranked)	Unknown future
System model [23]	A single system model	A single system model with a probabilistic parameterization	Several system models, one of which is most likely	Several system models, with different structures	Unknown system model, know we don't know
SLR treatment within coastal-hazard assessment	Present-day MSL, or modest SLR range for the next few decades ($\leq$2050)	Probabilistic SLR trajectories within a single RCP scenario, e.g., [18]	Rank one RCP SLR scenarios relative to each other, e.g., RCP2.6 now considered unlikely [42]	Treat all IPCC AR5 RCP scenarios as separate and equally plausible to test pathways	SLR rate at very long timeframes not considered in available literature, e.g., beyond 2150–2200
Other hazard source examples	Median, or "best estimate" of AEP, where calculable for non-SLR coastal hazards, e.g., storm-tide	Statistical probabilities, where calculable for non-SLR coastal hazards, e.g., storm-tide		An allowance for increased future storm-tide variability, e.g., $\pm$10%	Geomorphic response to SLR of tidal inlet/spit systems on sand or gravel shorelines
Coastal hazard assessment situation	Little uncertainty (or, uncertainty is inconsequential to decision being made, or deliberately ignored)	Statistical probabilities for storm-tide or coastal erosion based on historical observations	SLR scenarios added to storm-tide probability levels		High SLR scenarios added to present-day storm-tide or coastal erosion "best estimates"

Location of Uncertainty Within Coastal Hazard Assessment

Additional technical uncertainties are represented by the location dimension of the uncertainty matrix, and these can be highly scenario sensitive. These should also be made transparent to inform the DAPP process. That is, uncertainties located within the technical process of modeling the hazard, including the conceptual understanding of the processes, the data inputs, the structure and parameters used [21]. For example, coastal hazard assessment would ideally use a multi-hazard risk reduction approach, assessing risk from coastal erosion and flooding from tsunami as well as storm-tides and SLR. Furthermore, there can be complex interactions and responses to SLR between coastal flooding, erosion, and management interventions, e.g., [13,48,49]. For example, coastal erosion and flooding are connected [48], and beaches may erode, or may prograde under SLR when alongshore and sediment system coupling are considered, as up-drift cliffs supply sediment [13].

4. Using Uncertainty to Guide Coastal Hazard Assessment

A clear understanding of the uncertainties relevant to the decision being made can guide the approach to coastal hazard assessment. The framework shown in Figure 2 was developed to provide guidance to local government in New Zealand, when commissioning coastal hazard studies. It is designed to ensure that uncertainties are appropriately identified and managed, and for adequate hazard scenarios to be used in the DAPP process, which is a relatively new concept [16], and not yet commonly applied in New Zealand other than for a river delta flood situation [28]. Local government hazard analysts were concerned that without such a framework they may commission hazard studies that either do not meet the needs of the DAPP process, or are more complex and costly than required for the situation and decisions being considered (e.g., too many scenarios to model). Hence the framework in Figure 2 attempts to be specific about the hazard frequencies and magnitudes, statistical uncertainty and the SLR scenarios to consider, and for which types of activities to apply them, and thus to link back to the situational context and requirements. By also linking to the uncertainty types, the framework attempts to prevent the blurring between statistical and scenario uncertainty discussed in Section 3, or insufficient scenarios, which could lead to misinformed decision making and costly consequences if exposure increases and path dependency results.

The framework in Figure 2 shows relationships between the existing situation, the appropriate level of uncertainty that could be considered based on that situation, the coastal hazard assessment scenarios to match that level of uncertainty, and the associated hazard assessment modeling complexity. The framework attempts to provide logical flow paths from left to right, depicted by the arrows, that could guide the choice of hazard assessment scenarios, being cognizant of all stages within the hazard assessment; the land use situation, hazard modeling, and the decision-making process. For example, if the intention is to avoid a future hazard to a new development, then it is logical to follow the red boxes and account for recognized ignorance in the long-term rate of SLR beyond this century, by modeling high SLR values. However, there may be situations when it is practical to build new major infrastructure (e.g., highway) at low elevation, with staged adaptation planned to cope with future SLR, requiring a more comprehensive set of coastal hazard scenarios akin to adapting an existing development. Hence, Figure 2 draws a distinction, represented by the dashed arrows and dashed box, between the land use situation, the coastal hazard assessment modeling process, and the decision-making process. The eventual decision on whether and how to accept, adapt, or avoid the hazard, is informed by the coastal hazard assessment, but will also be influenced by other factors such as such as economic or social impact assessment [16] and the relevant statutory coastal planning policies (e.g., the NZCPS requires avoidance of increasing the risk from new development).

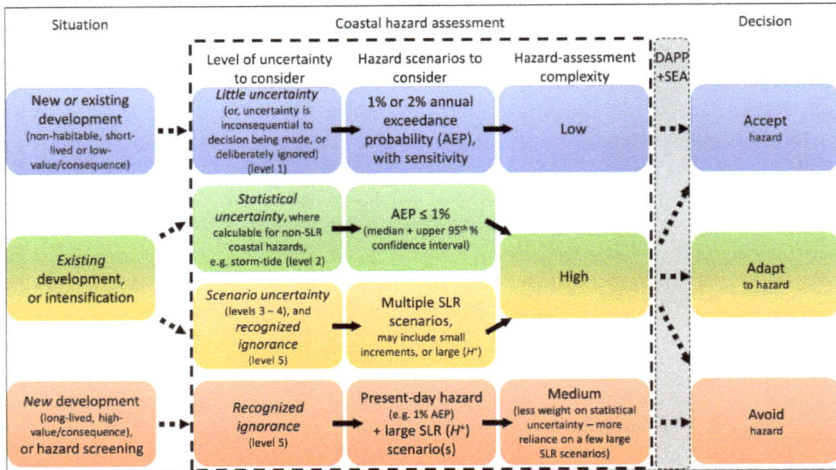

Figure 2. Uncertainty framework for coastal hazard assessments to support the DAPP process, showing a logical flow from the situation, to the related level of uncertainty as determined by the situation, the hazard scenarios to model, the likely hazard modeling complexity, and the possible decision type. A distinction is drawn (represented by the dashed arrows and dashed box) between the situation, the coastal hazard assessment process, the DAPP process and socio-economic assessment (SEA), and the decision type.

The framework in Figure 2 addresses three situations:

1. To avoid risk to new or existing development where, for non-habitable use, the risk of damage from coastal hazards and SLR is low, or the asset can be easily adapted to cope with future SLR. Although there may be high uncertainty around SLR in the long term, because the asset has a short life or low value, and has a functional need to be in the coastal margin, that uncertainty is inconsequential, or can be deliberately ignored. Examples might be a toilet block, a surf-lifesaving lookout, or a culvert supporting a minor access way. Such assets can be easily replaced or relocated, so modeling effort can be kept simple and low-cost. For example, using a simple "building block" model to allow for various coastal hazard sources, or relying on expert judgement or sensitivity testing to decide on an appropriate floor or culvert elevation or setback distance. The assumption in Figure 2 is that hazards are more likely to be accepted for non-habitable short-lived and/or low-value assets, although that decision will be influenced by the planning process, including socio-economic assessment.

2. The greatest demands on coastal hazard assessment are for existing, exposed developments, where ongoing adaptation will be required to cope with rising sea level. For avoiding risk to existing development, or for land use intensification or change in land use, the hazard assessment will require sufficient information to inform the decision(s) to be made, and, when intolerable or nuisance risks may emerge (if not already). This will require the use of both present-day statistical uncertainty (where calculable for non-SLR coastal hazards such as storm-tide), plus several SLR scenarios—thus, the hazard assessment is likely to be more complex and costly. Within the DAPP process, the hazard assessment will need to provide enough information to identify vulnerabilities and thresholds, to design adaptation pathways, and to identify trigger points for when to switch pathways before the threshold eventuates.

3. To avoid increasing the coastal risk exposure from new development and to test the longevity of the decision in establishing new developments on greenfield land where the logical and statutory requirement is to avoid future hazard (e.g., NZCPS); modeling effort can be kept relatively

straightforward, focusing on an upper-range hazard scenario of at least the maximum-likelihood 1% annual exceedance probability (AEP) hazard plus a higher SLR scenario, e.g., the H^+ SLR scenario (Section 2), or a higher percentile, e.g., [46].

The decision about whether to accept, adapt to, or avoid a hazard, can form a set of alternative pathways within the DAPP process. For example, a community might decide to accept a hazard in the short-term, until a trigger point is reached, after which they decide to adapt in some way, for example, by building a seawall, or shifting away from the coast. A range of adaptation pathways and a range of trigger points can be identified, e.g., [14]. There may be examples where a community must accept a hazard due to lack of resources or alternatives, or refuse to adapt, even though there may be considerable uncertainty surrounding future coastal hazard frequency and magnitude. The uncertainty framework provides guidance and supports assessment, but decision-makers make decisions based on many factors which they think appropriate. The hazard assessment and other evaluations such as economic and social impact assessments are only some of the decision inputs [16]. Nevertheless, all such decisions must still be made within the statutory framework operating in each jurisdiction or risk possible challenge.

Figure 3 provides a hypothetical example of three situations with different exposure to coastal hazards. We now use Figure 3 to explore how the framework in Figure 2 might apply to these three situations.

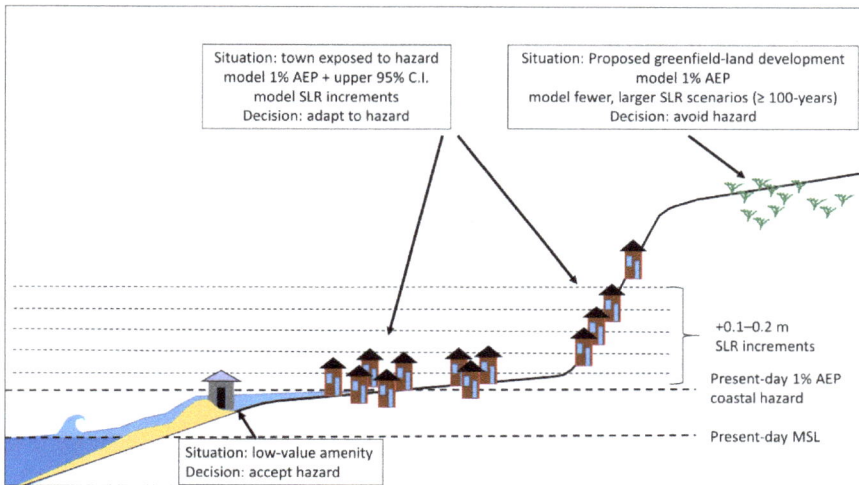

Figure 3. Choice of coastal hazard assessment model scenario based on hazard exposure. The degree of exposure indicates the level of uncertainty the coastal hazard assessment should address, the modeling scenarios required to assist decision making, and the likely complexity of the hazard assessment.

For communities that are already vulnerable to coastal hazards, it is likely that critical tipping points could be reached at relatively low SLR thresholds, such as for the town shown in Figure 3. The town is built on low-elevation land close to the coast and will need to adapt to coastal hazards with an early emergence of SLR impacts. The depth, extent, and frequency of the flooding and erosion hazards will grow incrementally with SLR, and tipping points (e.g., frequency of nuisance or damaging flooding or severe erosion events) may be reached well before 1 m of SLR occurs (which is often used as a single SLR scenario in hazard assessments). Areas on the hill slope will become progressively exposed as sea level rises incrementally. In this case, it would be useful to assess the impacts of a few regular small (e.g., 0.1–0.2 m) SLR height increments (on top of both the median and upper 95% of the

1% AEP flooding hazard) to identify potential tipping points and trigger points for input to the DAPP and the community engagement processes.

Where a new suburb is proposed to be built on a raised coastal platform approximately 1 m above present-day 1% AEP storm-tide level (Figure 3), hazard screening shows no exposure to 1% AEP coastal flooding or erosion at present-day MSL, but increasing hazard exposure after about 0.5 m or more of SLR from later this century. Coastal hazard assessment could instead focus on fewer SLR scenarios accounting for at least 100-year timeframes, such as 0.5 m, 1.0 m, H^+ SLR. Greenfields development in this suburb will require careful scrutiny to avoid increasing risk as the future unfolds.

The third situation involves accepting the hazard for a low-value public amenity, which was discussed in point 1 of this Section.

5. A Coastal Flood Assessment Case Study to Support Dynamic Adaptive Policy Pathways

This section provides a case study on different ways of presenting the coastal flood hazard. We then discuss the usefulness of the maps to the DAPP process.

Figure 4 presents an aerial photograph of Mission Bay, Auckland, New Zealand, along with results of a coastal flood assessment in the form of shaded areas representing the horizontal extent of the 1% AEP storm-tide plus wave setup elevation at: (i) present-day mean sea level (MSL); (ii) present-day MSL + 1 m SLR; and (iii) present-day MSL + 2 m SLR. These types of hazard maps provide a useful summary for planning authorities, showing both the present-day hazard, and identifying the potential future hazard for at least a 100-year timeframe, as required under the statutory NZCPS. Maps such as these formed the basis for development controls in the Auckland Region [50].

Although useful, such hazard-exposure maps in Figure 4 have several limitations:

- They show land either as 'in' or 'out' of the hazard area, but provide no information of the gradient in hazard magnitude away from the sea (e.g., a property at the landward edge of the 1% AEP + 1 m SLR area will only be affected towards the end of the 100-year planning timeframe);
- They provide no information on the timing of the emerging hazard;
- They provide no information on the increasing frequency of flooding with future SLR;
- The hazard analysis for the +1 m and +2 m SLR scenarios may not be useful for adaptation planning if flooding begins to occur frequently at lower SLR.

The analysis in Figure 4 also acts as a hazard screening tool, showing that much of Mission Bay could be affected by flooding after +1 m SLR, and, similar to the town in Figure 3, parts of Mission Bay are likely to reach tipping points before +1 m of SLR occurs.

The uncertainty framework in Figure 2 suggests that for such locations, the impacts of regular small (0.1–0.2 m) SLR height increments should be assessed (on top of both the median and upper 95% of the 1% AEP hazard) to identify potential trigger points for input to the DAPP and community engagement processes.

Figure 5 uses a static mapping technique to add 0.1 m SLR increments directly on top of the present-day median (maximum-likelihood) 1% AEP storm-tide elevation, working under the common assumption that the dominant effect changing the depth of flooding events will be SLR [11,51–53], and ignoring changing storm characteristics. Figure 5 provides more detail than Figure 4, and clearly indicates how flooding extent might change incrementally with SLR, depending on location. Similar mapping products are available in New Zealand[1] and overseas[2]. Properties on low-elevation land close to the sea will face flooding after a modest SLR, so will be affected sooner. Properties located further inland on higher elevation land are less exposed and will have longer to adapt to rising sea level. Such maps also assist councils to assess the emergence of risks to roads and other utilities

---

[1]   http://coastalinundation.waikatoregion.govt.nz/
[2]   https://www.coast.noaa.gov/floodexposure/

and services. The mapping of small SLR increments will be more useful for adaptation planning, as it relates a gradually increasing flooding extent to gradually increasing SLR. However, as with Figure 4, Figure 5 provides no information on the depth and frequency of flooding.

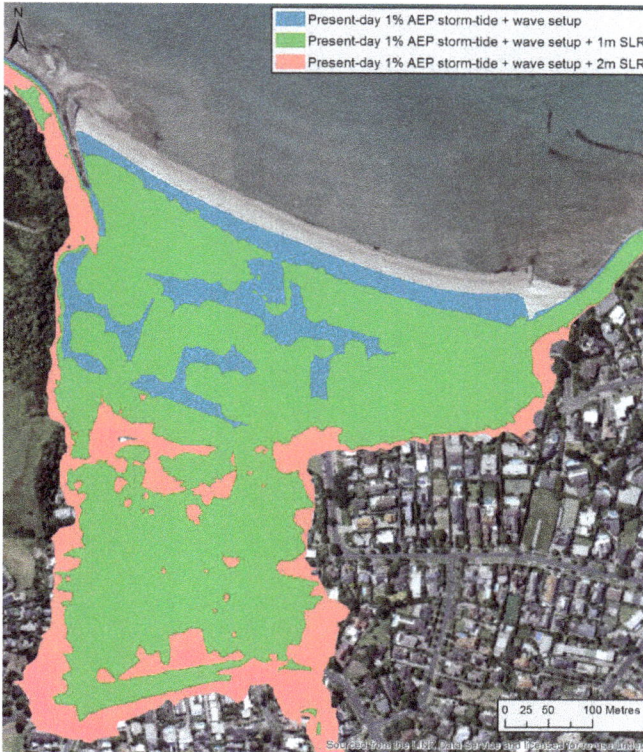

Figure 4. Coastal-storm flood mapping example at Mission Bay, Auckland. Aerial photograph of Mission Bay with present-day 1% AEP storm-tide plus wave setup elevation superimposed (blue shading), plus 1 m SLR (green shading), and plus 2 m SLR (pink shading). After: (50 @Engineers Australia, 2015).

Figures 6 and 7 provide even more information, mapping the expected depth and frequency of flooding for various SLR scenarios, again assuming that SLR will dominate the future increase in flooding frequency. Another example is NOAA's online sea-level rise viewer (https://coast.noaa.gov/slr), which maps flooding depth for various SLR scenarios.

Figure 6 shows the depth (and area) of flooding for a 1% AEP storm-tide at present-day MSL, plus 0.8 m SLR scenario. The map shows the increasing area and depth (severity) of future flooding as the sea rises. The changing frequency of flooding was determined by vertically translating the empirical sea-level distribution to account for SLR, e.g., [10]. The sea-level distribution was first merged with an extreme-value model to create a mixed-distribution model, which represented the full sea-level distribution [9].

Figure 7 shows the frequency (and area) of flooding for a 1% AEP storm-tide at present-day MSL, plus 0.8 m SLR scenario. The map shows that coastal-storm flooding becomes increasingly likely with SLR. In combination, Figures 6 and 7 show both the expected depth and frequency of future flooding.

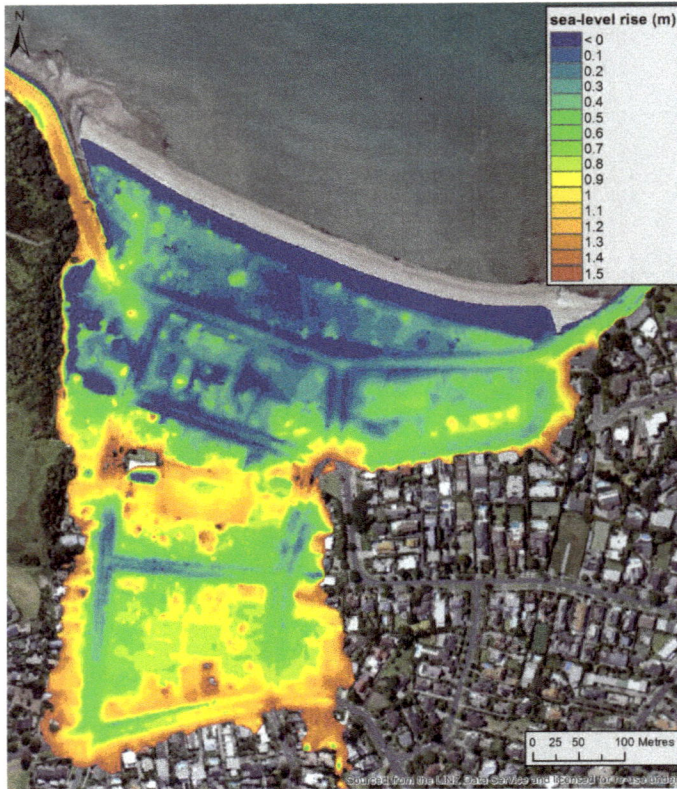

Figure 5. The effect of 0.1 m SLR increments on coastal-storm flood exposure at Mission Bay (Auckland). SLR increments have been added onto the 1% AEP storm-tide elevation, which was calculated for the present-day mean sea level.

The combination of these plots (Figures 6 and 7) provides information that is more useful for decision making than any of the other plots in isolation. For example, a property located beside the first street back from the sea is not presently exposed, but after 0.8 m SLR can be expected to be inundated by about 0.5 m or more of water, about 10 times per year. Clearly the owner of this property will face a tipping point before 0.8 m of SLR occurs, and certainly well before 1.0 or 2.0 m SLR occurs. Understanding the additional information that Figures 6 and 7 portray can enable decision makers to design trigger points ahead of intolerable damage occurring and design longer term strategies for managing the transition for the existing developments (NZCPS)[3].

Figures 6 and 7 were created using a static mapping technique within GIS, which does not consider the dynamic route of flooding nor connectivity to the sea. For example, there are red areas in Figure 7 showing "islands" of 365 daily exceedances per year, which may not occur. These "islands" might be connected to the sea through culverts, and can be identified and used to adjust the areas shown as flooded in the static maps. Dynamic models, which are more computationally expensive, could also

---

[3]   NZCPS Policy 27 I (e) identifying and planning for transition mechanisms and timeframes for moving to more sustainable approaches.

have been used to create more accurate maps. However, the spatially varying frequency and depth of flooding can be mapped irrespective of the mapping technique.

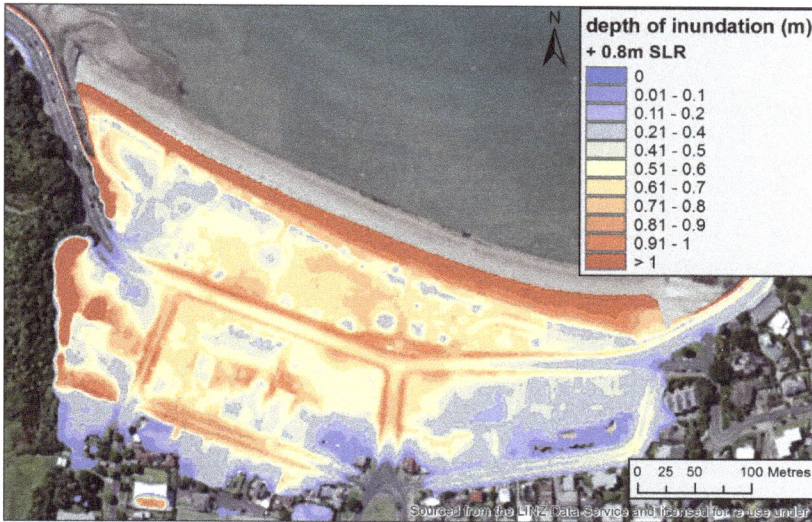

Figure 6. Depth of flooding at Mission Bay, Auckland, for a 1% AEP storm-tide at present-day MSL + 0.8 m SLR. Flooding was modelled using a static GIS technique. All areas below the modelled sea level are shown as inundated, regardless of connection to the sea—some inland areas may not become inundated as shown if such interconnections exist.

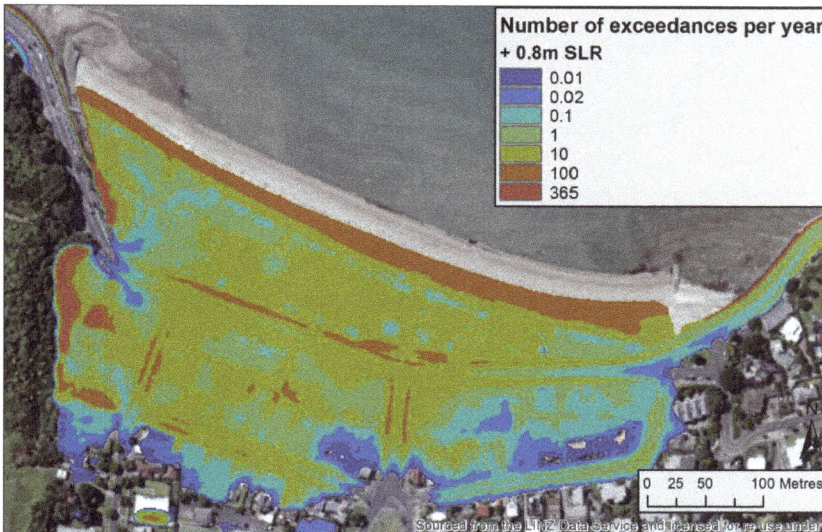

Figure 7. Frequency of flooding (exceedances per year) at Mission Bay, Auckland, for a 1% AEP storm-tide, at present-day MSL + 0.8 m SLR. Flooding was modelled using a static GIS technique. All areas below the modelled sea level are shown as inundated, regardless of connection to the sea—some inland areas may not become inundated as shown if such interconnections exist.

The final piece of the puzzle is to identify the likely timing of the various flooding scenarios mapped in Figures 6 and 7, which can be achieved using Table 1. In Section 6, we have provided a hypothetical example to demonstrate the intended use of the uncertainty framework (Figure 2), the maps (Figures 6 and 7), and possible SLR timing (Table 1) within a DAPP process.

6. Applying the Uncertainty Framework within the DAPP Process

This hypothetical example illustrates how the uncertainty framework might be used in a DAPP process:

1. A community living in the town shown in Figure 3 (or Figures 4–7) decides to proceed with a DAPP process before further development occurs. Such planning fulfils the requirement in the NZCPS for risk-based planning. There is also general agreement within the community (established through a community engagement process and council knowledge) that the planning is required, based on existing coastal flooding problems in some areas, plus an existing simple hazard assessment and expert opinion that show increasing flooding depth with SLR.

2. The local council, which is responsible for planning to reduce or avoid risk from climate change, commission a detailed coastal hazard study. Based on the uncertainty framework (Figure 2), the hazard study estimates the flood height from a storm-tide with a present-day likelihood of flooding of 1% AEP, plus the upper 95% confidence interval of the 1% AEP estimate. The effect of SLR is assessed by adding 0.1 m increments up to 0.5 m, onto the present-day 1% AEP estimate. A higher SLR of +1 m is also assessed to provide a longer-term scenario consistent with a 100-year planning timeframe, e.g., [50], and an H^+ SLR scenario of +1.9 m by 2150 (Table 1) is assessed for the purposes of risk avoidance for greenfields development within the town.

3. The hazard scenarios are mapped, and the maps show both the areal extent, the depth, and the expected frequency of flooding, as in Figures 6 and 7, and how the flooding area, depth and frequency change with the SLR scenarios.

4. The community then meets with the council, and the hazard maps for the various scenarios are presented and explained. The maps form the basis of a discussion whereby the community identifies vulnerable assets, and identifies tipping point scenarios where the depth and frequency of flooding of those assets (i.e., *consequences*) would become unacceptable if no action were taken, and therefore adaptation is required. Thus, when applying the framework, the *consequences* have been separated from the *likelihood* of occurrence, and the community initially makes decisions based primarily on *consequence*.

5. The possible timing of those scenarios is then assessed using Table 1. Thus, given that the *likelihood* of future SLR scenarios is unknown, Table 1 brackets the possible earliest and latest timing of *consequences*. There is a clear separation between the *statistical uncertainty* associated with the storm-tide estimates and the various SLR *scenarios*, which provides clarity to the decision-making process. Community understanding of the flooding risk can be further enhanced by using images of historical damaging coastal flooding (when available), to provide a visual representation of present-day statistical likelihood.

6. The community, with assistance from practitioners, uses the knowledge of the depth, frequency and timing to decide on several pre-determined courses of action (adaptation pathways). Those pathways could include staged alternative strategies such as coastal protection, building modifications, retreat from the coast, and avoidance of greenfields development. Planning provisions to control future development can form supporting strategies to avoid further lock-in of the current pathway. The community identifies potential trigger points, for example, based on a frequency of flooding of a given depth that is not tolerable, which identifies when a switch between pathways needs to occur. They then monitor and review the situation over time using the specified triggers in an iterative fashion as the physical and socio-economic conditions change.

Haasnoot et al. [14] and Lawrence and Haasnoot [28] provide specific examples of a DAPP process in action. The coastal hazard uncertainty framework provided here can add to such empirical examples in future.

7. Discussion and Conclusions

Coastal flooding has caused periodic damage, nuisance flooding or disruption in the past. On the back of rising sea levels, these hazards will greatly increase in frequency, depth and consequence in the future. Coastal hazard assessments require more clarity of hazard-exposure information in a way that uncertainty and dynamics of change are clearly understood, to better assist decision-making and community engagement processes.

The DAPP process involves the identification of trigger points, whereby communities decide ahead of time on potential courses of action or pathways for when those trigger points are reached. Coastal hazard assessments must therefore clearly assist communities and councils (in relation to specified levels of service) to decide what those trigger points are, by providing alternative scenarios along with the likely time range for those scenarios. This requires a careful treatment of uncertainty, because there are different levels of uncertainty that come into play when dealing with the long timeframes and progressive hazards associated with SLR. These sources of uncertainty need to be carefully separated out and communicated transparently.

For coastal hazard assessment, multiple uncertainty levels can be addressed by calculating *statistical* uncertainties for coastal hazards at present-day mean sea level, evaluating the additional hazard from various SLR increments (*scenario* uncertainty), and using high H^+ SLR scenarios to explore the implications of *deep* uncertainty about the hazard from possible upper-range SLR, and evaluating their consequence within the DAPP.

We developed an 'uncertainty framework for coastal hazard assessment', designed to guide local government when commissioning coastal hazard assessments to assist the dynamic adaptive policy pathways process. The framework provides a logical flow from the landuse situation, to the related level of uncertainty as determined by the situation, to which hazard scenarios to model, to the complexity of hazard modeling required, and to the possible decision type.

A case study illustrates how coastal hazard exposure can be mapped for small increments of SLR. Such increments represent a range of plausible future scenarios, which can be superimposed on high storm-tide elevations for which there is an estimated statistical likelihood. The mapping of small SLR increments will be useful for adaptation planning, as it relates a gradually increasing flooding extent to gradually increasing SLR. This can inform communities and councils on when intolerable hazard exposure and risk may emerge (in relation to a SLR and event frequency trigger, and using the bracketed time windows in Table 1).

Maps of coastal flooding typically show just the area of flooding, but we have demonstrated how these can be improved to also show the depth and expected frequency of flooding. This extra information is useful for decision making, showing the degree of exposure, where that exposure occurs, and as input to how much sea-level rise can be tolerated. When combined with information on the approximate bracketed timing of the incremental sea-level rise scenarios, the maps allow communities, stakeholders and councils to identify trigger points and expected earliest time for the emergence of intolerable flooding risk. The actual progression of SLR, and of the triggers before objectives are no longer met, can then be monitored and reviewed. The 'uncertainty framework for coastal hazard assessment' enhances adaptation practice by enabling more salient decision making, because there is greater clarity in the treatment of uncertainty and dynamic aspects of the future risks, which to date have become barriers to the implementation of long-term path dependency considerations in current planning practice.

Acknowledgments: The project was funded by the New Zealand Ministry of Business, Innovation and Employment under Strategic Investment Fund projects CAFS1703 and CAVA1704 of the National Institute of Water

J. Mar. Sci. Eng. **2017**, *5*, 40

and Atmospheric Research. Sanjay Wadhwa produced the flood maps. Aerial imagery sourced from LINZ Data Service and licensed for re-use under the Creative Commons Attribution 3.0 New Zealand license. Manuscript reviews by Mark Dickson, an anonymous reviewer, and the academic editor improved the final version.

Author Contributions: Stephens and Bell conceived the mapping concepts, which were applied within a national guidance document on which all three authors collaborated [54]. Lawrence introduced us to Walker's work on uncertainty [21], to Haasnoot's work on DAPP [16]. Bell developed Table 1. Stephens conceived the 'uncertainty framework for coastal hazard assessment', which was revised collaboratively by the authors. Stephens wrote the paper with input from Bell and Lawrence.

Conflicts of Interest: The authors declare no conflict of interest. The funding sponsors had no role in the design of the study; in the collection, analyses, or interpretation of data; in the writing of the manuscript, and in the decision to publish the results.

References

1. Schwartz, M.L. *Encyclopaedia of Coastal Science*; Springer: Dordrecht, The Netherlands, 2005.
2. Kopp, R.E.; Kemp, A.C.; Bittermann, K.; Horton, B.P.; Donnelly, J.P.; Gehrels, W.R.; Hay, C.C.; Mitrovica, J.X.; Morrow, E.D.; Rahmstof, S. Temperature-driven global sea-level variability in the Common Era. *Proc. Natl. Acad. Sci. USA* **2016**, *113*, E1434–E1441. [CrossRef] [PubMed]
3. Hinkel, J.; Lincke, D.; Vafeidis, A.T.; Perrette, M.; Nicholls, R.J.; Tole, R.S.J.; Marzeion, B.; Fettweis, X.; Ionescu, C.; Levermann, A. Coastal flood damage and adaptation costs under 21st century sea-level rise. *Proc. Natl. Acad. Sci. USA* **2014**, *111*, 3292–3297. [CrossRef] [PubMed]
4. Nicholls, R.J.; Marinova, N.; Lowe, J.A.; Brown, S.; Vellinga, P.; de Gusmão, D.; Hinkel, J.; Tol, R.S.J. Sea-level rise and its possible impacts given a 'beyond 4 °C world' in the twenty-first century. *Philos. Trans. R. Soc. A Math. Phys. Eng. Sci.* **2011**, *369*, 161–181. [CrossRef] [PubMed]
5. Church, J.A.; Clark, P.U.; Cazenave, A.; Gregory, J.M.; Jevrejeva, S.; Levermann, A.; Merrifield, M.A.; Milne, G.A.; Nerem, R.S.; Nunn, P.D.; et al. *Climate Change 2013: The Physical Science Basis. Contribution of Working Group I to the Fifth Assessment Report of the Intergovernmental Panel on Climate Change*; Stocker, T.F., Qin, D., Plattner, G.-K., Tignor, M., Allen, S.K., Boschung, J., Nauels, A., Xia, Y., Bex, V., Midgley, P.M., Eds.; Cambridge University Press: Cambridge, UK, 2013; pp. 1137–1216.
6. Pachauri, R.K.; Allen, M.R.; Barros, V.R.; Broome, J.; Cramer, W.; Christ, R.; Church, J.A.; Clarke, L.; Dahe, Q.; Dasgupta, P.; et al. *Climate change 2014: Synthesis Report. Contribution of Working Groups I, II and III to the Fifth Assessment Report of the Intergovernmental Panel on Climate Change*; Pachauri, R., Meyer, L., Eds.; IPCC: Geneva, Switzerland, 2014; 151p.
7. Nicholls, R.J.; Cazenave, A. Sea-Level Rise and Its Impact on Coastal Zones. *Science* **2010**, *328*, 1517–1520. [CrossRef] [PubMed]
8. Parlimentary Commissioner for the Environment (PCE). *Changing Climate and Rising Seas: Understanding the Science*; New Zealand Parlimentary Commissioner for the Environment Report; Parlimentary Commissioner for the Environment: Wellington, New Zealand, 2014; p. 56.
9. Stephens, S.A. *The Effect of Sea-Level Rise on the Frequency of Extreme Sea Levels in New Zealand*; NIWA Client Report to the Parlimentary Commissioner for the Environment; No. HAM2015-090; National Institute of Water and Atmospheric Research: Hamilton, New Zealand, 2015; p. 52.
10. Sweet, W.V.; Park, J. From the extreme to the mean: Acceleration and tipping points of coastal inundation from sea level rise. *Earths Future* **2014**, *2*, 579–600. [CrossRef]
11. Hunter, J.R. A simple technique for estimating an allowance for uncertain sea-level rise. *Clim. Chang.* **2012**, *113*, 239–252. [CrossRef]
12. Slangen, A.B.A.; van de Wal, R.S.W.; Reerink, T.J.; de Winter, R.C.; Hunter, J.R.; Woodworth, P.L.; Edwards, T. The Impact of Uncertainties in Ice Sheet Dynamics on Sea-Level Allowances at Tide Gauge Locations. *J. Mar. Sci. Eng.* **2017**, *5*, 21. [CrossRef]
13. Dickson, M.; Walkden, M.; Hall, J. Systemic impacts of climate change on an eroding coastal region over the twenty-first century. *Clim. Chang.* **2007**, *84*, 141–166. [CrossRef]
14. Haasnoot, M.; Schellekens, J.; Beersma, J.J.; Middelkoop, H.; Kwadijk, J.C.J. Transient scenarios for robust climate change adaptation illustrated for water management in The Netherlands. *Environ. Res. Lett.* **2015**, *10*, 105008. [CrossRef]

15. Werners, S.E.; Pfenninger, S.; van Slobbe, E.; Haasnoot, M.; Kwakkel, J.H.; Swart, R.J. Thresholds, tipping and turning points for sustainability under climate change. *Curr. Opin. Environ. Sustain.* **2013**, *5*, 334–340. [CrossRef]

16. Haasnoot, M.; Kwakkel, J.H.; Walker, W.E.; ter Maat, J. Dynamic adaptive policy pathways: A method for crafting robust decisions for a deeply uncertain world. *Glob. Environ. Chang.* **2013**, *23*, 485–498. [CrossRef]

17. Ministry for the Environment (MfE). *Coastal Hazards and Climate Change: A Guidance Manual for Local Government in New Zealand*, 2nd ed.; Ramsay, D., Bell, R., Eds.; Ministry for the Environment: Wellington, New Zealand, 2008.

18. Kopp, R.E.; Horton, R.M.; Little, C.M.; Mitrovica, J.X.; Oppenheimer, M.; Rasmussen, D.J.; Strauss, B.H.; Tebaldi, C. Probabilistic 21st and 22nd century sea-level projections at a global network of tide-gauge sites. *Earths Future* **2014**, *2*, 383–406. [CrossRef]

19. Bell, R.G.; Lawrence, J.; Stephens, S.A.; Allan, S.; Blackett, P.; Lemire, E.; Zwartz, D. Coastal Hazards and Climate Change: New Zealand Guidance. In Proceedings of the Australasian Coasts & Ports Conference 2017, Carins, Australia, 21–23 June 2017.

20. Lawrence, J.; Sullivan, F.; Lash, A.; Ide, G.; Cameron, C.; McGlinchey, L. Adapting to changing climate risk by local government in New Zealand: Institutional practice barriers and enablers. *Local Environ.* **2015**, *20*, 298–320. [CrossRef]

21. Walker, W.E.; Harremoës, P.; Rotmans, J.; van der Sluijs, J.P.; van Asselt, M.B.A.; Janssen, P.; Krayer von Krauss, M.P. Defining Uncertainty: A Conceptual Basis for Uncertainty Management in Model-Based Decision Support. *Integr. Assess.* **2003**, *4*, 5–17. [CrossRef]

22. Kwakkel, J.H.; Walker, W.E.; Marchau, V.A.W.J. Classifying and communicating uncertainties in model-based policy analysis. *Int. J. Technol. Policy Manag.* **2010**, *10*, 299–315. [CrossRef]

23. Walker, W.E.; Lempert, R.J.; Kwakkel, J.H. Deep Uncertainty. In *Encyclopedia of Operations Research and Management Science*; Gass, S., Fu, M.C., Eds.; Springer: New York, NY, USA, 2013; pp. 395–402.

24. Kwakkel, J.H.; Walker, W.E.; Marchau, V.A.W.J. Adaptive airport strategic planning. *Eur. J. Transp. Infrastruct. Res.* **2010**, *10*, 249–273.

25. Walker, W.E.; Rahman, S.A.; Cave, J. Adaptive policies, policy analysis, and policy-making. *Eur. J. Oper. Res.* **2001**, *128*, 282–289. [CrossRef]

26. Haasnoot, M.; Middelkoop, H.; Offermans, A.; van Beek, E.; van Deursen, W.P.A. Exploring pathways for sustainable water management in river deltas in a changing environment. *Clim. Chang.* **2012**, *115*, 795–819. [CrossRef]

27. Kwadijk, J.C.J.; Haasnoot, M.; Mulder, J.P.M.; Hoogvliet, M.M.C.; Jeuken, A.B.M.; van der Krogt, R.A.A.; van Oostrom, N.G.C.; Schelfhout, H.A.; van Velzen, E.H.; van Waveren, H.; et al. Using adaptation tipping points to prepare for climate change and sea level rise: A case study in the Netherlands. *Wiley Interdiscip. Rev. Clim. Chang.* **2010**, *1*, 729–740. [CrossRef]

28. Lawrence, J.; Haasnoot, M. What it took to catalyse uptake of dynamic adaptive pathways planning to address climate change uncertainty. *Environ. Sci. Policy* **2017**, *68*, 47–57. [CrossRef]

29. Jevrejeva, S.; Grinsted, A.; Moore, J.C. Upper limit for sea level projections by 2100. *Environ. Res. Lett.* **2014**, *9*, 104008. [CrossRef]

30. Ruckert, K.L.; Oddo, P.C.; Keller, K. Impacts of representing sea-level rise uncertainty on future flood risks: An example from San Francisco Bay. *PLoS ONE* **2017**, *12*, e0174666. [CrossRef] [PubMed]

31. Slangen, A.B.A.; Adloff, F.; Jevrejeva, S.; Leclercq, P.W.; Marzeion, B.; Wada, Y.; Winkelmann, R. A Review of Recent Updates of Sea-Level Projections at Global and Regional Scales. *Surv. Geophys.* **2017**, *38*, 385–406. [CrossRef]

32. DeConto, R.M.; Pollard, D. Contribution of Antarctica to past and future sea-level rise. *Nature* **2016**, *531*, 591–597. [CrossRef] [PubMed]

33. Ackerley, D.; Bell, R.G.; Mullan, A.B.; McMillan, H. Estimation of regional departures from global-average sea-level rise around New Zealand from AOGCM simulations. *Weather Clim.* **2013**, *33*, 2–22.

34. Horton, R.; Bader, D.; Kushnir, Y.; Little, C.; Blake, R.; Rosenzweig, C. New York City Panel on Climate Change 2015 Report. Chapter 1: Climate Observations and Projections. *Ann. N. Y. Acad. Sci.* **2015**, *1336*, 18–35. [CrossRef] [PubMed]

35. Cowell, P.J.; Thorn, B.G.; Jones, R.A.; Everts, C.H.; Simanovic, D. Management of uncertainty in predicting climate-change impacts on beaches. *J. Coast. Res.* **2006**, *22*, 232–245. [CrossRef]

36. Ramsay, D.L.; Gibberd, B.; Dahm, J.; Bell, R.G. *Defining Coastal Hazard Zones and Setback Lines. A Guide to Good Practice*; Envirolink Tools Report R3-2 NIWA: Hamilton, New Zealand, 2012; 91p. Available online: http://www.envirolink.govt.nz/Envirolink-tools/ (accessed on 26 August 2017).

37. Tonkin and Taylor Ltd. *Coastal Hazard Assessment. Stage 2*; Client Report to Christchurch City Council; 851857.001.v2.1; Tonkin and Taylor Ltd.: Auckland, New Zealand, 2015.

38. Tonkin and Taylor Ltd. *Coastal Flood Hazard Zones for Selected Northland Sites*; Client Report to Northland Regional Council; 30524.v1; Tonkin and Taylor Ltd.: Auckland, New Zealand, 2016.

39. Kenderdine, S.E.; Hart, D.E.; Cox, R.J.; de Lange, W.P.; Smith, M.H. *Peer Review of the Christchurch Coastal Hazards Assessment Report*; Review report produced for the Christchurch City Council; Christchurch City Council: Christchurch, New Zealand, 2016; 74p.

40. Buchanan, M.K.; Michael, O.; Robert, E.K. Amplification of flood frequencies with local sea level rise and emerging flood regimes. *Environ. Res. Lett.* **2017**, *12*, 064009. [CrossRef]

41. Lempert, R.J.; Popper, S.W.; Bankes, S.C. *Shaping the Next One Hundred Years: New Methods for Quantitative, Long-Term Policy Analysis*; Rand Corporation: Santa Monica, CA, USA, 2003.

42. City and County of San Francisco Sea Level Rise Committee. *Guidance for Incorporating Sea Level Rise into Capital Planning in San Francisco: Assessing Vulnerability and Risk to Support Adaptation*; San Francisco Department of the Environment: San Francisco, CA, USA, 2015.

43. Magnan, A.K.; Colombier, M.; Billé, R.; Joos, F.; Hoegh-Guldberg, O.; Pörtner, H.-O.; Waisman, H.; Spencer, T.; Gattuso, J.-P. Implications of the Paris agreement for the ocean. *Nat. Clim. Chang.* **2016**, *6*, 732–735. [CrossRef]

44. Le Cozannet, G.; Manceau, J.-C.; Rohmer, J. Bounding probabilistic sea-level projections within the framework of the possibility theory. *Environ. Res. Lett.* **2017**, *12*, 014012. [CrossRef]

45. Le Cozannet, G.; Rohmer, J.; Cazenave, A.; Idier, D.; van de Wal, R.; de Winter, R.; Pedreros, R.; Balouin, Y.; Vinchon, C.; Oliveros, C. Evaluating uncertainties of future marine flooding occurrence as sea-level rises. *Environ. Model. Softw.* **2015**, *73*, 44–56. [CrossRef]

46. Sweet, W.V.; Kopp, R.E.; Weaver, C.P.; Obeysekera, J.; Horton, R.M.; Thieler, E.R.; Zervas, C. Global and regional sea level rise scenarios for the United States. In *NOAA Technical Report NOS CO-OPS 083*; National Oceanic and Atmospheric Administration: Washington, DC, USA, 2017; 75p.

47. Parliamentary Commissioner for the Environment (PCE). *Preparing New Zealand for Rising Seas: Certainty and Uncertainty*; Parliamentary Commissioner for the Environment: Wellington, New Zealand, 2015; p. 92.

48. Dawson, R.J.; Dickson, M.E.; Nicholls, R.J.; Hall, J.W.; Walkden, M.J.A.; Stansby, P.K.; Mokrech, M.; Richards, J.; Zhou, J.; Milligan, J.; et al. Integrated analysis of risks of coastal flooding and cliff erosion under scenarios of long term change. *Clim. Chang.* **2009**, *95*, 249–288. [CrossRef]

49. Stive, M.J.F. How Important is Global Warming for Coastal Erosion? *Clim. Chang.* **2004**, *64*, 27–39. [CrossRef]

50. Stephens, S.A.; Bell, R.G. Planning for coastal-storm inundation and sea-level rise. In Proceedings of the Australasian Coasts & Ports Conference 2015, Auckland, New Zealand, 15–18 September 2015.

51. Tebaldi, C.; Strauss, B.H.; Zervas, C.E. Modelling sea level rise impacts on storm surges along US coasts. *Environ. Res. Lett.* **2012**, *7*, 014032. [CrossRef]

52. Hunter, J.R. Estimating sea-level extremes under conditions of uncertain sea-level rise. *Clim. Chang.* **2010**, *99*, 331–350. [CrossRef]

53. Buchanan, M.K.; Kopp, R.E.; Oppenheimer, M.; Tebaldi, C. Allowances for evolving coastal flood risk under uncertain local sea-level rise. *Clim. Chang.* **2016**, *137*, 347–362. [CrossRef]

54. Ministry for the Environment (MfE). *Coastal Hazards and Climate Change: Guidance for Local Government*; Bell, R.G., Lawrence, J., Allan, S., Blackett, P., Stephens, S.A., Eds.; New Zealand Ministry for the Environment Publication: Wellington, New Zealand, under review.

Journal of
Marine Science and Engineering

MDPI

Article

Choosing a Future Shoreline for the San Francisco Bay: Strategic Coastal Adaptation Insights from Cost Estimation

Daniella Hirschfeld *and Kristina E. Hill

College of Environmental Design, University of California Berkeley, Berkeley, CA 94720, USA;
kzhill@berkeley.edu
* Correspondence: daniellah@berkeley.edu; Tel.: +1-510-642-2962

Received: 31 May 2017; Accepted: 26 August 2017; Published: 4 September 2017

Abstract: In metropolitan regions made up of multiple independent jurisdictions, adaptation to increased coastal flooding due to sea level rise requires coordinated strategic planning of the physical and organizational approaches to be adopted. Here, we explore a flexible method for estimating physical adaptation costs along the San Francisco Bay shoreline. Our goal is to identify uncertainties that can hinder cooperation and decision-making. We categorized shoreline data, estimated the height of exceedance for sea level rise scenarios, and developed a set of unit costs for raising current infrastructure to meet future water levels. Using these cost estimates, we explored critical strategic planning questions, including shoreline positions, design heights, and infrastructure types. For shoreline position, we found that while the shortest line is in fact the least costly, building the future shoreline at today's transition from saltwater to freshwater vegetation is similar in cost but allows for the added possibility of conserving saltwater wetlands. Regulations requiring a specific infrastructure design height above the water level had a large impact on physical construction costs, increasing them by as much as 200%. Finally, our results show that the costs of raising existing walls may represent 70% to 90% of the total regional costs, suggesting that a shift to earthen terraces and levees will reduce adaptation costs significantly.

Keywords: sea level rise; coastal flooding; cost estimation; adaptation; coastal realignment; climate change; coastal planning; coastal management; San Francisco Bay; levees; seawalls

1. Introduction

Flooding and its associated consequences for human development are a major threat to coastal communities [1,2]. These impacts are exacerbated by human development in flood-prone areas [3,4]. Scientific evidence shows that climate change is intensifying these risks by accelerating relative sea level rise, elevating water tables in coastal areas, and increasing the incidence of extreme precipitation [5–7]. Estuaries provide a uniquely valuable setting for human settlement, but these urban regions are very vulnerable to sea level rise. Specifically, sea level rise in estuaries will result in landward movement of both the average and storm-driven high water lines; landward migration of the salinity gradient in surface and groundwater; changes in sediment transport and deposition; and coastal "squeeze", resulting in a loss of inter-tidal wetlands as well as damage to conventional urban districts and infrastructure [4,8]. Globally, some urban estuaries are already grappling with these threats, including the Thames Estuary [9] and the Wash region of the United Kingdom [10], the Elbe in Germany [11], and the Chesapeake Bay in the United States [12]. The San Francisco Bay urban region is also constructed in an estuary context, and presents an opportunity to gain insights about various physical adaptation strategies for shoreline realignment by estimating their adaptation costs and systematically varying key drivers of those costs. It is likely that some of these strategic insights will be applicable to other urban estuary regions.

Recent research shows that in the San Francisco Bay area, increased urban flooding and wetland habitat loss are among the greatest concerns. Currently, there are 140,000 people at risk from a 1 percent chance flood (commonly referred to as a 100 year flood), and with 1.4 m of sea level rise this number increases to 270,000 [13]. Estimates suggest that without adaptation actions, the cost of the impacts to buildings alone would be $49 billion from a 1 m rise in sea level under a 1 percent chance storm [14]. The current projections for marsh habitat along the Bay edge are similarly dramatic. Model projections suggest that sea level rise will produce significant losses of high marsh, which is particularly valuable habitat in the region [15]. In an effort to address these threats, the State of California adopted a new law to require the incorporation of climate change into local planning [16], and is in the process of updating its guidance to local communities. As part of the State's process, the California Ocean Protection Council Science Advisory Team released a report indicating that planners should consider up to 3 m of sea level rise by 2100 [17]. This report is based on the probabilistic assessment approach of Kopp et al. [18].

Addressing the threats of climate change will require complex system-based approaches that allow decision-makers to gain a strategic understanding of relationships between environmental trends and adaptation pathways [19,20]. Adaptation planning guidance calls for the need to address critical strategic questions [21]. For example, thresholds of change in system states may be used as decision points that initiate the shift from one adaptation pathway to another pathway [9]. Similarly, a vulnerability assessment could trigger further analysis and ultimately a change to a set of adaptation actions [22].

Despite this need, certain key planning questions remain unanswered for developed estuaries such as the San Francisco Bay. In particular, while it is clear that the shoreline position will tend to move inland unless humans intervene with new infrastructure, it is not clear what variables planners should use in making shoreline realignment decisions that maximize the benefits of investments in walls, levees, dunes, and wetlands. A recent survey of local government officials identified the need for cost benefit analysis guidance to help them adapt to climate change [23]. Efforts underway in the San Francisco Bay area suggest that raising existing structures is likely to be the most common response to current and predicted flooding [24,25].

In this study, we attempt to answer the question: how do different aspects of a cost estimate for coastal protective infrastructure (including earthen levees, concrete walls, and wetlands) reveal strategic opportunities for adaptation? We specifically explored physical, economic, and regulatory issues related to the adaptation costs for such infrastructure.

In Section 2, we provide details on the data and methods used in our analysis. Section 3 presents our results for the entire San Francisco Bay Area, with a focus on identifying the cost estimates for specific types of coastal infrastructure. In Section 4, we present a discussion of these results as they relate to strategic planning questions and propose further analysis. In Section 5, we provide key conclusions from our work.

2. Materials and Methods

The first challenge of describing a regional shoreline is the selection of categories that organize the diversity of existing conditions in a way that is useful to adaptation decision-making. Our research is guided by the analytical framework originally presented in an earlier paper by Hill, which is summarized in Figure 1 [26]. This framework was derived from concepts related to evolutionary landscapes. Shorelines are categorized as either landforms or walls, and either static or dynamic. The typology emphasizes transformability, in terms of the feasibility of raising the infrastructure over time, and the potential for a coastal infrastructure type to provide multiple benefits. For example, the typology treats walls as single-purpose structures that do not provide multiple benefits, such as recreation, wildlife habitat, or other ecosystem services. We use this typology to describe the current condition of the San Francisco Bay Edge, and to assess the potential costs of adapting to future sea level rise threats.

Our initial observation, from a review of the currently planned and built projects in the San Francisco Bay region, is that the legal, economic, and political challenges of land use change will drive public agencies to respond by seeking to raise existing shore zone structures. Therefore, we designed our cost estimation method to calculate the physical project costs associated with raising the height of existing structures at various positions within the shore zone.

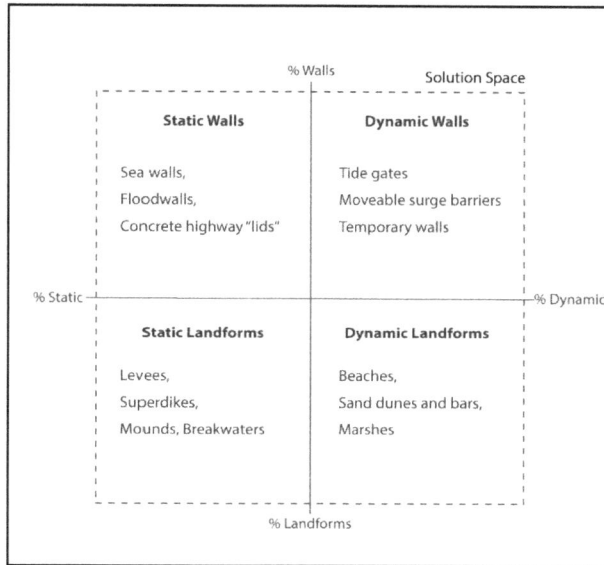

Figure 1. This four-quadrant diagram presents a typology of coastal protective shoreline structures [26]. The vertical axis is defined by the percentage of shoreline that is wall, versus the shoreline that is built with loose materials, such as sand and gravel (landform). The horizontal axis is defined by the percentage of shoreline that is dynamic (able to move, mechanically or by natural processes) versus static (fixed in position). We used this typology in our study because it allows us to differentiate among coastal structures by their approximate initial construction cost, the cost of raising the structures over time, and the variety of ecosystem services they can offer. Structures in the upper right quadrant are typically the most expensive, and structures in the lower right can offer the widest range of services.

2.1. Data Sources

Our analysis is based on the following two groups of data: (1) shore zone conditions and future sea level rise projections, and (2) cost estimation data. Three individual data sets, which are publically available, underlie our characterizations of the shore zone and its inundation under different sea level rise scenarios. The first dataset is a geographic information system (GIS) vector-based inventory of shoreline infrastructure along the San Francisco Bay, developed by the San Francisco Estuary Institute (SFEI) [27]. This dataset contains over 169,000 30-m line segments of linear shore structures (berms, levees, walls, etc.) that occur between mean higher high water (MHHW) and an elevation of 3 m above MHHW (NAVD88). These line segments include complex embayment shapes, resulting in large variations in the measured length of the Bay shoreline. For each 30-m line segment, the dataset describes four characteristics: the type of coastal structure, whether it is accredited as a protective structure, whether it is fronted by natural features (i.e., wetlands and beaches), and its current elevation relative to NAVD88.

SFEI also developed the second dataset [28]. This is a vector-based GIS inventory of aquatic features called the Bay Area Aquatic Resource Inventory (BAARI). It includes features such as wetlands, open water, and riparian areas.

The third dataset [29] is a product of the Coastal Storm Modeling System (CoSMoS) [30], developed at the US Geological Survey (USGS). Staff at the regional flood-mapping project by Point Blue, called Our Coast, Our Future (OCOF) (http://www.ourcoastourfuture.org), provided the data. This dataset provides water heights relative to NAVD88 under 40 different scenarios: 4 storm surge (mean water level, annual high water level (i.e., King Tide), and 20 year and 100 year storm-driven water levels) and 10 sea level rise scenarios (0–200 in increments of 25 and one extreme scenario of 500 cm).

The second group of datasets we used allowed us to develop a regional estimate of raising coastal protective infrastructure. First, we developed a database of unit costs after an extensive review of the costs of engineered structures relevant to the San Francisco Bay region, which is summarized in Table 1. Due to limited data, we do not present cost estimates for dynamic landforms, such as wetlands, or dynamic walls, such as tide gates. Next, we created a dataset of parcel-scale land costs, collected using a web data-scraping method developed by Chris Muir [31] for use with online real-estate data (https://www.zillow.com/). This method enabled us to download all of the data for the San Francisco Bay Area. All of our data can be accessed as described in Section 2.3.

Table 1. Sources and range of associated unit costs converted to thousands of 2016 USD$.

Cost Range: Thousands of 2016 $ (per Linear Kilometer per Meter of Elevation)	Design Type	Study Type and Source
$5.3–$13.2	Landform	SF Bay: Engineering Study (USACE, 2015)
$3.9–$12.4	Landform	Academic Publication from Planning Work (Jonkman, 2013)
$2.5–$5.5	Landform	SF Bay: Technical Report (Lowe, 2013)
$0.400–$33.0	Wall	International Technical Report (Linham and Nicholls 2010)
$5.8–$18.3	Wall	Academic Publication from Planning Work (Jonkman, 2013)
$24.5–$495.5	Wall	SF Bay: Engineering Study (GHD-GTC Joint Venture, 2016)

2.2. Shore Zone and Shoreline Positions, Sea Level Rise, and Cost Assessments

In this first section of our study, we sought to determine where sea level rise will likely cause saltwater flooding, how much higher the existing coastal structures would have to be to prevent exceedance by rising water levels, and how much it might cost to raise the existing structures around the entire San Francisco Bay edge. We identified three baseline scenarios of shoreline re-alignment for the sake of making cost comparisons.

In the shore zone and sea level rise assessment portion of our work, we took three specific steps. First, we used the reclassification scheme shown in Table 2 to align the shoreline data from SFEI to our shoreline typology. Additionally, we used Google Earth and site visits to identify small walls not previously detected by SFEI. Second, we conducted a rapid assessment of water exceedance levels by calculating the difference in height between the floodwater and the structure. We subtracted the height of every shoreline segment in the SFEI data from the USGS CoSMoS model's projected future water levels. In the third step, we generated in GIS three potential shoreline alignments for cost comparison. The most bayward line, referred to here as "Shoreline A", was mapped using SFEI's shoreline infrastructure data. We used the "Bayshore_Defense" category, with the values "First line of shoreline defense" or "Wetland on Bay shore" to designate "Shoreline A". We designated the other two shorelines using an intersection of SFEI's shoreline infrastructure data and the BAARI data to distinguish the saltwater and freshwater habitat zones [28]. The shoreline referred to as "Shoreline B" uses SFEI's mapping of saltwater habitat. The shoreline referred to as "Shoreline C" is the most landward of the three we designated, and is located on the landward side of freshwater wetland habitat as mapped by SFEI.

Table 2. Reclassification scheme used to match SFEI data to our analysis framework.

SFEI Class	Landform or Wall	Static or Dynamic
Berm	Landform	Static
Channel or opening	Landform	Static
Embankment	Landform	Static
Engineered Levee	Landform	Static
Shoreline Protection Structure [1]	Landform	Static
Natural Shoreline	Landform	Dynamic
Wetland	Landform	Dynamic
Floodwall	Wall	Static
Transportation Structure	Wall	Static
Water control structure	Wall	Dynamic

[1] This class included some smaller walls that we reclassified based on Google Earth and site visits.

Next, we calculated potential costs based on our review of unit costs specific to projects in the San Francisco Bay Area, and supplemented these data with cost calculations from the literature where needed. In Table 1, we show the range of unit costs for the two different protective infrastructure types considered in this project: landforms and walls. In this table, we present the information from five different technical and academic publications converted into the same units: thousands of 2016 USD$ for each linear kilometer, and for each meter of raised infrastructure height [32–36].

Using this information, we compared four different approaches to calculating an approximate cost of raising the existing coastal structures: (1) simple without parcel costs, (2) simple with parcel costs, (3) complex without parcel costs, and (4) complex with parcel costs. The "simple" approaches (#1 and #2) used a linear relationship when calculating the cost of raising the height of a levee or a wall. Based on the work of Jonkman et al. [33], we anticipate that the size of levees will increase approximately linearly for relatively low levels of sea level rise (0.5–1.5 m), as shown in Figure 2 and Table 3. However, as levee height is raised beyond 1.5 m, the relationship between height and cost is influenced more by the volume of the material in the levee. In our "complex" cost estimation approaches (#3 and #4), we applied a levee-growth cost factor that incorporates the geometric component of levee size, as shown in Table 3. In Approaches 3 and 4, we also used a more complex approach for estimating wall costs that incorporates the additional requirements of a seismically active region. We assumed that walls could be retrofitted if they need to be raised 0.5 m or less. Beyond 0.5 m of additional height, we assumed that the loadings on the wall would require it to be rebuilt.

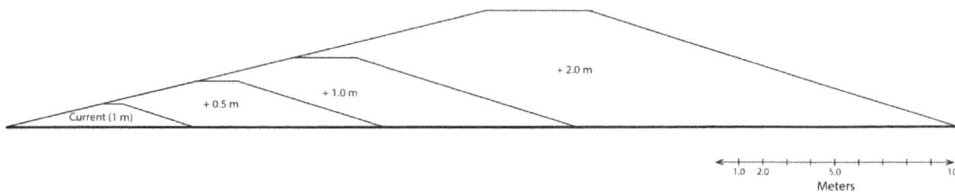

Figure 2. Cross-section of a levee depicted under different sea level rise scenarios.

Table 3. Implications for levee design under different sea level rise scenarios.

Sea Level Rise Scenario (m)	Levee Dimensions [1]		
	Height (m)	Width (m)	Cross-Sectional Area (m^2)
0	Basis = 1	Basis = 7.9	Basis = 4.4
0.5	2	15.8	17.6
1.0	3	23.7	39.6
2.0	5	39.5	110.0
5.0	11	86.9	532.4

[1] Based on levee design with a seaward ratio of 1:4 and a landward ratio of 1:3. The relationship between sea level rise and levee height assumes that wave breaking is depth limited.

We hypothesized that land costs are likely to be a major driver of the cost of higher levees, since more land must be acquired as the width of the levee increases to support its height (see Table 3) [33,35]. To calculate this cost, we obtained parcel-scale market cost estimates from an online real estate database (https://www.zillow.com/) on two separate dates, and calculated the average land costs for all parcels in each county. Prior research in the field of urban economics has shown these estimates to be accurate [37,38]. We increased these base costs slightly to reflect costs associated with eminent domain purchases in the American legal context. We then multiplied these average costs by the number of parcels needed to accommodate higher levees for each sea level rise scenario. This parcel calculation was applied to both the simple and complex levee approaches resulting in Approaches #2 and #4.

In Table 4, we present the final numbers we used for these four different cost estimation approaches. We used the median value of the data we found in the literature for either landforms or walls to define our "typical" unit cost. We used the high and low cost number from the literature for our range. The cost of purchasing land to address the width of the levee varies based on county, shoreline, and sea level rise scenario. In Table 4, we provide the range; however, the actual numbers can be accessed as described in Section 2.3.

Table 4. Cost estimates by infrastructure type and analysis approach.

Cost Approach	Static Landforms Typical (Range)	Static Walls Typical (Range)	Land Cost (Billions of 2016 USD$)
(1) Simple, without parcels (2) Simple, with parcels	$8.0 (±$4.0)	$218.0 (±$75.0)	NA 1.4 to 22.0
(3) Complex, without parcels (4) Complex, with parcels	Raising >3 m: Times 3 Raising >1.5 m: Times 2 Raising <1.5 m: Times 1	Raising >0.5 m: Times 4 Raising <0.5 m: Times 0.5	NA 1.4 to 22.0

2.3. Data and Methods Access

We made our datasets and detailed methods available through UC Berkeley's online data archive, known as DASH. The first dataset contains the water exceedance calculations for the three different shorelines we describe, as well as the full shore zone dataset, and can be accessed at this URL: https://doi.org/10.6078/D1W30C. The second dataset contains the cost calculations for the three different shorelines we describe, as well as the full shore zone dataset, and can be accessed at this URL: https://dx.doi.org/10.6078/D1KK59. Additionally at each URL the code and models we used can be downloaded. Finally, the three publicly available datasets used can be accessed from their original sources.

3. Results

3.1. Description of the Current Shore Zone and Alternative Future Shorelines

Using Hill's classification system [26], we defined four broad types of coastal infrastructure: (1) static landforms; (2) dynamic landforms; (3) static walls; and (4) dynamic walls. We evaluated all 169,000 30-m shoreline segments within the shore zone, all as described in Table 2. Each shoreline segment was only classified as a single one of the four types, and our analyses of the shorezone (as an area within which structures are placed or already exist) capture dual protection scenarios. In this area, we found that the San Francisco (SF) Bay shore zone is predominantly comprised of static landforms (69% of the entire SF Bay edge). The remainder contains 18% dynamic landforms, 12% static walls, and less than 1% dynamic walls. These structures, which comprise the set of existing structures, are not necessarily connected to each other, and are often not certified as effective flood protection structures.

In order to define organized future "shorelines" from this set of existing structures and compare their estimated costs, we defined three alternative shoreline positions relative to today's MHHW. We used ecological boundaries to generate our Shorelines A, B, and C, as described in Section 2.2. If flood protection structures are built along the most bayward line, Shoreline A, the structures would eliminate saltwater wetlands. The most landward future shoreline in this study, Shoreline C, would allow the beneficial flooding of wetland habitats by saltwater to continue, while still protecting most of the developed land. The wetlands that remain exposed would likely require additional sediment to keep pace with sea level rise [15]. Shoreline B is defined by the boundary between saltwater wetland vegetation and freshwater wetland vegetation, and lies between Shorelines A and C.

The map in Figure 3 shows these three alternative future shorelines for the SF Bay, which we used as the basis for our cost comparisons. In places with limited wetland vegetation, such as the Central Bay, the three shorelines converge. However, in portions of the Bay with expansive wetland vegetation and more gradual slopes, such as San Pablo and Suisun Bay, the three shorelines are very different.

These alternative shoreline alignments are different both in length and in the percentage of the different shoreline types we have defined. As shown in Figure 4, the more landward shorelines are longer. Shoreline C, which is the most landward, is 2154 km long; Shoreline B, which is at the boundary of freshwater and saltwater wetlands, is 1340 km long; and Shoreline A, at approximately the MHHW line, is 967 km long. Figure 4 also shows the composition of each shoreline in terms of our four shoreline types. Shoreline A is unique because of a more even split between the two landform categories, static and dynamic. Dynamic walls are 0.1% of all three shorelines and thus do not appear in the figure.

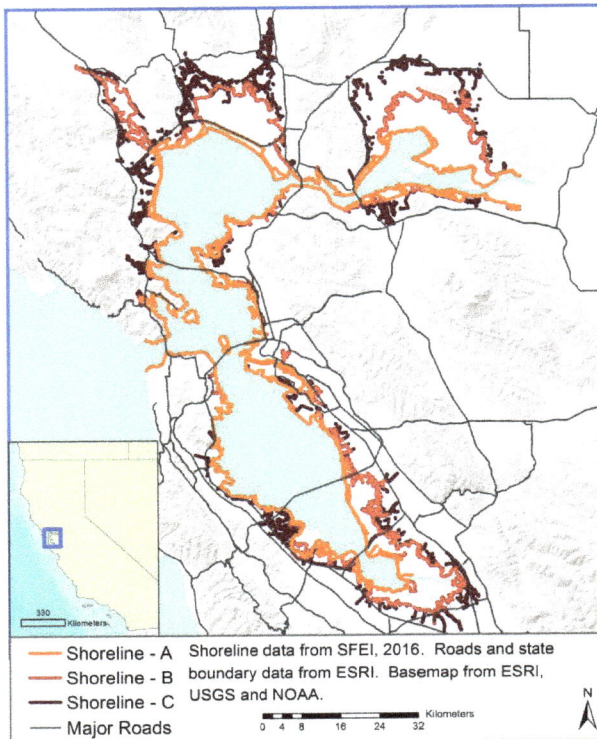

Figure 3. Map showing the full San Francisco Bay study area and the location of the three different shorelines—A, B, and C—we designated for comparative analysis purposes.

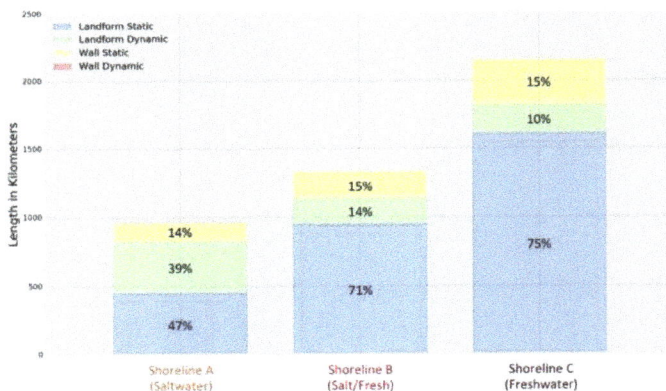

Figure 4. Chart showing each of the three shorelines A, B, and C. For each, the bar chart shows their total length, the length of each of the four coastal infrastructure types, and the percentage of each coastal infrastructure type. Note that dynamic walls are less than 1%, and therefore cannot be seen clearly in this figure.

3.2. Rapid Assessment of Water Exceedance Levels

Here, we used the CoSMoS projections of future water levels [30] to evaluate the potential costs of raising current coastal protective infrastructure to meet future sea level rise scenarios. We calculated the water height exceedance level by calculating the difference in height between the projected water levels and the structure for all four types of shoreline structures in our defined shore zone. Figure 5 shows the summary median values for four different sea level rise scenarios, without including storm surges. Note that in Figures 5 and 6, we present the "no-storm-surge" condition. We do this to separate the impacts of a permanently higher mean sea level from the impacts of temporary storm events, in which floodwaters recede when the event is over. Both types of flooding are important, but the adaptation cost and response to each might be different. All four types of shoreline structures see an increase in exceedance as sea levels rise; however, dynamic landforms (typically wetlands) face the greatest amount of water level exceedance as sea levels rise.

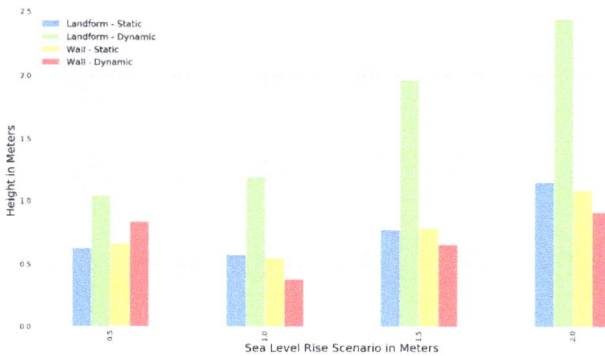

Figure 5. Figure showing the median height of the water that would exceed the top of the four types of protective infrastructure—static landforms, dynamic landforms, static walls, or dynamic walls—with sea level rise scenarios between 0.5 and 2 m.

Next, we calculated exceedance heights for each of the three specific shorelines we designated for cost comparison purposes: Shorelines A, B and C. Figure 6 shows the exceedance heights for the no-storm-surge condition under four different sea level rise scenarios. Here, we see an overall trend of an increase in exceedance as sea levels rise for all three shorelines.

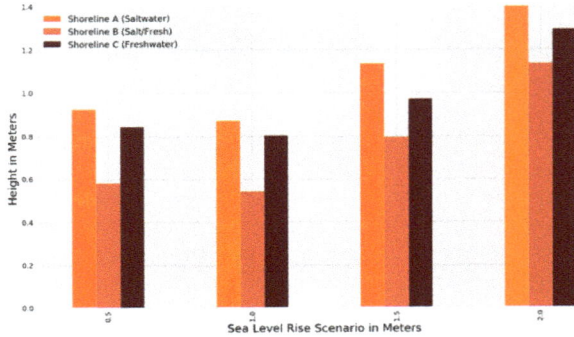

Figure 6. Figure showing the median height of the water that would exceed the height of thousands of line segments that represent the top of the three shorelines—A, B, and C—with sea level rise scenarios between 0.5 and 2 m.

3.3. Comparing Cost Estimation Approaches

Here, we present our cost estimation results based on the different cost estimation approaches described in Section 2.2 and summarized in Table 4.

Figure 7 shows our results from applying all four cost estimation approaches to Shoreline B, which sits on the boundary of saltwater vegetation and freshwater vegetation. This figure shows the total regional cost for the no-storm-surge condition under four different sea level rise scenarios. All four of the cost estimation approaches we used indicate that the costs of raising walls dominate the total cost estimate, constituting between 70 and 90 percent of the overall cost.

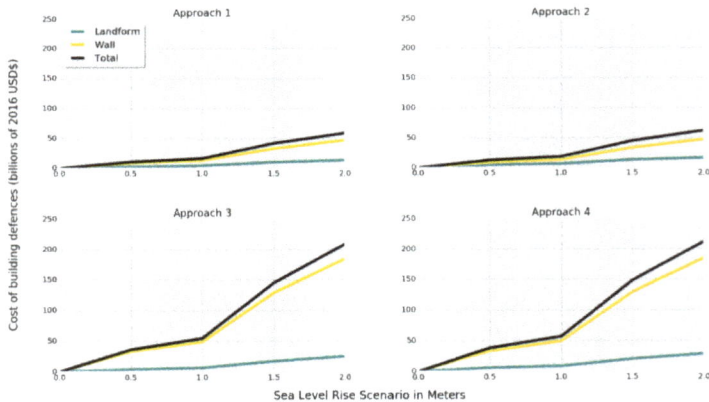

Figure 7. Figure comparing the four different cost calculation approaches for Shoreline B in billions of 2016 USD\$. The two approaches on the left side (#1 and #3) do not include land purchase costs for the landform cost calculation, whereas the two approaches on the right side (#2 and #4) include a land purchase cost. The upper two approaches (#1 and #2) use a linear calculation for landforms and wall cost calculation, whereas the two approaches on the bottom (#3 and #4) have a more nuanced approach to levee and wall cost calculations. Each subplot shows landform, wall, and total costs in billions of dollars starting at no sea level rise and extending to 2.0 m of sea level rise.

Cost increases are relatively moderate below 1 m of relative sea level rise, but increase much more rapidly beyond that. We found the average cost increase among all four approaches to be \$37 billion

2016 USD$ in order to raise existing structures to prevent overtopping from 0 m of sea level rise to 1 m of sea level rise. In contrast, we found the average increase to be $99 billion 2016 USD$ for the total costs of raising the structures to prevent overtopping caused between 1 m and 2 m of sea level rise. The cost of adapting to the first meter of sea level rise is predicted to be $62 billion 2016 USD$ lower than the cost of adapting to the second meter of sea level rise. In addition, Figure 7 shows that using a cost estimate that reflects changes in levee volume, and that uses wall replacement thresholds appropriate for a seismic region, has a large impact on the total cost (Approaches #3 and #4), whereas adding the cost of purchasing land (Approaches #2 and #4) produces a smaller impact on total costs.

Figure 8 shows the impact of land purchase costs alone (i.e., excluding wall costs) on all three shorelines by comparing our cost estimates from Approaches #1 and #2. Adding the estimated cost of purchasing private land increases the cost of adaptation. However, the impact of the land purchases is much greater for Shoreline C, which is the most landward of the shorelines. For Shorelines A and B, the difference between including estimated land costs and not including land costs is $1.3 billion 2016 USD$ at the lowest, and $3.02 billion 2016 USD$ at the highest. For Shoreline C, the longest and most landward shoreline, including land costs produces a difference between $7.8 and $11.2 billion 2016 USD$. The reason for this difference is that Shoreline C contains significantly more private parcels that would need to be purchased, compared with Shorelines A and B.

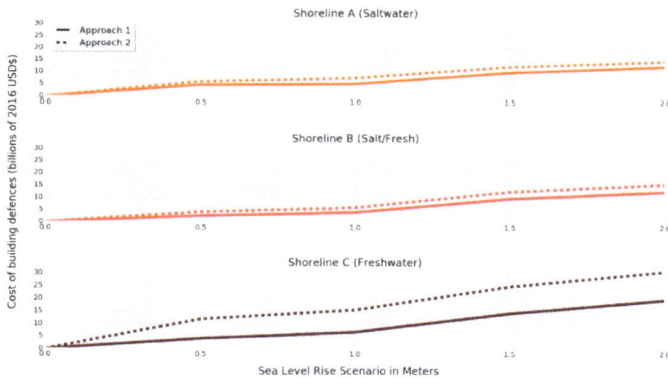

Figure 8. Figure comparing the impact of the land purchase costs across the three different shorelines—A, B, and C—in billions of 2016 USD$. Approach 1 (shown as a solid line) does not include land purchase costs, whereas Approach 2 (shown as a dashed line) includes a land purchase cost. Each subplot shows only the landform costs in billions of dollars.

Finally, we wanted to test the sensitivity of our cost estimation approach to traditional structural requirements for freeboard, defined as the distance between the normal water line and the top of a flood protection structure. While there are important functional reasons for these requirements, different types of infrastructure (wetlands vs. levees, for example) may use different engineering standards [39]. The design of resilient urban districts that tolerate some flooding can also reduce freeboard requirements for some types of shoreline structures.

Figure 9 shows the sensitivity of our cost estimation method to additional freeboard by displaying five different freeboard scenarios. For Shoreline B, we initially examined four different sea level rise scenarios with no storm surge and no freeboard. Then, we calculated the cost impact of requiring up to 1 m of freeboard, in increments of 0.25 m. This calculation uses Approach #3, which includes volume-based and seismically influenced cost calculations, but does not include land purchase costs. Figure 9 shows that requiring more freeboard has a large impact on overall costs. In our final modeling of estimated costs, presented in the next section, we used a freeboard of 0.61 m (2 feet) to represent current Federal Emergency Management Agency (FEMA) requirements [32].

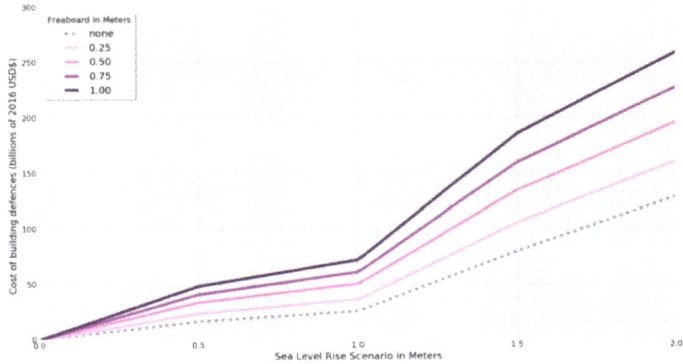

Figure 9. Figure showing the impact of different freeboard heights on the total cost for Shoreline B in billions of 2016 USD$. This calculation is based on Approach #3.

3.4. Overall Cost Estimates

In this section, we present our final cost estimates for the full Bay-wide scale of physical shoreline adaptation within the constraints of our assumptions. All of the results here reflect the application of our methods using Approach #4 (complex with land costs), with the freeboard requirement set at 0.61 m as noted above.

Table 5 and Figure 10 show a comparison of the three different shorelines we designated: A, B, and C. For each shoreline, we show a typical cost, which is based on an intermediate value from the data summarized in Table 1. The range is based on the distribution and the high and low end of the data summarized in Table 1. This comparison shows that Shorelines A and B are similar to each other, while Shoreline C stands out as significantly more expensive to protect.

Table 5. Estimated costs for raising coastal protective infrastructure to meet future sea level rise scenarios for the three designated potential shorelines in billions of USD$.

Sea Level Rise Scenario	Shoreline A (Saltwater)			Shoreline B (Salt/Fresh)			Shoreline C (Freshwater)		
	Range Low	Typical	Range High	Range Low	Typical	Range High	Range Low	Typical	Range High
0.5 m	$24	$39	$53	$25	$38	$51	$43	$63	$83
1.0 m	$33	$51	$70	$37	$57	$77	$69	$103	$137
1.5 m	$81	$126	$172	$95	$148	$200	$157	$240	$323
2.0 m	$116	$182	$248	$136	$212	$287	$217	$335	$453

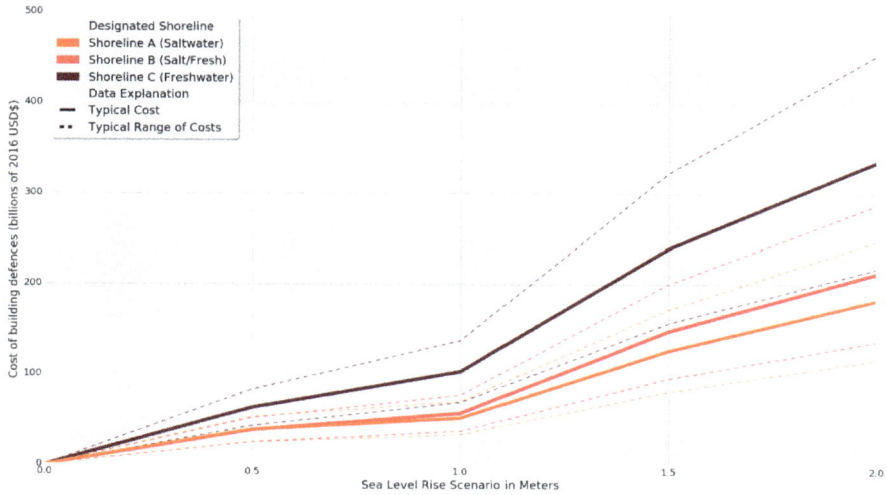

Figure 10. Figure comparing the average and the range of potential costs to raise the protective infrastructure along each of the three different shorelines. Costs are based on current infrastructure type—landform or wall—being raised to meet future water levels in a no-storm-surge condition and sea level rise scenarios. The cost estimates are done using Approach #4 and are shown as billions of 2016 USD$.

We also compared the Bay-wide costs of raising coastal infrastructure high enough to prevent flooding from new sea levels combined with a storm surge scenario (the 100-year or 1% storm surge). We display the "storm-vs.-no storm" comparison for our cost estimate for Shoreline B in Figure 11. In a 1-m sea level rise scenario, the cost goes from an average of $56.8 billion 2016 USD$ in a no-storm-surge condition to $98.9 billion 2016 USD$ in a one-percent chance storm condition. It is significant to note that raising infrastructure to prevent flooding from temporary storm events along with sea level rise is almost twice as expensive as raising the same structures to prevent flooding from permanent sea level rise only, while adapting to temporary flooding on the landward side.

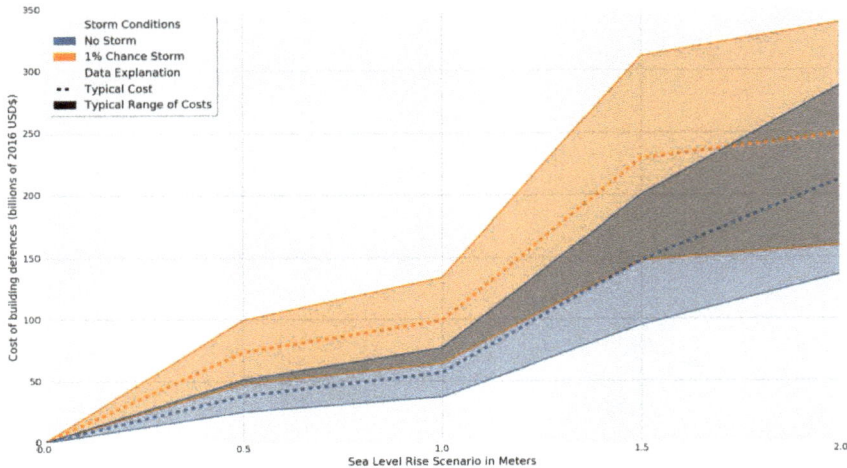

Figure 11. Figure showing the average and the range of potential costs under two different design scenarios for Shoreline B. The lines in blue show a scenario that assumes protective infrastructure would be designed to address only future sea level rise and not storm events. The lines in orange assume that protective infrastructure would be designed to address the 1% chance storm event as well as sea level rise. Costs are based on current infrastructure type—landform or wall—being raised to meet future water levels for sea level rise scenarios between 0 m and 2 m. The cost estimates are done using Approach 3: complex cost approximations without parcel values and are shown as billions of 2016 USD$.

4. Discussion: Strategic Implications of Results

4.1. Implications for Alternative Shorelines and Design Strategies

Coastal planning and engineering practitioners often suggest that the vulnerability of a shoreline can be reduced by shortening the overall length of the shoreline [40,41]. Our results show a lower cost for the shortest shoreline in Figure 10. However, our research raises specific concerns about this strategy. While we do find that the longest line (Shoreline C) represents significantly more cost to adapt to future sea level rise, the other two shorelines (A and B) are likely to be similar to each other in cost despite their different lengths. If we weigh these costs against the value of ecosystem services provided by the intertidal zone, the middle shoreline (Shoreline B) appears to be optimal. In San Francisco Bay today, site managers and public agency planners are proposing projects to raise coastal structures along this middle shoreline (our Shoreline B) [25,32].

Prior research suggests that raising wall infrastructure types will be an order of magnitude more expensive than raising the landform infrastructure types. Our results, as shown in Figure 7, support this conclusion. Figure 7 shows that the cost of raising the height of existing walls, which currently make up only 15% of Shorelines B and C, has a significantly greater impact on total cost than the cost of raising landforms. If we re-calculate our estimated cost for Shoreline B using only walls, the total cost increases by 7.5 to 8.9 times the cost of a combined landform-and-wall coastal shoreline. From a cost perspective, landform-based shoreline infrastructure may be a better investment over time. Switching to wall-based adaptation where there are no walls today introduces the risk of locking future generations into a much more expensive strategy.

There is significant debate both in the literature and in practice as to the ideal height of protective shoreline infrastructure, and the associated amount of temporary flooding that urban development would need to accommodate. On the one hand, reduction in risk is critical and protective infrastructure should be designed to prevent disastrous flooding events. On the other hand, researchers using models find that exposure to some flooding can reduce overall vulnerability [42], and may even promote the development of more resilient urban districts (sometimes called "floodable development"). Planners in the SF Bay region are currently trying to determine an appropriate height for raised structures in locations where the cost of providing coastal protection is outweighed by the multiple benefits that the protection provides [32,36]. Figure 11 shows that designing infrastructure to protect against infrequent, extreme flood events would be significantly more expensive than protecting against the permanent landward movement of the mean water line. While protecting against the 10 or 20 year storm may make financial sense, designing urban districts that are adapted to temporary flooding may be a better way to share costs and risks. Our results present important strategic implications for the design of vulnerability studies, and for setting goals in adaptation plans. If planners seek to accept some flooding while protecting against permanent inundation, physical plans might include floodable urban districts [43]. Adaptation costs could then be shared and optimized in new ways.

Actual adaptation strategies are likely to be complex in their spatial patterns and mix of structures. Previous land use is likely to influence realignment [8]. Rivers will contribute additional flooding problems, exacerbated by the increased incidence of extreme precipitation [7]. Coastal structures will also have upstream impacts. In other regions, even movable barriers, such as tide gates, have been shown to produce negative impacts on upstream wetlands and fish populations [44–47]. These, and a host of other specific considerations, would need to be taken into account when selecting a site-specific adaptation strategy. Further research is required to address all of these unique challenges. However, we hope that our work can highlight the cost-specific limitations of a regional strategy that relies on walls and static landforms, such as levees and terraces.

4.2. Benefits and Limitations of Our Cost Estimation Method

The ability to easily calculate the costs of protective infrastructure needed to address sea level rise would be a powerful tool for state and regional agencies, coastal planners, and local officials. The rapid assessment approach we present in this study is especially useful where relatively accurate regional datasets are available. Additionally, our method allows planners to customize key policy components, such as the desired freeboard height, to conduct sensitivity analyses. This may also allow our approach to be used in other geographic regions with different wave energy regimes, and therefore different structural design requirements. Our approach currently does not account for other major project costs, such as conducting required planning studies, managing soil contamination, and maintaining project contingency reserves. In addition, we have not included the capital, operations, and maintenance costs of pumps that will be required to remove stormwater when levees or other structures block the flow of runoff and tributaries, and to lower future groundwater levels. These very significant costs can also be estimated, but would require hydrologic flow data and groundwater modeling along shoreline segments to approximate the volumes of pumping that will be required.

4.3. Potential Future Research

Three important questions for future research arose during this project. One is whether and how coastal protective infrastructure can be paired with adjacent land uses and ecosystems to provide the greatest combined level of resilience [26]. Specifically, areas with high groundwater and riverine settings will require significant analysis to ensure that precipitation-based flooding and storm surge are both accounted for in designs and plans. Studies of possible pairings would allow planners to evaluate specific sites and develop a strategy for districts within regions.

Second, our results provide only the first step in the kind of full cost benefit analysis that is required for proper adaptation planning. The costs associated with potential damages as well as the significant impacts to physical assets and human populations will be a key factor in determining the levels of protection required. Moreover, estuary regions must consider how adaptation will affect ecosystem services, and seek to achieve greater connectivity for species and habitat. Without this advance planning, key functions can be completely lost as new or raised shoreline structures block hydrologic flows, wetland migration, and the movement of animals. Our research represents only a first step in estimating the costs of protective infrastructure using different spatial shoreline positions. It is essential to compare these costs to the value of assets that would be protected, including the value of key ecosystem services.

Finally, this study raised the question of whether it will be possible to identify thresholds associated with sea level rise and other drivers of coastal flooding. What thresholds in rates or magnitudes of processes would create unsafe conditions for occupying coastal developments? What thresholds might push coastal adaptation onto different adaptation pathways, for example, away from protection and towards realignment? Our method provides a starting point for exploring these questions from a cost perspective. The method would be enhanced by incorporating both additional capital and planning costs, as well as secondary costs such as those associated with the operation and maintenance of pumps. Including operations and maintenance costs would enrich the current method and provide for an evaluation that could identify different preferences among the alternative strategies.

5. Conclusions

5.1. Strategic Physical Design Implications

Given our assumptions about replacement thresholds, walls are significantly more expensive than levees and should only be used as a last resort approach to protection. Building them may create liabilities rather than assets for future generations, who will bear the costs of removing or replacing them. In general, creating structures that can fail catastrophically is unlikely to contribute to resilience over multiple decades when the rate of environmental change is accelerating, creating a greater likelihood of failure. Moreover, walls are typically single-purpose structures that do not provide benefits of recreation, carbon sequestration, and other ecosystem services. Walls, even movable barriers such as tide gates, produce negative impacts on upstream wetlands and fish populations [44–47].

When it comes to identifying an appropriate shoreline, our work suggests that defending the shortest shoreline may not be the best adaptation solution. Based on our approximation of the relative costs, Shoreline B (at the boundary between saltwater habitat and freshwater habitat) may be preferable to the shortest possible shoreline in this study (Shoreline A). Shoreline B has similar costs, but produces multiple benefits from ecosystem services. Shoreline C, which was the most expensive overall, also raises issues of unique flooding problems that occur where freshwater or brackish tributaries meet a saltwater body. Freshwater flooding driven by rainfall can combine with rising groundwater and saltwater flooding to create extreme flood conditions. These points of intersection among flows may require the realignment of shorelines, with the relocation of urban districts and infrastructure. Or, they could be locations where new strategies for human occupation of the shore zone are tested, between Shoreline B and Shoreline C. Floating concrete roadways, stormwater ponds, floating urban blocks, and freshwater habitat could form a mosaic of new development strategies that protect existing sewage treatment plants, which are often located at the intersections of fresh and salt water. This is a geographic zone that needs extensive further study in order to avoid the expansion or introduction of maladaptive strategies.

5.2. Strategic Regulatory Implications

We suggest that shoreline adaptation would be more successful if the shoreline structures continue to allow flooding in extreme events. While there is significant debate both in the literature and in practice as to the ideal design height for shoreline protective infrastructure, our results show clearly that creating requirements to design coastal structures with certain freeboard heights or storm surge expectations has a large impact on costs. Freeboard heights and other storm-specific design requirements should be set to reflect the intended purpose of the shoreline infrastructure, and its specific context within the shore zone. If coastal structures are designed to allow some overtopping without structural failure, higher sea levels can be allowed to produce temporary flooding from storm events. Periodic, temporary flooding can drive urban development to become more resilient by being re-designed to accommodate some floodwaters [42,43].

5.3. Strategic Economic and Overall Approach Implications

Most importantly, this paper illustrates the clear benefits of using a simplified typology to categorize shoreline types and calculate initial cost estimates. Working with a combination of Python and GIS enables our approach to handle very large datasets that are crucial for a large regional area. Moreover, our unit cost approach allows us to explore and compare different components of adaptation costs. Specifically, we find that the land purchase costs for levees contribute less to the overall regional cost than expected. Additionally, we see that triggers for replacing walls are a key driver of the total regional costs. We think this approach shows great promise for helping regional planners address critical strategic climate adaptation questions.

Acknowledgments: This study was performed as part of Daniella Hirschfeld's Ph.D. thesis work. Funding was provided by the McQuown Fellowship. SFEI and Point Blue provided data and assistance with data interpretation that was incredibly helpful to our work. Ellen Plane assisted by editing map data and constructing the shoreline alternatives, and helped to compile case studies of costs. Their support was greatly appreciated.

Author Contributions: D.H. and K.E.H. conceived and designed the key research questions and general methods collaboratively; D.H. performed the data collection and detailed model design; D.H. and K.E.H. analyzed the data; D.H. developed the figures and wrote Sections 1–4; K.E.H wrote Section 5 and contributed significant editing to the full paper.

Conflicts of Interest: The authors declare no conflict of interest.

References

1. Aerts, J.; Botzen, W.; Bowman, M.; Dircke, P.; Ward, D.I. *Climate Adaptation and Flood Risk in Coastal Cities*; Routledge: Abingdon, UK, 2013; ISBN 978-1-136-52893-4.
2. Brody, S.D.; Zahran, S.; Maghelal, P.; Grover, H.; Highfield, W.E. The Rising Costs of Floods: Examining the Impact of Planning and Development Decisions on Property Damage in Florida. *J. Am. Plan. Assoc.* **2007**, *73*, 330–345. [CrossRef]
3. Hill, K. Climate-Resilient Urban Waterfronts. In *Climate Adaptation and Flood Risk in Coastal Cities*; Routledge: Abingdon, UK, 2013; ISBN 978-1-136-52893-4.
4. Nicholls, R.J.; Wong, P.P.; Burkett, V.R.; Codignotto, J.O.; Hay, J.E.; McLean, R.F.; Ragoonaden, S.; Woodroffe, C.D. Coastal Systems and Low-Lying Areas. In *Climate Change 2007: Impacts, Adaptation and Vulnerability. Contribution of Working Group II to the Fourth Assessment Report of the Intergovenmental Panel on Climate Change*; Perry, M.L., Canziani, O.F., Palutikof, J.P., van der Linden, P.J., Hanson, C.E., Eds.; Cambridge University Press: Cambridge, UK, 2007; pp. 315–356.
5. IPCC. Summary for Policymakers. In *Climate Change 2013: The Physical Science Basis. Contribution of Working Group I to the Fifth Assessment Report of the Intergovernmental Panel on Climate Change*; Stocker, T.F., Qin, D., Plattner, G.-K., Tignor, M., Allen, S.K., Boschung, J., Nauels, A., Xia, Y., Bex, V., Midgley, P.M., Eds.; Cambridge University Press: Cambridge, UK; New York, NY, USA, 2013.
6. Rosenzweig, C.; Solecki, W.D.; Blake, R.; Bowman, M.; Faris, C.; Gornitz, V.; Horton, R.; Jacob, K.; LeBlanc, A.; Leichenko, R.; et al. Developing coastal adaptation to climate change in the New York City infrastructure-shed: Process, approach, tools, and strategies. *Clim. Chang.* **2011**, *106*, 93–127. [CrossRef]
7. Wahl, T.; Jain, S.; Bender, J.; Meyers, S.D.; Luther, M.E. Increasing risk of compound flooding from storm surge and rainfall for major US cities. *Nat. Clim. Chang.* **2015**, *5*, 1093–1097. [CrossRef]
8. French, P.W. Managed realignment? The developing story of a comparatively new approach to soft engineering. *Estuar. Coast. Shelf Sci.* **2006**, *67*, 409–423. [CrossRef]
9. Reeder, T.; Ranger, N. How Do You Adapt in an Uncertain World? Lessons from the Thames Estuary 2100 Project. Available online: http://www.worldresourcesreport.org/ (accessed on 15 April 2015).

10. Doody, J.P. Coastal squeeze and managed realignment in southeast England, does it tell us anything about the future? *Ocean Coast. Manag.* **2013**, *79*, 34–41. [CrossRef]

11. Nicholls, R.J.; Klein, R.J.T. Climate change and coastal management on Europe's coast. In *Managing European Coasts*; Vermaat, J., Salomons, W., Bouwer, L., Turner, K., Eds.; Environmental Science; Springer: Berlin/Heidelberg, Germany, 2005; pp. 199–226, ISBN 978-3-540-23454-8.

12. Ezer, T.; Corlett, W.B. Is sea level rise accelerating in the Chesapeake Bay? A demonstration of a novel new approach for analyzing sea level data. *Geophys. Res. Lett.* **2012**, *39*. [CrossRef]

13. Knowles, N. *Potential Inundation Due to Rising Sea Levels in the San Francisco Bay Region*; California Climate Change Center: Sacramento, CA, USA, 2009.

14. Heberger, M.; Cooley, H.; Moore, E.; Herrera, P.; Pacific Institute. *The Impacts of Sea Level Rise on the San Francisco Bay*; California Energy Commission: Sacramento, CA, USA, 2012.

15. Stralberg, D.; Brennan, M.; Callaway, J.C.; Wood, J.K.; Schile, L.M.; Jongsomjit, D.; Kelly, M.; Parker, V.T.; Crooks, S. Evaluating Tidal Marsh Sustainability in the Face of Sea-Level Rise: A Hybrid Modeling Approach Applied to San Francisco Bay. *PLoS ONE* **2011**, *6*, e27388. [CrossRef] [PubMed]

16. California Ocean Protection Council. Updating the State of California Sea-Level Rise Guidance Document. Available online: http://www.opc.ca.gov/climate-change/updating-californias-sea-level-rise-guidance/ (accessed on 18 April 2017).

17. Griggs, G.; Arvai, J.; Cayan, D.; DeConto, R.; Fox, J.; Fricker, H.A.; Kopp, R.E.; Tebaldi, C.; Whiteman, L.; California Ocean Protection Council Science Advisory Team Working Group. *Rising Seas in California: An Update on Sea Level Rise Science*; California Ocean Science Trust: Oakland, CA, USA, 2017.

18. Kopp, R.E.; Horton, R.M.; Little, C.M.; Mitrovica, J.X.; Oppenheimer, M.; Rasmussen, D.J.; Strauss, B.H.; Tebaldi, C. Probabilistic 21st and 22nd century sea-level projections at a global network of tide-gauge sites. *Earths Future* **2014**, *2*, 2014EF000239. [CrossRef]

19. Brown, S.; Nicholls, R.J.; Hanson, S.; Brundrit, G.; Dearing, J.A.; Dickson, M.E.; Gallop, S.L.; Gao, S.; Haigh, I.D.; Hinkel, J.; et al. Shifting perspectives on coastal impacts and adaptation. *Nat. Clim. Chang.* **2014**, *4*, 752–755. [CrossRef]

20. Hill, K. Climate Change: Implications for the Assumptions, Goals and Methods of Urban Environmental Planning. *Urban Plan.* **2016**, *1*, 103. [CrossRef]

21. Bierbaum, R.; Smith, J.B.; Lee, A.; Blair, M.; Carter, L.; Iii, F.S.C.; Fleming, P.; Ruffo, S.; Stults, M.; McNeeley, S.; et al. A comprehensive review of climate adaptation in the United States: more than before, but less than needed. *Mitig. Adapt. Strateg. Glob. Chang.* **2013**, *18*, 361–406. [CrossRef]

22. Haasnoot, M.; Kwakkel, J.H.; Walker, W.E.; ter Maat, J. Dynamic adaptive policy pathways: A method for crafting robust decisions for a deeply uncertain world. *Glob. Environ. Chang.* **2013**, *23*, 485–498. [CrossRef]

23. Nordgren, J.; Stults, M.; Meerow, S. Supporting local climate change adaptation: Where we are and where we need to go. *Environ. Sci. Policy* **2016**, *66*, 344–352. [CrossRef]

24. Hirschfeld, D.; Hill, K.; Plane, E. *SanFrancisco Bay—Adapt2SeaLevelRise—Case Studies*; UC Berkeley Library DASH: Berkeley, CA, USA, 2017.

25. San Francisquito Creek Joint Powers Authority. *SAFER Bay Project—Strategy to Advance Flood Protection, Ecosystems and Recreation along San Francisco Bay*; San Francisquito Creek Joint Powers Authority: East Palo Alto/Menlo Park, CA, USA, 2016; p. 91.

26. Hill, K. Coastal infrastructure: A typology for the next century of adaptation to sea-level rise. *Front. Ecol. Environ.* **2015**, *13*, 468–476. [CrossRef]

27. San Francisco Estuary Institute (SFEI). *San Francisco Bay Shore Inventory: Mapping for Sea Level Rise Planning GIS Data*; San Francisco Estuary Institute: Richmond, CA, USA, 2016; Available online: http://www.sfei.org/data/sf-bay-shore-inventory-gis-data#sthash.ctGdURWD.dpbs (accessed on 2 February 2015).

28. San Francisco Estuary Institute (SFEI). *Bay Area Aquatic Resource Inventory (BAARI)*; San Francisco Estuary Institute: Richmond, CA, USA, 2009. Available online: http://www.sfei.org/baari#sthash.MKGpl13z.dpbs (accessed on 13 May 2015).

29. United States Geological Survey (USGS). *Coastal Storm Modeling System (CoSMoS) Version 2.1: San Francisco Bay Area. Flood Extent and Depth*; Point Blue Conservation Science: Petaluma, CA, USA, 2014. Available online: www.ourcoastourfuture.org (accessed on 31 August 2017).

30. Barnard, P.L.; van Ormondt, M.; Erikson, L.H.; Eshleman, J.; Hapke, C.; Ruggiero, P.; Adams, P.N.; Foxgrover, A.C. Development of the Coastal Storm Modeling System (CoSMoS) for predicting the impact of storms on high-energy, active-margin coasts. *Nat. Hazards* **2014**, *74*, 1095–1125. [CrossRef]

31. ChrisMuir. Zillow Scraper for Python Using Selenium. 2017. Available online: https://github.com/ChrisMuir/Zillow/ (accessed on 30 March 2017).

32. United States Army Corps of Engineers. *South San Francisco Bay Shoreline Phase I Study - Final Integrated Document*; United States Army Corps of Engineers: San Francisco, CA, USA, 2015.

33. Jonkman, S.N.; Hillen, M.M.; Nicholls, R.J.; Kanning, W.; van Ledden, M. Costs of Adapting Coastal Defences to Sea-Level Rise—New Estimates and Their Implications. *J. Coast. Res.* **2013**, *29*, 1212–1226. [CrossRef]

34. Lowe, J.; Battalio, B.; Brennan, M. *Analysis of the Costs and Benefits of Using Tidal Marsh Restoration as a Sea Level Rise Adaptation Strategy in San Francisco Bay*; The Bay Institute: San Francisco, CA, USA, 2013.

35. Linham, M.M.; Nicholls, R.J. *Technologies for Climate Change Adaptation: Coastal Erosion and Flooding*; UNEP Riso Centre on Energy, Climate and Sustainable Development: Roskilde, Denmark, 2010; ISBN 978-87-550-3855-4.

36. GHD-GTC Joint Venture. *Recommendations for Hazard Mitigation for the Seawall Earthquake Vulnerability Study—Phase 3 Draft Report*; Port of San Francisco: San Francisco, CA, USA, 2016; p. 154.

37. Huang, H.; Tang, Y. Residential land use regulation and the US housing price cycle between 2000 and 2009. *J. Urban Econ.* **2012**, *71*, 93–99. [CrossRef]

38. Mian, A.; Sufi, A. The Consequences of Mortgage Credit Expansion: Evidence from the U.S. Mortgage Default Crisis. *Q. J. Econ.* **2009**, *124*, 1449–1496. [CrossRef]

39. Williams, S.; Ismail, N. Climate Change, Coastal Vulnerability and the Need for Adaptation Alternatives: Planning and Design Examples from Egypt and the USA. *J. Mar. Sci. Eng.* **2015**, *3*, 591–606. [CrossRef]

40. Hillen, M.; Jonkman, S.; Kanning, W.; Kok, M.; Geldenhuys, M.; Stive, M. *Coastal Defense Cost Estimates*; Delft University of Technology, Royal Haskoning: Delft, The Netherlands, 2010; ISSN 0169-6548.

41. Jonkman, S.N.; Kok, M.; van Ledden, M.; Vrijling, J.K. Risk-based design of flood defence systems: A preliminary analysis of the optimal protection level for the New Orleans metropolitan area. *J. Flood Risk Manag.* **2009**, *2*, 170–181. [CrossRef]

42. Di Baldassarre, G.; Viglione, A.; Carr, G.; Kuil, L.; Salinas, J.L.; Blöschl, G. Socio-hydrology: Conceptualising human-flood interactions. *Hydrol. Earth Syst. Sci.* **2013**, *17*, 3295–3303. [CrossRef]

43. Restemeyer, B.; Woltjer, J.; van den Brink, M. A strategy-based framework for assessing the flood resilience of cities—A Hamburg case study. *Plan. Theory Pract.* **2015**, *16*, 45–62. [CrossRef]

44. Elkema, M.; Wang, Z.B.; Stive, M.J.F. Impact of Back-Barrier Dams on the Development of the Ebb-Tidal Delta of the Eastern Scheldt. *J. Coast. Res.* **2012**, *28*, 1591–1605. [CrossRef]

45. Giannico, G.R.; Souder, J.A. *Tide Gates in the Pacific Northwest: Operation, Types, and Environmental Effects*; Oregon Sea Grant, Oregon State University: Corvallis, OR, USA, 2005; Volume 5.

46. Gordon, J.; Arbeider, M.; Scott, D.; Wilson, S.M.; Moore, J.W. When the Tides Don't Turn: Floodgates and Hypoxic Zones in the Lower Fraser River, British Columbia, Canada. *Estuar. Coasts* **2015**, *38*, 2337–2344. [CrossRef]

47. Louters, T.; Mulder, J.P.; Postma, R.; Hallie, F.P. Changes in coastal morphological processes due to the closure of tidal inlets in the SW Netherlands. *J. Coast. Res.* **1991**, *7*, 635–652.

Journal of
Marine Science and Engineering

MDPI

Article

Impact of North Atlantic Teleconnection Patterns on Northern European Sea Level

Léon Chafik [1],*,†, Jan Even Øie Nilsen [2] and Sönke Dangendorf [3]

[1] Geophysical Institute, University of Bergen, and Bjerknes Centre for Climate Research, 5020 Bergen, Norway
[2] Nansen Environmental and Remote Sensing Center, and Bjerknes Centre for Climate Research,
    5006 Bergen, Norway; jan.even.nilsen@nersc.no
[3] Research Institute for Water and Environment, University of Siegen, 57076 Siegen, Germany;
    Soenke.Dangendorf@uni-siegen.de
*   Correspondence: leon.chafik@uib.no; Tel.: +47-55-20-58-00
†   Current address: Universitetet i Bergen, Geofysisk institutt, Postboks 7803, 5020 Bergen, Norway.

Received: 31 May 2017; Accepted: 30 August 2017; Published: 6 September 2017

Abstract: Northern European sea levels show a non-stationary link to the North Atlantic Oscillation (NAO). The location of the centers of the NAO dipole, however, can be affected through the interplay with the East Atlantic (EAP) and the Scandinavian (SCAN) teleconnection patterns. Our results indicate the importance of accounting for the binary combination of the NAO with the EAP/SCAN for better understanding the non-stationary drivers inducing sea level variations along the European coasts. By combining altimetry and tide gauges, we find that anomalously high monthly sea levels along the Norwegian (North Sea) coast are predominantly governed by same positive phase NAO+/EAP+ (NAO+/SCAN+) type of atmospheric circulation, while the Newlyn and Brest tide gauges respond markedly to the opposite phase NAO−/EAP+ combination. Despite these regional differences, we find that coherent European sea level changes project onto a pattern resembling NAO+/SCAN+, which is signified by pressure anomalies over Scandinavia and southern Europe forcing winds to trace the continental slope, resulting in a pile-up of water along the European coasts through Ekman transport. We conclude that taking into consideration the interaction between these atmospheric circulation regimes is valuable and may help to understand the time-varying relationship between the NAO and European mean sea level.

Keywords: European sea level; atmospheric circulation; teleconnections; satellite altimetry; tide gauges

1. Introduction

Global mean sea level has been accelerating from rates of 1.1 to 2 mm/year before 1990 [1] towards unprecedented high rates of 3.3 ± 0.4 mm/year afterwards [2] and is expected to continue its acceleration in a warming climate primarily as a result of melting ice and ocean thermal expansion [3,4] giving rise to significant socioeconomic and environmental consequences on the coastal zones [5]. Rising seas can, however, deviate immensely on regional scale relative to the global mean owing to several processes acting at different locations on a range of timescales [6–8].

Wind forcing over the North Atlantic, for instance, can cause regional and coastal mean sea level variations acting from months to decades [9–11]. However, since the North Atlantic meridional dipole of the mean sea level pressure field is subject to spatial displacements (non-stationarity), the relationship between northern European sea level and atmospheric forcing, as typically defined by the North Atlantic Oscillation, is found to be temporally variable (e.g., Wakelin et al. [12]). This non-stationarity of the centers of action is presumed to induce notable impacts on northern European sea level (e.g., Dangendorf et al. [13]). The present study therefore focuses on understanding the response of northern European sea level to this non-stationarity of the centers of action through a simple approach that combines the three prominent North Atlantic teleconnection patterns [14].

Strong and weak periods of atmospheric forcing in the North Atlantic region are usually linked to the strength of the prevailing westerly winds as measured by the meridional mean sea-level pressure (MSLP) gradient between the Icelandic low and the Azores high, i.e., the North Atlantic Oscillation (NAO); a major mode of North Atlantic climate and ocean circulation variability [15–18]. During its positive phase, the westerlies are anomalously strong and more intense storms are generated. While during a negative NAO phase, the winds are weaker than normal and the jet stream and hence the storm tracks are shifted southward [15]. However, several recent studies have indicated that the NAO index alone is not sufficient to explain the climatic variations over both ocean [19,20] and land [21–24], and that other teleconnections, typically the second and third leading mode of atmospheric variability in the North Atlantic/European sector [25–27], are important to take into consideration. The reasoning is that North Atlantic climate variability is influenced by spatial movements of the centers of action that can only be captured when combining the NAO with the other two leading modes [14,28]. Already in 1939, Rossby [29] recognized that displacements of the semi-permanent centers of action are able to influence the climate sytem. Woollings et al. [30], for example, showed that the latitudinal position of the eddy-driven jetstream can be explained by a combination of the NAO and the East-Atlantic Pattern (EAP, the second leading mode). The latter was recently suggested by Comas-Bru and McDermott [23] to influence temperature and precipitation patterns over Europe, often also together with the Scandinavian pattern (SCAN, the third leading mode [27]).

From a sea-level perspective, a number of studies have reported on the impact of the NAO on the mean sea level, its variability, sensitivity and related processes [12,31–39]. Yan et al. [33] and Jevrejeva et al. [34] find that the NAO co-varies strongly with the sea level in northwest Europe using tide gauges dating back to the 19th century, but they also point out that this relationship is not stable over time. In particular, it was strongly enhanced during the second half of the 20th century. This finding is in agreement with that of Wakelin et al. [12], where the 1909–1954 period was reported to exhibit lower correlations with the NAO from 17 tide gauges in the North Sea region as compared to higher correlations during the 1955–2000 period (see also Andersson [31]). But although this intermittency in the relationship of the sea level to the NAO is argued to have been induced by shifts in the centers of action (see, e.g., Hilmer and Jung [40], Jung et al. [41]), the influence of this non-stationarity of the atmospheric circulation on the sea level along the European coasts remains poorly quantified. Perhaps with the exception of the studies by Kolker and Hameed [42] and Ullmann and Monbaliu [43]. The former assessed the influence of shifts in the position of the centers of action on sea level change at five tide gauges in the Atlantic, while the latter focused on daily sea surges at a single station along the Belgian coast and their relationship to weather regimes.

In this study, we utilize satellite altimetry, tide gauges and atmospheric reanalysis to investigate the influence of the migrations of the North Atlantic meridional pressure dipole on the sea level along the European shelves and coasts through the combination of the canonical NAO, EAP and SCAN patterns. Our findings suggest that the NAO alone is not always sufficient as a forcing mechanism (see, e.g., Dangendorf et al. [13], Kolker and Hameed [42], Ullmann and Monbaliu [43]), and that it is essential to take into account all three leading modes of North Atlantic atmospheric variability to explain the anomalous periods of the European sea level, not least since there are regional differences with a preference toward a specific atmospheric pattern that can only be formed by the linear combination of these teleconnections [14].

The paper is structured as follows. Section 2 describes the data and methodology. Section 3 shows the North Atlantic teleconnection patterns, their linear combinations, and how they influence the sea level along the European shelves and coasts consistently using both altimetry and tide gauges. These results are followed by a summarizing discussion and outlook in Section 4.

2. Data and Methodology

2.1. Satellite Altimetry

We use the DUACS DT2014 multi-mission satellite altimetry [44] to study the sea-level variability on the European continental shelf, Figure 1. An important change in the DT2014 is the referencing of the sea level anomaly products to the 20 years of available measurements as compared to the previous 7-year reference period. The new reprocessed sea level anomaly products demonstrate multiple significant improvements in, e.g., mesoscale signals, eddy kinetic energy, geostrophic currents (now more comparable to surface drifter observations) and sea level at higher latitudes as well as in coastal areas [44]. The DT2014 demonstrates a better consistency with sea levels from tide gauge stations. The geophysical corrections pertaining to inverse barometer, tides and dry/wet tropospheric effects have been applied beforehand. We here utilize the monthly mean absolute dynamic topography (the sum of sea-level anomalies and the MDT_CNES/CLS 2013 ocean mean dynamic topography) averaged from daily data, on a $1/4°$ cartesian grid, between January 1993 and December 2015.

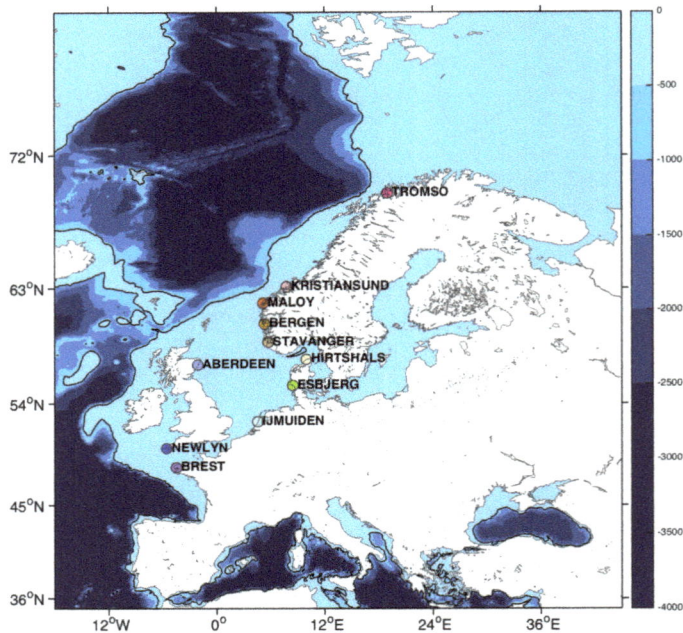

Figure 1. Bathymetric map (m) including the location of the 11 tide gauges used in this study. The 500 m isobath is indicated in black.

2.2. Tide Gauges

Monthly data from 11 tide gauges (TGs) along the European coasts are used, Figure 1. This data was downloaded from the webpage of the Permanent Service for Mean Sea Level (PSMSL) [45], and have been corrected for the inverse barometer effect using the National Center for Environmental Prediction/National Center for Atmospheric Research (NCEP/NCAR) reanalysis [46], linearly detrended, and deseasonalized by removing the mean seasonal cycle. The TG period under investigation stretches from January 1950 until December 2014. Note, however, that although a weak relationship during the 1909–1954 period between sea levels from tide gauges and the NAO has been reported (e.g., Wakelin et al. [12]), using these data since 1950 is not an issue for our results.

2.3. Atmospheric Data

We use the monthly NCEP/NCAR reanalysis [46] monthly mean sea level pressure (MSLP) and winds at 10 m spanning the 1950–2015 period. The grid resolution of this data is 2.5° × 2.5°. The monthly data have been linearly detrended, and deseasonalized by removing the mean seasonal cycle. The MSLP is used to calculate the three leading empirical orthogonal functions (EOFs) of atmospheric variability in the North Atlantic/European sector (80° W–50° E, 30° N–80° N) following the method described by Hannachi et al. [47], which is based on singular value decomposition. The data are weighted by the cosine of its latitude at every grid point.

2.4. Linear Combination of the Leading EOFs

The three leading EOFs together explain more than 60% of the monthly atmospheric variability in the North Atlantic/European sector. During wintertime, they explain, however, more than 90% of the variability in the subpolar North Atlantic and northern Europe [14]. The reasoning behind the linear combination is that although the north-south meridional MSLP dipole associated with the NAO is the dominant pattern, the atmospheric circulation, at any given time, is not solely composed of a pure NAO phase but rather involves the other two leading teleconnections, i.e., the East Atlantic Pattern (EAP) and Scandinavian Pattern (SCAN), which result in horizontal displacements of the centers of action. The linear combination of MSLP can thus be written, following [14], as:

$$MSLP = \alpha \times NAO + \beta \times \left\{ \begin{smallmatrix} EAP \\ SCAN \end{smallmatrix} \right\}$$
$$where \left\{ \begin{smallmatrix} \alpha \\ \beta \end{smallmatrix} \right\} = [-1, 0, 1]$$

The two phases of the NAO (EAP/SCAN) occur when $\alpha = \pm 1$ ($\alpha = 0$) and $\beta = 0$ ($\beta = \pm 1$), while the other four possible linear combinations occur when α and β are $\neq 0$. Moore et al. [14] show that this simple technique is able to capture the fundamental characteristics associated with the mobility of the centers of action; atmospheric variability is not a result of a single EOF, but instead results from a continuum of EOFs that are acting simultaneously, and can thus be reconstructed by the linear combination of these climate modes [28].

3. Results

3.1. North Atlantic Teleconnections and Their Interaction

The three leading modes of atmospheric variability in the North Atlantic/European sector are the NAO, EAP and SCAN. These three EOFs explain 32%, 17% and 15% of the MSLP variability over the 1950–2015 period based on the NCEP/NCAR reanalysis, Figure 2. The North Atlantic meridional MSLP dipole, i.e., between the region near Iceland/western Nordic Seas and the Azores, is a distinctive pattern of the first leading EOF, i.e., the NAO. The second leading EOF, i.e., the EAP, is primarily characterized by a monopole pressure encompassing the subpolar North Atlantic region with a node south of Iceland. The third EOF, i.e., the SCAN, is centered over the southern parts of Norway with a node close to Bergen (cf. Figure 1). Note, however, that the EAP and SCAN have another node over Eastern Europe and Greenland, respectively, albeit rather weak. It should also be mentioned that the EAP and SCAN are associated with the known weather regimes Atlantic ridge and Scandinavian blocking, respectively (see Cassou et al. [48]).

We now look into the atmospheric patterns when the NAO is linearly combined with the EAP (Figure 3) and SCAN (Figure 4). Figure 3 shows that during same (opposite) sign NAO-EAP events, the centers of action shift southwestward (northeastward). Due to these displacements, the zero line during, e.g., NAO+/EAP+ (NAO+/EAP−) conditions follows the North European coast (becomes more zonally-oriented).

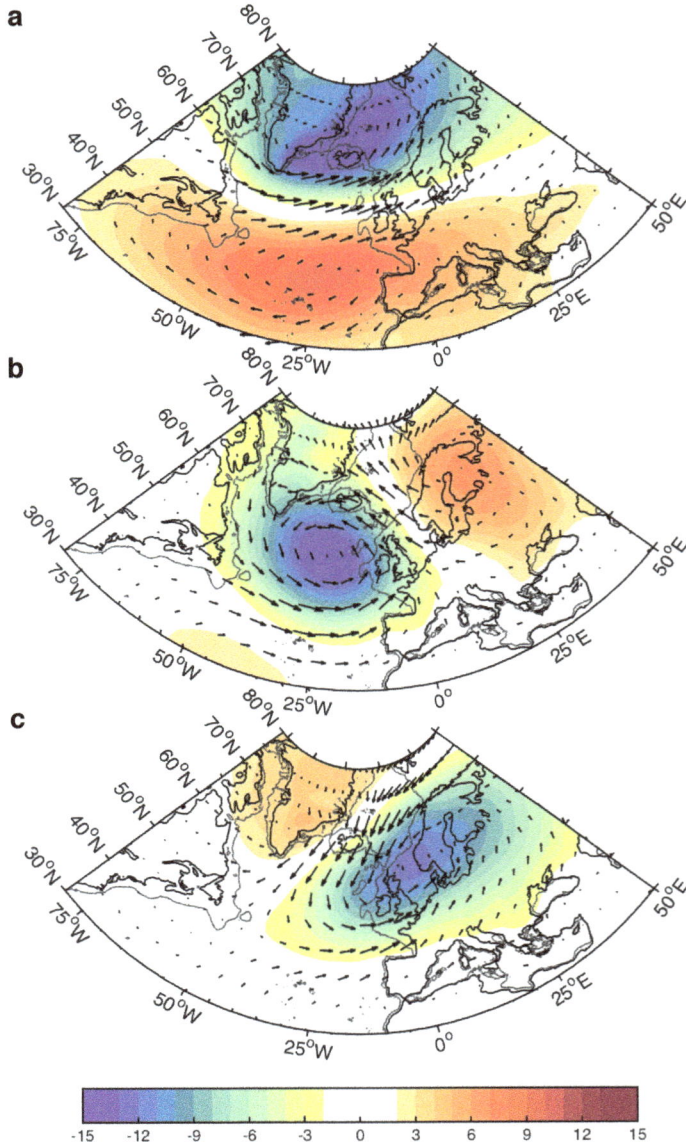

Figure 2. Composite analysis of MSLP (hPa) and 10 m winds based on the (a) NAO, (b) EAP and (c) SCAN teleconnection indices calculated using NCAR/NCEP for the 1950–2015 period. The composite difference is based on anomalously high (>0.75 std) and low (<−0.75 std) periods of the three leading principal components of MSLP.

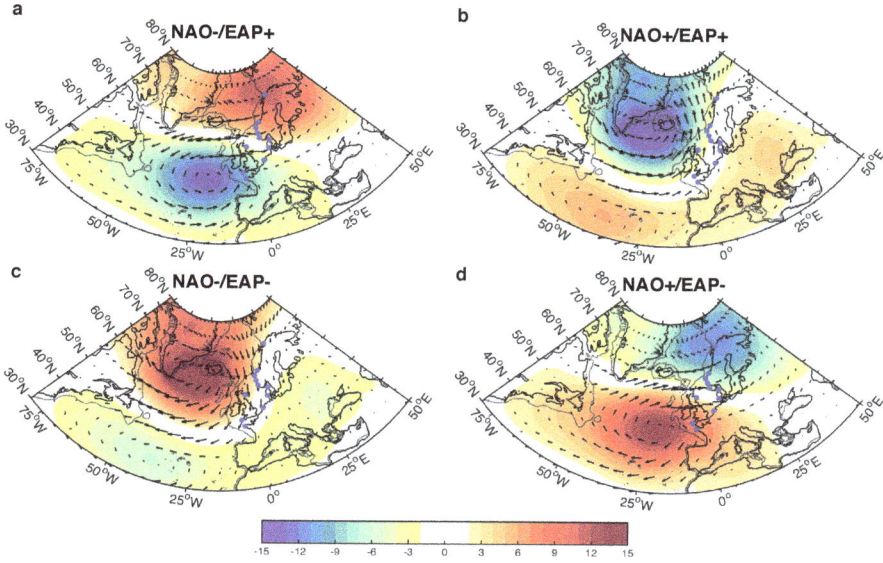

Figure 3. Meridional mean sea-level pressure (hPa, shading) and 10 m wind (vectors) patterns based on the four linear NAO-EAP combinations. (**a**) Negative and (**b**) positive NAO combined with the positive phase of the EAP. (**c**) Negative and (**d**) positive NAO combined with the negative phase of the EAP. The TG stations are overlaid to help with the orientation.

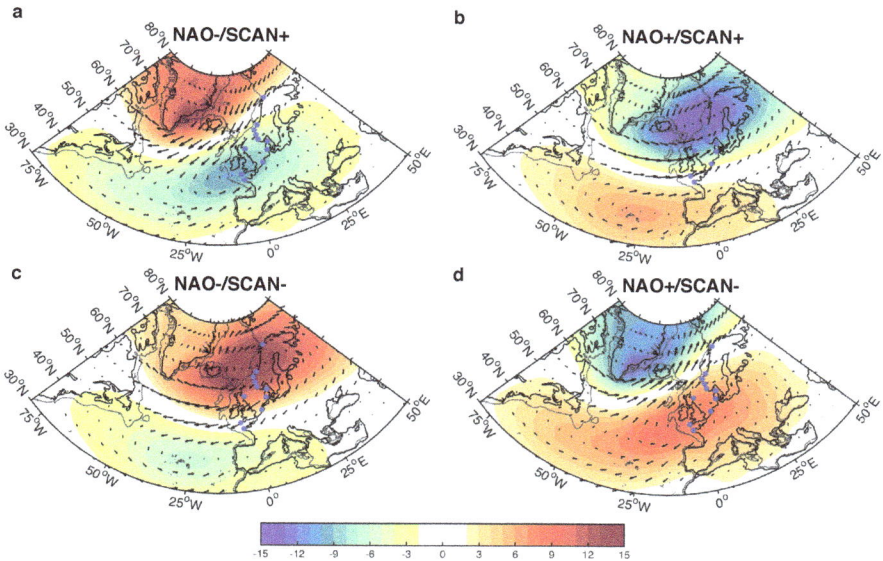

Figure 4. Meridional mean sea-level pressure (hPa, shading) and 10 m wind (vectors) patterns based on the four linear NAO-SCAN combinations. (**a**) Negative and (**b**) positive NAO combined with the positive phase of the SCAN. (**c**) Negative and (**d**) positive NAO combined with the negative phase of the SCAN. The TG stations are overlaid to help with the orientation. The TG stations are overlaid to aid with the orientation.

During same sign NAO-SCAN conditions (Figure 4), the northern node moves to the east over the Norwegian Sea, and the southern one moves to the west (and slightly to the south), now being more localized over the Azores. While during opposite sign NAO-SCAN events the northern anomaly is more displaced to the west over eastern Greenland, and the southern anomaly migrates northwards and to the east with a node just south off the British Isles resulting in an overall slanted spatial pattern. In general, the interplay between the NAO and the SCAN induces clockwise (anticlockwise) movements of the centers of action when having the same (opposite) sign. The impact of the NAO-EAP combination on the winter MSLP field reported by Moore et al. [14] are similar to our results shown in Figures 3 and 4.

3.2. The Combined Impact on European Sea Level as Seen in Altimetry

Before diagnosing the impact of the combined EOFs on the sea level along the European shelves and coasts, we show in Figure 5 how the sea level responds to the three leading teleconnection patterns separately. Figure 5a shows a coherent positive relationship to sea level over both the Norwegian and North Seas as a response to the NAO and its associated winds flowing parallel to these coasts (Figure 2a). Note that the magnitudes are larger along the coasts with the highest sea levels over the German Bight and the Baltic Sea (>0.15 m) [31]. The sea-level response to the EAP is dominated by a dipole pattern between the sea level in the North/Baltic Sea and elsewhere on the European shelves (Figure 5b). The negative and weakly positive sea level anomalies in the North Sea and Norwegian Sea, respectively, is mainly due to the large-scale winds being offshore-directed as a result of the EAP monopole structure over the subpolar North Atlantic region (Figure 2b). The response to positive SCAN periods is anomalously low sea level in the interior of the North Sea, along the Norwegian coast and the Barents Sea, and anomalously high sea level along a narrow coastal zone stretching from the Bay of Biscay to the northern tip of Denmark close to the Hirtshals TG station, as well as in the Baltic Sea (Figure 5c). Anomalously low MSLP in the outer part of the North Sea associated with the SCAN (cf. Figure 2c) induces alongshore winds in the inner part that ultimately lead to this sea level pattern.

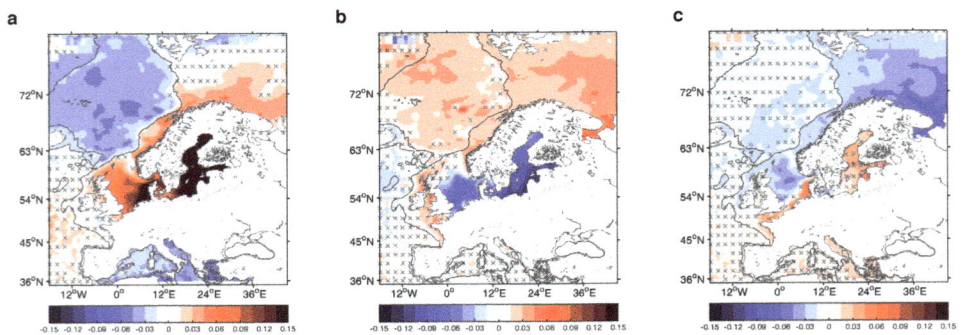

Figure 5. Sea level composite difference (m) from altimetry during the 1993–2015 period based on the (**a**) NAO, (**b**) EAP and (**c**) SCAN indices. The threshold value for the composites is 0.75 std of the teleconnection indices. The non-significant regions are indicated by gray crosses calculated using a two-sided *t* test.

We diagnose the effect of the binary interaction between the NAO with the EAP (Figure 6) and SCAN (Figure 7) on the sea level during the altimetry period, i.e., 1993–2015. The same and opposite phase combinations are observed to affect the northern shelf seas differently. Events with NAO-EAP values of the same sign show a more coherent spatial sea level pattern along the Norwegian shelf and the Barents Sea. There is little (only weak and regionalized as for the EAP alone) influence on the North Sea associated with same sign NAO-EAP events. In contrast, NAO-EAP values of the opposite sign show a strong influence on the sea level in the North and Baltic Seas. The phases of the EAP thus have different impacts on the North Sea and the Norwegian Sea, which can be attributed to the displacements of the centers of action and the orientation of the large-scale wind field. In the case of same sign NAO-EAP events, the winds closely trace the Norwegian coast (cf. Figure 3b,c) ultimately leading to anomalous (higher for NAO+/EAP+ situations) sea levels on the shelf, Figure 6b,c. While, during opposite sign situations, the modulation of the meridional pressure gradient enhances the westerlies across the North Sea (Figure 3a,d; note that the zero line crosses the North Sea), which plays an important role for the observed sea-level pattern in Figure 6a,d. These results are consistent with the findings of Dangendorf et al. [11], where anomalous sea-level situations between 1871 and 2011 from 14 tide gauges (located along the coasts of Belgium, Netherlands, Germany and Denmark, see their Figure 1 and regions 1–2) are shown to be strongly influenced by this westerly flow due to a low-pressure anomaly located over northern Scandinavia, which resembles the position of the northern center of action reflected by opposite sign NAO-EAP events.

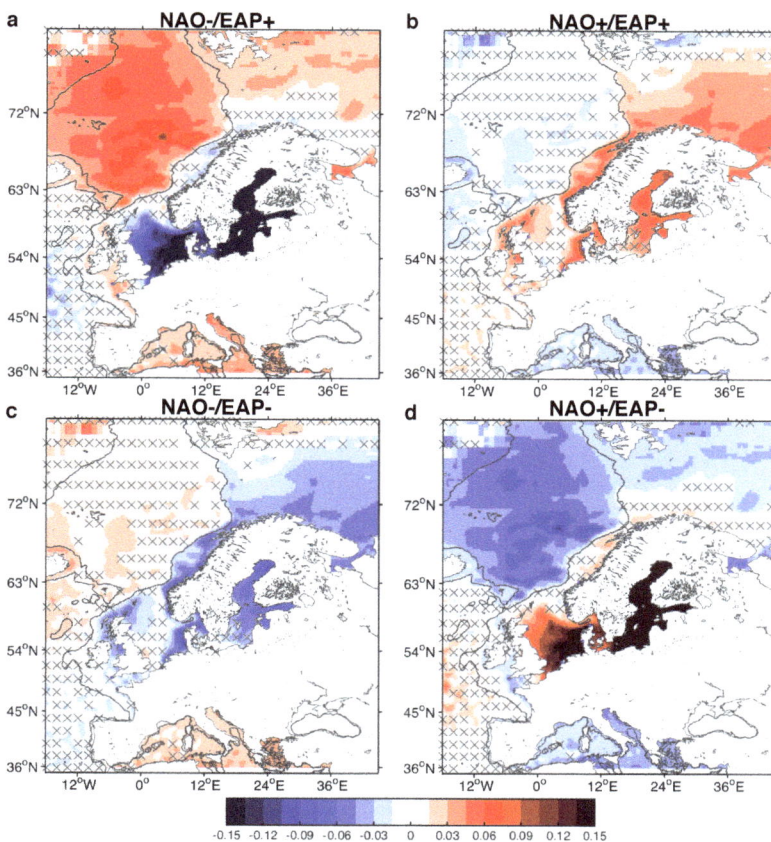

Figure 6. Sea level composite difference (m) from altimetry during the 1993–2015 period based on the four linear combinations of the NAO with the EAP. (**a**) Negative and (**b**) positive NAO combined with the positive phase of the EAP. (**c**) Negative and (**d**) positive NAO combined with the negative phase of the EAP. The threshold value for the composites is 0.75 std. The TG stations are overlaid to aid with the orientation. The non-significant regions below the 95% confidence level are indicated by gray crosses calculated using a two-sided *t* test.

The maximum (minimum) sea level associated with same sign NAO+/SCAN+ (NAO−/SCAN−) events appears to be locally concentrated to the German Bight and the Baltic Sea, cf. Figure 7. However, the opposite phases of the NAO-SCAN show a coherent sea level pattern dominating the North and Norwegian Seas, which is an immediate response to the anticlockwise movement and the slanting behavior of the centers of action ultimately reorgranizing the wind field to be parallel to these coasts (Figure 4a,d).

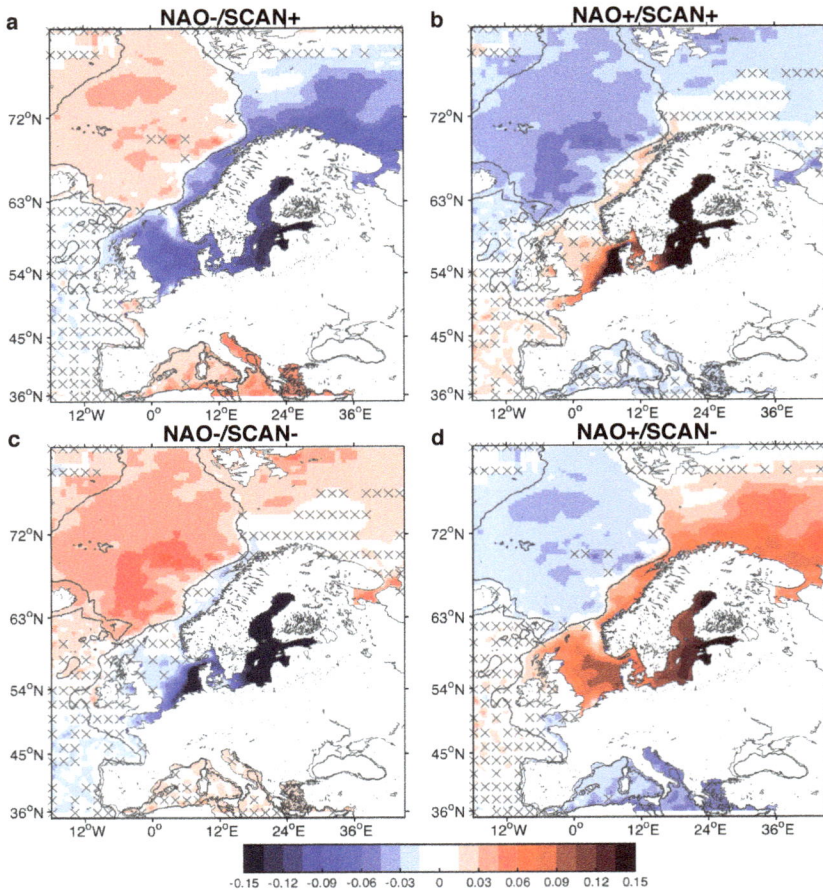

Figure 7. Sea level composite difference (m) from altimetry during the 1993–2015 period based on the four linear combinations of the NAO with the SCAN. (**a**) Negative and (**b**) positive NAO combined with the positive phase of the SCAN. (**c**) Negative and (**d**) positive NAO combined with the negative phase of the SCAN. The threshold value for the composites is 0.75 std. The TG stations are overlaid to aid with the orientation. The non-significant regions below the 95% confidence level are indicated by gray crosses calculated using a two-sided *t* test.

3.3. The Combined Impact on the European Coasts as Seen in Tide Gauges

We have shown in the previous section that anomalous periods of the teleconnection patterns and their interaction induce regional sea-level differences along the European shelves and coasts as observed during the altimetry period (1993–2015). We now investigate whether these regional differences are also present in TG records during the 1950–2014 period at the 11 European stations shown in Figure 1. We find that the sea level response at these tide gauges mirror to great detail the regional sea level structures manifested in altimetry and related to the combined effects of the NAO and either the EAP or SCAN (Figure 8).

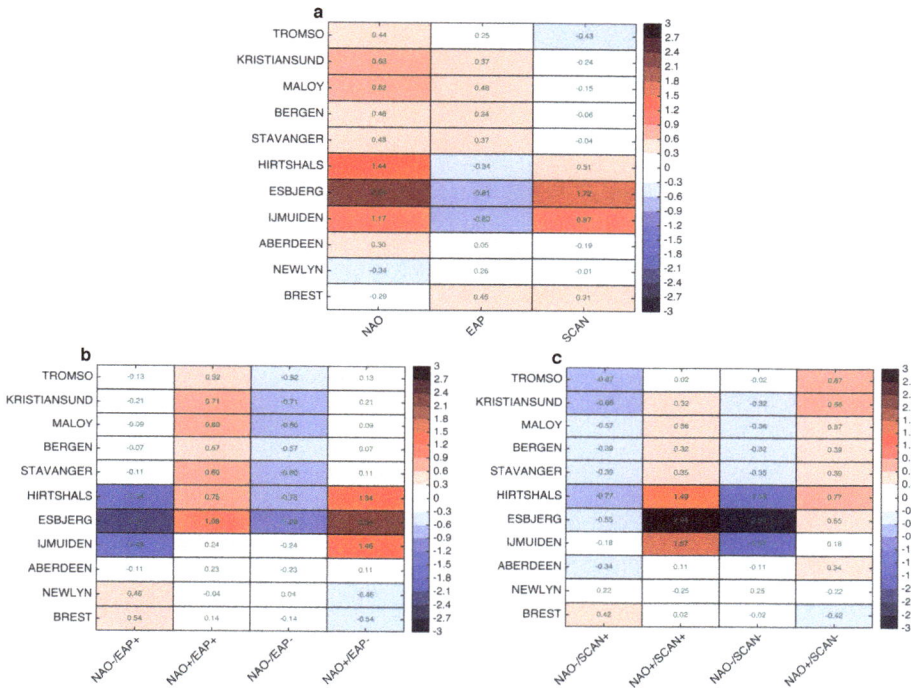

Figure 8. Composite difference of sea level (m) from the TGs based on the (**a**) three leading principal components of MSLP separately, (**b**) NAO-EAP and (**c**) NAO-SCAN combinations.

At the Norwegian TGs (Tromsø–Stavanger), the sea-level response is stronger during periods of same sign NAO-EAP combination as compared to those induced by the NAO or the EAP only (Figure 8a,b). For Brest and Newlyn, opposite sign NAO-EAP events appear to result in more anomalous sea levels as compared to any other mode or sign combination. Concerning the North Sea tide gauges, the NAO-EAP combinations (either same or opposite sign) do not show any more influence on the sea level than the NAO alone. However, events of same sign NAO-SCAN combinations are very likely to induce significantly stronger sea-level anomalies than solely through the NAO, especially for Esbjerg and Ijmuiden stations.

3.4. Teleconnection Space and Anomalous Sea Level at the European Coasts

Another approach to illustrate the preferred atmospheric state leading to anomalous sea-level periods at each TG station is to project these onto a space spanned by the NAO and the EAP (Figure 9) or SCAN (Figure 10) indices. Doing so, we study the spatial behavior and count how many of these anomalous months actually fall in the different quadrants spanned by the teleconnection indices. We can thus generalize the preferred combinations driving anomalous (>0.75 std) as well as extreme (>1.5 std) monthly sea levels as recorded at the TG stations along the European coasts.

Figure 9 demonstrates that both anomalously and extremely high monthly sea levels recorded from tide gauges along the Norwegian coast (Stavanger, Bergen, Måløy, Kristiansund and Tromsø) are dominated by the NAO+/EAP+ type of pattern, i.e., the corresponding same sign NAO-EAP values are clustered towards the upper-right quadrant. On average, about 45% (54%) of the anomalously (extremely) high monthly sea levels, respectively, are found in this quadrant. Anomalously high monthly sea levels recorded at tide gauge stations in the North Sea (Ijmuiden, Esbjerg, Hirtshals) are strongly dominated by NAO+/EAP−. On average, about 51% (59%) of the anomalous (extreme) monthly sea levels are clustered in the NAO+/EAP− quadrant. The tide gauges located in the south, i.e., Newlyn and Brest are also dominated by opposite sign NAO-EAP values, however, NAO−/EAP+ conditions are more likely to induce anomalously (extremely) high monthly sea levels, with an average of 46% (54%) of these months being clustered in this quandrant. As shown in Figure 3a, the NAO−/EAP+ pattern is characterized by a low-pressure anomaly outside the Bay of Biscay, which induces cyclonic winds parallel to the continental slope resulting in coastal convergence and anomalously high sea level.

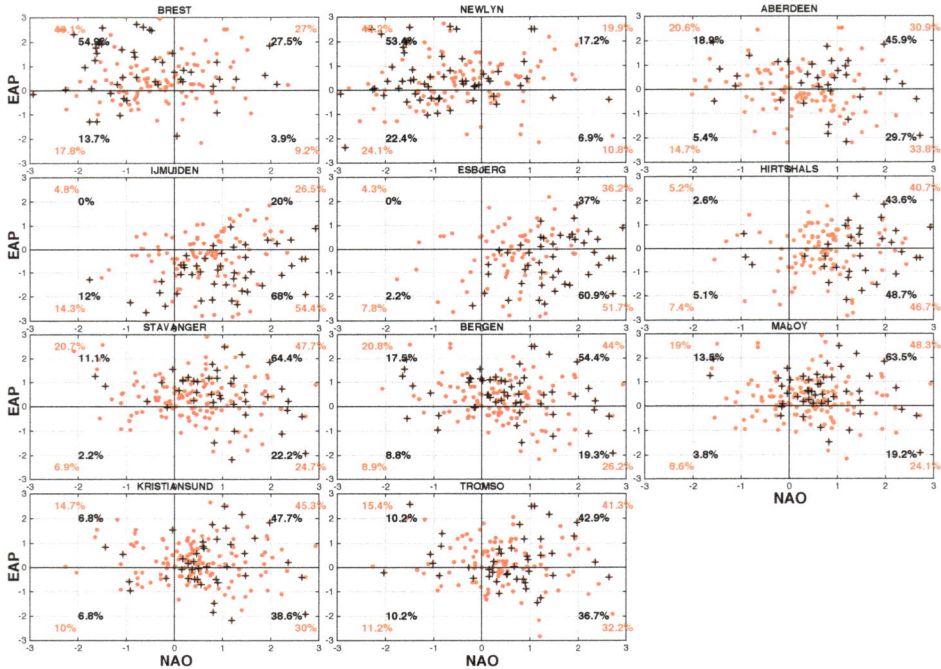

Figure 9. Anomalously/extremely high (red circles/black crosses) monthly sea levels recorded at the different TGs along the European coasts projected onto the NAO-EAP teleconnection space. The threshold value for the anomalous/extreme monthly sea levels is 0.75/1.5 std. The percentages represent the number of anomalously/extremely high (in red/black) months in every quadrant divided by the total number of anomalously high/extreme months in all quadrants.

Figure 10 underlines that opposite sign NAO-SCAN events are more likely to induce anomalous (extreme) monthly sea levels along the Norwegian coast (but also Aberdeen), with an average of 42% (48%) of these months being clustered in the NAO+/SCAN− quadrant. A notable feature, however, is that this relationship becomes more distinct as a function of latitude. For example, the number of anomalously (extremely) high monthly sea levels recorded at Tromsø and clustered in the NAO+/SCAN− quadrant is 51% (59.2%) compared to 36.3% (38.6%) in Bergen. A possible explanation is that since the NAO+/SCAN− leads to anticlockwise movements of the centers of action, the winds are thus better aligned with the continental slope in the northern parts of Norway (cf. Figure 4). The anomalously (extremely) high monthly sea levels observed at the North Sea tide gauges, however, are strongly dominated by NAO+/SCAN+, i.e., same sign values. Ijmuiden, Esbjerg and Hirtshals have 57.8% (68%), 65.5% (80.4%) and 51.9% (59%) of their anomalously (extremely) high monthly sea levels in the NAO+/SCAN+ quadrant. These high numbers are not surprising considering the strong cyclonic wind anomalies over the northern Seas induced as a result of the clockwise movement and intensification of the low-pressure node (cf. Figure 4).

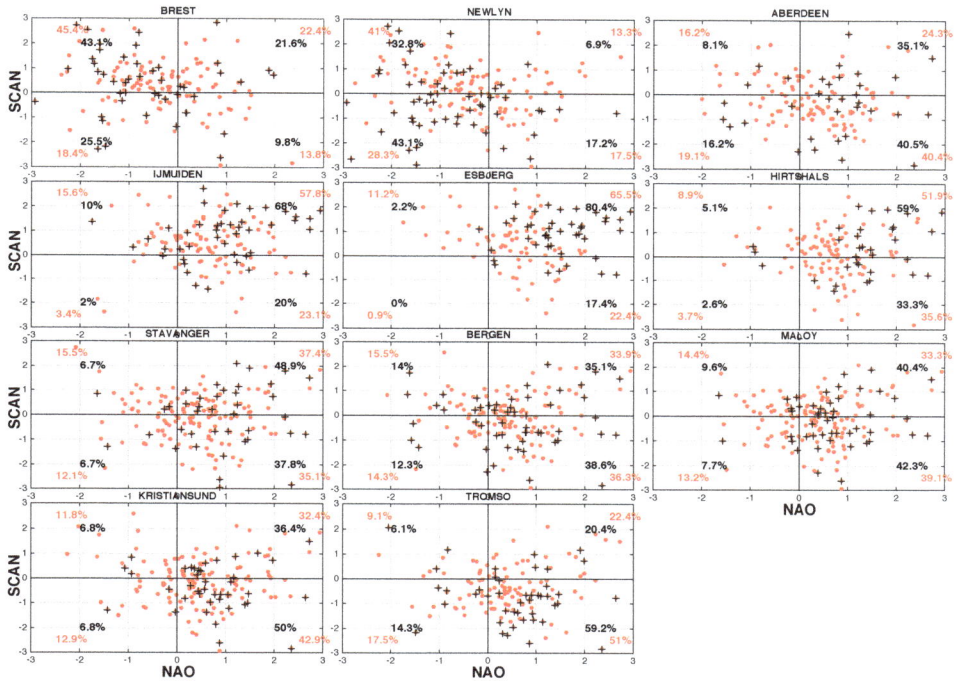

Figure 10. Anomalously/extremely high (red circles/black crosses) monthly sea levels recorded at the different TGs along the European coasts projected onto the NAO-SCAN teleconnection space. The threshold value for the anomalous/extreme monthly sea levels is 0.75/1.5 std. The percentages represent the number of anomalously/extremely high (in red/black) months in every quadrant divided by the total number of anomalously high/extreme months in all quadrants.

It is evident that anomalous sea-level variations at every TG is associated with a unique atmospheric state and pattern combination. We have, however, in Figure 11 identified three TG regions with common large-scale circulation patterns. At the Norwegian tide gauges, the anomalous monthly sea levels are dominated by an MSLP gradient between the Irminger Sea and central Europe. The North Sea tide gauges, apart from Aberdeen, project onto a north-south pressure gradient with nodes over northern Scandinavia and in the vicinity of the Bay of Biscay. Newlyn and Brest are associated with markedly poleward shifted centers of action, but are clearly dominated by the low-pressure (southern center of action) encompassing the subpolar North Atlantic. It is, however, important to keep in mind that at this stage the inverted barometric effect has been removed, so that this solely represents wind forcing. A key point to conclude with from this analysis is that the patterns driving the anomalous sea level variability at each tide gauge are different from the NAO dipole, and can only be understood in terms of the combinations with the other two teleconnection patterns, EAP and SCAN.

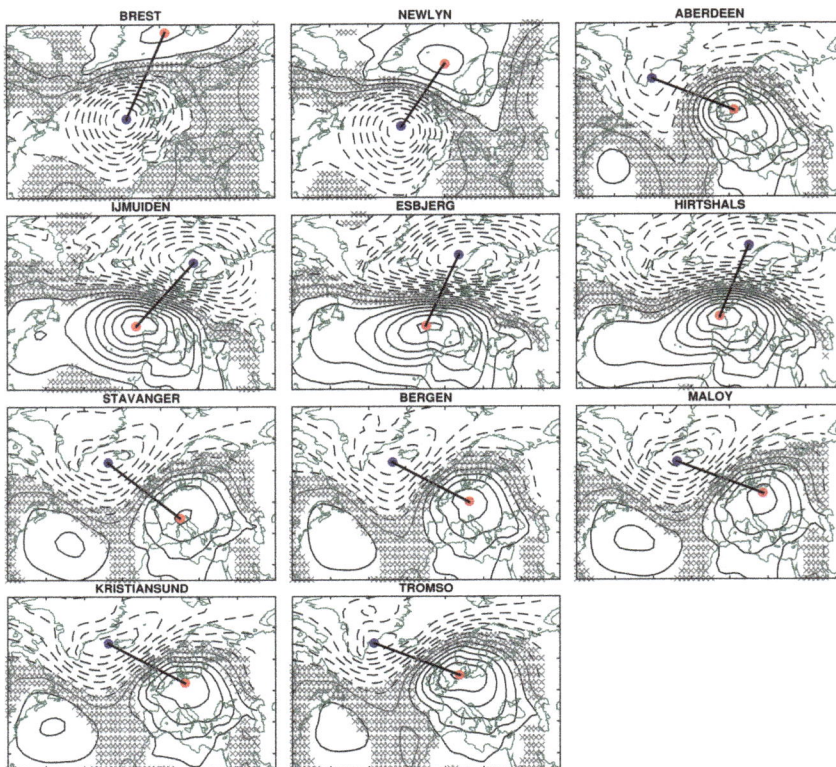

Figure 11. Composite difference of MSLP based on anomalous monthly sea levels from different TGs along the European coasts. The threshold value for the anomalous monthly TG sea levels is 0.75 std. Anomalously positive and negative MSLP is indicated by solid and dashed contours with a spacing of 1 hPa. The red and blue dots are assigned to the lowest and highest MSLP anomaly, respectively. The non-significant regions below the 99% confidence level are indicated by gray crosses calculated using a two-sided *t* test.

3.5. Coherent European Sea Level Variability

We have so far dealt with understanding anomalous monthly sea levels on a regional and local scale, however, in this section we make an attempt to represent the European shelves/coasts using a single index. Figure 12a shows the first leading EOF of monthly sea level variability on the European shelf only using altimetry. The spatial pattern of the leading EOF is notably a monopole with a node over the North Sea and explains about 51% of the variance. The associated first Principal Component (PC1) is shown in Figure 12b, where the anomalous months (using threshold value of 0.75 std) are highlighted. Figure 12c shows the PC1 based on the 11 European TGs displayed in Figure 1. The comparison between PC1 based on TGs and that from altimetry is good with a correlation of 0.71. This result suggests that the on-shelf PC1 is a good proxy that captures the sea-level variations along the European coast from TGs.

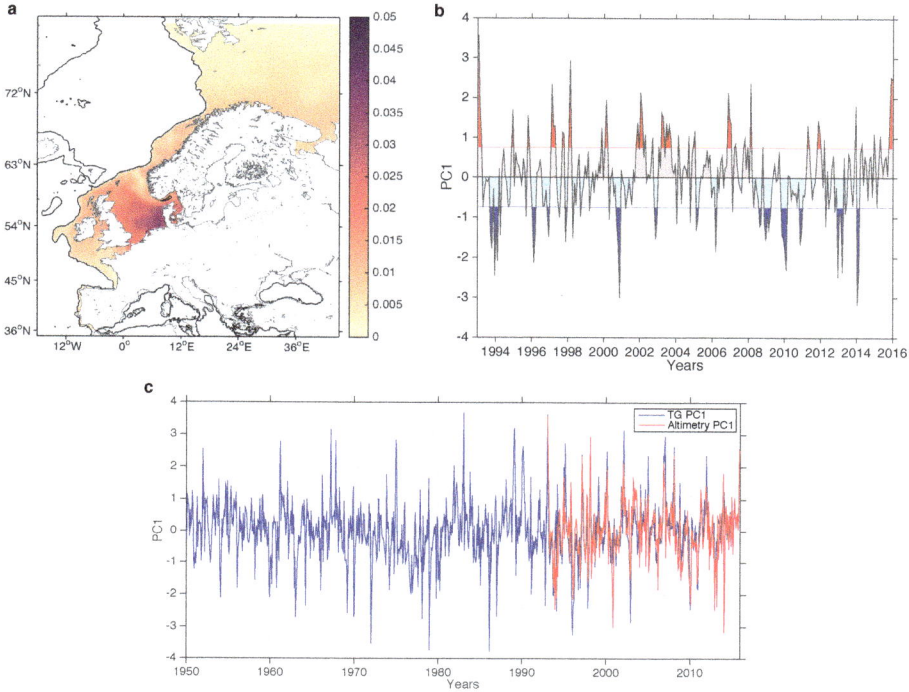

Figure 12. (**a**) First EOF of the monthly on-shelf sea level from altimetry (1993–2015). (**b**) The PC1 associated with the first EOF of the on-shelf sea level. (**c**) PC1 (blue) derived based on the 11 tide gauges (1950–2014) displayed in Figure 1 and overlaid by the on-shelf PC1 (red). The on-shelf and TG PC1 explains 51% and 61% of the variance, respectively. The correlation between the on-shelf and TG PCs is 0.71.

We will now explore the sea level and the atmospheric circulation patterns associated with anomalous monthly sea levels based on the on-shelf and TG PC1s. Figure 13a shows a composite difference between anomalously high and low on-shelf sea level. The highest amplitudes are found in the Barents, Norwegian shelf and North Sea and by the British Isles, and comparatively smaller amplitudes in the Bay of Biscay. Figure 13a further shows that the atmospheric centers of action are centered over Brest and the Barents Sea, respectively. The sea level and atmospheric circulation patterns projecting on the TG PC1 (Figure 13b) show similar features and are hence consistent with those based on the on-shelf PC1 (not surprising considering the strong correlation between these PC1s). These results suggest that due to this poleward and eastward reorganization (relative to the climatological NAO position) of the atmospheric circulation, the winds are nearly parallel to the continental slope all the way from the Bay of Biscay to northern Norway (longshore winds), which through Ekman transport and convergence towards the coast increases the sea level coherently along the European coast, as portrayed in Figure 13c. Coherent coastal sea level variability is thus dependent on the strength and direction of the winds controlled by migrations of the North Atlantic meridional pressure dipole.

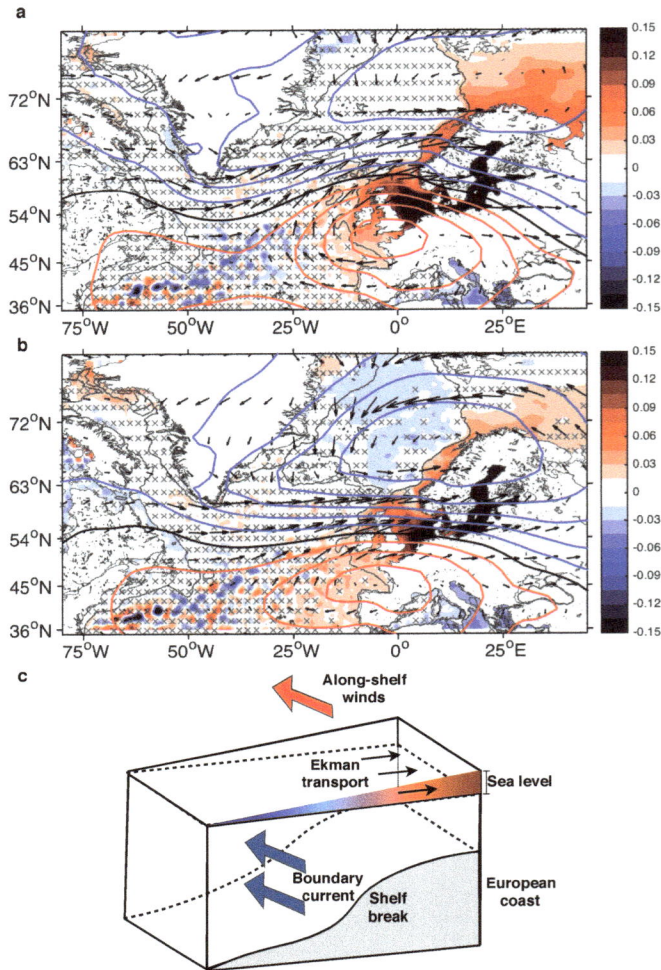

Figure 13. Composite difference of sea level from altimetry (m, shading), MSLP (hPa, contours) and 10 m winds (arrows) based on (**a**) the on-shelf and (**b**) the TG PC1s. The sea level, MSLP and winds have deseasoned and detrended before the composite analysis, which is based on the difference between anomalously high (>0.75 std) and low (<−0.75 std) periods in the PC1s. The black contour denotes the zero MSLP anomaly, while the blue/red contours denote negative/positive MSLP anomaly with a spacing of 2 hPa. (**c**) Schematic diagram including the main dynamical processes involved on monthly timescales. This case reflects southerly along-shelf/shore winds that through Ekman transports increase the coastal sea level. The non-significant regions below the 95% confidence level are indicated by gray crosses calculated using a two-sided *t* test.

The overall pattern driving the sea level variability at the European coasts is reminiscent of that associated with NAO+/SCAN+ (Figure 4b). We project, as above, these anomalous months onto the space spanned by the NAO and SCAN indices to investigate their distribution, cf. Figure 14. The main conclusion is that the anomalously high monthly sea levels of the on-shelf and TG PC1 are predominantly governed by NAO+/SCAN+ pattern, more than the NAO interplay with the

EAP (about ~9% difference between NAO+/EAP− and NAO+/EAP+ for both the anomalously and extremely high monthly sea levels, not shown).

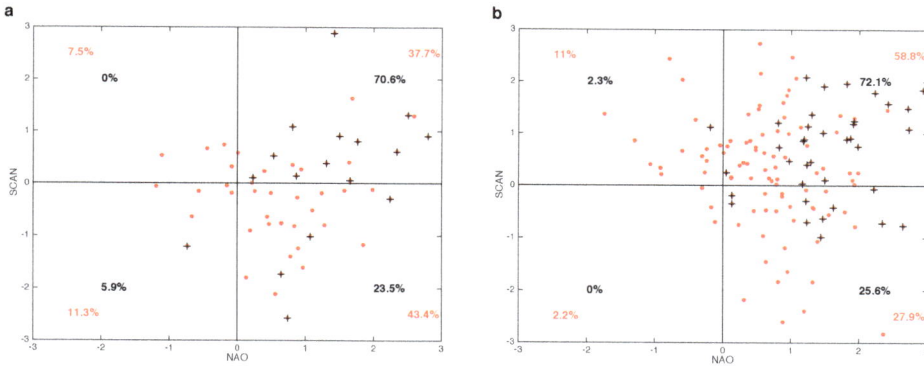

Figure 14. Anomalously/extremely high (red circles/black crosses) monthly sea levels based on (**a**) the on-shelf first Principal Component (PC1) constructed from altimetry, and (**b**) TG PC1, projected onto the NAO-SCAN teleconnection space. The percentages represent the number of anomalously/extremely (>0.75/1.5 std) high (in red/black) months in every quadrant divided by the total number of anomalously high/extreme months in all quadrants.

4. Summarizing Discussion and Outlook

Atmospheric forcing has long been recognized as the dominant forcing factor of sea level variability over the Northern European Shelf. The physical connection between atmospheric forcing and sea level expresses itself through a strong, but non-stationary link to the NAO. The location of the centers of action associated with the North Atlantic meridional pressure dipole, however, can be affected through the interplay with the EAP and SCAN teleconnections, potentially explaining the temporally-varying relationship between sea level variability and the NAO. Our results indicate that it is indeed important to account for the binary interaction of the NAO with EAP/SCAN to better understand the drivers inducing anomalous sea level periods along the European coast.

Our analysis was divided into three main parts. First, we demonstrated that the spatial movements of the centers of action can be captured through the combination of the NAO with the second or third leading EOF of atmospheric variability in the North Atlantic/European sector [14,28]. Apart from explaining the spatial migrations of the North Atlantic meridional pressure dipole, the binary patterns are able to modulate the strength of the MSLP at the centers of action and hence the wind speeds, which, in turn, further influences the sea level. Second, we showed that these combinations can explain the regional sea level differences along the European shelves and coasts, with consistent results from altimetry and tide gauges. It should, however, be mentioned that these regional differences are likely first induced by the NAO, but they are greatly enhanced under the influence of the other two teleconnection patterns. Tsimplis and Shaw [37] note, for example, that if the effect of the NAO and hence westerly winds are removed (regressed out) from the mean sea level at tide gauges around Europe, a weakened relationship between the regions is found. Third, we constructed an index that represent the entire European shelves/coasts, which is found to project on an MSLP dipole pattern with nodes over northwestern Scandinavia and the Bay of Biscay. This modified dipole pattern organizes the winds such that they trace the continental shelf break from the Bay of Biscay all the way to northern Norway. The teleconnection space revealed that the majority of anomalously high sea levels in the tide gauge based index (TG PC1) corresponded to positive phases of both the NAO and the SCAN and the combination of these teleconnections can thus be concluded to be the atmospheric pattern that most strongly induces coherent sea level variability along the European shelves and coasts (Figure 13).

The relationship between the NAO and sea level from tide gauges in the North Sea has been suggested by several studies to be highly variable with time, which is presumed to be due to the movements of the centers of action dominating different periods [12,31,33,34]. As shown by Hilmer and Jung [40], the centers of action was shifted eastwards during the 1978–1997 period as compared to the 1958–1977 period. Although not mentioned by these investigators, a close inspection of their results suggests that the mean 1978–1997 atmospheric state mirrors the NAO+/SCAN+ combination. This is corroborated by Moore et al. [14], where both the NAO and SCAN indices are shown to have been on the positive side over low-frequency timescales (cf. their Figure 8). An NAO+/SCAN+ situation leads to a clockwise shift of the centers of action with the low-pressure node being located slightly west of Norway, which strongly resembles the mean 1978–1997 atmospheric state shown by Hilmer and Jung [40] (see also Comas-Bru and McDermott [23] and Kolker and Hameed [42]). With respect to our results, we note that the majority of the anomalously high monthly sea levels at the North Sea TGs are induced by the NAO+/SCAN+ type of pattern (Figure 10), which may further suggest that the persistency of a same sign NAO-SCAN pattern is likely to reinforce this time-varying relationship between the atmospheric circulation and sea level. We conclude that the NAO+/SCAN+ combination may have induced this multidecadal shift in the location of the centers of action and as a result the correlation with the sea level in the North Sea and Baltic Seas was diagnosed to be higher. Therefore, our results could potentially be valuable if applied to sea level variations on longer timescales.

We note that the atmospheric circulations patterns inducing anomalous or extreme monthly sea levels along the European coast (when represented by a single index) based on both altimetry and TGs were similar (Figures 12 and 13). However, a close inspection of these patterns reveals some differences, although as shown in Figures 13 and 14, both the on-shelf and TG PCs are primarily associated with NAO+/SCAN+ type of pattern (Figure 4). The major difference is that the altimetric on-shelf PC1 projects onto an MSLP dipole pattern that is more shifted polewards (as compared to the pattern based on the TG PC1) with a low-pressure node close to the Barents Sea opening and a high-pressure node more-or-less close to Brest. This overall shift polewards can be seen to drive alongshelf/shore winds following the 500 m isobath from the Bay of Biscay all the way the Barents Sea opening. The TG PC1, however, projects onto an MSLP pattern that organizes the isobars in a zonal manner across the North Sea. One reason causing this pattern difference may in fact lie in our altimetric EOF analysis which, unlike the TG EOF analysis (note that northernmost station is Tromsø), accounts for the Barents Sea region. Furthermore, the atmospheric pattern based on TGs bears a strong resemblance to that reported to correlate well with the monthly mean sea level variability from 13 TGs in the German Bight region [13]. Dangendorf et al. [13] constructed a proxy based on MSLP over Scandinavia and the Iberian Peninsula, and found that it explained about 80% (as compared to the 30–35% explained by the NAO) of the variability and concluded, in addition, that this proxy is capable of reproducing the decadal mean sea-level fluctuations in the German Bight. The present study thus provides a better understanding of the pattern in Dangendorf et al. [13] by revealing that it emerges as a result of the interplay between the NAO and the SCAN (Figure 14b).

Anomalous and extreme monthly sea levels along the Norwegian coast evidently respond to the EAP, but there are some differences within this coastal zone. This applies, in particular, to the northernmost TGs, which are found to be more sensitive to the SCAN than to the EAP (cf. Figures 9 and 10). In contrast, for the North Sea, although dominated by the SCAN, we note that a non-negligible portion of anomalous and extreme monthly sea levels responds to the negative phase of the EAP as well. Due to the clockwise displacement and poleward shift of the North Atlantic meridional pressure dipole, the pattern associated with NAO+/EAP− has its zero-line crossing the North Sea, resulting in a situation that enhances the westerly winds and thereby induces anomalously high/extreme sea levels there. Thus, the North Sea TGs analyzed here are sensitive to both the phase of the SCAN and the EAP. Based on this conclusion, we asked ourselves whether it would not have been feasible to combine all three EOFs. That we did not conduct this type of analysis is, however, not a limitation since we found that the mobility of the North Atlantic meridional pressure dipole due to either the EAP or SCAN separately provides a dynamically clearer picture regarding their impact on sea level, which further acts a simple guideline of what to expect when two EOFs are combined (Figures 3, 4 and 8).

The schematics in Figure 13c depicts that the eastern boundary current locked to the continental slope is further enhanced by the same atmospheric circulation that induce anomalously high sea levels along the European shelves and coasts through geostrophic dynamics. Richter et al. [49] modelled the boundary current at one location (Svinøy) using the across-slope sea level gradient from tide gauges between the Norwegian west coast and the Faroe Islands and found a good correlation of 0.6. We think that the altimetric on-shelf or TG PC1 may be useful to reconstruct the variability of this boundary current, and that these binary combinations may further elucidate the drivers, which have solely been investigated in terms of the NAO (see, e.g., Skagseth et al. [50], Sandø et al. [51], Chafik et al. [52]). Furthermore, the European shelves respond on decadal timescales to open-ocean steric height variations [11,53], which may impact the across-slope sea level gradient and hence regulate the strength of the boundary current. Thus, the on-shelf or TG PC1 may further be useful to better understand the low-frequency variability of the nearly-barotropic shelf-edge current that carries Atlantic water to the Arctic, especially in the Nordic Seas region.

Our objective has been to understand the impact of North Atlantic teleconnections on northern European sea levels with less focus on the forcing mechanisms of these teleconnections, and the reader is referred to, e.g., Cassou et al. [48] for details (see also [54–58]). But, to shed some light on this issue, we apply a similar composite analysis as in Figure 13 on additional key variables and show the hemispheric-scale linkages (Figure 15). The main picture that emerges is that anomalously high northern European monthly sea levels (Figure 15a) coincide with a tripolar pattern of sea surface temperature (SST) anomalies extending across the North Atlantic and a warm SST anomaly in the central tropical Pacific (Figure 15b). These pronounced SST anomalies vary simultaneously with the upper-troposphere geopotential height anomalies (Figure 15c). A recent study by Schemm et al. [57] demonstrates that warm SST anomalies over these two key regions, i.e., central tropical Pacific and Gulf Stream, promote a more northeastward North Atlantic storm track (as anticipated during anomalous NAO+/SCAN+ periods, cf. Figure 13). In particular, they report that synchronously with central (eastern) Pacific El Niño winters, the cyclogenesis location over the Gulf Stream region is located north of its climatological position, which, in turn, steers the storm tracks into a northeastward (zonal) path across the Atlantic. In fact, they calculate that ~20–25% (2–3%) of the cyclones that originate from the Gulf Stream traverse the northern Nordic Seas region during active central (eastern) Pacific El Niño winters. Furthermore, the warm SST anomaly that normally accompanies a northward shift of the Gulf Stream current system has recently been suggested to be a key factor for triggering the SCAN teleconnection pattern through enhanced transient eddy vorticity flux [58]. Taken together, these results suggest that the North Atlantic tripolar SST pattern and its associated Rossby wave train (Figure 15b) emanating from the Gulf Stream region and connecting the western North Atlantic with the European sector is a potential source, likely also in combination with central Pacific El Niños, for triggering the teleconnection pattern driving anomalous monthly European sea levels. More research along these lines is thus warranted to better understand the global-scale drivers of European sea level change.

In a warming climate, the position of the centers of action associated with the North Atlantic meridional pressure dipole are projected to shift slightly northeastward and amplify by the end of the twenty-first century [59]. Such a change would dynamically push the European mean sea level to a higher state as well as increase the extreme monthly sea levels as this northeastward shift is reported to be associated with deeper cyclones under increasing anthropogenic climate change conditions (see, e.g., Pinto et al. [60], Feser et al. [61]). Therefore, even a future small shift in the position of the North Atlantic meridional pressure dipole may lead to large effects on European sea levels and hence enhanced flood risk, further underlining that a correct representation of these atmospheric teleconnections in climate models is a prerequisite for projecting sea level change, in particular, on the regional scale. Not least since the variability caused by the interaction of the teleconnection patterns demonstrated herein are on the order of magnitude as the centennial global mean sea level change [1]. Furthermore, on a national level these combinations may act as a guideline of which atmospheric patterns and concomitant teleconnection states are important. As we have demonstrated in the present study, it may not be sufficient to only predict the NAO (see, e.g., Scaife et al. [55]). Seasonal forecasts of the EAP and SCAN, which have been suggested to be linked to tropical Pacific and North Atlantic sea surface temperature variability [62,63], would be valuable to the European coastal communities.

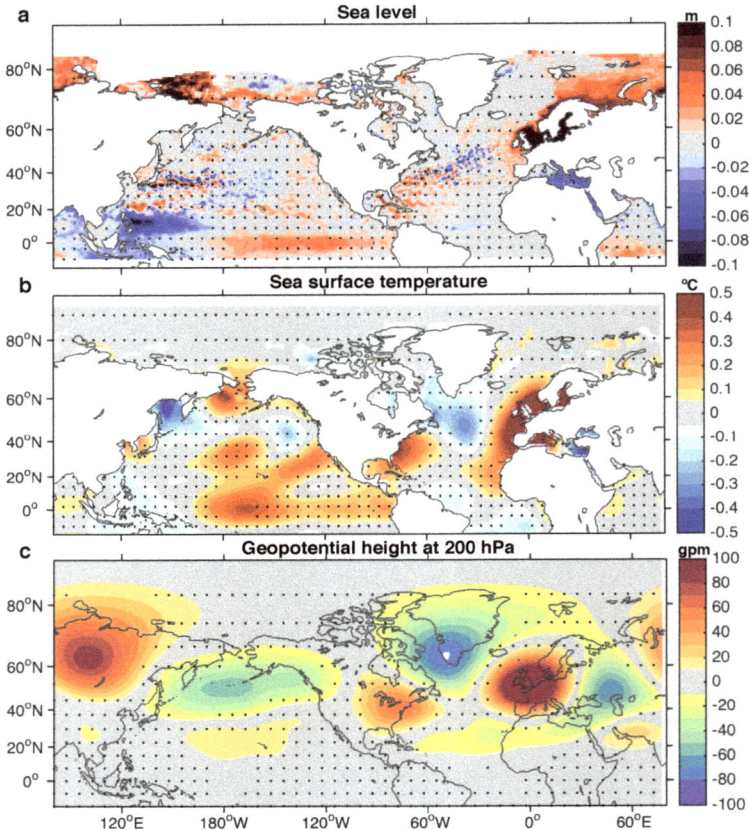

Figure 15. Composite difference of (**a**) sea level (m), (**b**) sea surface temperature (SST) (°C) and (**c**) non-zonal geopotential height (at 200 hPa) anomalies (gpm) based on the difference between anomalously high (>0.75 std) and low (<−0.75 std) periods of the on-shelf PC1 (1993–2015). The SST data is based on the extended reconstructed sea surface temperature version 4 (ERSSTv4) [64]. The geopotential height is from the NCEP/NCAR reanalysis [46]. The data have been deseasoned and detrended before the analysis. The stipplings indicate non-significant regions below the 95% confidence level using a two-sided *t* test.

Acknowledgments: This work was funded by the Centre for Climate Dynamics at the Bjerknes Centre, through the project iNcREASE. S.D. would like to thank the University of Siegen. The altimeter products were produced by Ssalto/Duacs and distributed by Aviso, with support from Cnes (http://www.aviso.altimetry.fr/duacs/). We thank PSMSL for the tide gauge data. We would like to thank three anonymous reviewers for valuable comments that improved the paper.

Author Contributions: L.C., J.E.O.N. and S.D. conceived and designed the analysis; L.C. conducted the analysis and wrote the paper; all authors contributed to the writing of the paper.

Conflicts of Interest: The authors declare no conflict of interest.

Abbreviations

The following abbreviations are used in this manuscript:

NAO	North Atlantic Oscillation
EAP	East Atlantic Pattern
SCAN	Scandinavian Pattern
DUACS	Data Unification and Altimeter Combination System
DT2014	Delayed Time 2014
MDT	Mean Dynamic Topography
CNES/CLS	Centre national d'études spatiales/Collecte Localisation Satellites
NCEP/NCAR	National Center for Environmental Prediction/National Center for Atmospheric Research
EOF	Empirical Orthogonal Functions
PC1	First Principal Component
TG	Tide Gauges
PSMSL	Permanent Service for Mean Sea Level
SST	Sea Surface Temperature
ERSSTv4	Extended Reconstructed Sea Surface Temperature version 4

References

1. Dangendorf, S.; Marcos, M.; Wöppelmann, G.; Conrad, C.P.; Frederikse, T.; Riva, R. Reassessment of 20th century global mean sea level rise. *Proc. Natl. Acad. Sci. USA* **2017**, *114*, doi:10.1073/pnas.1616007114.
2. Cazenave, A.; Dieng, H.B.; Meyssignac, B.; Von Schuckmann, K.; Decharme, B.; Berthier, E. The rate of sea-level rise. *Nat. Clim. Chang.* **2014**, *4*, 358–361.
3. Bindoff, N.L.; Willebrand, J.; Artale, V.; Cazenave, A.; Gregory, J.M.; Gulev, S.; Hanawa, K.; Le Quéré, C.; Levitus, S.; Nojiri, Y.; et al. Observations: Oceanic Climate Change and Sea Level. In *Climate Change 2007: The Physical Science Basis*; Cambridge University Press: Cambridge, UK; New York, NY, USA, 2007.
4. Church, J.A.; Clark, P.U.; Cazenave, A.; Gregory, J.M.; Jevrejeva, S.; Levermann, A.; Merrifield, M.A.; Milne, G.A.; Nerem, R.S.; Nunn, P.D.; et al. Sea-level rise by 2100. *Science* **2013**, *342*, 1445.
5. Nicholls, R.J.; Cazenave, A. Sea-level rise and its impact on coastal zones. *Science* **2010**, *328*, 1517–1520.
6. Kopp, R.E.; Horton, R.M.; Little, C.M.; Mitrovica, J.X.; Oppenheimer, M.; Rasmussen, D.; Strauss, B.H.; Tebaldi, C. Probabilistic 21st and 22nd century sea-level projections at a global network of tide-gauge sites. *Earth's Future* **2014**, *2*, 383–406.
7. Carson, M.; Köhl, A.; Stammer, D.; Slangen, A.; Katsman, C.; Van de Wal, R.; Church, J.; White, N. Coastal sea level changes, observed and projected during the 20th and 21st century. *Clim. Chang.* **2016**, *134*, 269–281.
8. Slangen, A.; Adloff, F.; Jevrejeva, S.; Leclercq, P.; Marzeion, B.; Wada, Y.; Winkelmann, R. A review of recent updates of sea-level projections at global and regional scales. *Surv. Geophys.* **2016**, *38*, 385–406.
9. Sturges, W.; Douglas, B.C. Wind effects on estimates of sea level rise. *J. Geophys. Res. Oceans* **2011**, *116*, doi:10.1029/2010JC006492.
10. Calafat, F.; Chambers, D.; Tsimplis, M. Mechanisms of decadal sea level variability in the eastern North Atlantic and the Mediterranean Sea. *J. Geophys. Res. Oceans* **2012**, *117*, doi:10.1029/2012JC008285.
11. Dangendorf, S.; Calafat, F.M.; Arns, A.; Wahl, T.; Haigh, I.D.; Jensen, J. Mean sea level variability in the North Sea: Processes and implications. *J. Geophys. Res. Oceans* **2014**, *119*, doi:10.1002/2014JC009901.
12. Wakelin, S.; Woodworth, P.; Flather, R.; Williams, J. Sea-level dependence on the NAO over the NW European Continental Shelf. *Geophys. Res. Lett.* **2003**, *30*, doi:10.1029/2003GL017041.

13. Dangendorf, S.; Wahl, T.; Nilson, E.; Klein, B.; Jensen, J. A new atmospheric proxy for sea level variability in the southeastern North Sea: Observations and future ensemble projections. *Clim. Dyn.* **2014**, *43*, 447–467.

14. Moore, G.W.K.; Renfrew, I.A.; Pickart, R.S. Multidecadal Mobility of the North Atlantic Oscillation. *J. Clim.* **2012**, *26*, 2453–2466.

15. Hurrell, J.W. Decadal Trends in the North Atlantic Oscillation: Regional Temperatures and Precipitation. *Science* **1995**, *269*, 676–679.

16. Stephenson, D.B.; Pavan, V.; Bojariu, R. Is the North Atlantic Oscillation a random walk? *Int. J. Clim.* **2000**, *20*, 1–18.

17. Nilsen, J.; Gao, Y.; Drange, H.; Furevik, T.; Bentsen, M. Simulated North Atlantic-Nordic Seas water mass exchanges in an isopycnic coordinate OGCM. *Geophys. Res. Lett.* **2003**, *30*, doi:10.1029/2002GL016597.

18. Furevik, T.; Nilsen, J.E.Ø. Large-Scale Atmospheric Circulation Variability and its Impacts on the Nordic Seas Ocean Climate—A Review. In *The Nordic Seas: An Integrated Perspective*; American Geophysical Union: Washington, DC, USA, 2005; pp. 105–136.

19. Moore, G.; Pickart, R.; Renfrew, I. Complexities in the climate of the subpolar North Atlantic: A case study from the winter of 2007. *Q. J. R. Meteorol. Soc.* **2011**, *137*, 757–767.

20. Castelle, B.; Dodet, G.; Masselink, G.; Scott, T. A new climate index controlling winter wave activity along the Atlantic coast of Europe: The West Europe Pressure Anomaly. *Geophys. Res. Lett.* **2017**, *44*, 1384–1392.

21. Fereday, D.; Knight, J.; Scaife, A.; Folland, C.; Philipp, A. Cluster analysis of North Atlantic–European circulation types and links with tropical Pacific sea surface temperatures. *J. Clim.* **2008**, *21*, 3687–3703.

22. Moore, G.W.K.; Renfrew, I.A. Cold European winters: Interplay between the NAO and the East Atlantic mode. *Atmos. Sci. Lett.* **2012**, *13*, 1–8.

23. Comas-Bru, L.; McDermott, F. Impacts of the EA and SCA patterns on the European twentieth century NAO–winter climate relationship. *Q. J. R. Meteorol. Soc.* **2013**, *140*, 354–363.

24. Bastos, A.; Janssens, I.; Gouveia, C.; Trigo, R.; Ciais, P.; Chevallier, F.; Peñuelas, J.; Rödenbeck, C.; Piao, S.; Friedlingstein, P.; et al. European land CO_2 sink influenced by NAO and East-Atlantic Pattern coupling. *Nat. Commun.* **2016**, *7*, 10315.

25. Wallace, J.M.; Gutzler, D.S. Teleconnections in the Geopotential Height Field during the Northern Hemisphere Winter. *Mon. Weather Rev.* **1981**, *109*, 784–812.

26. Barnston, A.G.; Livezey, R.E. Classification, Seasonality and Persistence of Low-Frequency Atmospheric Circulation Patterns. *Mon. Weather Rev.* **1987**, *115*, 1083–1126.

27. Bueh, C.; Nakamura, H. Scandinavian pattern and its climatic impact. *Q. J. R. Meteorol. Soc.* **2007**, *133*, 2117–2131.

28. Franzke, C.; Feldstein, S.B. The continuum and dynamics of Northern Hemisphere teleconnection patterns. *J. Atmos. Sci.* **2005**, *62*, 3250–3267.

29. Rossby, C.G. Relation between variations in the intensity of the zonal circulation of the atmosphere and the displacements of the semi-permanent centers of action. *J. Mar. Res.* **1939**, *2*, 38–55.

30. Woollings, T.; Hannachi, A.; Hoskins, B. Variability of the North Atlantic eddy-driven jet stream. *Q. J. R. Meteorol. Soc.* **2010**, *136*, 856–868.

31. Andersson, H.C. Influence of long-term regional and large-scale atmospheric circulation on the Baltic sea level. *Tellus A* **2002**, *54*, 76–88.

32. Woolf, D.K.; Shaw, A.G.; Tsimplis, M.N. The influence of the North Atlantic Oscillation on sea-level variability in the North Atlantic region. *Glob. Atmos. Ocean Syst.* **2003**, *9*, 145–167.

33. Yan, Z.; Tsimplis, M.N.; Woolf, D. Analysis of the relationship between the North Atlantic oscillation and sea-level changes in northwest Europe. *Int. J. Clim.* **2004**, *24*, 743–758.

34. Jevrejeva, S.; Moore, J.; Woodworth, P.; Grinsted, A. Influence of large-scale atmospheric circulation on European sea level: Results based on the wavelet transform method. *Tellus A* **2005**, *57*, 183–193.

35. Papadopoulos, A.; Tsimplis, M. Coherent coastal sea-level variability at interdecadal and interannual scales from tide gauges. *J. Coast. Res.* **2006**, *22*, 625–639.

36. Woodworth, P.; Flather, R.; Williams, J.; Wakelin, S.; Jevrejeva, S. The dependence of UK extreme sea levels and storm surges on the North Atlantic Oscillation. *Cont. Shelf Res.* **2007**, *27*, 935–946.

37. Tsimplis, M.N.; Shaw, A.G. The forcing of mean sea level variability around Europe. *Glob. Planet. Chang.* **2008**, *63*, 196–202.

38. Dangendorf, S.; Wahl, T.; Hein, H.; Jensen, J.; Mai, S.; Mudersbach, C. Mean sea level variability and influence of the North Atlantic Oscillation on long-term trends in the German Bight. *Water* **2012**, *4*, 170–195.

39. Chen, X.; Dangendorf, S.; Narayan, N.; O'Driscoll, K.; Tsimplis, M.N.; Su, J.; Mayer, B.; Pohlmann, T. On sea level change in the North Sea influenced by the North Atlantic Oscillation: Local and remote steric effects. *Estuar. Coast. Shelf Sci.* **2014**, *151*, 186–195.

40. Hilmer, M.; Jung, T. Evidence for a recent change in the link between the North Atlantic Oscillation and Arctic sea ice export. *Geophys. Res. Lett.* **2000**, *27*, 989–992.

41. Jung, T.; Hilmer, M.; Ruprecht, E.; Kleppek, S.; Gulev, S.K.; Zolina, O. Characteristics of the recent eastward shift of interannual NAO variability. *J. Clim.* **2003**, *16*, 3371–3382.

42. Kolker, A.S.; Hameed, S. Meteorologically driven trends in sea level rise. *Geophys. Res. Lett.* **2007**, *34*, doi:10.1029/2007GL031814.

43. Ullmann, A.; Monbaliu, J. Changes in atmospheric circulation over the North Atlantic and sea-surge variations along the Belgian coast during the twentieth century. *Int. J. Clim.* **2010**, *30*, 558–568.

44. Pujol, M.I.; Faugère, Y.; Taburet, G.; Dupuy, S.; Pelloquin, C.; Ablain, M.; Picot, N. DUACS DT2014: The new multi-mission altimeter data set reprocessed over 20 years. *Ocean Sci.* **2016**, *12*, 1067–1090.

45. Holgate, S.J.; Matthews, A.; Woodworth, P.L.; Rickards, L.J.; Tamisiea, M.E.; Bradshaw, E.; Foden, P.R.; Gordon, K.M.; Jevrejeva, S.; Pugh, J. New data systems and products at the permanent service for mean sea level. *J. Coast. Res.* **2013**, *29*, 493–504.

46. Kalnay, E.; Kanamitsu, M.; Kistler, R.; Collins, W.; Deaven, D.; Gandin, L.; Iredell, M.; Saha, S.; White, G.; Woollen, J.; et al. The NCEP/NCAR 40-Year Reanalysis Project. *Bull. Am. Meteorol. Soc.* **1996**, *77*, 437–471.

47. Hannachi, A.; Jolliffe, I.T.; Stephenson, D.B. Empirical orthogonal functions and related techniques in atmospheric science: A review. *Int. J. Clim.* **2007**, *27*, 1119–1152.

48. Cassou, C.; Terray, L.; Hurrell, J.W.; Deser, C. North Atlantic winter climate regimes: Spatial asymmetry, stationarity with time, and oceanic forcing. *J. Clim.* **2004**, *17*, 1055–1068.

49. Richter, K.; Segtnan, O.H.; Furevik, T. Variability of the Atlantic inflow to the Nordic Seas and its causes inferred from observations of sea surface height. *J. Geophys. Res. Oceans* **2012**, *117*, C04004.

50. Skagseth, Ø.; Orvik, K.A.; Furevik, T. Coherent variability of the Norwegian Atlantic Slope Current derived from TOPEX/ERS altimeter data. *Geophys. Res. Lett.* **2004**, *31*, L14304.

51. Sandø, A.; Nilsen, J.; Eldevik, T.; Bentsen, M. Mechanisms for variable North Atlantic–Nordic seas exchanges. *J. Geophys. Res. Oceans* **2012**, *117*, doi:10.1029/2012JC008177.

52. Chafik, L.; Nilsson, J.; Skagseth, Ø.; Lundberg, P. On the flow of Atlantic water and temperature anomalies in the Nordic Seas toward the Arctic Ocean. *J. Geophys. Res. Oceans* **2015**, *120*, 7897–7918.

53. Frederikse, T.; Riva, R.; Slobbe, C.; Broerse, T.; Verlaan, M. Estimating decadal variability in sea level from tide gauge records: An application to the North Sea. *J. Geophys. Res. Oceans* **2016**, *121*, 1529–1545.

54. Li, Y.; Lau, N.C. Impact of ENSO on the atmospheric variability over the North Atlantic in late winter—Role of transient eddies. *J. Clim.* **2012**, *25*, 320–342.

55. Scaife, A.; Arribas, A.; Blockley, E.; Brookshaw, A.; Clark, R.; Dunstone, N.; Eade, R.; Fereday, D.; Folland, C.; Gordon, M.; et al. Skillful long-range prediction of European and North American winters. *Geophys. Res. Lett.* **2014**, *41*, 2514–2519.

56. Drouard, M.; Rivière, G.; Arbogast, P. The link between the North Pacific climate variability and the North Atlantic Oscillation via downstream propagation of synoptic waves. *J. Clim.* **2015**, *28*, 3957–3976.

57. Schemm, S.; Ciasto, L.M.; Li, C.; Kvamstø, N.G. Influence of tropical Pacific sea surface temperature on the genesis of Gulf Stream cyclones. *J. Atmos. Sci.* **2016**, *73*, 4203–4214.

58. Jung, O.; Sung, M.K.; Sato, K.; Lim, Y.K.; Kim, S.J.; Baek, E.H.; Jeong, J.H.; Kim, B.M. How does the SST variability over the western North Atlantic Ocean control Arctic warming over the Barents–Kara Seas? *Environ. Res. Lett.* **2017**, *12*, 034021.

59. Ulbrich, U.; Christoph, M. A shift of the NAO and increasing storm track activity over Europe due to anthropogenic greenhouse gas forcing. *Clim. Dyn.* **1999**, *15*, 551–559.

60. Pinto, J.G.; Zacharias, S.; Fink, A.H.; Leckebusch, G.C.; Ulbrich, U. Factors contributing to the development of extreme North Atlantic cyclones and their relationship with the NAO. *Clim. Dyn.* **2009**, *32*, 711–737.

61. Feser, F.; Barcikowska, M.; Krueger, O.; Schenk, F.; Weisse, R.; Xia, L. Storminess over the North Atlantic and northwestern Europe—A review. *Q. J. R. Meteorol. Soc.* **2015**, *141*, 350–382.

62. Iglesias, I.; Lorenzo, M.N.; Taboada, J.J. Seasonal predictability of the East atlantic pattern from sea surface temperatures. *PLoS ONE* **2014**, *9*, e86439.

63. King, M.P.; Herceg-Bulić, I.; Kucharski, F.; Keenlyside, N. Interannual tropical Pacific sea surface temperature anomalies teleconnection to Northern Hemisphere atmosphere in November. *Clim. Dyn.* **2017**, 1–19, doi:10.1007/s00382-017-3727-5.

64. Huang, B.; Thorne, P.W.; Smith, T.M.; Liu, W.; Lawrimore, J.; Banzon, V.F.; Zhang, H.M.; Peterson, T.C.; Menne, M. Further exploring and quantifying uncertainties for extended reconstructed sea surface temperature (ERSST) version 4 (v4). *J. Clim.* **2016**, *29*, 3119–3142.

Journal of Marine Science and Engineering

MDPI

Article

Cost and Materials Required to Retrofit US Seaports in Response to Sea Level Rise: A Thought Exercise for Climate Response

Austin Becker [1,*] , Ariel Hippe [2] and Elizabeth L. Mclean [1]

[1] Department of Marine Affair, University of Rhode Island, Kingston, RI 02881, USA; elmclean@uri.edu
[2] DOWL, Anchorage, AK 99503, USA; ahippe@dowl.com
* Correspondence: abecker@uri.edu; Tel.: +1-401-874-4192

Received: 27 April 2017; Accepted: 4 September 2017; Published: 14 September 2017

Abstract: Climate changes projected for 2100 and beyond could result in a worldwide race for adaptation resources on a scale never seen before. This paper describes a model for estimating the cost and materials of elevating coastal seaport infrastructure in the United States to prevent damage from sea level rise associated with climate change. This study pilots the use of a generic port model (GenPort) as a basis from which to estimate regional materials and monetary demands, resulting in projections that would be infeasible to calculate on an individual port-by-port basis. We estimate the combined cost of adding two meters of additional fill material to elevate the working surface and then reconstructing the generic port. We use the resulting unit area cost to develop an estimate to elevate and retrofit 100 major United States commercial coastal ports. A total of $57 billion to $78 billion (2012 US dollars) and 704 million cubic meters of fill would be required to elevate the 100 ports by two meters and to reconstruct associated infrastructure. This estimation method and the results serve as a thought exercise to provoke considerations of the cumulative monetary and material demands of widespread adaptations of seaport infrastructure. The model can be adapted for use in multiple infrastructure sectors and coastal managers can use the outlined considerations as a basis for individual port adaptation strategy assessments.

Keywords: seaports; resilience; climate adaptation; estimating; elevation

1. Introduction

Climate changes projected for 2100 and beyond could result in a race for adaptation-related construction resources on a scale never seen before. Many adaptation projects will need to utilize the same types of construction resources simultaneously because local adaptation requirements are driven by global phenomena. Calculating potential global demand for several construction resources poses challenges, due to the site-specific nature of adaptation designs. In this research, we created a model to estimate resources required to adapt just one coastal use for the United States (US): major seaports. The US seaport system supports $4.6 trillion in economic activity annually and 23 million jobs [1]. As important hubs of commerce, damage to ports can cripple economies both locally and regionally, triggering far-reaching impacts to economic systems and supply chains [2]. In 2013, Chambers et al. reported that freight vessels transported 53% of US imports and 38% of exports by value [3]. Infrastructure construction takes years to plan, design, and build. Often outliving its 30–50-year design life, much port infrastructure built today will likely continue to function at the end of the century [4]. Climate change over the next century could force multiple adaptations in many locations [5]. To plan in advance, government and port authorities need to implement strategies to protect infrastructure for future environmental conditions [4] as well as forecast the cost and demand of necessary construction [5].

The location of coastal ports places them in the path of ocean storms and rising sea levels [6]. As the impacts of climate change become more evident, port infrastructure is likely to be hit first and hit hardest, unless adaptation is undertaken [7]. New projects for adaptation could place high demands on construction materials for the protection of major coastal seaports, not only in the US, but worldwide [8]. A quantitative analysis to aggregate such potential costs to adapt ports has not yet been conducted [9].

In this paper, we examine the cumulative regional and national costs of adapting 100 Hawaiian, Alaskan, West Coast, Gulf Coast and East Coast US coastal ports to sea level rise (SLR) through one potential adaptation solution: elevating the port footprint and reconstructing associated infrastructure. Scientists report the global SLR to range from 0.6 to 2.0 m by 2100 [10–12], with an upper bound rise of 4.3 m by 2200 [13,14]. However, even a small amount of SLR can have major impacts on storm surge heights and associated flooding [15]. Category 3–5 hurricanes in the Atlantic basin may double in frequency by 2100 [16]. In 2012, Becker et al. found that a two-meter SLR was the threshold at which all seaport managers they surveyed felt that they would be required to act and protect their facilities [17]. Some measure of coastal infrastructure protection can be achieved through the construction of dikes, relocation, or through elevation of ports [8]. Each of these strategies will have costs and benefits to be considered by local decision makers. Dikes, for example, will protect some areas, but leave others exposed. Flooding could be worse for those areas just outside the dike, as surge waters that once flooded inland would be now displaced to the areas outside of a new structure. Financial considerations, as well as social and environmental issues, will all need to be considered carefully. Green solutions, such as the Room for the River project in the Netherlands, can also help to offset flooding to infrastructure [18]. Ultimately, decision makers will implement different strategies for different infrastructure assets, but all strategies will be cost and materials intensive [19–21].

Here, we explore only one strategy: elevating the footprint and infrastructure of coastal ports in the United States two meters based on high-end projections for SLR by 2100 [11]. We do not advocate this as the appropriate or best solution for any specific port and note that elevating a port alone may create additional challenges for multi-modal connections and surrounding areas. For example, if a port is elevated, but critical rail and road connections remain below flood elevations, the port would be rendered inoperable for the movement of cargo during flood events. We also do not assess the risk or probability of sea level rise and surge for individual ports, which have already been addressed [22–24]. Rather, we present this estimation as a thought exercise to provoke consideration of the cumulative resource demands of widespread adaptations to maintain functioning seaports and for consideration of design and planning of future port expansion. These calculations can serve as an "upper-bound estimate" of potential investment for one particular strategy. Analyzing the cumulative demands can present a clearer picture of the challenges inherent in any strategy for protection, not only in the procurement of funding, but also in the procurement of materials such as fill.

Researchers have estimated costs to adapt coastal structures at a global scale [9] and for the US [25,26] for a variety of adaptation strategies. Hinkel et al. [27], for example, calculate costs to protect coastal areas by utilizing a theoretical dike design [28] at the coastline. The height and cost of the structure increases relative to the asset value and population requiring protection, amounting to an annual investment and maintenance cost of $12–71 billion in 2100 to protect coastal areas [27]. Nicholls et al. [9] calculated that adapting ports worldwide would cost approximately $0.21 billion per year. By leveraging local information and engineering knowledge, Becker et al. in their "minimum assumption credible design" (MACD) found that in order to protect 221 of the worlds seaports using a dike and berm design, approximately 436 million cubic meters of construction material would be required [5]. Except for this last example, most studies are not based on detailed design specifications that incorporate materials and costs of various components. In practice, every seaport is different: they handle different cargos, are of different sizes, and face a range of environmental conditions. As an in-depth assessment of each individual port and its infrastructure would take thousands of hours to complete, developing a national estimate requires a simplified method. The method in this paper serves as a pilot for the development of such aggregate estimates; this method could be used in other sectors (e.g., power plants, sewage treatment facilities, and airports) to develop an understanding of the potential cost and resources required for adaptation to climate change.

This study used a generic container port model, "GenPort" to extrapolate aggregate costs using empirical data for 100 coastal port land areas throughout the US. Typically, estimation for such projects is conducted on a case-by-case basis and utilizes estimation tools such as the Construction Specifications Institute [29] *MasterFormat* model which breaks construction costs into divisions of work such as concrete, plumbing and finishes. This approach is too specific for the purposes of creating aggregate estimates. Instead, the GenPort cost estimations were based on ground elevation and a simplified approach that categorizes the major types of port infrastructure requiring reconstruction. Methods on the development of our cost and materials estimates are summarized below (full details and equations may be found in the Appendixs A and B).

2. Methodology

We generated aggregate estimates of costs and material requirements to elevate and retrofit US ports by designing a retrofit for a "generic container port" model (GenPort). Using GenPort as a basis to estimate costs of land elevation and infrastructure retrofit to mitigate SLR simplifies what, in practice, would require complex site-specific projected calculations [6]. GenPort contains the typical components of a container shipping port in the US. It comprises a 0.4 square kilometer (100 acre) two-berth marine container terminal (Pers. comm. T. Ward) such as one that might be retrofitted from an existing functioning port. We calculated the cost of infrastructure reconstruction per square meter of the GenPort as well as the cost to elevate by two meters in order to develop an equation for calculating the full cost to elevate and retrofit GenPort (see Appendix A). We use RSMeans cost data, allowing for unit-based estimations (e.g., square meter of infrastructure) to estimate costs for administration buildings and warehouses. RSMeans provides cost information used for estimates and projections in the construction industry (see www.rsmeans.com). Then, we calculated the land area for 100 major coastal commercial ports in the US by region (see Appendix B). Finally, we combined the costs to reconstruct and elevate GenPort with the aggregate land area of the 100 coastal ports in the US to obtain a national cost estimate. True costs for the retrofit would of course differ for liquid bulk, dry bulk, or general cargo terminals and in many cases would likely be higher than retrofitting a container terminal. However, we used the GenPort design as a proxy, as most major ports are container facilities. The model was run using a two-meter elevation design for this pilot study, but could easily be run with higher or lower elevations. Recent SLR projections suggest two meters as an upper bound projection for SLR by 2100 [12,14,30,31]. In addition, port operators suggested that a two meter rise would prompt them to take significant measures to adapt their infrastructure [17].

2.1. GenPort Retrofit Design

The GenPort design assumed that existing port infrastructure is currently at an elevation that protects it from present day storm surge and sea levels. Thus, elevation of the primary and secondary container yard areas increases the level of protection by providing for an additional two meters of rise for future climate-driven projections. Changes in storm surge height are dependent on local bathymetry and the geography of the local coastline [32]. Therefore, the change in storm surge height could vary significantly along a fairly short distance of coastline [33,34]. Though two meters is a global average, local and regional variations would result in different levels of rise and different future storm surge levels for any given port. Some of the projected future impacts will be due to an increase in storm intensity [35]. Other potential threats are explained by anthropogenic subsidence that can lead to the rapid sinking of ground levels. For example, the Long Beach/Los Angeles port experienced at least two meters of subsidence due to oil pumping in the past century [36]. These variations remain difficult to accurately project [37], thus the single SLR value of two meters was used for this pilot study.

We based the technical design of GenPort's layout on reconstruction plans for the Port of Gulfport (MS) after it was destroyed by a nine meter storm surge during Hurricane Katrina [38] and on the general plan for container ports outlined by Thoresen in 2003 [39]. The Port of Gulfport adopted a plan to elevate the entire container port from three to nine meters as a strategy to build the port's resilience to Katrina-magnitude storms. Their design called for the elevation of the primary and secondary yards, where containers are stored, while leaving the berth and apron at a height suitable for offloading cargo from container ships. Over time, the apron area would also need elevation, but retrofitted incrementally to keep pace with sea level rise. The difference in height between the apron area and main yard areas would be accommodated through a retaining wall and ramps to allow for the movement of containers between the two yards. Though the Port of Gulfport ultimately did not move forward with this plan [40], it stands as an example of how a port retrofit design could feasibly elevate a portion of the port's footprint to accommodate storm surges associated with sea level rise. Outside of this plan from the Port of Gulfport, we found no other "real world examples" of designs for elevating an entire port. While the GenPort layout would be appropriate for many terminals and was used as a basis for analysis, large modern terminals would likely require variations in design to facilitate the transport of cargo between the apron and the primary yard. For example, one alternative design would place the retaining wall behind the landward crane rail. This would allow for the transfer of containers between the apron and the yard via the crane, eliminating the need for cargo transport via ramps (though perhaps requiring taller cranes). This design variation would not significantly impact the cost estimate generated in this investigation, but does illustrate that actual designs would vary from port to port based on their needs.

GenPort consists of general use areas and generic building types in order to develop equations to calculate the cost of reconstruction on a square meter basis. The costs to elevate appropriate portions of the terminal and to construct ramps and a retaining wall were also developed into equations. The land use at GenPort was divided into three area components: apron, primary yard and secondary yard (Figure 1). For the elevated primary and secondary yard areas, we include a range of land use combinations to accommodate a variety of port configurations. Infrastructure costs vary with land use type, and using a range of land use ratios allows for the integration of this variation into the cost estimate.

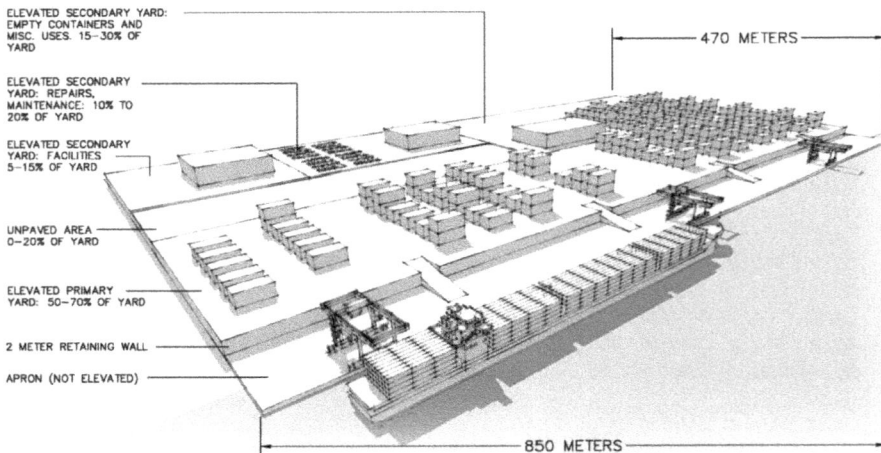

Figure 1. Schematic layout of GenPort with three area components included in the cost estimate (image by authors).

2.1.1. Apron

The apron is the area immediately behind the berth where ships tie up and the cargo is loaded and unloaded (Figure 2a). At a container port, the cranes sit on the apron and lift cargo on or off the ship to or from trucks or rail, which move the containers to other secondary yard locations at the port or to an off-port destination (Figure 2b). This area typically has a width of 15 to 50 m [39]. The GenPort layout assumes a width of 35 m. In the GenPort design, the apron remains at its pre-retrofit elevation (generally approximately 2–3 m above mean high tide levels) while the primary and secondary yards are elevated.

Figure 2. (**a**) Photograph of the "Apron" area at the Port of Boston; and (**b**) photograph of primary (to right) and secondary (to left) yards in Wilmington, North Carolina (photos by A. Becker).

2.1.2. Primary and Secondary Yards

The primary yard for container and cargo storage is located behind the apron and takes up 50–70% of the total yard area [39]. Cargo and facilities on the primary yard are exposed to storm surge impacts at many ports. For this exercise, we elevated the primary yard in the GenPort design by two meters to accommodate future SLR and storm surges. It incorporates the construction of a new retaining wall between the apron and the primary yard.

The secondary yard contains four main sections [39]: (1) the area for facilities such as office buildings, customs facilities and parking, which takes up 5–15% of the total yard area; (2) an area that holds empty containers and the container freight station which takes up 15–30% of the total yard area; (3) an area for repairs, storage and maintenance, which takes up 10–20% of the total yard area; and (4) an area that is unpaved, which takes up 0–20% percent of the total area (Figure 2). The secondary yard was also elevated two meters. In the next sections, we use the GenPort model to develop the cost estimate calculations for both elevation (e.g., fill procurement and placement) and reconstruction of GenPort's major facilities.

2.2. Cost Estimate Calculation to Elevate and Retrofit GenPort

The total cost to elevate one square meter the GenPort was presented as the Total Port Adaptation Cost (TPC). This value resulted from calculating the estimates for the total fill placement costs (FC) combined with retaining wall cost (RWC), ramp cost (RC), yard cost (YC) and geotechnical assessment cost (GC). Here, geotechnical assessment refers to engineering, physical, and environmental surveys of the construction site with a particular focus on soils and foundation conditions. Engineering and administration costs were assumed to be equal to 10% and 8% of the construction costs, respectively. Administration costs include miscellaneous insurance, bonding, construction management, and permitting. We incorporated these costs by applying a 1.18 multiplication factor to the base construction costs (see Equation (A1)).

The fill cost to raise the elevation of the port assumes the availability of clean, soft dredged material from the surrounding area. The unit cost of dredging was taken from the Army Corps of Engineers annual analyses of dredging costs [41]. Use of dredged material was generally more cost effective than trucking in fill from offsite. Dredging will not always be possible due to lack of suitable dredging material or environmental restrictions. In some locations, contaminated dredge material, rock bottoms requiring blasting, or other circumstances make the use of dredged fill infeasible. The cost of overland fill in these situations is likely to be within the same order of magnitude as the total cost presented in Table 1. Calculations are based on clean dredge materials.

Table 1. Average cost of overland fill to elevate land areas (2012 US dollars). Six years of cost for new work dredging (clean dredge material) are averaged and multiplied by a dozing and compaction factor of 20% to estimate the unit cost of adding fill to elevate land areas [42].

Item	Unit Cost/m^3
2007 New Work Dredging	$15.11
2008 New Work Dredging	$14.08
2009 New Work Dredging	$24.16
2010 New Work Dredging	$19.97
2011 New Work Dredging	$22.98
2012 New Work Dredging	$21.71
6-year New Work Dredging Average	$20.19
Dozing and Compaction (20%)	$4.04
Total Cost	$24.23

Fill cost calculations included dredging costs from 2007 through 2012. New work dredging costs were converted to 2012 dollars and averaged to incorporate year-to-year cost fluctuations. A 20% dozing and compaction factor accounted for the cost of placing the fill (Table 1). The calculations excluded the costs of erosion control or potential environmental protection requirements.

The cost of fill (FC) per square meter for elevated portions of the GenPort utilizes the total fill cost (Table 1) (see also Equation (A2)) multiplied by the height of the elevation increase and by the fraction of the port area to be elevated, represented by the variable EA. This was equal to the elevated area of the port divided by the total port area (for the GenPort model, EA equals 0.926 because 92.6% of the total port area is elevated). The GenPort design assumed the construction of a retaining wall at the boundary between the apron and the primary yard as a means of holding back the fill in the elevated port area. To calculate the cost per linear meter of the retaining wall, the cost of a cast-in-place level concrete retaining wall from RSMeans was used. Because the square meter cost of the wall increased with the height of the wall, the height of the wall and the square meter cost of the wall were treated as a linear relationship (Figure 3). To calculate the retaining wall cost per square meter of port (RWC), the retaining wall square meter cost (see trend line equation in Figure 3) was multiplied by the elevation increase to calculate the linear cost of the retaining wall and divided by 470 m, the width of the GenPort model (Equation (A3)).

The ramp cost (RC) variable represents the cost of ramp construction per square meter of port. Ramps enable access between the apron and the yard of the port. The ramps used in our calculations are 10 m in width and have a five percent incline. We accounted for the construction of two ramps per berth, totaling four ramps in the GenPort model. We assumed this construction uses the same dredged fill material used to elevate the yard, topped with 15 cm of crushed 2.5–1.3 cm stone base and 20 cm thick concrete paving (Table 2) [42,43]. For these unit costs, the price associated with constructing a single ramp increases as a function of the elevation increase. We calculated the total ramp construction cost per square meter of port by multiplying one ramp cost by four ramps per model port and then dividing by the total GenPort area (see Equation (A4)).

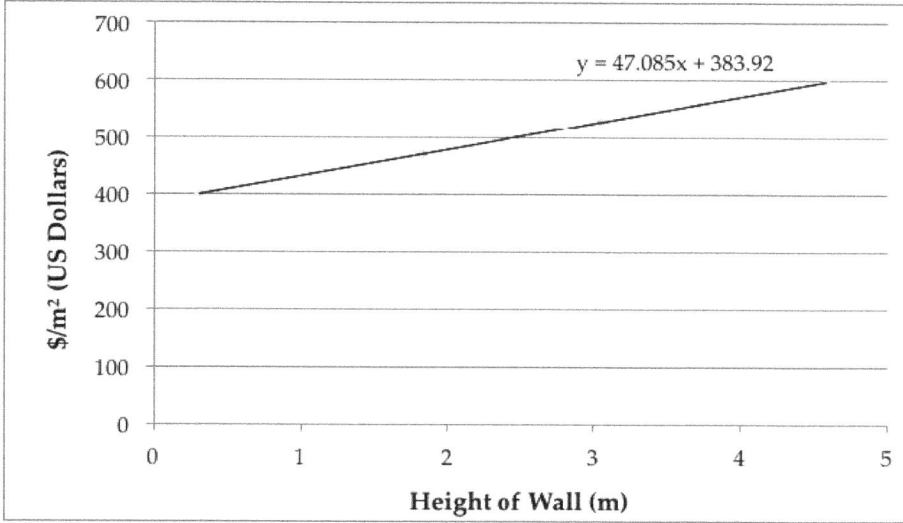

Figure 3. Equation and trend line for square meter costs of the retaining wall as a function of wall height (2012 US dollars).

Table 2. Unit cost of ramp materials, including fill, concrete pavement, and crushed stone (2012 US dollars).

Item	Cost
Dredged fill	$24.23/m^3
20 cm concrete pavement	$54.36/m^2
Crushed 2.5–1.3 cm stone base, 15 cm deep	$10.17/m^2

The yard surface and structures costs (YC) incorporate two major components: a primary yard where most container storage and movements on and off the port takes place and a secondary yard that holds empty containers, a container freight station, and space for repairs, storage and maintenance. Facilities such as office buildings, customs facilities and parking are also located on the secondary yard. Unpaved areas may also provide space for freight laydown or storage but do not contribute to the yard cost.

The primary yard, which requires paving, uses concrete block pavers because they function well in areas where heavy equipment is in use. Asphalt has a lower load capacity than concrete and there is a higher possibility of container supports penetrating asphalt, especially when its bearing capacity has been further reduced by warm weather [39]. For this exercise, no distinction was made between traffic lanes inside the yard and the actual stacking areas, as concrete block pavers were assumed for both. The presented calculations assumed that the subgrade soil condition was good because the new construction would take place on the site of the existing port. *The Port Designer's Handbook* recommends the use of 8–10 cm thick inter-locking concrete pavers or 10–12 cm thick rectangular concrete pavers [39]. Similar to the Dundalk and Seagirt marine terminal repaving projects [42], we used concrete pavers on top of a bed of sand, a bituminous-stabilized base course, and a crushed stone subbase (Table 3).

Table 3. Primary yard cost components total cost value (2012 US dollars) [42,43].

Item	Cost/m^2
100 mm thick, 100 mm × 200 mm rectangular concrete paver + 25 mm sand + stabilizer	$40.47
20 cm bituminous-stabilized base course	$22.78
Crushed 2.5–1.3 cm stone subbase, 15 cm deep	$10.17
Total	$73.42

The secondary yard paving costs were calculated using the 20 cm concrete pavement and 15 cm crushed stone base costs used for the ramps (Table 2). Pavers were not included in secondary yard costs. For the portion of the secondary yard dedicated to repairs and maintenance, the cost of the maintenance building was calculated using the cost of a concrete block warehouse. The overflow area of the secondary yard dedicated to empty containers, container repair, and freight handling used the same square meter costs for warehouses and concrete paving. The cost of a one story office building with exterior insulation and finish systems on metal studs was used to calculate the cost of office space (Table 4) [43].

Table 4. Secondary yard cost components per percentage coverage (2012 US dollars).

Components	Percent Coverage	Cost/m^2
Repairs, Maintenance Area		
Concrete Block Warehouse	10%	$1051.63
Paving	90%	$64.52
Repairs and Maintenance Area Weighted Average		$163.23
Facilities Area		
Administration Building	15%	$1929.43
Paving	85%	$64.52
Facilities Area Weighted Average		$344.26
Overflow Area (Empty Containers, Container Freight Station, Misc.)		
Concrete Block Warehouse	3%	$1051.63
Paving	97%	$64.52
Overflow Area Weighted Average		$94.13

We consolidated the costs of the primary and secondary yards into a single average square meter cost for reconstructing all yard infrastructure at GenPort (Table 5). Table 5 shows the range of expected land use mixes for a typical port. The most expensive percentage mix of uses and the least expensive mix of uses are presented by using an upper bound (UYC), representing the most expensive mix of uses, and a lower bound yard cost (LYC), representing the least expensive mix of uses (Equations (A5) and (A6)). The most expensive mix of uses maximizes the area devoted to facilities (15% of the total yard area) and to repairs and maintenance (20% of the total yard area). The least expensive mix of uses maximizes the unpaved areas (20% of the total yard area) (Table 5).

Finally, the GenPort adaptation cost estimate assumed that there was sufficient geotechnical stability of the existing port site to accommodate the required additional fill. In some locations, the addition of fill may overload the waterside containment significantly, increasing construction costs. The additional cost required to upgrade waterside containment structures in such cases was not included in the cost estimate. The geotechnical assessment cost (GC) assumes deep boring tests within 50 m of the berth and shallow boring tests in the areas of the port over 50 m from the berth. Geotechnical assessment costs were calculated using the total weighted average cost per square meter of port area (Table 6).

Table 5. Unit costs to reconstruct the yard for each land use type were used to calculate the weighted total yard cost per square meter, including surface and structure reconstruction cost components (2012 US dollars). Most expensive mix of uses maximizes land use devoted to facilities, repairs, and maintenance. Least expensive mix of uses maximizes unpaved areas.

Land Use	Yard Coverage	Cost/m^2	Least Expensive Mix of Uses	Most Expensive Mix of Uses
Primary Yard (from Table 3)	50–70%	$73.42	50%	50%
Facilities Areas	5–15%	$344.26	5%	15%
Repairs, Storage and Maintenance Areas	10–20%	$163.23	10%	20%
Overflow Areas (Empty Containers, Container Freight Station, Misc.)	15–30%	$94.13	15%	15%
Unpaved Areas	0–20%	$0	20%	0%
Total Weighted Yard Cost/m^2			$84.37	$135.11

Table 6. Cost of geotechnical assessment components (2012 US dollars) [43].

Component	Linear Bore Cost ($/m)	Bore Depth (m)	Bore Spacing (m^2)	Assessment Cost ($/m^2)
Shallow Bore	$184.22	9	900	$1.87
Deep Bore	$184.22	23	500	$8.42
Weighted Average				$2.57

2.3. Apron and Other Costs Not Included

The GenPort retrofit design assumes that the waterfront portion of the apron remains at the pre-retrofit existing elevation without alteration. This allows for ship accessibility, as significantly raising the apron height over the existing sea level could impede the loading and unloading of cargo. In addition to reducing construction costs, maintaining the existing apron and berth elevation would prevent difficulties in loading and unloading cargo. Ramps allow vehicles access to the apron from the elevated portions of the port in order to transport containers from sea level up to the new raised laydown area [44]. However, some cranes, utilities and other waterfront port infrastructure would not be protected by this design and would still be subjected to increased threat of flooding due to storm surge. In practical application, the apron area would be elevated incrementally to keep in step with SLR.

Other costs were considered, but ultimately not included in the cost calculations due to their high level of variability between different locations and uses. For example, in some ports, years of port activities may have left traces of fuels and other hazardous materials that would have to be remediated. Likewise, environmental permitting, changes to utilities (e.g., electrical cabling, frames for reefers, lighting posts, and drainage facilities), and the cost of demolition and rubble removal were not included in the estimation.

2.4. Calculation of Port Area for All US Ports

To generate an aggregate estimate for the cost and materials to retrofit US ports using the GenPort design, we manually calculated the port land area values for the 100 major US East Coast, Gulf Coast, West Coast, Hawaiian and Alaskan coastal ports (see Appendix B). Using satellite imagery available through Google Earth, we traced a polygon around the infrastructure at each port that was clearly associated with port activity, including yard storage areas and associated structures. We included all United States coastal commercial ports that handle freight [8], excluding marinas, fishing harbors, or other surrounding infrastructure that are not apparently tied directly to port activities. The polygon overlays are exemplified below for Oakland, CA, where area recorded as "coastal port infrastructure" in red was used in the aggregate calculation estimates (Figure 4). Although not all port areas captured through this process are container operations, the GenPort model is used in this exercise as a proxy for a variety of port operations. The summary of the ports and the total square kilometers of port infrastructure used in this study are presented in Table 7; see further details in Appendix B.

Figure 4. Example of a port polygon (Oakland, CA) determined manually using satellite imagery available from ESRI (Sources: ESRI, DigitalGlobe, GeoEye, Earthstar Geographics, CNES/Airbus DS, USDA, USGS, AeroGRID, IGN, and the GIS User Community).

Table 7. Number of ports and port areas (in square kilometers) in each region of the US.

Region	Number of Ports	Total Port Area (km^2)
Hawaii	8	5.7
Alaska	27	5.9
West Coast	22	110.0
Gulf Coast	17	129.5
East Coast	26	129.1
Total	100	380.2

3. Results

Applying the cost calculations from GenPort to the total port area calculated through the digitization of 100 US coastal port footprints generated an estimate for cost and materials to elevate and reconstruct the sampled ports. Although not all 100 ports handle containers exclusively, this method provided a rough estimate of costs and materials based on a generic layout. In this section we summarize key findings, including the costs to elevate and rebuild port infrastructure for GenPort on a per square kilometer basis. We apply these calculations to estimate costs for ports in the US.

3.1. Cost to Elevate GenPort per Square Kilometer

The cost per kilometer to retrofit and elevate one square kilometer of GenPort two meters varied between $206 million and $151 million (Figure 5). The variance between the upper and lower bound costs reflect the range of values found in the total yard cost (YC) relative to how the yard was used. Figure 5 presents upper and lower bound costs per square kilometer as a function of increase in elevation.

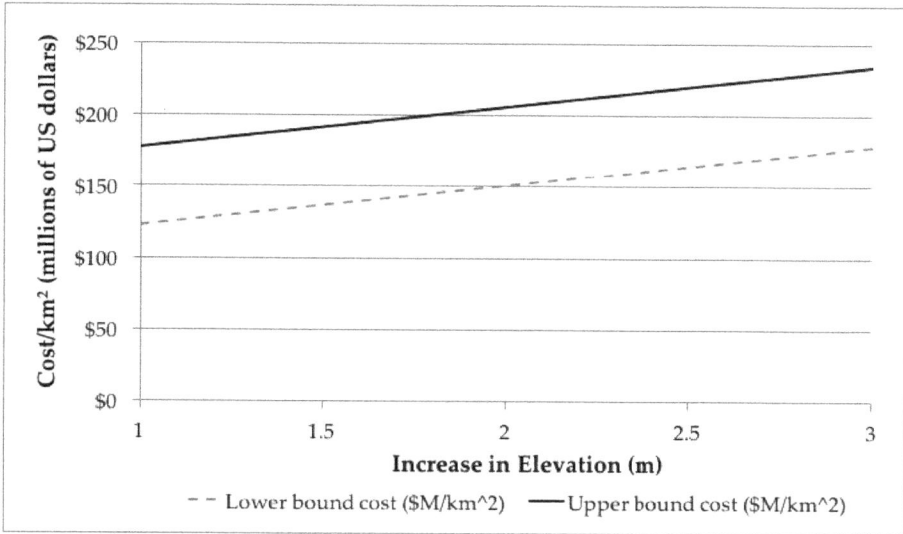

Figure 5. Upper and lower bound cost in US dollars per square kilometer of port area to increase port elevation and rebuild associated infrastructure as a function of increase in port elevation in meters.

The majority of the cost for adaptation was in the reconstruction of the yard (Figure 6) due to the high cost of facilities reconstruction (e.g., offices, repair sheds, paving). The incremental cost to increase a port's elevation was relatively low in comparison to the total cost per square kilometer of the full renovation project (Table 8). The cost of the components and total cost to elevate port infrastructure two meters are presented in Table 9. The cost components include the 1.18 engineering and administration multiplication factor.

Table 8. Total cost in millions of US dollars per square kilometer of port area to increase port elevation one to three meters and retrofit port.

Elevation Increase (m)	Lower Bound Cost ($millions/km^2)	Upper Bound Cost ($millions/km^2)
1	$123	$178
2	$151	$206
3	$179	$235

Table 9. Cost components and total cost (US dollars) to elevate and retrofit GenPort two meters.

1.18 × RWC	1.18 × RC	1.18 × FC	1.18 × GC	1.18 × LYC	1.18 × UYC	Lower Bound Total Cost	Upper Bound Total Cost
		($/m^2)				($/km^2)	
$2.40	$0.46	$52.92	$3.03	$92.14	$147.56	$150,939,700	$206,354,257

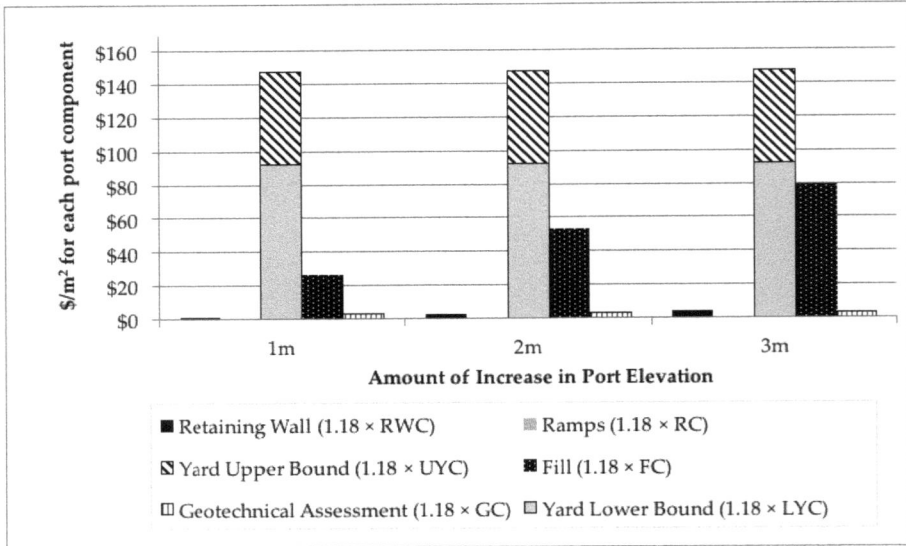

Figure 6. Cost to elevate and retrofit per square meter of port area by cost of component.

3.2. Cost to Elevate All US Ports

Elevating 100 commercial US coastal ports by two meters sums to between $57,379 million and $78,444 million (Table 10). A cost breakdown by geographic region is presented in Table 10.

Table 10. Total cost to elevate all US coastal port infrastructures by two meters.

Region	Number of Ports	Port Area (km^2)	Total Cost ($millions) to Elevate Two Meters and Retrofit (Lower Bound)	(Upper Bound)
Hawaii	8	5.7	$860	$1176
Alaska	27	5.9	$890	$1217
West Coast	22	110.0	$16,600	$22,694
Gulf Coast	17	129.5	$19,540	$26,714
East Coast	26	129.1	$19,489	$26,644
Total		380.1	$57,379	$78,444

Combined, these projects would require about 704 million cubic meters of fill to elevate all 100 ports by two meters (Table 11). This amounts to approximately four years of dredging material, based on all 2012 USACE projects [41].

Table 11. Amount of fill necessary to elevate 100 US commercial ports compared to US Army Corps 2012 dredging quantities [41].

Dredging	Cubic Meters (Millions m^3)	Years Needed to Provide Required Fill Volume
Total US commercial port fill requirement	704	
US Army Corps 2012 total dredging	182	4
US Army Corps 2012 new work dredging	17	42

4. Discussion

As noted by Nicholls et al. [9] and others, few assessments of regional port adaptation to climate change have been conducted, largely due to a lack of comprehensive physical data. The use of the GenPort model simplified the estimation process to allow for consideration of regional or national scale costs and material requirements. The GenPort model estimated a cost of $151 to $206 million per square kilometer to elevate and reconstruct the 0.4 square kilometer (100 acre) GenPort. This included procurement and placement of two meters of fill, geotechnical assessment, construction of the ramps and retaining walls, and reconstruction of the yard. The estimate also included a standard engineering and administration multiplication factor of 1.18. The use of GenPort enables the creation of an aggregate estimate for the costs and the quantity of fill necessary for this type of design that can be extrapolated to develop an estimate for ports on a regional or national scale, based on current land area occupied by the sample of ports. This pilot study found that elevating and retrofitting 100 major US ports to accommodate a two-meter sea level rise using the GenPort design would result in a total cost of between $57 and $78 billion (US). The GenPort model simplifies data requirements for elevation and reconstruction for port operation; actual port elevation and retrofit would involve additional costs, including but not limited to permits and environmental remediation assessments. This cost estimate could be refined in a number of ways, including the development of a variety of GenPort type port designs for large modern ports, bulk and liquid bulk terminals, and general cargo terminals, each of which would have unique requirements for retrofit. This could be accomplished by introducing additional infrastructure and land use types, such as tank farms or rail facilities, to the total yard cost calculations. Further, elevating a port is a large-scale project; environmental impacts are compounded by the immediate proximity of sensitive aquatic habitats for many ports. Environmental permitting and remediation would likely contribute significantly to the final price tag.

In another analysis of the cost of elevating port ground levels, Nicholls et al. [9] compensated for a lack of port area data by applying a traffic-to-area conversion. In their analysis, they used a cost of $15 million per square kilometer to raise ground elevation by one meter based on a 1990 IPCC report, which equates to $26 million (US) in 2012 when adjusted for inflation. This value, "based on Dutch procedures including design, execution, taxes, levies and fees and the assumption that the operation would take place as one event" [9,28], excluded the cost of adapting buildings and infrastructure. Comparatively, our calculations estimate $31 million per square kilometer to raise a port one meter (including geotechnical assessment and retaining wall) and an additional $92 to $148 million per square kilometer to reconstruct the yard infrastructure (see Table 9 and Figure 6). Our estimate also aligns with the budget developed for an actual elevation project in Gulfport (MS) for which the Mississippi State Port Authority at Gulfport proposed a 4.6 m increase in the elevation of its western pier [45]. While the scope of that project was ultimately reduced and the elevation component eventually eliminated [40], the port initially budgeted approximately $250 million per square kilometer to raise the terminal by 4.6 m [44]. Our model predicts a $225 to $280 million per square kilometer cost for the same elevation increase.

Port elevation represents one of a number of preventative measures that can be taken against the damages associated with climate change, along with construction of seawalls, hardening of structures, or the relocation of the port, though this last option is unlikely to be the most economical adaptation strategy as it requires the acquisition of a large area of land and the restructuring of the surrounding transportation network. Any design to retrofit will also require compromise. GenPort's ramps, for example, would slow operations both on the apron and between the apron and the elevated container storage area. Individual ports would need to design a retrofit suited to their specific needs. Ports exposed to large storm surges may opt for dikes and sea walls, as opposed elevation of the port infrastructure. Such decisions will of course be made on a case-by-case basis. Many ports are likely to elevate their infrastructure more gradually as old structures need to be replaced. This will allow for a more incremental capital investment strategy that will only be increased to accommodate the marginal costs of filling to increase port elevation, as opposed to the full costs represented in this study. Though the cost to elevate ports would be immense, the cost of doing nothing to protect ports would very likely far exceed the investment in adaptation. Just one storm, Hurricane Katrina in 2005, caused $100 million in damages to Mississippi's port infrastructure [46] and Hurricane Sandy resulted in between $40 million and $55 million of damage to the maritime assets of the Port Authority of New York and New Jersey [47]. Due to the indirect and intangible costs associated with damaged infrastructure, the costs of doing nothing are thought to far exceed the costs of adaptation for coastal infrastructure [27].

In addition to capital costs, obtaining the necessary fill material to raise infrastructure would likely be a limiting factor. This study shows that elevating these 100 US ports alone would require 704 million cubic meters of dredged fill, roughly 42 times more than the volume of material generated through new work dredging in the United States by the Army Corps of Engineers in 2012 and four times more than all material dredged my the Army Corps of Engineers in 2012, including maintenance and emergency dredging [48].

Finally, we note that sea level rise does not (at this point) have a fixed end point and the rate of rise will likely only increase over time [14,49]. Thus, a single retrofit project to elevate infrastructure does not result in a final solution for sea level rise. Rather, it simply buys time and protects infrastructure for a time until sea levels rise enough to overcome once again. Depending on the rate of rise, a piece of infrastructure may need to be elevated numerous times (along with its surrounding network of road, rail, and utilities). This study did not consider options such as designing a project to accommodate multiple (incremental) elevations over time. However, such ideas could and should be considered.

Ports are only one of many types of coastal infrastructure at risk from SLR and storm surge. We focused on ports for this study, as ports are relatively self-contained and conducive to the use of a generic design. In addition to ports, sewage treatment plans, airports, roads, rail, power plants, and other uses will require protection or relocation. We propose that the methodology described herein could be adapted for other such uses in order to develop a better understanding of costs and materials required to protect such uses on a national, or even global, scale.

5. Conclusions

Few ports have begun to invest in adaptation measures in preparation for the levels of SLR expected by the end of the century [50]. That said, according to American Association of Port Authorities, US ports currently anticipate spending $2.1 billion on capital upgrades annually [1]. Even if most this budget was put toward climate change adaptations, funding is likely to be insufficient to accommodate all US ports efforts to protect against major sea level rise. However, many ports will eventually require major modifications, such as new sea walls and storm gates or elevation, that require significant material and financial resources on a scale not seen in the past. Simultaneous implementation of large scale construction projects can increase resource demand. However, estimating such projected demand on a regional, national, or global scale remains a challenge.

Using the GenPort model to develop square meter equations for elevating and retrofitting a generic container port, we estimate a cost of $57 to 78 billion to elevate all major US seaports by two meters. Naturally, each port would need to determine the cost-effectiveness of raising infrastructure (once) to an anticipated final height against implementing two (or more) such capital improvement projects carried out incrementally as sea level rise. We suggest that using a generic design model, such as GenPort, as an estimation tool can help inform the emerging dialogue about climate adaptation and building coastal resilience. Most designs are generated on a site-specific basis and would not be appropriate to base estimates of demand at a national scale. Decision makers must consider this increase in demand as they weigh the costs and benefits of various adaptation strategies. In order to understand the impact that such projects will have on a national scale, new methods must be developed to estimate the potential aggregate demand as many projects move from concept to design and build. This research suggests the use of a generic port model to aid in the development of such estimates for one particular climate adaptation strategy: elevation of port facilities. Ports represent as little as 5% of coastal land uses that would compete globally for construction resources under this climate scenario. We note that climate change will generate new demands for many resources on both the national and global scales. Our estimation method represents a novel approach to help understand this new level of demand that will be driven by implementation of many preferred adaptation strategies.

Supplementary Materials: The Supplementary Materials are available online at www.mdpi.com/2077-1312/5/3/44/s1.

Acknowledgments: This study was performed as part of a research program at Stanford University and at the University of Rhode Island. The authors wish to acknowledge Martin Fischer and Ben Schwegler at Stanford University for their role in the initial development of these ideas. Zachary Driver assisted in the development of the schematic image of GenPort. The authors also thank the anonymous reviewers for their edits and insightful suggestions.

Author Contributions: Austin Becker and Ariel Hippe conceived, designed, and undertook the research associated with this study. Elizabeth L. Mclean provided additional input, updated, formatted and substantially edited the manuscript.

Conflicts of Interest: The authors declare no conflict of interest.

Appendix A. Equations to Calculate the Total Port Adaptation Cost (From Section 2.2 Cost Estimate Calculations)

A.1. Equation (A1)—The Total Port Adaptation Cost (TPC)

$$TPC \ (\$/m^2) \ = \ 1.18 \times (FC + RWC + RC + YC + GC) \tag{A1}$$

- TPC represents the total port adaptation cost for GenPort per square meter of port area.
- 1.18 is a multiplication factor that incorporates engineering and administration costs into the total port adaptation cost. Engineering costs are assumed to be equal to 10% of the construction costs and administration costs are assumed to be equal to 8% of the construction costs. Administration costs include miscellaneous insurance, bonding, construction management, and permitting.
- FC represents the fill cost equation (Equation (A2)).
- RWC represents the retaining wall cost equation (Equation (A3)).
- RC represents the ramp cost equation (Equation (A4)).
- YC represents the yard surface cost (Equations (A5) and (A6)).
- GC represents the geotechnical assessment cost (see main text).

A.2. Equation (A2)—The Cost of Fill (FC) for Elevated Portions of the GenPort

$$FC \ (\$/m^2) = (\$24.23/m^3) \times height \times EA \tag{A2}$$

- $\$24.23/m^3$ represents the total fill unit cost (Table 1).
- Height represents the elevation increase.
- EA represents the elevated area, which for the GenPort is equal to 92.6%.

A.3. Equation (A3)—Retaining Wall Cost (RWC)

Retaining walls in the GenPort serve as the boundary between the apron and the primary yard. The square meter cost of the wall increases with the height of the wall. This increase was treated as a linear relationship. To convert to per square meter of port area, the square meter cost of retaining wall is multiplied by the elevation increase and divided by 470 m, the width of the GenPort model (see Figure 3).

$$RWC \ (\$/m^2) \ = \ (\$47.09/m^3 \times height + \$383.92/m^2) \times height \times \frac{1}{470 \, m} \tag{A3}$$

- $47.09/m^3 \times$ height + $383/m^2$ represents the square meter cost of retaining wall from RSMeans as a function of wall height see (Figure 3, trend line equation).
- Height represents the elevation increase.
- 470 m represents the width of the GenPort model.

A.4. Equation (A4)—Ramp Costs (RC)

The ramps allow access between the apron and the yard of the port. The ramps used in our calculations are 10 m in width and have a five percent incline. Four ramps are calculated for the GenPort model. Construction of the ramps is based on the use of the same dredged fill material used to elevate the yard as well as a stone base and paving. The price associated with constructing each increases as a function of the elevation increase. To calculate the total ramp construction cost per square meter of port area, this cost is multiplied by four ramps per model port and divided by the total GenPort area:

$$RC\left(\$/m^2\right) = \frac{4 \times \left(\$2423/m^2 \times height^2 + \$12,922/m \times height + \$1632\right)}{399,500\ m^2} \tag{A4}$$

- $2423/m^2 \times height^2$ represents the dredged fill cost as a function of elevation increase for a single ramp.
- $12,922/m \times$ height represents the paving and stone base cost as a function of elevation increase for a single ramp (Table 2).
- $1632 represents additional paving costs independent of ramp height for a single ramp.
- 399,500 m^2 represents the total GenPort area.

A.5. Equation (A5)—Upper Bound Costs (UYC) and Equation (A6)—Lower Bound Yard Costs (LYC)

The upper bound total yard cost (UYC) is based on the most expensive yard land use mix presented in Table 5 and is calculated with Equation (A5); the lower bound yard cost (LYC) is based on the least expensive yard land use mix presented in Table 5 and is calculated with Equation (A6). Both of these values are attained by multiplying the total weighted yard cost by the elevated area (EA) variable:

$$UYC\ (upper\ bound\ yard\ cost\ )\ (\$/m^2) = \$135.11/m^2 \times EA \tag{A5}$$

$$LYC\ (lower\ bound\ yard\ cost)\ (\$/m^2) = \$84.37/m^2 \times EA \tag{A6}$$

Appendix B. Cost and Materials for Ports by Region

Table A1. Cost and materials required to elevate and retrofit Hawaii Seaports in response to two meters of sea level rise.

Port Name	Area (km²)	Lower Bound Total Cost ($)	Upper Bound Total Cost ($)	Fill (m³)
Hilo	0.29	$43,800,326	$59,880,759	537,150
Kawaihae	0.16	$24,512,311	$33,511,526	300,609
Kaumalapau Harbor	0.02	$2,822,494	$3,858,717	34,614
Barber's Point	0.78	$117,865,331	$161,137,281	1,445,453
Port Allen	0.07	$10,612,278	$14,508,368	130,145
Nawiliwili Bay	0.36	$54,952,192	$75,126,813	673,912
Honolulu	3.62	$545,949,407	$746,384,049	6,695,304
Kahului	0.39	$59,572,937	$81,443,976	730,579
TOTAL	5.70	$860,087,276	$1,175,851,490	10,547,765

Table A2. Cost and materials required to elevate and retrofit Gulf Coast Seaports in response to two meters of sea level rise.

Port Name	Area (km²)	Lower Bound Total Cost ($)	Upper Bound Total Cost ($)	Fill (m³)
Sabine Pass	0.06	$9,088,981	$12,425,823	111,464
Sabine	0.26	$39,952,994	$54,620,954	489,967
Palacios	0.03	$4,185,822	$5,722,565	51,333
Port Lavaca	0.08	$11,684,427	$15,974,135	143,293
Rockport	0.06	$8,636,288	$11,806,932	105,912
Port Aransas	0.11	$15,982,254	$21,849,825	196,000
Port Ingleside	1.33	$201,121,677	$274,959,566	2,466,475
Mobile	9.08	$1,370,901,741	$1,874,201,491	16,812,188
New Orleans	17.19	$2,595,239,835	$3,548,031,359	31,826,979
Corpus Christi	16.28	$2,456,785,892	$3,358,746,761	30,129,035
Galveston	20.09	$3,032,030,718	$4,145,181,470	37,183,606
Houston	53.03	$8,004,880,735	$10,943,716,067	98,168,641
Gulfport	1.23	$185,009,384	$252,931,959	2,268,881
Baytown	0.68	$102,527,020	$140,167,809	1,257,350
Pensacola	0.35	$52,915,522	$72,342,420	648,935
Port Manatee	0.53	$79,656,114	$108,900,297	976,871
Tampa-St. Petersburg	9.07	$1,369,375,484	$1,872,114,899	16,793,471
TOTAL	129.46	$19,539,974,887	$26,713,694,332	239,630,400

Table A3. Cost and materials required to elevate and retrofit East Coast Seaports in response to two meters of sea level rise.

Port Name	Area (km²)	Lower Bound Total Cost ($)	Upper Bound Total Cost ($)	Fill (m³)
Fort Pierce	0.10	$14,614,872	$19,980,436	179,231
Port Everglades	4.10	$618,925,576	$846,151,990	7,590,255
New York/New Jersey	34.26	$5,171,066,231	$7,069,522,016	63,415,878
Canaveral	3.84	$579,060,752	$791,651,576	7,101,368
Charleston	16.19	$2,444,180,843	$3,341,514,015	29,974,452
Baltimore	21.28	$3,211,517,304	$4,390,563,043	39,384,758
Boston	12.64	$1,908,617,107	$2,609,328,532	23,406,513
Sayreville	0.14	$21,333,290	$29,165,390	261,623
South Amboy	0.13	$19,734,547	$26,979,699	242,017
Cape Charles	0.28	$42,117,385	$57,579,959	516,511
Newport News	3.69	$556,464,483	$760,759,530	6,824,257
Morehead City	0.37	$55,222,725	$75,496,669	677,229
New Bedford	0.17	$25,881,285	$35,383,092	317,398
Fall River	0.72	$108,135,121	$147,834,815	1,326,126
Quonset Point	0.55	$82,395,569	$112,645,490	1,010,466
Davisville Depot	0.32	$48,362,192	$66,117,424	593,094
Providence	1.92	$289,057,394	$395,179,160	3,544,884
Savannah	12.01	$1,813,337,833	$2,479,069,337	22,238,047
Miami	2.87	$433,603,204	$592,792,137	5,317,535
Jacksonville	11.12	$1,678,951,740	$2,295,346,020	20,589,990
New Haven	1.08	$163,013,122	$222,860,201	1,999,127
Bridgeport	0.52	$78,593,556	$107,447,643	963,840
Elizabethport	0.27	$40,191,408	$54,946,897	492,891
Stapleton SI	0.14	$20,978,788	$28,680,740	257,275
Port Richmond SI	0.06	$8,490,935	$11,608,215	104,129
Mariners Harbor	0.36	$54,826,264	$74,954,653	672,367
TOTAL	129.12	$19,488,673,526	$26,643,558,680	239,001,261

Table A4. Cost and materials required to elevate and retrofit West Coast Seaports in response to two meters of sea level rise.

Port Name	Area (km²)	Lower Bound Total Cost ($)	Upper Bound Total Cost ($)	Fill (m³)
Los Angeles	33.0	$4,980,464,045	$6,808,943,967	61,078,410
Portland	3.13	$472,524,057	$646,002,019	5,794,845
San Diego	17.71	$2,672,969,813	$3,654,298,378	32,780,228
San Francisco	7.19	$1,085,692,895	$1,484,283,798	13,314,501
Seattle	10.42	$1,572,279,356	$2,149,510,957	19,281,802
Tacoma	18.95	$2,860,487,259	$3,910,659,186	35,079,866
Olympia	0.25	$37,968,754	$51,908,239	465,634
Point Wells	0.17	$24,923,495	$34,073,669	305,652
Everett	0.69	$103,435,488	$141,409,804	1,268,491
Anacortes	0.19	$29,004,415	$39,652,819	355,698
Bellingham	0.57	$85,494,333	$116,881,904	1,048,468
Eureka	0.99	$150,063,752	$205,156,722	1,840,322
Samoa	0.07	$9,904,636	$13,540,929	121,466
North Bend	0.08	$12,688,131	$17,346,330	155,602
Coos Bay	0.37	$56,197,468	$76,829,269	689,183
Shelton	0.39	$59,450,628	$81,276,762	729,079
Port Angeles	0.13	$19,690,362	$26,919,293	241,475
Hueneme	3.22	$485,459,747	$663,686,795	5,953,483
Redwood City	0.56	$83,927,726	$114,740,148	1,029,256
Port Gamble	0.06	$8,733,822	$11,940,274	107,108
Point Richmond	1.06	$159,849,560	$218,535,198	1,960,331
Virginia Beach	10.79	$1,628,636,435	$2,226,558,435	19,972,943
TOTAL	109.98	$16,599,846,176	$22,694,154,894	203,573,843

Table A5. Cost and Materials Required to Elevate and Retrofit Alaska Seaports in Response to two meters of Sea Level Rise.

Port Name	Area (km²)	Lower Bound Total Cost ($)	Upper Bound Total Cost ($)	Fill (m³)
Anchorage	1.22	$184,215,925	$251,847,197	2,259,150
Ketchikan	0.18	$26,718,747	$36,528,012	327,668
Ward Cove	0.18	$26,711,261	$36,517,778	327,576
Wrangell	0.02	$3,019,830	$4,128,502	37,034
Craig	0.01	$1,463,596	$2,000,926	17,949
Klawok	0.16	$23,569,285	$32,222,287	289,044
Sitka	0.06	$9,230,978	$12,619,950	113,205
Hoonah	0.03	$5,110,972	$6,987,365	62,679
Juneau	0.30	$45,491,259	$62,192,485	557,887
Tanani Point	0.20	$29,580,547	$40,440,466	362,764
Cordova	0.49	$74,591,251	$101,975,969	914,757
Valdez	0.20	$29,686,207	$40,584,918	364,060
Whittier	0.17	$25,955,088	$35,483,991	318,303
Seward	0.35	$52,394,937	$71,630,713	642,550
Port Graham	0.02	$2,325,579	$3,179,370	28,520
Seldovia	0.01	$1,922,694	$2,628,574	23,579
Homer	0.26	$39,130,550	$53,496,565	479,881
Nikiski	1.36	$204,776,336	$279,955,961	2,511,295
Kodiak	0.09	$13,352,114	$18,254,081	163,745
Womens Bay	0.21	$31,945,282	$43,673,368	391,764
Sand Point	0.04	$6,023,583	$8,235,024	73,871
Baralof Bay	0.03	$4,856,861	$6,639,962	59,563
King Cove	0.04	$5,346,994	$7,310,038	65,573
Dillingham	0.06	$8,581,510	$11,732,044	105,240
Prudhoe Bay	0.13	$19,225,426	$26,283,664	235,773
Minturn	0.01	$829,798	$1,134,441	10,176
Naknek	0.09	$14,022,708	$19,170,871	171,969
TOTAL	5.90	$890,079,316	$1,216,854,521	10,915,575

References

1. AAPA (American Association of Port Authorities). Talking Points: United States Seaports and Job Creation. Available online: http://aapa.files.cms-plus.com/PDFs/USSeaportsandJobCreation5-12-2015.png (accessed on 16 August 2016).
2. Burkett, V.; Davidson, M. *Coastal Impacts, Adaptation, and Vulnerabilities: A Technical Input to the 2013 National Climate Assessment*; Island Press: Washington, DC, USA, 2012.
3. Chambers, M.; Liu, M. Maritime Trade and Transportation by the Numbers. In *Bureau of Transportation Statistics*; BTS Publications: Washington, DC, USA, 2013.
4. Savonis, M.J.; Potter, J.R.; Snow, C.B. Continuing Challenges in Transportation Adaptation. *Curr. Sustain. Renew. Energy Rep.* **2014**, *1*, 27–34. [CrossRef]
5. Becker, A.; Chase, N.T.; Fischer, M.; Schwegler, B.; Mosher, K. A method to estimate climate-critical construction materials applied to seaport protection. *Glob. Environ. Chang.* **2016**, *40*, 125–136. [CrossRef]
6. Becker, A.; Toilliez, J.; Mitchell, T. Considering Sea Level Change When Designing Marine Civil Works: Recommendations for Best Practices. In *Handbook of Coastal Disaster Mitigation for Engineers and Planners*; Esteban, M., Takagi, H., Shibayama, T., Eds.; Elsevier: Waltham, MA, USA, 2015.
7. Becker, A.; Acciaro, M.; Asariotis, R.; Carera, E.; Cretegny, L.; Crist, P.; Esteban, M.; Mather, A.; Messner, S.; Naruse, S.; et al. A Note on Climate change adaptation for seaports: A challenge for global ports, a challenge for global society. *Clim. Chang.* **2013**, *120*, 683–695. [CrossRef]
8. NGIA (National Geospatial-Intelligence Agency). *World Port Index*, 23rd ed.; National Imagery and Mapping Agency: Springfield, VA, USA, 2014.
9. Nicholls, R.J.; Brown, S.; Hanson, S.; Hinkel, J. *Economics of Coastal Zone Adaptation to Climate Change*; World Bank Discussion Papers, 10; International Bank for Reconstruction and Development/World Bank: Washington, DC, USA, 2010.
10. Intergovernmental Panel on Climate Change (IPCC). *Climate Change 2013: The Physical Science Basis: Working Group I Contribution to the Fifth Assessment Report of the Intergovernmental Panel on Climate Change*; 1107415322; Intergovernmental Panel on Climate Change: Cambridge, UK; New York, NY, USA, 2013; p. 1552.
11. Rahmstorf, S. A new view on sea level rise. *Nat. Rep. Clim. Chang.* **2010**, 44–45. [CrossRef]
12. Jevrejeva, S.; Jackson, L.P.; Riva, R.E.; Grinsted, A.; Moore, J.C. Coastal sea level rise with warming above 2 degrees C. *Proc. Natl. Acad. Sci. USA* **2016**, *113*, 13342–13347. [CrossRef] [PubMed]
13. Vellinga, P.; Katsman, C.; Sterl, A.; Beersma, J.; Hazeleger, W.; Church, J.; Kopp, R.; Kroon, D.; Oppenheimer, M.; Plag, H. *Exploring High-End Climate Change Scenarios for Flood Protection of The Netherlands*; KNMI: De Bilt, The Netherlands, 2008.
14. Sweet, W.; Kopp, R.; Weaver, C.; Obeyskera, J.; Horton, R.; Thieler, E.; Zervas, C. *Global and Regional Sea Level Rise Scenarios for the United States*; NOAA: Silver Spring, MD, USA, 2017.
15. NRC (National Research Council). *America's Climate Choices: Adapting to the Impacts of Climate Change*; American Geophysical Union: Washington, DC, USA, 2010.
16. Bender, M.A.; Knutson, T.R.; Tuleya, R.E.; Sirutis, J.J.; Vecchi, G.A.; Garner, S.T.; Held, I.M. Modeled impact of anthropogenic warming on the frequency of intense Atlantic hurricanes. *Science* **2010**, *327*, 454–458. [CrossRef] [PubMed]
17. Becker, A.; Inoue, S.; Fischer, M.; Schwegler, B. Climate change impacts on international seaports: Knowledge, perceptions, and planning efforts among port administrators. *Clim. Chang.* **2012**, *110*, 5–29. [CrossRef]
18. Room for the Rivers. Room for the River Information. Available online: https://www.ruimtevoorderivier.nl/history/ (accessed on 14 June 2017).
19. Lonsdale, K.G.; Downing, T.E.; Nicholls, R.J.; Parker, D.; Vafeidis, A.T.; Dawson, R.; Hall, J. Plausible responses to the threat of rapid sea-level rise in the Thames Estuary. *Clim. Chang.* **2008**, *91*, 145–169. [CrossRef]
20. Blodget, H.; Wile, R. Hey, It Will only Cost $7 Billion to Build a Storm Surge Barrier for New York—Whaddya Say? Available online: http://www.businessinsider.com/new-york-storm-surge-barrier-2012-11 (accessed on 15 June 2017).

21. Dronkers, J.; Gilbert, T.; Butler, L.W.; Carey, J.J.; Campbell, J.; James, E.; McKenzie, C.; Misdorp, R.; Quin, N.; Ries, K.L.; et al. *Strategies for Adaptation to Sea Level Rise*; Intergovernmental Panel on Climate Change: Geneva, Switzerland, 1990.

22. Ghile, Y.B.; Taner, M.Ü.; Brown, C.; Grijsen, J.G.; Talbi, A. Bottom-up climate risk assessment of infrastructure investment in the Niger River Basin. *Clim. Chang.* **2013**, *122*, 97–110. [CrossRef]

23. Hanson, S.; Nicholls, R.; Ranger, N.; Hallegatte, S.; Corfee-Morlot, J.; Herweijer, C.; Chateau, J. A global ranking of port cities with high exposure to climate extremes. *Clim. Chang.* **2010**, *104*, 89–111. [CrossRef]

24. Hallegatte, S.; Green, C.; Nicholls, R.J.; Corfee-Morlot, J. Future flood losses in major coastal cities. *Nat. Clim. Chang.* **2013**, *3*, 802–806. [CrossRef]

25. Neumann, J.; Hudgens, D.; Herter, J.; Martinich, J. The economics of adaptation along developed coastlines. *Wiley Interdiscip. Rev. Clim. Chang.* **2011**, *2*, 89–98. [CrossRef]

26. Aerts, J.; Botzen, W.J.; Eamanuel, K.; Lin, N.; de Moel, H.; Michael-Kerjan, E. Evaluating Flood Resilience Strategies for Coastal Megacities. *Science* **2014**, *344*, 473–476. [CrossRef] [PubMed]

27. Hinkel, J.; Lincke, D.; Vafeidis, A.T.; Perrette, M.; Nicholls, R.J.; Tol, R.S.J.; Marzeion, B.; Fettweis, X.; Ionescu, C.; Levermann, A. Coastal flood damage and adaptation costs under 21st century sea-level rise. *Proc. Natl. Acad. Sci. USA* **2014**, *119*, 3292–3297. [CrossRef] [PubMed]

28. Hoozemans, F.; Marchand, M.; Pennekamp, H. *A Global Vulnerability Analysis: Vulnerability Assessment for Population, Coastal Wetlands and Rice Production on a Global Scale*; Public Works and Water Management: Emmeloord, The Netherlands, 1993.

29. IHS Markit. Construction Specifications Institute. Available online: http://www.ihs.com/products/industry-standards/organizations/csi/index.aspx (accessed on 15 July 2015).

30. Rahmstorf, S. A Semi-Empirical Approach to Projecting Future Sea-Level Rise. *Science* **2007**, *315*, 368–370. [CrossRef] [PubMed]

31. Muis, S.; Verlaan, M.; Winsemius, H.C.; Aerts, J.C.; Ward, P.J. A global reanalysis of storm surges and extreme sea levels. *Nat. Commun.* **2016**, *7*, 11969. [CrossRef] [PubMed]

32. Weisse, R. *Marine Climate and Climate Change: Storms, Wind Waves and Storm Surges*; Springer Science & Business Media: Berlin/Heidelberg, Germany, 2010.

33. Shepard, C.C.; Agostini, V.N.; Gilmer, B.; Allen, T.; Stone, J.; Brooks, W.; Beck, M.W. Assessing future risk: Quantifying the effects of sea level rise on storm surge risk for the southern shores of Long Island, New York. *Nat. Hazards* **2012**, *60*, 727–745. [CrossRef]

34. Tebaldi, C.; Strauss, B.H.; Zervas, C.E. Modelling sea level rise impacts on storm surges along US coasts. *Environ. Res. Lett.* **2012**, *7*, 014032. [CrossRef]

35. Kirshen, P.; Watson, C.; Douglas, E.; Gontz, A.; Lee, J.; Tian, Y. Coastal flooding in the Northeastern United States due to climate change. *Mitig. Adapt. Strateg. Glob. Chang.* **2007**, *13*, 437–451. [CrossRef]

36. City of Long Beach. Subsidence History. Available online: http://www.longbeach.gov/lbgo/about-us/oil/subsidence/ (accessed on 10 March 2015).

37. Hanson, S.; Nicholls, R.J. Extreme flood events and port cities through the twenty-first century. In *Maritime Transport and the Climate Change Challenge*; Asariotis, R., Benemara, H., Eds.; Earthscan/Routledge: New York, NY, USA, 2012; p. 243.

38. MSPA (Mississippi State Port Authority). *Gulfport Master Plan Update 2007—Final Report*; DMJM Harris and AECOM: MS, USA, 2007.

39. Thoresen, C.A. *Port Designer's Handbook: Recommendations and Guidelines*; Thomas Telford Services Ltd.: London, UK, 2003.

40. MSPG (Mississippi State Port at Gulfport). Port Commission Nixes 25' Elevation Plan. Available online: http://www.portofthefuture.com/news-headlines/port-authority-nixes-25-feet-elevation-for-gulfport/ (accessed on 12 November 2012).

41. USACE (U.S. Army Corps of Engineers). *Fiscal Year 2012 Analysis of Dredging Costs*; United States of America Corps of Engineers: Washington, DC, USA, 2012.

42. Shafer, T.J. A Case History of Concrete Block Papers at Dundalk and Seagirt Marine Terminals. In Proceedings of the International Conference on Concrete Block Paving, San Francisco, CA, USA, 6–8 November 2006; pp. 757–765.

43. RSMeans Corporation. *RSMeans Building Construction Cost Data 2013*; Reed Construction: Norwell, MA, USA, 2012.

44. Becker, A. Interview with Joseph Conn, Director of Disater Recovery at Mississippi State Port Authority at Gulfport. Unpublished work. 2010.
45. MSPG (Mississippi State Port at Gulfport). *Port of Gulfport Restoration Project*; MSPG: Gulfport, MI, USA, 2013.
46. PEER (Joint Legislative Committee on Performance Evaluation and Expenditure Review). *The Impact of Hurricane Katrina on Mississippi's Commercial Public Ports and Opportunities for Expansion of the Ports*; Peer Report # 487; Mississippi Legislature: Jackson, MS, USA, 2006.
47. Recovery Support Strategy. New York Recovers: Hurricane Sandy Federal Recovery Support Strategy—Version 1. June 2013. Available online: https://portal.hud.gov/hudportal/documents/huddoc?id=rssnewyorkrecovers09132013.png (accessed on 10 June 2017).
48. USACE (U.S. Army Corps of Engineers). Analysis of Dredging Costs. Available online: http://www.navigationdatacenter.us/dredge/ddcosts.htm (accessed on 15 January 2014).
49. Vitousek, S.; Barnard, P.L.; Fletcher, C.H.; Frazer, N.; Erikson, L.; Storlazzi, C.D. Doubling of coastal flooding frequency within decades due to sea-level rise. *Sci. Rep.* **2017**, *7*, 1399. [CrossRef] [PubMed]
50. Ng, A.; Becker, A.; Cahoon, S.; Chen, S.-L.; Earl, P.; Yang, Z. *Climate Change and Adaptation Planning for Ports*; Routledge: New York, NY, USA, 2016.

Journal of
Marine Science and Engineering

MDPI

Article

Biorock Electric Reefs Grow Back Severely Eroded Beaches in Months

Thomas J. F. Goreau [1,2,*] **and Paulus Prong** [2,3]

1   Global Coral Reef Alliance, 37 Pleasant Street, Cambridge, MA 02139, USA
2   Biorock Indonesia, Bali 80361, Indonesia; paulus.prong@yahoo.co.id
3   Pulau Gangga Dive Resort, Sulawesi 95253, Indonesia
*   Correspondence: goreau@globalcoral.org; Tel.: +1-617-864-4226

Received: 18 June 2017; Accepted: 27 September 2017; Published: 11 October 2017

Abstract: Severely eroded beaches on low lying islands in Indonesia were grown back in a few months—believed to be a record—using an innovative method of shore protection, Biorock electric reef technology. Biorock shore protection reefs are growing limestone structures that get stronger with age and repair themselves, are cheaper than concrete or rock sea walls and breakwaters, and are much more effective at shore protection and beach growth. Biorock reefs are permeable, porous, growing, self-repairing structures of any size or shape, which dissipate wave energy by internal refraction, diffraction, and frictional dissipation. They do not cause reflection of waves like hard sea walls and breakwaters, which erodes the sand in front of, and then underneath, such structures, until they collapse. Biorock reefs stimulate settlement, growth, survival, and resistance to the environmental stress of all forms of marine life, restoring coral reefs, sea grasses, biological sand production, and fisheries habitat. Biorock reefs can grow back eroded beaches and islands faster than the rate of sea level rise, and are the most cost-effective method of shore protection and adaptation to global sea level rise for low lying islands and coasts.

Keywords: beach restoration; erosion protection; corals; sea grass; reefs; Biorock electric technology; sea level rise; porous & permeable breakwaters; sand production; climate change

1. Introduction

Accelerating global sea level rise is now causing almost all beaches worldwide to erode [1]. The current rate of sea level rise, now 3 mm/year [2], will accelerate greatly in the future as the melting of ice caps increases, masked by shorter term regional fluctuations driven by local weather [3].

IPCC projections of sea level rise are often thought by the public to represent the end point of sea level rise response to fossil fuel CO_2, but in fact they are merely points along the first 5, 10, 20—or at most 100—years, of the initial rise of a curve that will continue to increase for thousands of years. The time horizons for IPCCC climate change projections were chosen for political purposes, not for scientific ones, and therefore miss the vast bulk of the real world long-term sea level and temperature responses to increased greenhouse gases [4].

Since the ocean holds nearly 93% of the heat in the Earth ocean-atmosphere-soil-vegetation-rock-ice system [5] and it takes around 1500 years for the ocean to mix and turn over [6], Earth's surface will not fully warm up until after the deep ocean waters, now about 4 degrees above freezing, heat up. Global temperatures and sea levels lag thousands of years behind CO_2 increases because of ocean mixing, so we have not yet really begun to feel the inevitable temperature and sea level responses. Because of these politically chosen time horizons, IPCC projections do NOT include more than 90% of the long-term climate response to changing CO_2 [4,7–9]. By greatly underestimating the all too real long-term responses of temperature and sea level, they have lulled political decision makers into a false sense of complacency about the magnitude and duration of human-caused climate change or the urgency of reversing them before the really serious impacts hit future generations [4].

Improved estimates of long-term global climate impacts are made from actual paleoclimate records of changes in global CO_2, temperature, and sea levels from the Antarctic ice cores, deep sea sediments, and fossil coral reefs over the last few million years [7,9]. The last time that global temperatures were 1–2 °C above today's level, sea levels were about 8 m higher, crocodiles and hippopotamuses lived in swamps where London, England, now stands (Rhodes Fairbridge, 1987, personal communication) [10,11], and CO_2 levels were around 270 ppm, around 40% lower than today [7,9]. Comparison of long-term global climate change records suggests that the steady state climate for the present (2017) CO_2 concentration of 400 ppm, once the climate system has fully responded, are about 17 °C and 23 m above today's levels [9]. We are committed to such changes even if there is no further CO_2 increase starting right now because of the excess already in the atmosphere, unless that is reduced. No amount of emissions reduction can reduce excess atmospheric CO_2, only increased natural carbon sinks with storage in soil and biomass carbon can draw down the dangerous excess in time to avert extreme long term changes [4,8,12], which would last for hundreds of thousands to millions of years. Perhaps the largest cost of adaptation to climate change will be the cost of protecting low lying islands and coasts from being flooded by global sea level rise.

Beach erosion is largely controlled by refraction of offshore waves by bottom topography [13]. The reflection of waves by steep cliffs prevents any accumulation of sand at their bases. In contrast, shallow sandy beach fore-shores are almost always protected from waves by reefs, Waves are refracted as water passes through porous and permeable reef structures, without being reflected.

Conventional methods of shore protection rely on "hard" solid structures like sea walls and breakwaters that are designed to reflect waves, like rock cliffs. This concentrates all the energy of the wave at the hard, reflecting surface, and the force on the structure itself is twice the momentum of the wave due to the reversal of the wave direction vector [14,15]. The inevitable result of this energy focusing is that first all the sand is washed away in front of the structure, and then is scoured away underneath it until the structure settles, cracks, falls apart, and needs to be rebuilt. These structures protect what is behind them until they fall down, but they cause erosion in front of them and guarantee loss of sediment. All such structures are consequently ephemeral and will fall down sooner or later, depending on how large and strong they are.

This is well known to coastal engineers but most feel there is no alternative to impermeable solid walls, even though so called "porous" or "permeable" breakwaters, made of small distributed modules shaped like coral reefs, with holes within structures and passages between them, seem to protect shores with much less material and with greatly reduced reflection. But we could find no experimental or theoretical modeling literature on porous permeable structures like natural coral reef structures found by searching on Google Scholar. Most search results for porous breakwaters were for solid rock walls with crevices between stones rather than reef-like structures with a much greater range of pore and spacing sizes, capable of interacting with waves over larger wavelength ranges.

Coral reefs provide the most perfect natural shore protection, dissipating around 97% of incident wave energy by frictional dissipation [16]. Healthy reefs produce sand as well as protect it, and rapidly build beaches behind them. They are sand factories, generating vast amounts of new sand, largely remains of calcareous green and red algae. Every grain of white limestone sand on a tropical beach is the skeletal remains of a living coral reef organism. Once corals die from high temperatures, pollution, or disease, the previously growing and self-repairing reef framework starts to deteriorate and crumble from boring organisms that excavate the rock [17]. Because of the mass mortality of corals around the world caused by global warming [18–23], tropical beaches that were growing until recently have begun rapid erosion, and islands are washing away because of global sea level rise.

Here we describe the results of a novel method of beach restoration—Biorock electric reef shore protection—which avoids the intrinsic physical flaws of hard reflective structures, and which grows beaches back at record rates at a lower cost, with less materials, and with much greater environmental benefits than seawalls. Biorock electric reefs are grown by low voltage electrolysis of sea water, which causes growth of limestone rock minerals dissolved in sea water over steel surfaces, which are completely protected from corrosion [24,25]. Biorock reef structures can be any size or shape, and are the only marine construction material that gets stronger with age and is self-repairing [26]. When grown slowly, less than 1–2 cm/year, this material is several times harder than Portland Cement concrete [25].

Biorock shore protection reefs are open mesh frameworks designed to permit water to flow through them, like coral reefs. The size and shape of the structures, and of the holes in them, determine their performance dissipating wave energy. The electricity needed for electrolysis is safe extremely low voltage (ELV) direct current provided by transformers, chargers, batteries, solar panels, wind mills, ocean current generators, or wave energy generators, depending on which source is most cost effective at the site [27].

Biorock reefs in Grand Turk survived the two worst hurricanes in the history of the Turks and Caicos Islands, which occurred three days apart and damaged or destroyed 80% of the buildings on the island [28]. Sand was observed to build up around the bases of Biorock reef structures. In contrast, concrete reef balls nearby caused such severe sand scour around and under them that they buried themselves into the sand, digging their own graves. Solid objects, by forcing bottom currents to accelerate as they diverge around them, cause erosion to a depth of about half the height of the structure, and about as wide as the structure height [29]. Biorock reefs, being permeable and porous to waves, had the opposite effect than reef balls, baffling waves, lowering their velocity, and causing sand deposition instead of erosion [30].

The first Biorock shore protection reef was built in front of a beach that had washed away at Ihuru Island, North Male Atoll, the Maldives, in 1997. Sand bags were being piled in front of trees and buildings that were falling into the sea, which the hotel thought they had no chance of saving. The Biorock reef was a linear structure parallel to the shore, 50 m long, about 5 m wide, and about 1.5 m high, built on eroded reef bedrock. The structure cemented itself solidly to the limestone bedrock with mineral growth. Waves were observed to slow down as they were refracted through the structure, dissipating energy by surface friction. Sand immediately began to accumulate on the shore line and under and around the reef, and the beach grew back naturally and rapidly in a few years, and stabilized with no further erosion, even though the 2004 Tsunami passed right over it [30]. Corals growing on the Biorock reef had 50 times (5000%) higher coral survival than the adjacent natural coral reef after the 1998 coral bleaching event [25]. For a decade after the bleaching event this resort had the only healthy reef full of corals and fishes in front of their beach in the Maldives. The hotel whose reef and beach were saved by the Biorock project turned the power off, with the result that the corals, no longer protected from bleaching by the Biorock process [31,32], suffered severe mortality in the 2016 bleaching event.

The second group of Biorock shore protection reefs were built at three eroding beach locations at Gili Trawangan, Lombok, Indonesia around 2010. These consisted of 4 to 6 separate reef modules designed to break waves up by slowing down separate portions of the wave front and driving the incoming wave front out of coherence, using less structural materials. The Biorock structures are shaped like an upside-down wave, which is optimal for dissipating wave energy, with no vertical surfaces to cause reflections, and so are called Biorock Anti-Wave structures (BAW). Although these structures were small, new beach growth was clearly visible at all sites within 8 months on Google Earth images [30]. One set of structures was of, and creation of a gap underneath it, so that it was on the way to falling down. One year later the gap underneath had completely filled in, and the sand had risen by about a meter to cover half the vertical wall height. The seawall on the neighboring property, built at the same time, but not protected by Bioorck, completely collapsed within a year [30]. The hotels whose beaches had been restored by the projects then turned the power off. Because the structures, no longer maintained, growing, or protected from rusting, are now collapsing, beach erosion has now resumed, with new sea walls being constantly built and falling down.

2. Materials and Methods

Pulau Gangga, North Sulawesi, Indonesia, has suffered progressive beach erosion. The index maps show its location on various scales (Figure 1a–d), and Google Earth images show the rapid erosion of the beach between 2013 and 2014 (Figure 2a,b). The site is outside the typhoon belt because it is close to the Equator, but it is affected by both the Australian and Indo-China Monsoons. From around December through May the winds and waves are usually from the southeast. A strong southward tidally-modulated current normally sweeps sand from north to south at the site.

The formerly wide sand beach had largely washed away by late 2015, leaving an erosion cliff about 1.36 m high along the shore, with trees falling into the sea, and beach pavilion buildings have had to be repeatedly torn down and moved inland. In front of the 200 m of severely eroded beach we built 48 Biorock Anti Wave reefs in a staggered design to dissipate wave energy before it hits the beach. The time of installation, January 2016, was just before the monsoon season when erosion takes place on this beach, and was done as fast as possible before waves made installation difficult.

48 Biorock Anti Wave reefs were deployed in the sea grass beds in front of the eroding beach in January 2016. They were arrayed in twelve groups of four, each group powered by a single power supply located 100 m away on land, connected by electrical cables dug into the sand (Figures 3 and 4a–d). They grow thickest at the bottom, and thinnest on top. The bottom 10–20 cm of the structures were always submerged at low tide, but above this they were exposed to the air for various periods, depending on the tidal cycle. Since structures grow only when and where submerged in salt water, the bottoms are always growing, while the top grows only when submerged, about half of the time. The gabion baskets were deployed in a grid pattern (visible from aerial images below) at low tide in shallow sand, seagrass, and coral rubble.

The resort had previously purchased gabion wire baskets for stones to make a breakwater, not knowing that these would quickly rust and fall apart. Because these were already available, they were incorporated into the core of the Biorock Anti Wave structures, because the Biorock electrolysis process prevents any rusting of the steel. The rocks are bound in place by growth of minerals over the mesh and by prolific growth of barnacles, oysters, and mussels, preventing rock shifting in heavy surge from breaking the gabion. There are both advantages and disadvantages to the use of rock gabions in Biorock shore protection structures, as they can cause scour by acting as a near solid wall, but they cause more rapid initial results slowing wave erosion than a more open structure that does not incorporate them. Gabions are not an essential part of the design, in fact not using them makes BAW units much faster and cheaper to construct and deploy. In this study gabions were used only for convenience as they had been previously bought and were already on site. Not to include rocks at all relies purely on the growth of a biological reef to provide long-term growing shore protection, instead of that provided by the rocks.

The core of each structure was a double gabion basket, 1 m × 1 m × 2 m with the long dimension parallel to the shore. These were placed on the shallow sea floor at predetermined sites, with the long axis parallel to the shore, and filled in with rounded river stones (largely in the 20–50 cm size range). At the start, the rocks had clean surfaces with nothing growing on them. The gabion baskets were overlain with standard welded steel mesh bars used in local construction, spacing 15 cm, dimensions about 2.1 m by 5.4 m. These sheets were curved into an arc to fit over the top of the gabion basket, with the long dimensions at right angles, so that the long axis of the arc was perpendicular to the shore, oriented into the waves coming over the shelf edge coral reef. They were welded across at the base with support rebars and vertical bars to strengthen them. Each unit was carried by four people at low tide, placed over a gabion basket and wired to it with hose clamps and binding wire. The growth of minerals over the steel cement it firmly to any hard rock bottom, and cement sediment around the bases on sand or mud, firmly attaching the structure to the bottom. The structures sit on the bottom under their own weight and that of the rocks they contain.

Beach profiles were measured using the U-tube water level measuring method [32]. They are estimated to be accurate vertically to about a millimeter, and horizontally to about a centimeter, by repeated measurements. The beach profile before the start of the experiment was estimated from photographs taken before and after from the same positions with common objects of known size in the images for scale. The accuracy of the pre-project beach profile estimate is thought to be about 10 cm vertically and one meter horizontally. Unfortunately, the apparatus for making rapid and accurate beach profiles was not built until September 2016, eight months after the start of the project. A second set of measurements was made in January 2017, a year after the start of the project, after a severe storm and at several more intervals since then. Initial beach profiles were estimated by measuring the height of the erosion cliff at the start of the project using the measured height of concrete foundations as a scale.

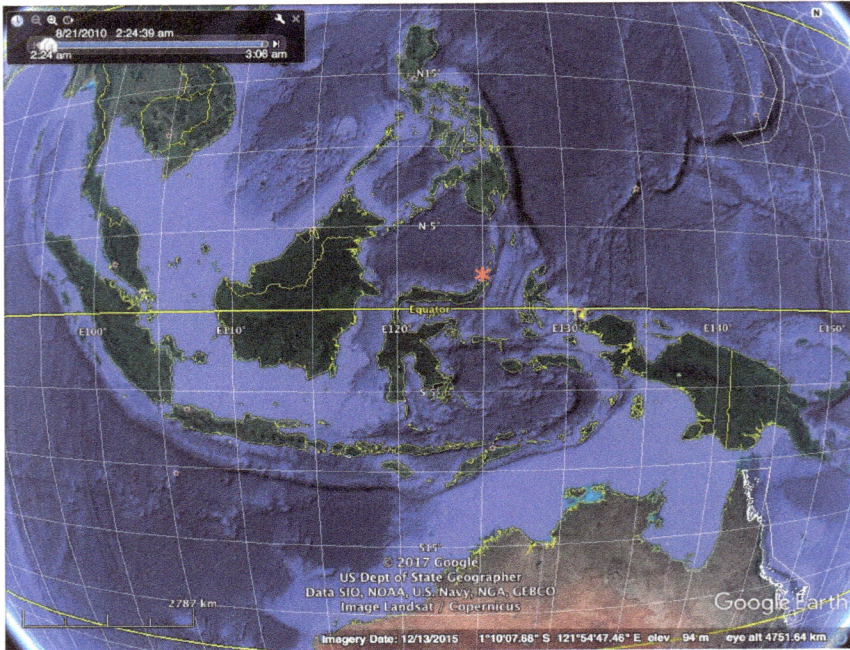

(a)

Figure 1. *Cont.*

(b)

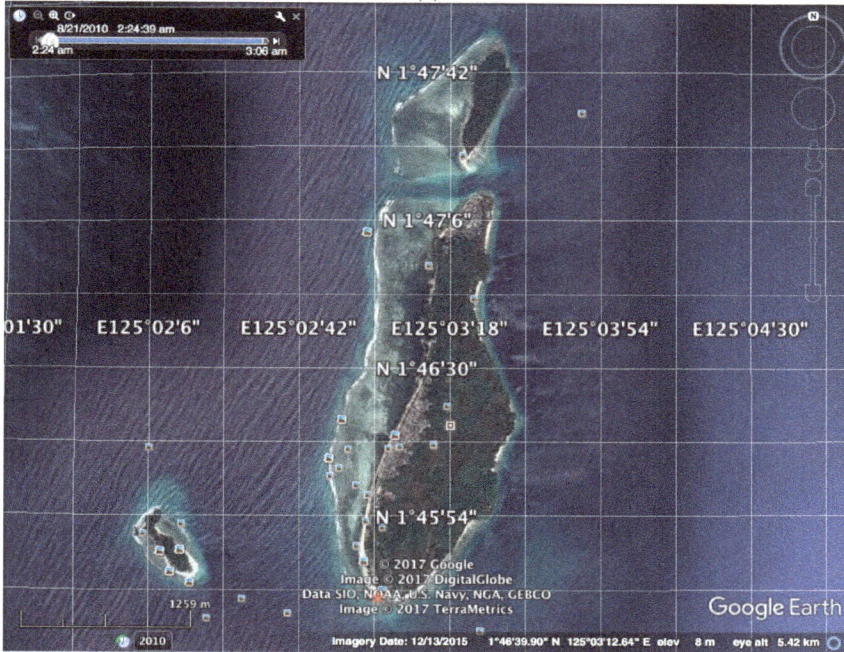

(c)

Figure 1. The site shown on Google Earth Images of the location of the project (red star) on decreasing scales: (**a**) Indonesia, (**b**) North Sulawesi, (**c**) Pulau Gangga.

(a) 23 May 2013

(b) 4 January 2014

Figure 2. *Cont.*

(c) 15 December 2014

(d) 4 August 2017

Figure 2. Erosion of the beach prior to the project. Google Earth images from 2013 (**a**) and 2014 (**b**) showing short term changes in the beach before the project, the last available image taken near low tide a year before the project began in early 2016 (**c**), and the first image after the project, taken near high tide 1.5 years after (**d**). Notice the cores of the Biorock Anti Wave modules as white spots in the seagrass off the regenerated beach. Beach erosion beyond the project near the pier at the south was caused by storm waves from the southeast and longshore drift sand blockage by the solid rock pier. Waves from the northwest shown in (**c**) are typical during erosion of this westward facing beach. Note the row of pavilions (dark spots) on the edge of the beach erosion scarp and brown dying trees with roots exposed in (**c**) and how in (**d**) the roofs are now hidden by new leaf canopies due to prolific tree leaf regrowth after sand buildup around their roots.

Figure 3. Soon after start of the project, at low tide (wide angle image). The tops of the reef restoration structures are clearly visible. The eroded curve is the area where the beach grew back.

(a)

(b)

Figure 4. *Cont.*

(c)

(d)

Figure 4. Aerial views, oblique at low tide (**a**) and vertical at high tide (**b**) on 12 March 2017, fourteen months after the installation, showing the Biorock Anti Wave structures (dark spots in sea grass beds). The northern edge of the project, on the same day at high (**c**) and low (**d**) tides. The Red spot in the last image two images is the northern limit of the project.

3. Results

The speed of beach regrowth astonished local residents. The formerly concave beach, ending in an eroding cliff, is now convex in its profile and growing. Large trees that had been dying after sand washed away, exposing their roots to sea water, have leafed out new canopies since the sand built up around their roots.

By the time of the first beach profile measurements the eroded beach had almost completely grown back and the erosion scarp, 1.36 m tall, was reduced to about 10 cm. About 80% of the beach grew back in less than 3 months and has continued to grow ever since, even during strong storm conditions that in the past caused severe erosion. Over the course of a year the beach has increased in height by more than a meter, and in width by more than 15 m, over a two hundred meter length, a conservative estimate of an increase of beach sand volume of 3000 cubic meters. Most of the gains occurred in the first two months, but have continued since (Figure 5a–g).

The beach growth in the first year was wider at the south than at the north. But there were interesting differences after a severe storm in early January, which caused some erosion of the southern end of the beach, while there was substantial growth of the central and northern sections. Since then the center and north have continued to grow, while the south has continued to erode slowly (Figure 6a–c).

(a) November 2015

(b) December 2015

Figure 5. *Cont.*

(c) December 2015

(d) May 2016

Figure 5. *Cont.*

(e) August 2016

(f) November 2016

(g) January 2017

Figure 5. Before & after photos of beach taken at various times: (**a**) November 2015, beach looking north two months before start of project, tree falling into the sea and roots exposed at beach erosion scarp, (**b**) December 2015, one month before project, 1.36 m high erosion scarp and foundations of beach pavilions about to collapse near center of beach, (**c**) December 2015, one month before project, large old tree collapsing into the sea, leaves dying, roots exposed, (**d**) May 2016, 4 months after project installation, lower branches of fallen tree buried in new beach sand growth, roots buried, new growth, (**e**) August 2016, seven months after, (**f**) November 2016, ten months after, (**g**) January 2017, Twelve months after, soon after a severe storm, looking south. Most of the beach grew after the storm, even though this was the sort of event that had caused heavy erosion in the past. The south end, shown here, was worst affected.

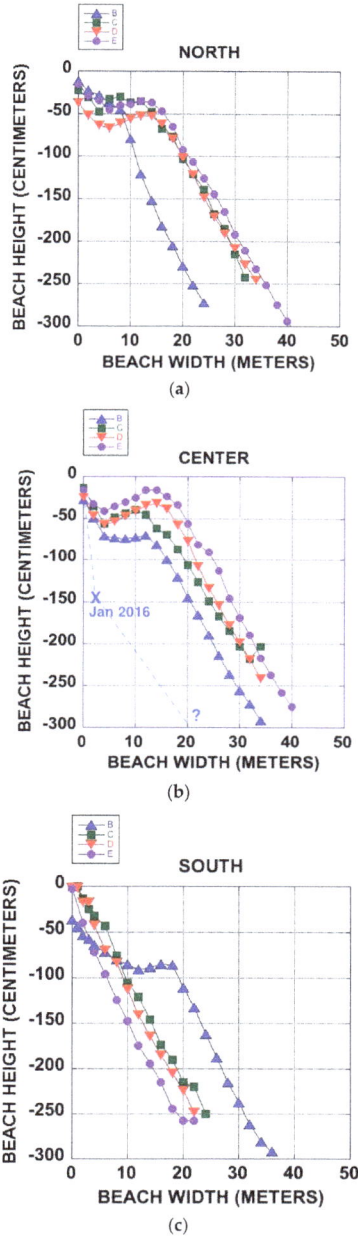

Figure 6. Beach profiles, measured at different times for the north (**a**), center (**b**), and south (**c**). The Zero reference datum for the central profile is the top of the concrete pillar foundation at the left of the December 2015 photograph. A 1.36 m vertical cliff stood at what is now the zero distance, based on measurements from photographs. Dates of measurements: X and dashed line estimate from January 2016; KEY to symbols: B upward triangle, 21 September 2016; C square, 11 January 2017; D downward triangle, 11 March 2017, E circle, 10 July 2017.

There has been prolific growth of hard and soft corals all over the bases of the structures and in intervening areas (Figure 7a–e), prolific growth of sea grasses all around them, dense settlement and growth of barnacles, mussels, and clams on the rocks in the core of the BAW structures, and a rapid increase in juvenile fish and echinoderm populations. In addition, there has been prolific growth of sand-producing calcareous green and red algae around and between the structures.

(a)

(b)

Figure 7. *Cont.*

(c)

(d)

(e)

Figure 7. Underwater images: One end of a Biorock Anti Wave structures at high tide soon after installation (**a**), growth of barnacles over smooth river stones inside the gabions (**b**), examples of rapid growth of corals, sea grasses, and other organisms under, over, and around their bases (**c**–**e**).

Wave fronts are dissipated as they pass through the structures, breaking up the coherence of incoming wave fronts (Figure 8a–c). This dissipation behaves as refraction through permeable structures. But the wave interaction can include a reflective component once structures have grown to be solid with all pores filled in, or if the rock fill is too impermeable to transmit wave pressure through intervening spaces. In that case, sand-eroding scour will be caused, while purely refractive dissipation increases sand accumulation underneath and behind the structure. There is also a diffractive component caused by wave interaction with the metal structure spacing and lattice spacing, which appeared to damp waves at spatial scales that range from the spacing of the metal grid used (0.1 m), the size of the modules (1–5 m), and the spacing of the modules (roughly 10 m).

(a)

(b)

Figure 8. *Cont.*

(c)

Figure 8. Diffraction of energy. Incoming wave energy dissipated by interaction with structure these photographs were taken in rapid succession, the middle picture appears brighter because of break in the clouds, and the photographer moved between the first two images. (**a**) approaching wave, (**b**) wave starting to interact with mesh, (**c**) wave energy dissipated by friction.

4. Discussion

We are not aware of any other case in which a badly eroded beach has been grown back so quickly and naturally. In addition, the Biorock reefs have caused prolific growth of corals, barnacles, oysters, mussels, and seagrass, and created a juvenile fish habitat [30].

Biorock reefs can be any size or shape. Mineral growth extends up to the high tide mark. They can be built entirely subtidal, as at Ihuru, entirely exposed at low tide as at Gili Trawangan, or have only the tops exposed at low tide, as at Pulau Gangga. Since the structures grow only when submerged, those in the intertidal grow most on the bottom, and least on the top. The structures attach themselves solidly to bedrock, and cement loose sand around their bases. Whether the structures are entirely submerged or partially exposed affects their wave mitigating performance. Those that are fully submerged are never visible from shore and generate real coral reef communities or oyster and mussel reefs in muddier or colder waters. Those reefs located in the intertidal zone are visible at low tide, which may be aesthetically objectionable to those who want a clear ocean view from the beach, but they are more effective in protecting the shore if there is a storm during high tide, which would pass over deeper reef structures.

The size, shape, and spacing of the modules affect their performance, and cost. What is astonishing is that record beach growth was achieved with far less material and far lower cost than sea wall or a breakwater. Conventional reinforced concrete structures are made by first building a reinforcing bar frame, and the steel is a very small part of the total cost compared to cement, stone, wooden forms, and labor. Since steel in reinforced concrete structures invariably rusts, expands, and cracks the concrete, such structures have finite lifetimes, especially in salty coastal air.

Biorock structural steel is completely protected from rusting, and the continuous growth of hard minerals makes it constantly stronger, and able to grow back first in areas that are physically damaged. The cost is far less than a reinforced concrete structure of the same size and shape, because instead of cement, rocks, labor, and wooden forms we simply provide an electrical supply instead. Estimates of Biorock reef costs range from $20–1290/m of shoreline depending on the size of the reef grown, while other methods range from $60–155,000/m, or 3–120 times more expensive [16]. The amount of electricity used is small, the entire Pulau Gangga beach restoration project uses about one air conditioner worth of electricity.

Shore protection provided by Biorock includes both production of new sand by prolific growth of calcareous algae around them, and physical protection from sand erosion by wave energy dissipation. The results of beach growth after the project was installed indicate that beach sand accumulation is a very dynamic function of wave and current interactions with reef structures. Biorock reef structures physically dissipate wave energy and reduce erosion at the shoreline, while also generating new sand. They should slow down transport of sand by north to south tidal currents at the site. That, and the interplay of the structures with waves coming from different directions, may explain why the unusual January 2017 storm seems to have transported sand in the reverse of the usual direction. Further measurements will reveal if this trend continues, or is reversed by sand production from increased calcareous algae growth around the structures.

Wave energy dissipation due to friction at the surface of the growing limestone minerals produced by the Biorock process is very quickly exceeded by the much larger, and rougher, surfaces provided by the prolific growth of corals, barnacles, sea grass, and all other living marine organisms as the structure becomes rapidly overgrown. The size, shape, and spacing of the structures, as well as the reef organisms growing on them, affect their performance as wave absorbers, and their designs can be readily changed by adding, or removing, sections as needed. Such structures can be designed to be oyster, mussel, clam, lobster, or fish habitat for highly productive and sustainable mariculture.

Biorock structures interact with waves over a very broad range of wavelengths, and are expected to produce wave diffraction on wavelength scales similar to the spacing of the structures—about 10 m—and over the spacing of the mesh—about 10 cm. Since the growing, self-repairing Biorock structures cannot be modelled by conventional hydrodynamic modeling schemes, it will be important to make physical measurements of wave energy around such structures to evaluate actual performance, and to optimize them for beach growth purposes.

The initial results of this project resulted in extraordinary beach growth, which could be improved with further experimentation under a wider range of conditions, and should be much more widely applied as a cost-effective beach restoration solution that uniquely restores marine ecosystem services.

5. Conclusions

We have grown back severely eroded beaches naturally in months by growing Biorock electric reefs in front of them. These structures cost far less than sea walls or breakwaters and work on entirely different physical principles. They restore marine ecosystems as well as beaches. The exceptionally rapid growth of corals on them [31,33] provides additional shore protection, and the rapid growth of sand-producing calcareous algae on and around them produces new biological sand supplies. These structures can easily keep pace with global sea level rise because solid hard electrochemical minerals can be grown upwards at rates up to 2 cm/year—around 5 times faster than sea level rise—and grow still much faster when corals, oysters, mussels, and other calcareous organisms cover them. As a result, Biorock shore protection reefs quickly turn eroding beaches into growing ones, can protect entire islands and even grow new ones. Biorock is the most cost-effective technology for protecting eroding coasts, for restoring fisheries habitat, and is critically needed to save the low-lying islands and coasts now threatened by global sea level rise, and the billions of coastal people who will become climate refugees if global warming is not rapidly reversed [4,12].

6. Dedication

This paper is dedicated to the memory of the late Wolf Hilbertz, who first invented the Biorock process of growing minerals in the ocean in 1976, and who foresaw all its applications, including shore protection.

Acknowledgments: We thank the entire management and staff of Pulau Gangga Dive Resort and its parent company, Lotus Resorts, for their willingness to try new, better, more natural and effective approaches to shore protection. Lotus Resorts paid for all materials, equipment, domestic travel, and time. The authors thank them deeply for their willingness to pioneer innovative methods of shore protection. We also thank Lori Grace for providing funds for a round trip ticket to Indonesia for Thomas J. F. Goreau to construct the device to measure beach profiles. We thank the anonymous reviewers for their constructive suggestions that have improved the original draft.

Author Contributions: T.J.F.G. and P.P. conceived, designed, built, and installed the first four Biorock Anti Wave modules and connected them to power; P.P. then built and installed the rest. T.J.F.G. analyzed the data; contributed reagents/materials/analysis tools; and wrote the paper. Although they are not listed as authors, because they did not work on the projects in the water, the entire management and staff of Lotus Resorts, operators of Pulau Gangga Dive Resort, played absolutely crucial roles in the design, logistics, funding, advice, information, and support for the work described. T.J.F.G. is a co-inventor of the Biorock electric technology of marine ecosystem restoration.

Conflicts of Interest: The authors declare no conflict of interest.

References

1. Pilkey, O.H.; Neal, W.J.; Bush, D.M. Coastal erosion. In *Coastal Zones and Estuaries. Encyclopedia of Life Support Systems (EOLSS)*; UNESCO: Paris, France, 1992.
2. Dieng, H.B.; Cazenave, A.; Meyssignac, B.; Ablain, M. New estimate of the current rate of sea level rise from a sea level budget approach. *Geophys. Res. Lett.* **2017**. [CrossRef]
3. Nieves, V.; Marcos, M.; Willis, J.K. Upper-Ocean Contribution to Short-Term Regional Coastal Sea Level Variability along the United States. *J. Clim.* **2017**, *30*, 4037–4045. [CrossRef]
4. Goreau, T.J. Global biogeochemical restoration to stabilize CO_2 at safe levels in time to avoid severe climate change impacts to Earth's life support systems: Implications for the United Nations Framework Convention on Climate Change. In *Geotherapy: Innovative Technologies for Soil Fertility Restoration, Carbon Sequestration, and Reversing Atmospheric CO_2 Increase*; Goreau, T.J., Larson, R.G., Campe, J.A., Eds.; CRC Press: Boca Raton, FL, USA, 2014.
5. Levitus, S.J.; Antonov, I.; Boyer, T.P.; Baranova, O.K.; Garcia, H.E.; Locarnini, R.A.; Mishonov, A.V.; Reagan, J.R.; Seidov, D.; Yarosh, E.S.; et al. World ocean heat content and thermosteric sea level change (0–2000 m), 1955–2010. *Geophys. Res. Lett.* **2012**, *39*, L10603. [CrossRef]
6. Gebbie, G.; Huybers, P. The Mean Age of Ocean Waters Inferred from Radiocarbon Observations: Sensitivity to Surface Sources and Accounting for Mixing Histories. *J. Phys. Oceanogr.* **2012**, *42*, 291–305. [CrossRef]
7. Goreau, T.J. Balancing atmospheric carbon dioxide. *Ambio* **1990**, *19*, 230–236.
8. Goreau, T.J. Tropical ecophysiology, climate change, and the global carbon cycle. In *Impacts of Climate Change on Ecosystems and Species: Environmental Context*; Pernetta, J., Leemans, R., Elder, D., Humphrey, S., Eds.; International Union for the Conservation of Nature: Gland, Switzerland, 1995; pp. 65–79.
9. Rohling, E.J.; Grant, K.; Bolshaw, M.; Roberts, A.P.; Siddall, M.; Hemleben, C.; Kucera, M. Antarctic temperature and global sea level closely coupled over the past five glacial cycles. *Nat. Geosci.* **2009**. [CrossRef]
10. Koenigswald, W.V. Mammalian faunas from the interglacial periods in Central Europe and their stratigraphic correlation. In *The Climate of Past Interglacials*; Sirocko, F., Claussen, M., Sanchez Goñi, M.F., Litt, T., Eds.; Elsevier: Amsterdam, The Netherlands, 2006; pp. 445–454.
11. Koenigswald, W.V. Discontinuities in the Faunal Assemblages and Early Human Populations of Central and Western Europe During the Middle and Late Pleistocene. In *Continuity and Discontinuity in the Peopling of Europe: One Hundred Fifty Years of Neanderthal Study, 101 Vertebrate Paleobiology and Paleoanthropology*; Condemi, S., Weniger, G.-C., Eds.; Springer Science + Business Media B.V.: Dordrecht, The Netherlands, 2011.
12. Goreau, T.J. The other half of the global carbon dioxide problem. *Nature* **1987**, *328*, 581–582. [CrossRef]
13. Munk, W.H.; Traylor, M.A. Refraction of Ocean Waves: A Process Linking Underwater Topography to Beach Erosion. *J. Geol.* **1947**, *55*, 1–26. [CrossRef]
14. Wiegel, R.L. *Oceanographical Engineering*; Dover: New York, NY, USA, 1992.
15. Schiereck, G.J. *Introduction to Bed, Bank, and Shore Protection: Engineering the Interface of Soil and Water*; VSSD: Delft, The Netherlands, 2006.
16. Ferrario, F.; Beck, M.W.; Storlazzi, C.D.; Micheli, F.; Shepard, C.C.; Airoldi, L. The effectiveness of coral reefs for coastal hazard risk reduction and adaptation. *Nat. Commun.* **2014**. [CrossRef] [PubMed]
17. Goreau, T.F.; Goreau, N.I.; Goreau, T.J. Corals and Coral Reefs. *Sci. Am.* **1979**, *241*, 124–136. [CrossRef]

18. Goreau, T.J. *Testimony to the National Ocean Policy Study Subcommittee of the United States Senate Committee on Commerce, Science, and Transportation, S. HRG. 101-1138: 30-37*; US Government Printing Office: Washington, DC, USA, 1991.

19. Hayes, R.L.; Goreau, T.J. The tropical coral reef ecosystem as a harbinger of global warming. *World Resour. Rev.* **1991**, *3*, 306–322.

20. Goreau, T.J.; Hayes, R.L.; Clark, J.W.; Basta, D.J.; Robertson, C.N. Elevated sea surface temperatures correlate with Caribbean coral reef bleaching. In *A Global Warming Forum: Scientific, Economic, and Legal Overview*; Geyer, R.A., Ed.; CRC Press: Boca Raton, FL, USA, 1993; pp. 225–255.

21. Goreau, T.J.; Hayes, R.L. Coral bleaching and ocean "hot spots". *Ambio* **1994**, *23*, 176–180.

22. Goreau, T.J.; Hayes, R.L. Global coral reef bleaching and sea surface temperature trends from satellite-derived Hotspot analysis. *World Resour. Rev.* **2005**, *17*, 254–293.

23. Goreau, T.J.; Hayes, R.L.; McAllister, D. Regional patterns of sea surface temperature rise: Implications for global ocean circulation change and the future of coral reefs and fisheries. *World Resour. Rev.* **2005**, *17*, 350–374.

24. Hilbertz, W.H.; Goreau, T.J. Method of Enhancing the Growth of Aquatic Organisms, and Structures Created Thereby. U.S. Patent No. 5,543,034, 6 August 1996.

25. Goreau, T.J.; Hilbertz, W. Marine ecosystem restoration: Costs and benefits for coral reefs. *World Resour. Rev.* **2005**, *17*, 375–409.

26. Goreau, T.J. Marine electrolysis for building materials and environmental restoration. In *Electrolysis*; Kleperis, J., Linkov, V., Eds.; InTech Publishing: Rijeka, Croatia, 2012; pp. 273–290.

27. Goreau, T.J. Marine ecosystem electrotherapy: Practice and theory. In *Innovative Technologies for Marine Ecosystem Restoration*; Goreau, T.J., Trench, R.K., Eds.; CRC Press: Boca Raton, FL, USA, 2012.

28. Wells, L.; Perez, F.; Hibbert, M.; Clervaux, L.; Johnson, J.; Goreau, T. Effect of severe hurricanes on Biorock coral reef restoration projects in Grand Turk, Turks and Caicos Islands. *Rev. Biol. Trop.* **2010**, *58*, 141–149. [PubMed]

29. Shyue, S.-W.; Yang, K.-C. Investigating terrain changes around artificial reefs by using a multi-beam echosounder. *ICES J. Mar. Sci.* **2002**, *59*, S338–S342. [CrossRef]

30. Goreau, T.J.; Hilbertz, W.; Azeez, A.; Hakeem, A.; Sarkisian, T.; Gutzeit, F.; Spenhoff, A. Restoring reefs to grow back beaches and protect coasts from erosion and global sea level rise. In *Innovative Technologies for Marine Ecosystem Restoration*; Goreau, T.J., Trench, R.K., Eds.; CRC Press: Boca Raton, FL, USA, 2012.

31. Goreau, T.J. Electrical stimulation greatly increases settlement, growth, survival, and stress resistance of marine organisms. *Nat. Resour.* **2014**, *5*, 527–537. [CrossRef]

32. Andrade, F.; Ferreira, M.A. A Simple Method of Measuring Beach Profiles. *J. Coast. Res.* **2006**, *22*, 995–999. [CrossRef]

33. Goreau, T.J.; Cervino, J.; Polina, R. Increased zooxanthellae numbers and mitotic index in electrically stimulated corals. *Symbiosis* **2004**, *37*, 107–120.

Review

Sea Level Change and Coastal Climate Services: The Way Forward

Gonéri Le Cozannet [1,*], Robert J. Nicholls [2], Jochen Hinkel [3], William V. Sweet [4],
Kathleen L. McInnes [5], Roderik S. W. Van de Wal [6], Aimée B. A. Slangen [7], Jason A. Lowe [8] and
Kathleen D. White [9]

1   Bureau de Recherches Géologiques et Minières (BRGM), French Geological Survey, Orléans 45060, France
2   Engineering and the Environment, University of Southampton, Highfield, Southampton SO17 1BJ, UK;
    R.J.Nicholls@soton.ac.uk
3   Global Climate Forum and Division of Resource Economics at Albrecht Daniel Thaer-Institute and Berlin
    Workshop in Institutional Analysis of Social-Ecological Systems (WINS), Humboldt-University,
    Berlin 10178, Germany; hinkel@globalclimateforum.org
4   Center for Operational Oceanographic Products and Services, National Oceanic and Atmospheric
    Administration (NOAA), Silver Spring, MD 20910, USA; william.sweet@noaa.gov
5   Commonwealth Scientific and Industrial Research Organisation (CSIRO), Aspendale,
    Victoria 3195, Australia; Kathleen.Mcinnes@csiro.au
6   Institute for Marine and Atmospheric research Utrecht (IMAU), University of Utrecht,
    Utrecht 3584 CC, The Netherlands; r.s.w.vandewal@uu.nl
7   NIOZ Royal Netherlands Institute for Sea Research, Department of Estuarine & Delta Systems, and Utrecht
    University, PO Box 140, Yerseke 4400 AC, The Netherlands; aimee.slangen@nioz.nl
8   Hadley Centre, UK MetOffice, Exeter EX1 3PB, United Kingdom and Priestley Centre, University of Leeds,
    Leeds LS2 9JT, UK; jason.lowe@metoffice.gov.uk
9   US Army Corps of Engineers Headquarters Washington, Washington, DC 20314, USA;
    Kathleen.D.White@usace.army.mil
*   Correspondence: g.lecozannet@brgm.fr; Tel.: +33(0)2-38-64-36-14

Received: 7 May 2017; Accepted: 3 October 2017; Published: 16 October 2017

Abstract: For many climate change impacts such as drought and heat waves, global and national frameworks for climate services are providing ever more critical support to adaptation activities. Coastal zones are especially in need of climate services for adaptation, as they are increasingly threatened by sea level rise and its impacts, such as submergence, flooding, shoreline erosion, salinization and wetland change. In this paper, we examine how annual to multi-decadal sea level projections can be used within coastal climate services (CCS). To this end, we review the current state-of-the art of coastal climate services in the US, Australia and France, and identify lessons learned. More broadly, we also review current barriers in the development of CCS, and identify research and development efforts for overcoming barriers and facilitating their continued growth. The latter includes: (1) research in the field of sea level, coastal and adaptation science and (2) cross-cutting research in the area of user interactions, decision making, propagation of uncertainties and overall service architecture design. We suggest that standard approaches are required to translate relative sea level information into the forms required to inform the wide range of relevant decisions across coastal management, including coastal adaptation.

Keywords: climate services; coastal zones; sea level projections

---

1. Introduction

The concept of climate services emerged some 15 years ago to support decision-making related to mitigation of and adaptation to climate change [1]. As a general principle, climate services transfer climate information from research to users in order to help users manage and communicate the risks and opportunities of climate variability and change [2–4]. The users of climate services span a variety of needs and have differing levels of understanding of scientific information and, crucially, the strengths and limitations of this information in supporting their particular decisions. Users include local, regional, and national government entities, business and industry, beneficiaries of coastal ecosystem services (e.g., fisheries, tourism), transportation providers, and members of the public. Their decisions span a wide spatio-temporal range, further complicating issues for different users and for those providing climate services. Climate services are not limited to the provision of research data and information. Rather, they refer to the translation of climate research into an operational delivery of services in support to adaptation and mitigation of climate change. Such services may be provided on either a fee-paying or a free-of-charge basis. However, a viable economic model involving private or public funding for both the research and it translation is required to ensure their sustainability.

Existing climate services have focused on different thematic areas that vary by region and country: in Europe, they provide essential climate projections relevant for mean and extreme temperatures and precipitation and their impacts, but much less information is made available to support coastal adaptation [5–7]. Conversely, in the US, more information is available for relative sea level changes, including subsidence or uplift components [8], because this topic was the focus of a national study published in 1987 [9] and follow-on national studies [10–12].

The need for coastal climate services (CCS) is becoming more apparent as coastal stakeholders require support to adapt to global and local sea level rise and increase their resilience to coastal hazards and risks such as flooding, erosion and saline intrusion in estuaries and aquifers (see definitions in Table 1). As sea levels continue to rise, the annual frequency of minor tidal flooding is growing rapidly in frequency in many world regions, such as dozens of U.S. coastal communities [10]. Flooding during extreme events (storm surges and tropical cyclones) will become stronger, shoreline erosion will become more severe, and human interventions more costly [13,14]. In the future, changing sea level is expected to affect economic activities associated with maritime and inland navigation and environmental goods and services upon which many coastal communities rely (e.g., fishing, tourism). Furthermore, because many decisions taken today in coastal zones have implications for decades or more, longer-term information on future sea level rise is required to avoid maladaptation and substantial economic losses. Indeed, sea level will continue to rise for centuries even under low greenhouse gas emissions, as ice sheets and ocean expansion are characterized by long response times [15,16]. Hence, there will be a continuing need for CCS for adaptation that consider future relative sea level rise (including subsidence or land uplift) and associated impacts, even if the ambitious climate goals of the Paris Agreement are met [17]. Furthermore, CCS are required by a large number of private and public stakeholders, with the result that a market for economic activities responding to this demand is emerging [18]. However, these services are often tailored very specifically to the particular hazards, consequences, and other needs of the locality or region implementing adaptation, and do not necessarily identify themselves as climate services. As a consequence, defining CCS and characterizing their users and providers remains challenging.

This article addresses the challenge of CCS based on information related to changes in observed and expected future sea levels. So far, this information has been provided primarily in the form of scenarios corresponding to plausible future sea level changes, or projections. These provide future sea level rise (SLR) given assumptions such as social and economic narratives [19], global temperature increase (e.g., 1.5 °C above preindustrial levels) [16], or representative concentration pathways [20,21]. They cover a range of timescales from the next decades to the coming centuries, can be continuous or discrete, and some estimate the related uncertainties. Furthermore, the best available sea level projections also consider regional variability in SLR [8,20,22–26]. Today, an increasing number of regional projections are being produced for specific regions or countries (e.g., northern Europe [27], The Netherlands [28,29], Canada [30], Norway [31]; Australia [32], United States [10,11], etc.). These include specific attention to relevant processes influencing relative sea level in the region of interest (e.g., Global Isostatic Adjustment in the case of Norway). In fact, future sea levels are only a part of the climate information needed for coastal adaptation: climate impacts to the warming of surface ocean waters and changes in ocean acidification [33,34] are especially important to anticipate how ecosystem services will be altered in the future [35]. However, coastal managers currently have limited access to specialised and tailored information on projections of future sea levels [7], and the significance of potential damages and losses motivates a specific study in this area [36–38].

This article addresses requirements for effective use of SLR information in coastal climate services (CCS). To do this, we review selected practices. This review leads to the definition of generic characteristics of CCS based on sea level projections to be drawn, in order to ultimately examine how the use of sea level projections in CCS could be improved. Specifically, we address the following questions:

- What is the current state-of-the-art in the area of CCS using SLR projections (Section 2)?
- What are the current technical barriers to satisfying the demand for CCS based on sea level projections (Section 3)?
- What is needed to overcome barriers and to facilitate the use of sea level information in CCS (Section 4)?

In the conclusion, we provide recommendations for stakeholders involved in designing the next generation of climate services.

Table 1. Terms used in this article and their definitions.

Term	Definition Used in This Article
Service	Economic activity characterized by the trade of intangible assets.
Climate service	Any type of service using climate information and supporting adaptation to and mitigation of climate change.
Coastal climate service (CCS)	Climate services in coastal areas. Note that this article focuses on CCS using sea level information.
Coastal services	Any type of service provided in coastal areas, not necessarily using climate information.

2. Current Coastal Climate Services Using Sea Level Information

This section examines the state of CCS based on sea level scenarios or projections today. To do so, we start with an analysis of existing CCS (Section 2.1). Then, based on these examples, we provide a generic picture characterizing current CCS, and the different stakeholders involved therein (Section 2.2).

2.1. Examples of Existing Coastal Climate Services

2.1.1. Example 1: USA Coastal Climate Services

The type of flooding events related to SLR range from frequent chronic flooding (also called tidal or nuisance flooding in the literature [10,39–42]), to rare, event-driven flooding, which occurs in most cases during storms or cyclones. In many places, the frequency of chronic flooding is rapidly increasing in annual frequency (Figure 1a) and accelerating in some coastal towns (Figure 1b) [10,39,43]. Such floods often occur during relatively calm, sunny conditions. These floods adversely affect ground-level and subsurface infrastructure in many U.S. coastal communities (e.g., roadways, storm/waste/fresh-water systems, and private/commercial property) that were not designed for repetitive salt-water exposure or inundation. Until long-term adaptation strategies are put into place, SLR related impacts will be experienced as more-frequent chronic flooding will occur at high-tide (Figure 1c).

In the U.S., the demand by decision makers for information about chronic coastal flooding is mounting: for example, the growing rise in chronic flooding in Norfolk is seen as posing long-term problems due to its cumulative toll and being monitored by credit-rating companies [44]. Afflicted communities are assessing locations and damages associated with chronic flooding and budgeting on an annual basis for anticipated costs required for mobilization of emergency responders to close streets and for temporary installation of pumps, sand bags and storm-water inflow preventers. In the US, such land use and coastal risk management decisions taken at local scale complement the state and federal government coastal planning strategies.

Access to seasonal/annual flood-frequency predictions allows for more effective preparedness and response. As a result, the National Oceanic and Atmospheric Administration (NOAA) has recently provided experimental annual flood predictions that consider past trends, and in some locations, interannual variability with the El Niño Southern Oscillation [45]. This allows for statistical-dynamical tidal-flood [46,47] and sea level anomaly [48] predictions, enabling uses to improve readiness for current and future flooding. These products complement mid- and longer-term climate services such as NOAA's SLR Viewer web mapping tool [49], that support community decision making around infrastructure plans and designs that consider performance and reliability for local relative SLR up to 100 years in the future.

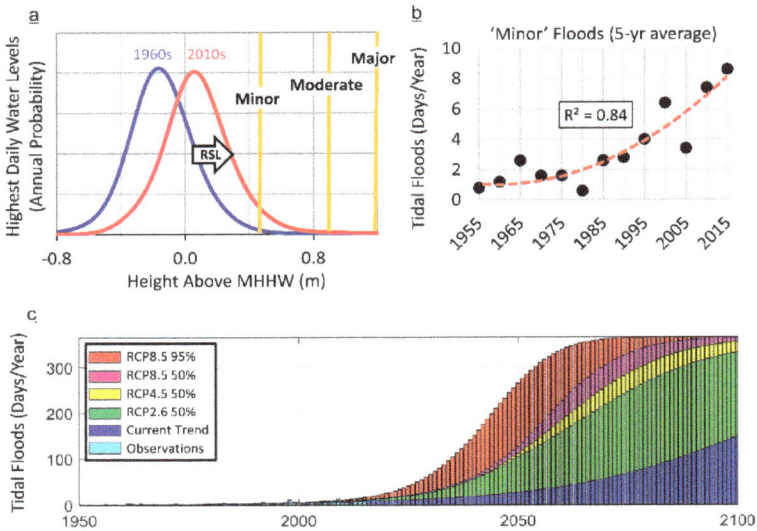

Figure 1. (**a**) Multi-year empirical (smoothed) distributions for daily highest water levels in Norfolk, VA, USA, for the 1960s and 2010s, showing extent that local relative sea level (RSL) rise has increased the flood probability relative to impact thresholds defined locally by NOAA's National Weather Service [50] for minor (~0.5 m: nuisance level), moderate (~0.9 m) and major (~1.2 m: local level of Hurricane Sandy in 2012) impacts, relative to mean higher high water (MHHW) tidal datum of the National Tidal Datum Epoch (1983–2001) and due to RSL rise; (**b**) annual flood frequencies (based upon 5-year averages) in Norfolk for recurrent tidal floods with minor impacts are accelerating, as shown by the quadratic trend fit (goodness of fit [R^2] = 0.84). From Sweet et al. [12]; In (**c**) are projections of local minor daily floods for Norfolk, VA of Sweet and Park [43] forced by the local SLR trend (navy blue) and those of Kopp et al. [8] based upon representation concentration pathways (RCP). The *y*-axis in c saturates at 365 days/year with a flood.

Several providers of CCS exist in the US, including academic institutions, nongovernmental organizations, and federal government agencies. Each organization acts at a different phases of the disaster management cycle, which include prevention, preparedness, crisis management and recovery, according to the disaster risk reduction terminology [51]. Each phase involves different actions such as: (1) flood monitoring, warning, and near-real-time local hazard mapping to improve preparedness and manage the crisis (role of NOAA); (2) hydrodynamic modelling and long-term hazard mapping to support prevention, preparedness, and recovery (Federal Emergency Management Agency—FEMA) [52]; (3) engineered nonstructural, structural, natural, and nature-based infrastructure adaptation to reduce risks from coastal hazards (US Army Corps of Engineers—USACE) [53,54]; and (4) evaluation of impacts to complex geomorphic systems such as coastal aquifers, shorelines and active coastal zones (US Geological Survey—USGS) [55–57]. In many other countries (e.g., France), a similar composite network of organizations is involved in the provision of climate services. From the perspective of this article, these CCS providers can be classified in two groups: those providing information on sea level observations, modelling and analysis, and those using this information to provide information on preparedness and response to coastal impacts (Figure 2).

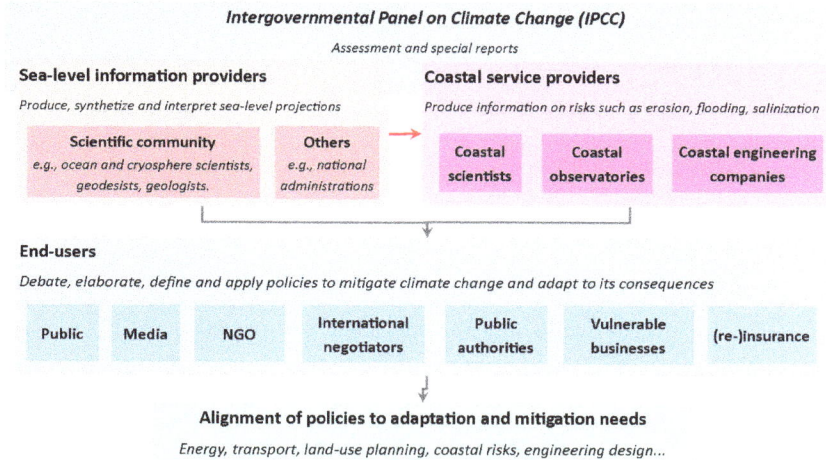

Figure 2. Current interactions between users and producers of sea level projections. Arrows denote information flows involving sea level projections or derived products.

2.1.2. Example 2: Australian Coastal Climate Services

Similar to the US, coastal planning at national to regional scale in Australia is shared by the federal and the state/territory governments, while local coastal planning and management is primarily the role of local governments under the direction of state policy. Each of the state/territory governments has established different SLR planning guidelines. For example, the state of Victoria has stipulated since 2008 (and reassessed in 2014) that planning authorities must plan for a SLR increase of 'not less than' 0.8 m by 2100, whereas in South Australia new developments should take into consideration 0.3 m SLR by 2050 and a further 0.7 m SLR between 2050 and 2100 [58]. To support local council planning and management, CCS in the form of regional sea level projections and their uncertainties up to 2100, including sea level allowances to support coastal defense upgrades, have been developed [32]. These are delivered as a national climate service along with other projected atmospheric and oceanic variables such as temperature, precipitation, wind, ocean and pH [59], but also as a more specific coastal climate service at the scale of individual Australian coastal councils on the 'CoastAdapt' web site [60]. In addition to SLR projections and allowances at the coastal council scale, CoastAdapt provides tools such as inundation mapping software, local coastline morphological information, coastal climate adaptation decision support guidance and Australian case studies on coastal adaptation.

2.1.3. Example 3: French Coastal Climate Services

CCS are developing at different paces depending on the region and country or state. Until 2010, the awareness and implementation of coastal adaptation was much more advanced in the UK and the Netherlands than elsewhere in Europe [61], including France, where local coastal risk prevention plans lagged. However, awareness can rise rapidly due to extreme events, as illustrated by the case of the Xynthia storm and surge in France in 2010, which caused more than 50 deaths and significant economic damages on the Atlantic coast [62]. After this event, the French risk prevention regulations were modified to improve coastal hazard maps [63] by defining standards to model coastal processes such as storm surge and wave setup [64] and to consider future SLR. To avoid heterogeneous responses along the coastlines driven by individual municipality selections, the scientific community concerned with SLR was consulted, and a standardized fixed sea level projection of +60 cm by the end of the 21st century was defined and included into the regulation [65]. This resulted in additional constraints on land use planning policies, while stimulating research and development on coastal flood modelling [66]. In addition, non-binding regulatory frameworks were implemented to facilitate land use planning in coastal zones, and the insurance and reinsurance industry is considering the introduction of new insurance products to anticipate and favour adaptation. Despite the real progress in the development of CCS since 2010, there are still concerns due to the limitations of sea level projections currently used, because uncertainties, temporal, regional and local variability are not considered. In this case, SLR projections are essentially used to limit further urbanisation in coastal zones, potentially creating a maladaptation trap in area where sea level will exceed the threshold defined by the regulation [67]. Recently, an additional law for adaptation has been discussed in the parliament to institute construction regulations that consider the expected lifetime of engineered infrastructure together with shoreline change predictions. However, providing shoreline erosion predictions with a sufficient degree of confidence remains a research challenge today. This is illustrative of a situation where the implementation of an adaptation policy is limited by research.

2.1.4. Example 4: Critical Settlements and Infrastructure

Sea level projections beyond 2100 are being developed for critical coastal assets, such as nuclear power plants [68], security infrastructure [11], and critical settlements such as atoll islands or low-lying deltas. However, in many vulnerable areas, long-term issues related to SLR are not considered due to the urgency of other risks, such as cyclones, earthquakes or industrial risks, and also because of institutional, social and economic barriers to their prevention and management [69]. For example, developing countries have large immediate development concerns, which inevitably consume the limited human, technical and financial resources at the expense of multi-decadal to multi-centennial issues. New management initiatives such as the Bangladesh Delta Plan 2100 [70] provide mechanisms to address this issue of balancing temporal scales. To cover all relevant timescales in such vulnerable areas, CCS will also require support from international projects involving international networks of scientists, as well as the strong background knowledge provided by the Intergovernmental Panel on Climate Change (IPCC). More general climate services, considering not only sea level, are based on IPCC results, for instance in the Netherlands by the KNMI [71]. In the UK, national climate projections combine international studies, including research reported in the IPCC, with domestic modeling approaches [72].

2.2. Generic Lessons from These Examples

2.2.1. The Status of Coastal Climate Services

The review has shown that coastal climate services already exist in some places. In the US, several services [49,50] are based on scenarios [9], while others [10,11] have built on the recently released set of sea level projections [8]. In many other countries, unlike other climate services relying on a core service providing essential climate variables such as mean and extreme temperatures, CCS are emerging in a scattered manner linked to local or national coastal management. In all cases, they remain mostly unconnected to the global framework of climate services. The range of examples here demonstrates that a one-size-fits-all approach is not warranted. On the other hand, these case studies enable the identification of common elements of coastal climate services, which could apply on a more global scale. Nevertheless, it can be also noted that the sea level projections considered in these examples are mostly consistent with those published by the IPCC.

2.2.2. The Purpose of Coastal Climate Services

The demand for services based on sea level projections is driven by three main user needs (Table 2; after Titus and Narayan, 1995 [73]):

(1) To analyse the benefits of mitigation of climate change, by comparing coastal impacts of sea level projections under different greenhouse gas emissions [74] (Row 1 in Table 2);
(2) To highlight research needs [75,76] (Row 2 in Table 2);
(3) To support adaptation to present and/or future sea level changes (Row 3a–d in Table 2).

The third need can be further divided into a range of more specific needs related to adaptation and disaster risk reduction strategies [77]. Global to national level users, for example, require assessments of impacts over varying spatio-temporal scales, in order to stimulate the implementation of adaptation plans or regulations, as in the French example (Row 3a in Table 2). Other users at regional to local scales require services to support preparedness to current sea level and flood hazards (Row 3b in Table 2) and to understand local to regional adaptation needs over multi-decadal timescales [78], including critical infrastructure where relevant [11] (Row 3c in Table 2). The U.S. example shows that coastal researchers and engineers have already partly responded to these needs, for example by providing annual predictions of high-water event frequencies to help budget appropriately in terms of costs of preparedness and crisis management (e.g., road closures, installation of pumps, sandbags, inflow preventers in storm water systems). Finally, Row 3d in Table 2 highlights the emergence of services aiming at evaluating the efficiency of adaptation measures and policies [79]. For example, building dikes to reduce the occurrence of overflow modifies flooding risks at local to regional scale: catastrophic flooding can still occur due to overtopping or breaching of dikes, and applying this measure systematically can result in amplifications of the tidal and surge maxima, ultimately increasing the flooding. Overall, Table 2 illustrates that CCS refer to a wide range of impacts and temporal and spatial scales, especially in the area of adaptation.

Table 2. Coastal climate services based on sea level scenarios or projections.

Need	Examples of Services	Key Challenge	Timescales of Sea Level Projections Required
1. To inform and encourage climate change mitigation efforts	21st and 22nd century sea level projections to evaluate benefits of mitigation for coastal areas [74], in particular in support to the negotiations revising the intended nationally determined contributions (INDCs)	To discriminate among the global impacts of different sea level projections corresponding to different greenhouse gas emissions pathways.	From 2050 onward
2. To highlight research needs	Research results highlighting needs for new SLR projections [75,80] or new coastal impacts assessment methods [55,81–84]	To demonstrate issues that are uncertain and sensitive requiring further research at local to regional scales.	From now to 2100 and beyond
3a. To understand global coastal adaption costs and benefits	Macro-scale studies demonstrating that adaptation is more cost-efficient than doing nothing [36,37,84] or evaluating the responsibilities of countries in SLR and their needs for adaptation [85]	To distinguish between the coastal impacts induced by future SLR from those induced by other processes at regional to global scales.	Coming decades to 2100 and beyond

Table 2. *Cont.*

Need	Examples of Services	Key Challenge	Timescales of Sea Level Projections Required
3b. To enhance preparedness for changing coastal hazards	Supporting preparedness, prevention and adaptation planning [10,12,43,47], according to the disaster risk reduction terminology [51,77]	To model impacts of sea level changes of a few 10's cm at local to regional scales (e.g., cities, estuaries) with improved confidence about when these effects will occur.	Near-term forecasts and projections, up to 2050, with a strong focus on the coming years to decade
3c. To understand local adaptation needs	Coastal vulnerability indicators [86] Detailed [66,81,87] to appropriate complexity modelling [88] Expected annual damages, adaptation needs [89–94] Critical infrastructures such as nuclear power plants [68,95] Critical settlements such as atoll islands [96,97]	To assess local to regional SLR, coastal environmental evolution and societal development within a single framework. To identify the timescales of local to regional changes for complex biophysical and human systems.	Coming decades to 2100 and beyond
3d. To evaluate local adaptation measures and policies	Robust decision making ([98] for an application of the approach in another context) Tipping points [54,98] Dynamic adaptive policy pathways or robustness approach [99,100]	To differentiate local to regional impacts of SLR according to different adaptation options.	Coming decades to 2100 and beyond

2.2.3. Users and Providers of Coastal Climate Services

The examples of Section 2.1 indicate that the climate services involves interactions among users and service providers (illustrated through arrows displaying information flows in Figure 2). This is suggestive of an evolution toward higher technology readiness levels, whereby CCS are entering in a phase of development, according to the research-to-operation scale of Brooks [101]. Figure 2 identifies three groups of users and providers of CCS, who all build upon the IPCC reports, and, when available, other expert groups:

- End-users of CCS, who ultimately benefit from them, and who are in charge of implementing adaptation and mitigation: this refers to a wide range of parties concerned with mitigating climate change and adapting to its consequences, who are generally involved in the process of decision making.

- Sea level information providers, such as the climate science community or government agencies who develop, use and interpret the models evaluating future sea level changes, and ultimately provide of mean and extreme sea level scenarios and projections.

- The coastal service providers, including coastal engineers, and consultants whose expertise is concerned with evaluating coastal hazards, such as coastal flooding, erosion and sedimentation, saline intrusions in estuaries, lagoons and coastal aquifers as well as their impacts on human activities, the environment and the economy. Traditionally, these coastal service providers have provided coastal information to end users. They are users of sea level projections.

2.2.4. The Business Case for Coastal Climate Services

The examples in Section 2.1 show that both public and private sectors are involved in the development of climate services [102]. However, economic activities supplying climate services remain today almost exclusively driven by a public demand (e.g., near-term investments for coastal protection or adaptation to chronic flooding vs. longer-term investments for larger infrastructure or relocation) or by public regulations [18]. Private organizations involved in coastal climate services are responding to this public demand, as in the case of France, where both private and public organization are assessing coastal hazards in support to local regulatory coastal risk prevention plans. Furthermore, despite the variety of the examples presented above, large vulnerable geographic regions are not covered by climate services today. Therefore, incentives and investments will be necessary to accelerate the development of coastal services.

3. Barriers to Coastal Climate Services Using SLR Scenarios or Projections

3.1. Common Barriers in the Development of Climate Services

Common barriers to the uptake of a variety of climate services have been extensively discussed in the literature [3,4,18,102–106]. Table 3 examines the extent to which these common barriers apply to the case of CCS. Among the four barriers identified during the first phase of early design of climate services, only one applies to CCS: as adaptation is now considered an urgent issue, users are requesting an increasing amount of information, which remains in the field of research. Wherever the research community is the only service provider able to respond to the demand while meeting the quality standards, the different time-scales involved in translating research into operations means that adaptation is often lagging behind the expected schedules. While this difficulty appears obvious in the French example (Section 2.1.3), it is not specific to the case of CCS and rather highlights the need to support innovation, as previously noted by Brooks [101].

In the second phase of climate services development, called the development phase, a range of barriers pertaining to difficulties in specifying and satisfying service requirements arise. Section 2 has already provided some empirical evidence of these barriers: when attempting to define CCS based on sea level projections, no single framework can be identified. Moreover, as previously noted by Hinkel et al. [7], sea level information provided so far has remained largely centered on science, while barely considering the workflows of users and their usages of sea level information. Overall, these barriers arise due to a lack of communication among stakeholders involved in the design and development of CCS. In the remainder of this section, we thus examine current interactions and information flows among stakeholders of coastal climate services (arrows in Figure 2), in order to identify specific barriers to their further development. Specifically, we successively examine three important barriers in the development phase (see Table 3):

- Lack of formalized requirements from end-users (Section 3.2)
- Lack of formalized requirements from translators of climate information into services (Section 3.3)
- Lack of salient sea level information (Section 3.4)

Table 3. Barriers to the uptake of CCS (after: Vaughan and Dessai [3]; Brasseur and Gallardo [4]; Cavelier et al. [18], Brooks [101]; Nuseibeh and Easterbrook [103], Cash et al. [104]; Monfray and Bley [105]).

	Generic Climate Services	Coastal Climate Services
Phase of Development	**Barriers Identified in Previous Studies**	**Relevance in the Case of CCS**
Early design	Lack of interactions among providers of climate information and end-users [3,101]	Partial (see Section 2): there exist examples where stakeholders have engaged in a loop of interactions to support coastal adaptation, but many potential end-users just do not have access to the expertise needed (e.g., developing countries)
	Insufficient awareness regarding vulnerability to climate change [4]	Partial (see Section 2): sea level projections beyond the likely range of IPCC are frequently used [7,78,100], but most users are unaware about long term SLR commitment.
	Lack of understanding of the decision-making context [3]	Partial (see Section 2): for example, sea level projections have been used in coastal engineering design at the municipality scale in Australia [32,106]
	Differences in working times for scientists and decision makers providing or using climate services [4]	Yes: the French example shows that by establishing a regulation in favour of adaptation, coastal stakeholders require operational products within six months or a year (e.g., multidecadal shoreline erosion predictions), while research has hardly provided with a satisfactory level of confidence so far in this area (Section 3.1).

Table 3. *Cont.*

	Generic Climate Services	Coastal Climate Services
Phase of Development	**Barriers Identified in Previous Studies**	**Relevance in the Case of CCS**
Development	Lack of formalized requirements from end-users [101,103]	Partial: requirements have been provided in many cases (e.g., defining setback lines, sea level allowances), but no global standards exist (see Section 3.2)
	Lack of formalized requirements from translators of climate information into services [101,103]	Partial: coastal service providers have hardly provided detailed formalized requirements for sea level information besides Nicholls et al. [78]; see Section 3.3)
	Limited ability of impact models to include climate information [105]	Yes: coastal evolution models have limitations over the time and space scales relevant for SLR [107,108] (see Section 3.3)
	Limited credibility and legitimacy of climate change impacts modelling frameworks [104,105]	Yes: coastal impact and adaptation modeling frameworks only cover part of the sea level, biophysical and socioeconomic uncertainty [37] (Section 3.3) and some key coastal datasets remain incomplete (e.g., information on subsidence and current shoreline changes [109]).
	Limited salience of current scientific results, including sea level information (relevance to the user needs) [3,4,104]	Yes: coastal impact and adaptation modeling frameworks are incomplete which may lead to maladaptation [7] (Section 3.3), and sea level information remains difficult to interpret for CCS providers (Section 3.4)
	Lack of awareness regarding the climate and sectorial information available [3,18]	Partial: there are informed users of coastal climate information, as shown by the Australian and UK examples.
	Lack of funding for innovation [101]	Country dependent
	Lack of evaluation and validation [101]	Country dependent
Operations	Limited societal benefits [3,18]	None: in general not relevant to developed coastal areas as shown by the large latent demand for CCS (Section 2)
	Lack of business model [4,18]	Yes: (Section 2): sea level projections are used in regulatory frameworks or in public or private procurements (e.g., World Bank projects in developing countries [110]). However, the long term impacts of SLR are often little addressed.
	Inadequate governance [3,18]	Yes (Section 2): UK continuously improve their use of sea level projections in CCS for more than a decade [78,111–113]

3.2. Lack of Formalized Requirements from End-Users

Difficulty in specifying end-user needs is a common barrier to the uptake of climate services [101,103]. In the area of SLR, a specific difficulty consists of understanding the extent to which users may accept higher damages than those implied by the likely range [7]. Indeed, knowledge about risk aversion and acceptability levels are difficult to obtain and to communicate [114] and can vary widely between users. Section 2 provided empirical evidence that this specific difficulty has not prevented the development of CCS in practice, as it presents examples where end-users have been able to include sea level information and coastal impact studies in their workflows. For example, coastal impacts studies are commonly used to justify "low-regret" strategies, such as maintaining ecological services and quality in coastal zones [115], relocating some buildings or activities, or to limit further urbanization in low lying or erodible coastal areas [116]. They are also used to anticipate the upgrade of defence works in coastal areas, which will be necessary to control safety levels despite sea level rise, demographic growth and land use pressure [32,36,89,90]. Moreover, other products are emerging: for example, coastal engineers not only design coastal defences that anticipate future upgrades [117], but also perform vulnerability assessments for existing infrastructure, which may have already experienced changing sea levels over a century or more [100].

For all those examples, precise products are defined (e.g., "setback lines", "sea level allowances", "elevation thresholds"), which all require knowledge of historic sea levels and SLR scenarios or projections. These examples demonstrate that the hypothesis that CCS systematically lack precise requirements can be rejected. Hence, other information flows in Figure 2 must be considered to explain why CCS emerge too slowly.

3.3. Lack of Formalized Requirements from Translators of Climate Information into Services

To obtain information on coastal impacts of SLR, end-users are turning to traditional coastal service providers such as coastal engineers, geologists, consultants and coastal scientists (upper right block in Figure 2). For these service providers, the main challenge is to choose an appropriate conceptual or physical modelling framework to evaluate coastal hazards, impacts and adaptation over time, in order to respond adequately to the questions raised by end-users. The multiple impacts of sea level rise (inundation/flooding, erosion, salinization, wetland loss and change, etc.) means that such assessments need to be comprehensive and consider a variety adaptation responses [118]. Coastal service providers use a range of approaches to tackle this challenge.

A first approach used by coastal service providers consists of classifying the coastal systems considered according to their vulnerability, instead of focusing on the details of SLR projections [86]. Such classifications translate some simple principles on a map: for example, regardless of the future sea level, eroding sand spits, former wetlands, unconsolidated cliffs and low-lying areas exposed to storms are known to be the most vulnerable areas. Once these simple rules have been identified, elaborating a coastal vulnerability index becomes a classical multi-criteria decision mapping problem [119], involving heterogeneous data regarding the physical and human coastal environment and expert opinion [86,120–123]. This approach considers rough SLR assumptions only (e.g., 1 m by 2100) and can identify critical areas where avoiding further urbanisation and development or planned retreat should be considered. This qualitative approach is unable to identify exactly when adaptation is required, though it can provide a range of time over which effects may be expected. For complex situations or cases where the consequences are high, more detailed approaches may be required. Hence, its use remains presently limited to assessments and evaluations other than for critical areas (Table 2, 1st Column).

A second approach applied to assess impacts and adaptation uses scenarios or sea level projections with coastal models of varying complexity to quantify possible impacts [88,124–127]. Hydrodynamic models of floods and salinization assuming constant morphology are reasonably skilful [66,93,128], but coastal evolution morphodynamic models still have limited predictive capabilities over the timescales relevant for decision making on coastal adaptation. Most of them rely on equilibrium profile assumptions, such as the Bruun rule, to model the impacts of SLR [108,109,129–131]. While flexible probabilistic modelling approaches for estimating setback lines from erosion are increasingly being developed [106,116,132,133], care is required in interpreting the uncertainties reported in IPCC SLR scenarios, particularly when representing the uncertainties by probability distributions (see Section 3.4.3 below). Similar cautions apply to the calculation of sea level allowances [32,90–92]. Furthermore, in many areas where active sediment processes are taking place, changing bathymetries modify coastal hydrodynamic processes and the related flooding hazards [55,134]. Similarly, neither socioeconomic uncertainties nor the impacts of human adaptation are fully addressed in existing modeling frameworks [37]. Because these complex and coupled processes are still poorly represented in integrated coastal impact models, large residual uncertainties remain.

Some end-users are already accustomed to making decisions in a context of uncertainties. However, Table 2 (Column 3) shows that several needs are very demanding for coastal impact models: for example, the decision-making requirement to discriminate between the efficiency of different adaptation strategies will not be satisfied if the different modelling results are dwarfed by the uncertainties of coastal impact models outcomes. Ultimately, these large uncertainties can hinder the ability of CCS providers to meet user needs for specific risk acceptability criteria, and at worst, they may mislead coastal decision-making [7].

We conclude from this subsection that coastal impact models do not always have the capabilities required to satisfy user needs and meet the challenges exposed in the 3rd Column of Table 2. Consequently, relying on providers to interpret user requirements alone is not sufficient. However, this practice remains quite common today as illustrated by the one-way information flow in Figure 2 between sea level and coastal impact information providers. Instead, there are still large research and development efforts to improve current coastal impact models, their use of sea level scenarios or projections, and their use in CCS. Hence, a challenge for CCS will be to promote two-way communication between sea level information providers and coastal service providers to ensure that sea level projections are fit-for-purpose for a larger range of next-user applications.

3.4. Lack of Salient Sea Level Information

3.4.1. Requirements for SLR Information

Both modelling approaches discussed above require sea level products considering the following aspects:

- Regional to local variability of sea level change, including coastal vertical land movement, irrespective of whether they are driven by climate change, tectonic change or direct human interventions, to downscale to local impacts assessments [109].
- Uncertainties, including likely, high-end and low-end scenarios, which are needed to examine impacts and adaptation responses [37,38,90], to test the robustness of adaptation measures, to identify minimum adaptation needs;
- The temporal evolution of sea level, which is important for estimating when to adapt or to define times of emergence for coastal impacts and adaptation needs.
- Most existing sea level scenarios and projections have addressed regional variability and their uncertainties (Sections 3.4.2 and 3.4.3 below), all following approximately the same methodology (Figure 3). However, much less information is available on temporal evolution (Section 3.4.4 below). Today, such sea level products are either used directly in coastal impact studies, or to define fixed standardized sea level scenarios can be defined by end-users, as is the case in France and some US agencies [53,54].

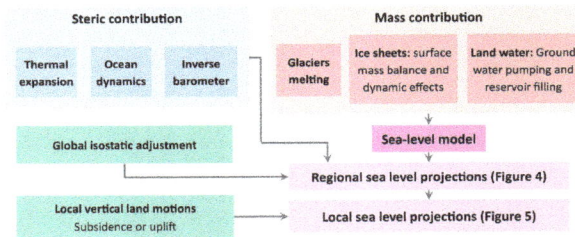

Figure 3. Methodology currently used to produce spatio-temporal sea level projections (Adapted from [23]).

3.4.2. Barriers to Providing Regional to Local Variability of Sea Level Changes

Scenarios and projections accounting for the oceanic, mass exchange and solid earth deformation processes causing regional sea level variability are now widely available [8,20–26] (Figure 4). However, coastal users in general still lack precise information to convert SLR from global models to a local on-shelf scale, including distortion of the SLR signal on continental shelves [135,136]. Furthermore, while coastal subsidence or uplift cause additional regional to local sea level variability (Figure 5), only the effects of the global isostatic adjustment and the response of the solid Earth to current large scale ice and water mass redistributions are usually included in sea level projections. Other vertical ground motions due to tectonic, volcanic, hydro-sedimentary and ground stability processes are often not available. However, they can have substantial effects at spatial scales ranging from a few meters to entire regions. Pointwise information regarding subsidence is included in the projections of Kopp et al. [8] and the U.S. scenarios of Sweet et al. [12], based on an analysis of tide gauge records. Alternatively, Global Navigation Satellite Systems (GNSS) measurements can be used, where possible, to infer vertical land motion trend rates [11,137]. These pointwise geodetic observations can be insufficient to characterize vertical ground motions where substantial and spatially variable subsidence rates exist [138–140]. Furthermore, these approaches assume that subsidence-related trends will persist through 2100, which could be invalidated if human activities are largely the cause of the measured subsidence and these activities cease in the future [140–145].

Figure 5A,B illustrate these issues in the case of Manila (Philippines), which has displayed vertical ground motions of >10 mm/year and net changes of several metres during the 20th Century. In fact, this city is prone to large subsidence due to groundwater withdrawal and/or drainage, similar to other coastal cities built on thick Holocene and Pleistocene deposits and where groundwater extractions exceed the recharge of aquifers [118,145,146]. The ground motion velocity fields shown in Figure 5A,B show that current pointwise measurements (Tide gauge, GPS, Doris) remain spatially too sparse and records too short, so that coastal users would have difficulties to quantify the contribution of vertical ground motions to relative SLR without complementary information from synthetic aperture radar interferometry [138–140,144]. This example shows that for subsidence and uplift processes unrelated to climate change, more observations would be extremely beneficial to CCS.

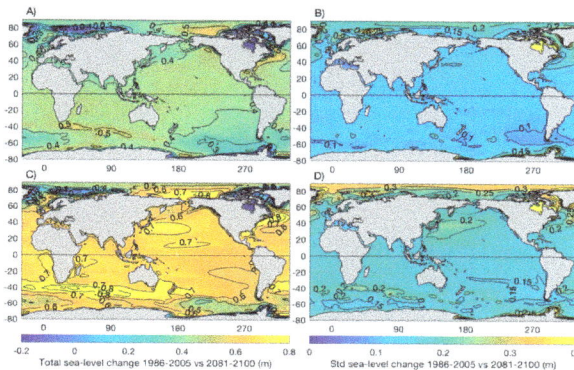

Figure 4. regional variability of SLR and their uncertainties (1-standard deviation) in the RCP 2.6 (**A,B**) and RCP 8.5 (**C,D**) scenarios (data: IPCC AR5 Ch13 [20]).

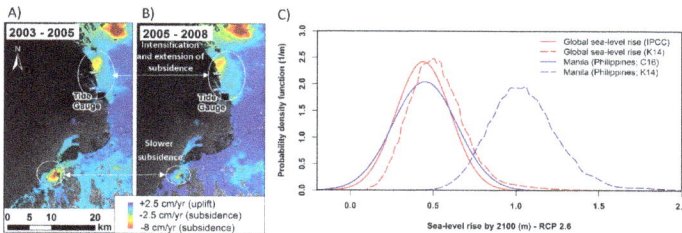

Figure 5. Vertical ground motions and SLR projections for Manila (The Philippines). (**A,B**) vertical ground motions estimated by Synthetic Aperture Radar Interferometry (InSAR), probably caused by groundwater extractions in Manila, displaying strong spatial variability and non-linear evolutions in time; (**C**) probabilistic representations of SLR global mean and near Manila, by 2100 and for the scenario RCP 2.6, based on the IPCC, Carson et al. (C16) [24] and Kopp et al. (K14) [8] data. The probability density functions (PDF) in Manila are slightly flatter than at global scale, reflecting the larger uncertainties caused by the large distance to glacial ice melting sources. Unlike Carson et al., Kopp et al. include a background subsidence estimated from sea level time series, so that the PDF is shifted to the right. However, future vertical ground motion will depend on future groundwater extraction, and is likely to be variable by location, increasing uncertainty. These examples illustrate the type of information coastal service providers need to analyse when designing sea level projections applicable at local scales. Data from IPCC AR5 Ch13 [20], Carson et al. [24], Kopp et al. [8], Raucoules et al. [140].

3.4.3. Barriers to Providing Information on Uncertainties of Future Sea Level Change

Before addressing the uncertainties, it is worth noting that even if climate change is mitigated to reach the Paris agreement objectives, sea level will continue rising for millennia [15], and that uncertainties remain about the speed of the process.

Users of sea level scenarios generally account for uncertainties by using several scenarios and assessing the robustness of their results to these scenarios (e.g., [54]). Several studies have attempted to provide coastal users with further details on uncertainties in future sea level projections, either by adjusting probability distributions to the median and likely range provided by the IPCC, or by designing non-parametric methods [8,89,90] (Figure 5C). This exercise is difficult because the processes involved in ice sheet melting are deeply uncertain. In the near future, the uncertainties in sea level projections are likely to be informed by two lines of research:

- Using one or several sets of probabilistic sea level projections, assuming a specific modelling framework for ice sheets. For example, the recent projections by Kopp et al. [147] are based on the modelling assumptions of DeConto and Pollard [148], but other projections could be based on the probabilistic projections of Ritz et al. [149].
- Using sea level projections based on expert judgement of ice sheets contributions to future SLR [150], or combining expert judgement with process-based models [151].

In the first case, deep uncertainties will be reflected by the different distributions resulting from different modelling frameworks. In the latter case, due to the lack of information to define each quantile of any probability distribution, it might not be possible to credibly quantify the precise probability of the tails of the SLR distribution [152]. However, this does not mean there is not useful information on these more extreme changes [100]. This raises the need for other theories of uncertainties able to convey differences among different estimates while minimizing the introduction of arbitrary information in uncertainty representations [153,154]. These latter approaches are complementary to probabilistic descriptions of uncertainties: in some coastal areas, users will require an optimal response to SLR through probabilistic projections [89,155], whereas others will use sea level scenarios or projections conveying minimum or maximum SLR estimates, in order to estimate minimum adaptation needs or to explore high end scenarios[1].

As shown by Figure 4B,D the uncertainties of local sea level projections display a regional variability. Furthermore, each source contributing to future SLR has a different probability distribution, with the longest tail probably being due to west Antarctic ice-sheet instability/melting. Due to the regional fingerprints of each contribution to sea level change, the regional probabilistic sea level projection will vary from place to place [8,25,26,156] (Figures 4B,D and 5C). This means that new estimates for each of the contributions will also have different consequences on a regional level.

However, this information has been only made available for coastal users since 2014 (in particular through the supplementary materials of Kopp et al. and of the IPCC report Ch13), thus limiting the number of coastal studies accounting for this spatial variability of uncertainties in SLR projections. From the perspective of coastal service providers, regional probabilistic sea level projections should also account for additional sources of uncertainties due to meso-scale coastal oceanic processes or vertical ground motions. However, existing regional sea level projections only partly take into account these sources of uncertainties, as this type information is often not available. It is made more difficult because the confidence in each term may be different. It is not clear at this time whether users are fully aware of the different uncertainties introduced by the use of projections beyond the uncertainties they are accustomed to dealing with in the scenario approach. This topic deserves further exploration by providers to fully understand how the differences in decisions taken using scenarios vary from

---

[1]   We avoid the term worst case, as it is impossible to define accurately and precisely.

those taken based on projections. This is required to be sure that the services do not inadvertently bias decisions makers in a way that increases the potential for adverse impacts.

3.4.4. Barriers to Providing Information on the Temporal Dynamics of Sea Level Changes

Beside the long-term sea level trend, the seasonal, interannual and multidecadal variability superimposed on these trends is of importance. For example, Slangen et al. (this volume, [92]) calculate the change in flooding risk due to the combined effect of sea level trends and variability by using the concept of allowances [89,90]. Users can find information on the temporal variability in the trends in SLR projections in the integrated Data Center of the University of Hamburg, which presents the timeseries of the IPCC AR5 projections [20,24,157]. However, this information only partly includes the meso-scale ocean processes causing interannual to decadal sea level variability such as El Niño lower frequency modes of variability [158–160], or interactions between tides and SLR [161–163]. Furthermore, vertical land motion (especially subsidence) can be highly non-linear in space and time, and may depend on non-predictable natural or anthropogenic processes such as tectonics or drainage/groundwater extraction [140]. In addition to the efforts to explicitly model these sources of temporal variability in sea level projections (see Section 2.1), coastal service providers have taken into account these processes either by analyzing their impacts on past sea level observations and assuming they will remain unchanged in the future [29,164], or by considering them as an additional source of uncertainty together with other processes causing deviations to a global or regional average [78,111]. While coastal service providers are aware of these issues, many of them require guidance to use these different components, design locally applicable sea level projections, and propagate their uncertainties and temporal dynamics into coastal impacts models. Finally, most sea level projections end in 2100, while as already noted in Table 2 there can be interest in longer-term projections for long-term design and planning issues.

4. Elements for Overcoming Barriers and Facilitating the Use of Sea Level Projections in Coastal Climate Services

4.1. A Framework for Coastal Climate Services

Due to the diversity of stakeholders, user needs and kinds of services required, future work on coastal climate services would greatly benefit from an overarching framework to guide their development. This subsection provides such a framework by extending the general climate service framework of Monfray and Bley (2016) [105] to the case of CCS based on sea level scenarios or projections (Figure 6). The starting point for developing CCS is the demand by user communities for services related to climate, sea level and coastal science. Neither observations and databases available today [137,165–169] nor the existing models and climate services (e.g., CMIP-5 [170], CORDEX [171], solid Earth deformation models [172–175]) directly respond to this demand. Hence, as in other areas of climate services, there is a need to strengthen the linkages between users and climate, sea level and coastal information providers [105]. In the context of the development of CCS, the scientific community is not only concerned with developing observations and models, but it is also expected to play a major role in the development of the boundary layers that translate climate, sea level and coastal information for users. Here, relevant approaches include building on best practices from the science community (such as the IPCC reports [13,20]) and exemplary case studies that combine both the science of SLR with applications (e.g., Section 2). For example, several countries such as the USA have moved beyond scenarios to planning and implementation for chronic and long-term effects of sea level changes (Section 2.1.1). Figure 6 also identifies boundary areas, which connect user requirements to climate, sea level and coastal science. This transitional layer includes eight topic areas, where applied research is needed, including (1) cross-cutting research (see Section 4.2 below) and (2) topical research in the area of mean and extreme sea level scenarios and projections, biophysical and socioeconomic impacts and adaptation (see Section 3). We argue that removing the barriers identified in Section 3 implies combining efforts in these interdisciplinary and topical research areas.

Importantly, Figure 6 escapes from the simplistic linear top-down model of "providing information to users", and recognizes the benefits of co-design and co-development between users and researchers, starting with the specific decision and governance context users are facing. A vital component of this is the longer term influence of the user needs on underpinning climate and impact science development.

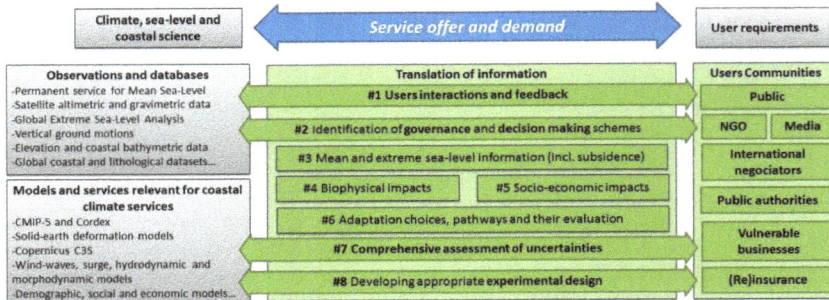

Figure 6. A framework for developing CCS (after: Monfray and Bley, 2016 [105]). This framework is valid as long as CCS based on sea level scenarios or projections are in a phase of research and development (Table 3).

4.2. Addressing Cross-Cutting Research Needs

Whatever the coastal impact considered, four cross-cutting lines of research can help to bridge the gap between the user demand and coastal and sea level science (Figure 6, shaded green arrows). They include all the disciplines needed to:

- Interact with users appropriately [105,176–179], in order to progressively settle on CCS meeting their needs.
- Identify where CCS can be mainstreamed into existing decision making frameworks [180–182].
- Address uncertainties in a consistent and comprehensive way all along the chain of disciplines involved, taking account of different levels of confidence in uncertainty estimates, and considering not only observations or models, but also users' differing needs and risk tolerance (following Hinkel et al. [7]).
- Develop appropriate experimental designs for decision support tools, combining all required components of CCS, addressing the challenges identified in the 3rd Column of Table 2, and allowing a comprehensive assessment of uncertainties from all relevant sources, including climate and coastal processes [83,183].

In the previous sections, we have shown that irrespective of the specific needs of end-users, a main concern of coastal service providers is to deliver robust, trusted and understood outputs (see Table 2 and Section 3). This leads to questions such as: are the conclusions suggested by coastal impact modelling experiments robust against the uncertainties in sea level projections and other uncertain parameters incorporated by the model and in agreement with existing data? Are the modelling results sensitive to different and equally appropriate modelling strategies [81,183]? Does the model itself have a bias or imprecision that could jeopardize the robustness of these conclusions? How are the residual uncertainties resulting from incomplete knowledge of the different processes generating coastal risks addressed [184]? If the uncertainties prevent the drawing of robust conclusions, is there a way to reduce them? How is expert judgement incorporated alongside the use of models?

A natural response to such concerns consists of propagating the uncertainties through the chain of models and conducting a sensitivity analysis [185,186]. Here, research is needed to guide users toward coastal modelling schemes that are sufficiently accurate, precise and resolved spatially and temporally to incorporate sea level scenarios or projections and to respond to their needs (Table 2, 3rd Column). At present, coastal impacts models do not always have the required accuracy and precision to meet these challenges [108], except in the case of coastal flood modelling during extreme events, where errors can be reduced to about 10 cm [66]. In the area of coastal evolution and erosion, further research efforts are needed to quantify modelling uncertainties for examples of mixed physical/empirical models able to incorporate precise sea level projections [56,133,187]. Furthermore, detailed salinization impact studies require high-complexity hydrogeological models able to represent groundwater fluxes through heterogeneous geological layers, and all models finally require evolving shorelines and bathymetries as boundary conditions. Finally, uncertainty theories are also useful to understand and model the residual uncertainties of coastal impact models, including those resulting from the high right hand tail of probabilistic sea level projections (see Section 3). This should be accompanied with similar efforts to model the uncertainties of other impact drivers, such as variable and changing waves and surge patterns, and to better understand the differences in decisions resulting from scenarios and those resulting from projections. Such research focused on the overall architecture of the experimental design can greatly benefit from existing frameworks developed at various scales [36,126,127].

4.3. Sustaining Public Finance and Developing Viable Private Business Model

When new services are supplied, a key question remains as to whether and how public demand can be satisfied before private demand has developed [101]. Presently, CCS are primarily demanded publicly, through research funding and public tenders, by agencies involved in government investments, by regulatory and permit requirements, or as terms in calls for tenders (Section 2). However, the demand is expected to grow as industries seek to assess their exposure to risk in response to the recommendations of the Task Force for Financial Disclosure [188] and as the Paris Agreement includes a strong adaptation focus. In developing countries such as small island developing states, CCS are increasingly supported by organizations such as the multilateral funds under the UNFCCC (e.g., the Green Climate Fund) and other national funds, donor organizations and the finance community concerned with the adaptation finance gap [189]. A prerequisite for sustaining and enhancing these financial flows is to identify viable business models for CCS in vulnerable developing countries that do not depend on public funding only.

4.4. Guidance and Capacity Building for Developing Countries

The information conveyed through IPCC reports is critically needed for the uptake of local CCS, especially in developing countries. They often constitute the most credible scientific reference for many end-users, but suffer from the major problem that new science emerges between reports, and waiting for the next round of IPCC assessment might lead to the most appropriate science being omitted from planning. At present the IPCC TGICA group (Task Group on Data and Scenario Support for Impact and Climate Analysis) provides some guidance on the use of IPCC scenarios. However, there is a need for guidance to be developed more closely with the working groups that develop the scenarios with more focus on the limitation of the scenarios and updating of guidance that is more closely tied to the time-scales of the major assessment reports. This could include improved access to datasets and tools for processing and extracting the relevant regional data. Additionally, there is a need to look to other initiatives, including the global framework on climate services [2] to combine IPCC information with that from other more frequently updated information. Together these can help users to conduct local coastal risks analysis and develop their own CCS.

5. Conclusions

Over the coming decade, climate services, supporting adaptation to and mitigation of climate change, are expected to become a viable market, driven by near-term user needs, by regulations, and by bi-and multilateral development institutes and financial mechanisms to support adaptation such us those established under the UNFCCC. Coastal zones are hotspots of climate change impacts, so that climate services will be increasingly needed to support adaptation. From this perspective, it seems important to identify how relevant information such as sea level scenarios or projections can be included in CCS. Given this context, this paper has provided preliminary answers to the following three questions:

- *What is the current state of the art in the area of coastal climate services using SLR scenarios or projections?* Section 2 has shown that there are some examples of early CCS based on SLR scenarios. The demand for these CCS is driven by decision makers interested in investments in coastal risk prevention and adaptation in the near-term future, by regulations and calls for tenders of international organisations addressing longer term issues, and, finally, by the demand of public and private decision makers concerned with critical infrastructure or settlements. However, there is large thematic and geographical diversity in the demanded CCS based on sea level projections. At least in Europe, this diversity explains why the development of coastal climate services is lagging behind those developments in other sectors. Overall, we find that CCS are emerging too slowly to meet the diversity of challenges posed by coastal climate change impacts, but there are already viable business models in place in some countries such as the USA and France.

- *What are the current technical barriers to satisfying the demand for coastal climate services based on sea level scenarios or projections?* Section 3 has identified two sets of barriers. The first one pertains to topical research needs, in order to increase confidence in coastal impact models and to respond to user's needs for sea level information; the latter includes sea level projections considering near-term (seasonal to decadal) and long term (beyond 2100) timescales, local to regional vertical ground motions, as well as the description (and reduction) of the related uncertainties. The second one pertains to barriers at the interface between sea level and coastal information providers: indeed, there are large gaps between what sea level science is able to provide (probabilistic, regional and time-evolving SLR projections often focusing on the open ocean) and methods of traditional coastal services providers (detailed flood or hydrodynamic modeling, probabilistic sizing of coastal defenses). We suggest that these difficulties prevent the integration of different components of CCS in a way that satisfies end user's needs. Hence, we argue that besides disciplinary research needs, what is lacking today is an accepted common methodology to elaborate CCS for adaptation.

- *What is needed to overcome barriers and to facilitate the use of sea level information in coastal climate services?* To overcome barriers identified in Section 3, we recommend defining a global framework for CCS. Section 4 proposes an integrated framework involving all stakeholders concerned with developing CCS. This framework addresses: (1) cross-cutting issues such as user interactions, decision making frameworks, uncertainties and overall architecture of the services to be developed; and (2) topical research on sea level science, coastal hydrodynamics, morphodynamics, biology, demography and economy, to ensure that current coastal modeling tools are able to include sea level rise information adequately. This framework could be useful to establish standards in the area of coastal climate services supporting adaptation to and mitigation of climate change.

Acknowledgments: This paper is a contribution to the Grand Challenge "Regional Sea level changes and coastal impacts" of the World Climate Research Programme (WCRP-GC). G.L.C., J.H. and R.v.d.W. acknowledge funding from the ERANET ERA4CS (INSeaPTION project). K.L.M. acknowledges funding from the NESP Earth Systems and Climate Change Hub. We thank all members of the WCRP GC, especially Detlef Stammer and Jonathan Gregory, for insightful discussions. We thank Daniel Raucoules for help in preparing Figure 5 and Olivier Douez for advice on coastal aquifer salinization. We thank Robert Kopp and Mark Carson for making their data available. These data are used to plot Figure 5C. Finally, we thank the two reviewers whose insightful comments led to significant improvements on this article. The scientific results and conclusions, as well as any views or opinions expressed herein, are those of the author(s) and do not necessarily reflect the views of USACE, NOAA or the Department of Commerce.

Author Contributions: This paper was drafted by G.L.C. and all authors based on contributions, meetings and exchanges of the WCRP GC (Work Package 5), led by R.J.N. K.L.M. and G.L.C. and involving J.H., R.S.W.V.d.W., K.D.W. and others. R.J.N. provided feedback on CCS development at the international and country levels. J.H. and G.L.C. proposed a common framework for CCS. W.V.S. and K.D.W. provided the US example. W.V.S. provided Figure 1. K.L.M. provided the Australian examples and insight into uncertainties. R.S.W.V.d.W. and A.B.A.S. provided knowledge of current sea level rise projections and their use. A.B.A.S. provided Figure 4. J.A.L. provided background regarding IPCC guidance, the global framework for climate services and UK examples. All authors wrote the final version of this article.

Conflicts of Interest: The authors declare no conflict of interest.

References

1. National Research Council of the National Academies (NRC); Board on Atmospheric Sciences and Climate. *A Climate Services Vision: First Steps Toward the Future*; The National Academies Press: Washington, DC, USA, 2001.

2. Hewitt, C.; Mason, S.; Walland, D. Commentary: The global framework for climate services. *Nat. Clim. Chang.* **2012**, *2*, 831–832. [CrossRef]

3. Vaughan, C.; Dessai, S. Climate services for society: Origins, institutional arrangements, and design elements for an evaluation framework. *Wiley Interdiscip. Rev.-Clim. Chang.* **2014**, *5*, 587–603. [CrossRef] [PubMed]

4. Brasseur, G.P.; Gallardo, L. Climate services: Lessons learned and future prospects. *Earths Future* **2016**, *4*, 79–89. [CrossRef]

5. Lémond, J.; Dandin, P.; Planton, S.; Vautard, R.; Pagé, C.; Déqué, M.; Moisselin, J.M. DRIAS: A step toward Climate Services in France. *Adv. Sci. Res.* **2011**, *6*, 179–186. [CrossRef]

6. Kjellstrom, E.; Nikulin, G.; Hansson, U.; Strandberg, G.; Ullerstig, A. 21st century changes in the european climate: Uncertainties derived from an ensemble of regional climate model simulations. *Tellus Ser. Dyn. Meteorol. Oceanogr.* **2011**, *63*, 24–40. [CrossRef]

7. Hinkel, J.; Jaeger, C.; Nicholls, R.J.; Lowe, J.; Renn, O.; Shi, P.J. Sea-level rise scenarios and coastal risk management. *Nat. Clim. Chang.* **2015**, *5*, 188–190. [CrossRef]

8. Kopp, R.E.; Horton, R.M.; Little, C.M.; Mitrovica, J.X.; Oppenheimer, M.; Rasmussen, D.J.; Strauss, B.H.; Tebaldi, C. Probabilistic 21st and 22nd century sea-level projections at a global network of tide-gauge sites. *Earths Future* **2014**, *2*, 383–406. [CrossRef]

9. National Research Council; Committee on Engineering Implications of Changes in Relative Mean Sea Level; Marine Board; Commission on Engineering and Technical Systems. *Responding to Changes in Sea Level, Engineering Implications*; National Academy Press: Washington, DC, USA, 1987. Available online: http://www.nap.edu/catalog.php?record_id=1006 (accessed on 10 October 2017).

10. Sweet, W.; Park, J.; Marra, J.; Zervas, C.; Gill, S. Sea-Level Rise and Nuisance Flood Frequency Changes around the United States. National Oceanic and Atmospheric Administration. NOAA Technical Report NOS CO-OPS 073, 2014. Available online: https://tidesandcurrents.noaa.gov/publications/NOAA_Technical_Report_NOS_COOPS_073.pdf (accessed on 4 May 2017).

11. Hall, J.A.; Gill, S.; Obeysekera, J.; Sweet, W.; Knuuti, K.; Marburger, J. Regional Sea Level Scenarios for Coastal Risk Management: Managing the Uncertainty of Future Sea Level Change and Extreme Water Levels for Department of Defense Coastal Sites Worldwide. U.S. Department of Defense, Strategic Environmental Research and Development Program, 2016; 224p. Available online: https://www.serdp-estcp.org/content/download/38961/375873/version/4/file/CARSWG+SLR+April+2016.pdf (accessed on 4 May 2017).

12. Sweet, W.V.; Kopp, R.E.; Weaver, C.P.; Obeysekera, J.; Thieler, E.R.; Zervas, C. Global and Regional Sea Level Scenarios for the United States. NOAA Technical Report NOS CO-OPS 083, 2017. Available online: https://tidesandcurrents.noaa.gov/publications/techrpt83_Global_and_Regional_SLR_Scenarios_for_the_US_final.pdf (accessed on 4 May 2017).

13. Wong, P.P.; Losada, I.J.; Gattuso, J.-P.; Hinkel, J.; Khattabi, A.; McInnes, K.L.; Saito, Y.; Sallenger, A. Coastal systems and low-lying areas. In *Climate Change 2014: Impacts, Adaptation, and Vulnerability*; Part A: Global and Sectoral Aspects. Contribution of Working Group II to the Fifth Assessment Report of the Intergovernmental Panel on Climate Change; Field, B.C., Barros, V.R., Dokken, D.J., Mach, K.J., Mastrandrea, M.D., Bilir, T.E., Chatterjee, M., Ebi, K.L., Estrada, Y.O., Genova, R.C., et al., Eds.; Cambridge University Press: Cambridge, UK; New York, NY, USA, 2014; pp. 361–409.

14. Vitousek, S.; Barnard, P.L.; Fletcher, C.H.; Frazer, N.; Erikson, L.; Storlazzi, C.D. Doubling of coastal flooding frequency within decades due to sea-level rise. *Sci. Rep.* **2017**, *7*, 1399. [CrossRef] [PubMed]

15. Clark, P.U.; Shakun, J.D.; Marcott, S.A.; Mix, A.C.; Eby, M.; Kulp, S.; Levermann, A.; Milne, G.A.; Pfister, P.L.; Santer, B.D.; et al. Consequences of twenty-first-century policy for multi-millennial climate and sea-level change. *Nat. Clim. Chang.* **2016**, *6*, 360–369. [CrossRef]

16. Lissner, T.K.; Fischer, E.M. Differential climate impacts for policy-relevant limits to global warming: The case of 1.5 °C and 2 °C. *Earth Syst. Dyn.* **2016**, *7*, 327–351.

17. United Nations Framework Convention on Climate Change (UNFCCC). Available online: http://unfccc.int/paris_agreement/items/9485.php (accessed on 4 May 2017).

18. Cavelier, R.; Borel, C.; Chareyron, V.; Chaussade, M.; Le Cozannet, G.; Morin, D.; Ritti, D. Condition for a market uptake of climate services for adaptation in France. *Clim. Serv.* **2017**, *6*, 34–40. [CrossRef]

19. Meehl, G.A.; Stocker, T.F.; Collins, W.D.; Friedlingstein, P.; Gaye, A.T.; Gregory, J.M.; Kitoh, A.; Knutti, R.; Murphy, J.M.; Noda, A.; et al. 2007: Global Climate Projections. In *Climate Change 2007: The Physical Science Basis*; Contribution of Working Group I to the Fourth Assessment Report of the Intergovernmental Panel on Climate Change; Solomon, S., Qin, D., Manning, M., Chen, Z., Marquis, M., Averyt, K.B., Tignor, M., Miller, H.L., Eds.; Cambridge University Press: Cambridge, UK; New York, NY, USA, 2007.

20. Church, J.A.; Clark, P.U.; Cazenave, A.; Gregory, J.M.; Jevrejeva, S.; Levermann, A.; Merrifield, M.A.; Milne, G.A.; Nerem, R.S.; Nunn, P.D.; et al. Sea Level Change. In *Climate Change 2013: The Physical Science Basis*; Contribution of Working Group I to the Fifth Assessment Report of the Intergovernmental Panel on Climate Change; Stocker, T.F., Qin, D., Plattner, G.-K., Tignor, M., Allen, S.K., Boschung, J., Nauels, A., Xia, Y., Bex, V., Midgley, P.M., Eds.; Cambridge University Press: Cambridge, UK; New York, NY, USA, 2013; pp. 1137–1216.

21. Mengel, M.; Levermann, A.; Frieler, K.; Robinson, A.; Marzeion, B.; Winkelmann, R. Future sea level rise constrained by observations and long-term commitment. *Proc. Natl. Acad. Sci. USA* **2016**, *113*, 2597–2602. [CrossRef] [PubMed]

22. Slangen, A.B.A.; Carson, M.; Katsman, C.A.; van de Wal, R.S.W.; Kohl, A.; Vermeersen, L.L.A.; Stammer, D. Projecting twenty-first century regional sea-level changes. *Clim. Chang.* **2014**, *124*, 317–332. [CrossRef]

23. Slangen, A.B.A.; Katsman, C.A.; van de Wal, R.S.W.; Vermeersen, L.L.A.; Riva, R.E.M. Towards regional projections of twenty-first century sea-level change based on ipcc sres scenarios. *Clim. Dyn.* **2012**, *38*, 1191–1209. [CrossRef]

24. Carson, M.; Kohl, A.; Stammer, D.; Slangen, A.B.A.; Katsman, C.A.; van de Wal, R.S.W.; Church, J.; White, N. Coastal sea level changes, observed and projected during the 20th and 21st century. *Clim. Chang.* **2016**, *134*, 269–281. [CrossRef]

25. Perrette, M.; Landerer, F.; Riva, R.; Frieler, K.; Meinshausen, M. A scaling approach to project regional sea level rise and its uncertainties. *Earth Syst. Dyn.* **2013**, *4*, 11–29. [CrossRef]

26. Jackson, L.P.; Jevrejeva, S. A probabilistic approach to 21st century regional sea-level projections using rcp and high-end scenarios. *Glob. Planet. Chang.* **2016**, *146*, 179–189. [CrossRef]

27. Grinsted, A.; Jevrejeva, S.; Riva, R.E.M.; Dahl-Jensen, D. Sea level rise projections for northern europe under rcp8.5. *Clim. Res.* **2015**, *64*, 15–23. [CrossRef]

28. Katsman, C.A.; Sterl, A.; Beersma, J.J.; van den Brink, H.W.; Church, J.A.; Hazeleger, W.; Kopp, R.E.; Kroon, D.; Kwadijk, J.; Lammersen, R.; et al. Exploring high-end scenarios for local sea level rise to develop flood protection strategies for a low-lying delta-the netherlands as an example. *Clim. Chang.* **2011**, *109*, 617–645. [CrossRef]

29. De Vries, H.; Katsman, C.; Drijfhout, S. Constructing scenarios of regional sea level change using global temperature pathways. *Environ. Res. Lett.* **2014**, *9*. [CrossRef]

30. Han, G.Q.; Ma, Z.M.; Chen, N.; Thomson, R.; Slangen, A. Changes in mean relative sea level around Canada in the twentieth and twenty-first centuries. *Atmos.-Ocean* **2015**, *53*, 452–463. [CrossRef]

31. Simpson, M.J.R.; Nilsen, J.E.Ø.; Ravndal, O.R.; Breili, K.; Sande, H.; Kierulf, H.P.; Steffen, H.; Jansen, E.; Carson, M.; Vestøl, O. Sea-level change for Norway. NCCS report n°1/2015, 2015. Available online: http://www.miljodirektoratet.no/Documents/publikasjoner/M405/M405.pdf (accessed on 4 May 2017).

32. McInnes, K.L.; Church, J.A.; Monselesan, D.; Hunter, J.R.; O'Grady, J.G.; Haigh, I.D.; Zhang, X. Sea-level Rise Projections for Australia: Information for Impact and Adaptation Planning. *Aust. Meteorol. Oceanogr. J.* **2015**, *65*, 127–149. [CrossRef]

33. Bates, N.R.; Astor, Y.M.; Church, M.J.; Currie, K.; Dore, J.E.; González-Dávila, M.; Lorenzoni, L.; Muller-Karger, F.; Olafsson, J.; Santana-Casiano, J.M. A time-series view of changing ocean chemistry due to ocean uptake of anthropogenic CO_2 and ocean acidification. *Oceanography* **2014**, *27*, 126–141. [CrossRef]

34. Mathis, J.T.; Cross, J.N.; Evans, W.; Doney, S.C. 2015: Ocean acidification in the surface waters of the Pacific–Arctic boundary regions. *Oceanography* **2015**, *28*, 122–135. [CrossRef]

35. Gattuso, J.P.; Magnan, A.; Bille, R.; Cheung, W.W.L.; Howes, E.L.; Joos, F.; Hoegh-Guldberg, O.; Kelly, P.; Pörtner, P.H.-O.; Rogers, A.D.; et al. Contrasting futures for ocean and society from different anthropogenic CO2 emissions scenarios. *Science* **2015**, *349*. [CrossRef] [PubMed]

36. Hallegatte, S.; Green, C.; Nicholls, R.J.; Corfee-Morlot, J. Future flood losses in major coastal cities. *Nat. Clim. Chang.* **2013**, *3*, 802–806. [CrossRef]

37. Hinkel, J.; Lincke, D.; Vafeidis, A.T.; Perrette, M.; Nicholls, R.J.; Tol, R.S.J.; Marzeion, B.; Fettweis, X.; Ionescu, C.; Levermann, A. Coastal flood damage and adaptation costs under 21st century sea-level rise. *Proc. Natl. Acad. Sci. USA* **2014**, *111*, 3292–3297. [CrossRef] [PubMed]

38. Hinkel, J.; Nicholls, R.J.; Tol, R.S.J.; Wang, Z.B.; Hamilton, J.M.; Boot, G.; Vafeidis, A.T.; McFadden, L.; Ganopolski, A.; Klein, R.J.T. A global analysis of erosion of sandy beaches and sea-level rise: An application of diva. *Glob. Planet. Chang.* **2013**, *111*, 150–158. [CrossRef]

39. Moftakhari, H.R.; AghaKouchak, A.; Sanders, B.F.; Feldman, D.L.; Sweet, W.; Matthew, R.A.; Luke, A. Increased nuisance flooding along the coasts of the united states due to sea level rise: Past and future. *Geophys. Res. Lett.* **2015**, *42*, 9846–9852. [CrossRef]

40. Moftakhari, H.R.; AghaKouchak, A.; Sanders, B.F.; Matthew, R.A. Cumulative hazard: The case of nuisance flooding. *Earths Future* **2017**, *5*, 214–223. [CrossRef]

41. Dahl, K.A.; Fitzpatrick, M.F.; Spanger-Siegfried, E. Sea level rise drives increased tidal flooding frequency at tide gauges along the us east and gulf coasts: Projections for 2030 and 2045. *PLoS ONE* **2017**, *12*. [CrossRef] [PubMed]

42. Ezer, T.; Atkinson, L.P. Accelerated flooding along the us east coast: On the impact of sea-level rise, tides, storms, the gulf stream, and the north atlantic oscillations. *Earths Future* **2014**, *2*, 362–382. [CrossRef]

43. Sweet, W.V.; Park, J. From the extreme to the mean: Acceleration and tipping points of coastal inundation from sea level rise. *Earths Future* **2014**, *2*, 579–600. [CrossRef]

44. Moodys' Flood Risk in Coastal Virginia. Available online: https://www.moodys.com/research/Moodys-Flood-risk-in-coastal-Virginia-supports-need-for-proactive--PR_328282 (accessed on 4 May 2017).

45. ENSO Forecasts at the International Research Institute of the University of Columbia. Available online: http://iri.columbia.edu/our-expertise/climate/forecasts/enso/current (accessed on 4 May 2017).

46. Sweet, W.V.; Marra, J.J. 2014 State of Nuisance Tidal Flooding, 2015. Available online: http://www.noaanews.noaa.gov/stories2015/2014%20State%20of%20Nuisance%20Tidal%20Flooding.pdf (accessed on 4 May 2017).

47. Sweet, W.V.; Marra, J.J. 2016: 2015 State of Nuisance Tidal Flooding, 2016. Available online: https://www.ncdc.noaa.gov/monitoring-content/sotc/national/2016/may/sweet-marra-nuisance-flooding-2015.pdf (accessed on 4 May 2017).

48. Widlansky, M.J.; Marra, J.J.; Chowdhury, M.R.; Stephens, S.A.; Miles, E.R.; Fauchereau, N.; Spillmanf, C.M.; Smithf, G.; Beardf, G.; Wells, J. Multi-model ensemble sea level forecasts for tropical Pacific islands. *J. Appl. Meteorol. Climatol.* **2017**. [CrossRef]

49. NOAA Sea Level Rise Viewer Web Mapping Tool. Available online: https://coast.noaa.gov/slr (accessed on 4 May 2017).

50. NOAA's National Weather Service. Available online: http://water.weather.gov/ahps (accessed on 4 May 2017).

51. Disaster Risk Reduction Terminology. Available online: https://www.unisdr.org/we/inform/terminology (accessed on 27 August 2017).

52. Federal Emergency Management Agency (FEMA). *Projected Impact of Relative SLR on the National Flood Insurance Program*; FEMA: Washington, DC, USA, 1991.

53. USACE. Engineer Regulation 1100-2-8162, Incorporating Sea Level Change in Civil Works Programs. Department of the Army, U.S. Army Corps of Engineers: Washington, DC, USA, 2013. Available online: http://www.publications.usace.army.mil/Portals/76/Publications/EngineerRegulations/ER_1100-2-8162.pdf (accessed on 4 May 2017).

54. U.S. Army Corps of Engineers (USACE). Engineer Technical Letter 1100-2-1, Procedures to Evaluate Sea Level Change: Impacts, Responses, and Adaptation. Department of the Army, U.S. Army Corps of Engineers: Washington, DC, USA, 2014. Available online: http://www.publications.usace.army.mil/Portals/76/Publications/EngineerRegulations/ER_1100-2-8162.pdf (accessed on 4 May 2017).

55. Lentz, E.E.; Thieler, E.R.; Plant, N.G.; Stippa, S.R.; Horton, R.M.; Gesch, D.B. Evaluation of dynamic coastal response to sea-level rise modifies inundation likelihood. *Nat. Clim. Chang.* **2016**, *6*, 696–700. [CrossRef]

56. Gutierrez, B.T.; Plant, N.G.; Thieler, E.R. A bayesian network to predict coastal vulnerability to sea level rise. *J. Geophys. Res.-Earth Surf.* **2011**, *116*. [CrossRef]

57. Masterson, J.P.; Fienen, M.N.; Thieler, E.R.; Gesch, D.B.; Gutierrez, B.T.; Plant, N.G. Effects of sea-level rise on barrier island groundwater system dynamics—Ecohydrological implications. *Ecohydrology* **2014**, *7*, 1064–1071. [CrossRef]

58. Good, M. Government Coastal Planning to Rising Sea Levels. Technical Report, 2011. Available online: http://acecrc.org.au/wp-content/uploads/2015/03/TR-Government-Coastal-Planning-Responses-to-Rising-Sea-Levels.pdf (accessed on 4 May 2017).

59. Climate Change in Australia. Available online: https://www.climatechangeinaustralia.gov.au/en/ (accessed on 4 May 2017).

60. CoastAdapt. Available online: https://coastadapt.com.au/ (accessed on 4 May 2017).

61. Tol, R.S.J.; Klein, R.J.T.; Nicholls, R.J. Towards successful adaptation to sea-level rise along europe's coasts. *J. Coast. Res.* **2008**, *24*, 432–442. [CrossRef]

62. Przyluski, V.; Hallegatte, S. *Gestion des Risques Naturels: Leçons de la Tempête Xynthia*; Editions Quae: Versailles, France, 2013.

63. Deboudt, P. Towards coastal risk management in france. *Ocean Coast. Manag.* **2010**, *53*, 366–378. [CrossRef]

64. Stockdon, H.F.; Holman, R.A.; Howd, P.A.; Sallenger, A.H. Empirical parameterization of setup, swash, and runup. *Coast. Eng.* **2006**, *53*, 573–588. [CrossRef]

65. Ministère de L'Ecologie, du Développement Durable, des Transports et du Logement (MEDDTL). Circulaire du 27 Juillet 2011 Relative à la Prise en Compte du Risque de Submersion Marine Dans les Plans de Prévention des Risques Naturels Littoraux, 2011. Available online: http://www.bulletin-officiel.developpement-durable.gouv.fr/fiches/BO201115/met_20110015_0100_0021.pdf (accessed on 4 May 2017).

66. Le Roy, S.; Pedreros, R.; Andre, C.; Paris, F.; Lecacheux, S.; Marche, F.; Vinchon, C. Coastal flooding of urban areas by overtopping: Dynamic modelling application to the johanna storm (2008) in gavres (france). *Nat. Hazards Earth Syst. Sci.* **2015**, *15*, 2497–2510. [CrossRef]

67. Magnan, A.K.; Schipper, E.L.F.; Burkett, M.; Bharwani, S.; Burton, I.; Eriksen, S.; Gemenne, F.; Schaar, J.; Ziervogel, G. Addressing the risk of maladaptation to climate change. *Wiley Interdiscip. Rev.-Clim. Chang.* **2016**, *7*, 646–665. [CrossRef]

68. Wilby, R.L.; Nicholls, R.J.; Warren, R.; Wheater, H.S.; Clarke, D.; Dawson, R.J. Keeping nuclear and other coastal sites safe from climate change. *Proc. Inst. Civ. Eng.-Civ. Eng.* **2011**, *164*, 129–136. [CrossRef]

69. Conway, D.; Mustelin, J. Strategies for improving adaptation practice in developing countries. *Nat. Clim. Chang.* **2014**, *4*, 339–342. [CrossRef]

70. Bangladesh Delta Plan 2100. Available online: http://www.bangladeshdeltaplan2100.org/ (accessed on 29 August 2017).

71. Van den Hurk, B.; van Oldenborgh, G.J.; Lenderink, G.; Hazeleger, W.; Haarsma, R.; de Vries, H. Drivers of mean climate change around the netherlands derived from cmip5. *Clim. Dyn.* **2014**, *42*, 1683–1697. [CrossRef]

72. UK Climate Reports Projections. Available online: http://ukclimateprojections.metoffice.gov.uk/21678 (accessed on 4 May 2017).

73. Titus, J.G.; Narayanan, V.K. *The Probability of Sea Level Rise*; US Environmental Protection Agency: Washington, DC, USA; Office of Policy, Planning, and Evaluation: Bethesda, MD, USA; Climate Change Division, Adaptation Branch: Washington, DC, USA, 1995; Volume 95.

74. Nicholls, R.J.; Lowe, J.A. Benefits of mitigation of climate change for coastal areas. *Glob. Environ. Chang.-Hum. Policy Dimens.* **2004**, *14*, 229–244. [CrossRef]

75. Rahmstorf, S. A semi-empirical approach to projecting future sea-level rise. *Science* **2007**, *315*, 368–370. [CrossRef] [PubMed]

76. Hansen, J.; Sato, M.; Hearty, P.; Ruedy, R.; Kelley, M.; Masson-Delmotte, V.; Russell, G.; Tselioudis, G.; Cao, J.J.; Rignot, E.; et al. Ice melt, sea level rise and superstorms: Evidence from paleoclimate data, climate modeling, and modern observations that 2 a degrees c global warming could be dangerous. *Atmos. Chem. Phys.* **2016**, *16*, 3761–3812. [CrossRef]

77. Romieu, E.; Welle, T.; Schneiderbauer, S.; Pelling, M.; Vinchon, C. Vulnerability assessment within climate change and natural hazard contexts: Revealing gaps and synergies through coastal applications. *Sustain. Sci.* **2010**, *5*, 159–170. [CrossRef]

78. Nicholls, R.J.; Hanson, S.E.; Lowe, J.A.; Warrick, R.A.; Lu, X.F.; Long, A.J. Sea-level scenarios for evaluating coastal impacts. *Wiley Interdiscip. Rev.-Clim. Chang.* **2014**, *5*, 129–150. [CrossRef]

79. Hallegatte, S. Strategies to adapt to an uncertain climate change. *Glob. Environ. Chang.* **2009**, *19*, 240–247. [CrossRef]

80. Hoffman, J.S.; Keyes, D.L.; Titus, J.G. *Projecting Future Sea Level Rise: Methodology, Estimates to the Year 2100, and Research Needs*; Strategic Studies Staff, Office of Policy Analysis: Washington, DC, USA; Office of Policy and Resource Management: Washington, DC, USA; US Environmental Protection Agency: Washington, DC, USA, 1983.

81. Bilskie, M.V.; Hagen, S.C.; Alizad, K.; Medeiros, S.C.; Passeri, D.L.; Needham, H.F.; Cox, A. Dynamic simulation and numerical analysis of hurricane storm surge under sea level rise with geomorphologic changes along the northern gulf of mexico. *Earths Future* **2016**, *4*, 177–193. [CrossRef]

82. Smith, M.D.; Murray, A.B.; Gopalakrishnan, S.; Keeler, A.G.; Landry, C.E.; McNamara, D.; Moore, L.J. *Geoengineering Coastlines? From Accidental to Intentional*; From Accidental to Intentional (June 2014); Duke Environmental and Energy Economics Working Paper EE, 14-02; Elsevier: Amsterdam, The Netherlands, 2014.

83. Le Cozannet, G.; Rohmer, J.; Cazenave, A.; Idier, D.; van de Wal, R.; de Winter, R.; Pedreros, R.; Balouin, Y.; Vinchon, C.; Oliveros, C. Evaluating uncertainties of future marine flooding occurrence as sea-level rises. *Environ. Model. Softw.* **2015**, *73*, 44–56. [CrossRef]

84. Hoshino, S.; Esteban, M.; Mikami, T.; Takagi, H.; Shibayama, T. Estimation of increase in storm surge damage due to climate change and sea level rise in the greater tokyo area. *Nat. Hazards* **2016**, *80*, 539–565. [CrossRef]

85. Hardy, R.D.; Nuse, B.L. Global sea-level rise: Weighing country responsibility and risk. *Clim. Chang.* **2016**, *137*, 333–345. [CrossRef]

86. Gornitz, V. Global coastal hazards from future sea-level rise. *Glob. Planet. Chang.* **1991**, *89*, 379–398. [CrossRef]

87. McInnes, K.L.; Walsh, K.J.E.; Hoeke, R.K.; O'Grady, J.G.; Colberg, F.; Hubbert, G.D. Quantifying storm tide risk in fiji due to climate variability and change. *Glob. Planet. Chang.* **2014**, *116*, 115–129. [CrossRef]

88. French, J.; Payo, A.; Murray, B.; Orford, J.; Eliot, M.; Cowell, P. Appropriate complexity for the prediction of coastal and estuarine geomorphic behaviour at decadal to centennial scales. *Geomorphology* **2016**, *256*, 3–16. [CrossRef]

89. Hunter, J. A simple technique for estimating an allowance for uncertain sea-level rise. *Clim. Chang.* **2012**, *113*, 239–252. [CrossRef]

90. Hunter, J.R.; Church, J.A.; White, N.J.; Zhang, X. Towards a global regionally varying allowance for sea-level rise. *Ocean Eng.* **2013**, *71*, 17–27. [CrossRef]

91. Buchanan, M.K.; Kopp, R.E.; Oppenheimer, M.; Tebaldi, C. Allowances for evolving coastal flood risk under uncertain local sea-level rise. *Clim. Chang.* **2016**, *137*, 347–362. [CrossRef]

92. Slangen, A.B.A.; van de Wal, R.S.W.; Reerink, T.J.; de Winter, R.C.; Hunter, J.R.; Woodworth, P.L.; Edwards, T. The impact of uncertainties in ice sheet dynamics on sea-level allowances at tide gauge locations. *J. Mar. Sci. Eng.* **2017**, *5*, 21. [CrossRef]

93. Dawson, R.J.; Dickson, M.E.; Nicholls, R.J.; Hall, J.W.; Walkden, M.J.A.; Stansby, P.K.; Mokrech, M.; Richards, J.; Zhou, J.; Milligan, J.; et al. Integrated analysis of risks of coastal flooding and cliff erosion under scenarios of long term change. *Clim. Chang.* **2009**, *95*, 249–288. [CrossRef]

94. Dawson, D.; Shaw, J.; Gehrels, W.R. Sea-level rise impacts on transport infrastructure: The notorious case of the coastal railway line at Dawlish, England. *J. Trans. Geogr.* **2016**, *51*, 97–109. [CrossRef]

95. Brown, J.M.; Ciavola, P.; Masselink, G.; McCall, R.; Plater, A.J. Preface: Monitoring and modelling to guide coastal adaptation to extreme storm events in a changing climate. *Nat. Hazards Earth Syst. Sci.* **2016**, *16*, 463–467. [CrossRef]

96. Nurse, L.A.; McLean, R.F.; Agard, J.; Briguglio, L.P.; Duvat-Magnan, V.; Pelesikoti, N.; Tompkins, E.; Webb, A. Small islands. In *Climate Change 2014: Impacts, Adaptation, and Vulnerability*; Part B: Regional Aspects; Contribution of Working Group II to the Fifth Assessment Report of the Intergovernmental Panel on Climate Change; Barros, V.R., Field, C.B., Dokken, D.J., Mastrandrea, M.D., Mach, K.J., Bilir, T.E., Chatterjee, M., Ebi, K.L., Estrada, Y.O., Genova, R.C., et al., Eds.; Cambridge University Press: Cambridge, UK; New York, NY, USA, 2014; pp. 1613–1654.

97. Storlazzi, C.D.; Elias, E.P.L.; Berkowitz, P. Many atolls may be uninhabitable within decades due to climate change. *Sci. Rep.* **2015**, *5*, 14546. [CrossRef] [PubMed]

98. Hall, J.W.; Lempert, R.J.; Keller, K.; Hackbarth, A.; Mijere, C.; McInerney, D.J. Robust climate policies under uncertainty: A comparison of robust decision making and info-gap methods. *Risk Anal.* **2012**, *32*, 1657–1672. [CrossRef] [PubMed]

99. Haasnoot, M.; Kwakkel, J.H.; Walker, W.E.; ter Maat, J. Dynamic adaptive policy pathways: A method for crafting robust decisions for a deeply uncertain world. *Glob. Environ. Chang.-Hum. Policy Dimens.* **2013**, *23*, 485–498. [CrossRef]

100. Ranger, N.; Reeder, T.; Lowe, J. Addressing 'deep' uncertainty over long-term climate in major infrastructure projects: Four innovations of the Thames Estuary 2100 Project. *EURO J. Decis. Process.* **2013**, *1*, 233–262. [CrossRef]

101. Brooks, M.S. Accelerating innovation in climate services: The 3 e's for climate service providers. *Bull. Am. Meteorol. Soc.* **2013**, *94*, 807–819. [CrossRef]

102. Tompkins, E.L.; Eakin, H. Managing private and public adaptation to climate change. *Glob. Environ. Chang.* **2012**, *22*, 3–11. [CrossRef]

103. Nuseibeh, B.; Easterbrook, S. Requirements engineering: A roadmap. In Proceedings of the Future of Software Engineering, Limerick, Ireland, 4–11 June 2000; pp. 35–46.

104. Cash, D.W.; Clark, W.C.; Alcock, F.; Dickson, N.M.; Eckley, N.; Guston, D.H.; Jager, J.; Mitchell, R.B. Knowledge systems for sustainable development. *Proc. Natl. Acad. Sci. USA* **2003**, *100*, 8086–8091. [CrossRef] [PubMed]

105. Monfray, P.; Bley, D. JPI Climate: A key player in advancing Climate Services in Europe. *Clim. Serv.* **2016**, *4*, 61–64. [CrossRef]

106. Kinsela, M.A.; Monis, B.D.; Daley, M.J.A.; Hanslow, D.J. A flexible approach to forecasting coastline change on wave dominated beaches. *J. Coast. Res.* **2016**, 952–956. [CrossRef]

107. Stive, M.J.F.; Aarninkhof, S.G.J.; Hamm, L.; Hanson, H.; Larson, M.; Wijnberg, K.M.; Nicholls, R.J.; Capobianco, M. Variability of shore and shoreline evolution. *Coast. Eng.* **2002**, *47*, 211–235. [CrossRef]

108. Ranasinghe, R. Assessing climate change impacts on open sandy coasts: A review. *Earth-Sci. Rev.* **2016**, *160*, 320–332. [CrossRef]

109. Cazenave, A.; Le Cozannet, G. Sea level rise and its coastal impacts. *Earths Future* **2014**, *2*, 15–34. [CrossRef]

110. World Bank Group Climate Change Action Plan. Available online: http://pubdocs.worldbank.org/en/677331460056382875/WBG-Climate-Change-Action-Plan-public-version.pdf (accessed on 5 May 2017).

111. Hulme, M.; Jenkins, G.J.; Lu, X.; Tumpenny, J.R.; Mitchell, T.D.; Jones, R.G.; Lowe, J.; Murphy, J.M.; Hassell, D.C. *Climate Change Scenarios for the United Kingdom: The UKCIP02 Scientific Report*; Tyndall Centre for Climate Change Research, University of East Anglia: Norwich, UK, 2002; 120p.

112. Lowe, J.A.; Howard, T.; Pardaens, A.; Tinker, J.; Holt, J.; Wakelin, S.; Milne, G.; Leake, J.; Wolf, J.; Horsburgh, K.; et al. *UK Climate Projections Science Report: Marine and Coastal Projections*; Met Office Hadley Centre: Exeter, UK, 2009.

113. UKCP18 (Forthcoming). Available online: http://ukclimateprojections.metoffice.gov.uk/24125 (accessed on 5 May 2017).

114. Idier, D.; Rohmer, J.; Bulteau, T.; Delvallée, E. Development of an inverse method for coastal risk management. *Nat. Hazards Earth Syst. Sci.* **2013**, *13*, 999–1013. [CrossRef]

115. Spalding, M.D.; McIvor, A.L.; Beck, M.W.; Koch, E.W.; Moller, I.; Reed, D.J.; Rubinoff, P.; Spencer, T.; Tolhurst, T.J.; Wamsley, T.V.; et al. Coastal ecosystems: A critical element of risk reduction. *Conserv. Lett.* **2014**, *7*, 293–301. [CrossRef]

116. Wainwright, D.J.; Ranasinghe, R.; Callaghan, D.P.; Woodroffe, C.D.; Jongejan, R.; Dougherty, A.J.; Rogers, K.; Cowell, P.J. Moving from deterministic towards probabilistic coastal hazard and risk assessment: Development of a modelling framework and application to narrabeen beach, new south wales, australia. *Coast. Eng.* **2015**, *96*, 92–99. [CrossRef]

117. Sergent, P.; Prevot, G.; Mattarolo, G.; Brossard, J.; Morel, G.; Mar, F.; Benoit, M.; Ropert, F.; Kergadallan, X.; Trichet, J.J.; et al. Adaptation of coastal structures to mean sea level rise. *La Houille Blanche* **2014**, 54–61. [CrossRef]

118. Nicholls, R.J. Impacts of and responses to sea-level rise. In *Understanding Sea-Level Rise and Variability*; Church, J.A., Woodworth, P.L., Aarup, T., Wilson, W.S., Eds.; Wiley-Blackwell: Chichester, GB, USA, 2010; pp. 17–51.

119. Malczewski, J. Gis-based multicriteria decision analysis: A survey of the literature. *Int. J. Geogr. Inf. Sci.* **2006**, *20*, 703–726. [CrossRef]

120. Boruff, B.J.; Emrich, C.; Cutter, S.L. Erosion hazard vulnerability of us coastal counties. *J. Coast. Res.* **2005**, *21*, 932–942. [CrossRef]

121. Hanson, S.; Nicholls, R.J.; Balson, P.; Brown, I.; French, J.R.; Spencer, T.; Sutherland, W.J. Capturing coastal geomorphological change within regional integrated assessment: An outcome-driven fuzzy logic approach. *J. Coast. Res.* **2010**, *26*, 831–842. [CrossRef]

122. Bagdanavičiūtė, I.; Kelpsaite, L.; Soomere, T. Multi-criteria evaluation approach to coastal vulnerability index development in micro-tidal low-lying areas. *Ocean Coast. Manag.* **2015**, *104*, 124–135. [CrossRef]

123. Le Cozannet, G.; Garcin, M.; Bulteau, T.; Mirgon, C.; Yates, M.L.; Mendez, M.; Baills, A.; Idier, D.; Oliveros, C. An APH-derived method for mapping the physical vulnerability of coastal areas at regional scales. *Nat. Hazards Earth Syst. Sci.* **2013**, *13*, 1209–1227. [CrossRef]

124. Hanson, H.; Aarninkhof, S.; Capobianco, M.; Jimenez, J.A.; Larson, M.; Nicholls, R.J.; Plant, N.G.; Southgate, H.N.; Steetzel, H.J.; Stive, M.J.F.; et al. Modelling of coastal evolution on yearly to decadal time scales. *J. Coast. Res.* **2003**, *19*, 790–811.

125. Mokrech, M.; Nicholls, R.J.; Richards, J.A.; Henriques, C.; Holman, I.P.; Shackley, S. Regional impact assessment of flooding under future climate and socio-economic scenarios for east anglia and north west england. *Clim. Chang.* **2008**, *90*, 31–55. [CrossRef]

126. Hinkel, J.; Klein, R.J.T. Integrating knowledge to assess coastal vulnerability to sea-level rise: The development of the diva tool. *Glob. Environ. Chang.-Hum. Policy Dimens.* **2009**, *19*, 384–395. [CrossRef]

127. Narayan, S.; Nicholls, R.J.; Clarke, D.; Hanson, S.; Reeve, D.; Horrillo-Caraballo, J.; le Cozannet, G.; Hissel, F.; Kowalska, B.; Parda, R.; et al. The spr systems model as a conceptual foundation for rapid integrated risk appraisals: Lessons from europe. *Coast. Eng.* **2014**, *87*, 15–31. [CrossRef]

128. Ketabchi, H.; Mahmoodzadeh, D.; Ataie-Ashtiani, B.; Simmons, C.T. Sea-level rise impacts on seawater intrusion in coastal aquifers: Review and integration. *J. Hydrol.* **2016**, *535*, 235–255. [CrossRef]

129. Cooper, J.A.G.; Pilkey, O.H. Sea-level rise and shoreline retreat: Time to abandon the bruun rule. *Glob. Planet. Chang.* **2004**, *43*, 157–171. [CrossRef]

130. Silva, P.A.; Bertin, X.; Fortunato, A.B.; Oliveira, A. Intercomparison of sediment transport formulas in current and combined wave-current conditions. *J. Coast. Res.* **2009**, *25*, 559–563.

131. Ranasinghe, R.; Stive, M.J.F. Rising seas and retreating coastlines. *Clim. Chang.* **2009**, *97*, 465–468. [CrossRef]

132. Ranasinghe, R.; Duong, T.M.; Uhlenbrook, S.; Roelvink, D.; Stive, M. Climate-change impact assessment for inlet-interrupted coastlines. *Nat. Clim. Chang.* **2013**, *3*, 83–87. [CrossRef]

133. Ranasinghe, R.; Callaghan, D.; Stive, M.J.F. Estimating coastal recession due to sea level rise: Beyond the bruun rule. *Clim. Chang.* **2012**, *110*, 561–574. [CrossRef]

134. Familkhalili, R.; Talke, S.A. The effect of channel deepening on tides and storm surge: A case study of Wilmington, NC. *Geophys. Res. Lett.* **2016**, *43*. [CrossRef]

135. Zhang, X.; Church, J.A.; Monselesan, D.; McInnes, K.L. Sea level projections for the Australian region in the 21st century. *Geophys. Res. Lett.* **2017**, *44*. [CrossRef]

136. Zhang, X.; Oke, P.; Feng, M.; Chamberlain, M.; Church, J.; Monselesan, D.; Sun, C.; Matear, R.; Schiller, A.; Fiedler, R. A near-Global Eddy-Resolving OGCM for Climate Studies. *Geosci. Model Dev. Discuss.* **2016**. [CrossRef]

137. Santamaria-Gomez, A.; Gravelle, M.; Collilieux, X.; Guichard, M.; Miguez, B.M.; Tiphaneau, P.; Woppelmann, G. Mitigating the effects of vertical land motion in tide gauge records using a state-of-the-art gps velocity field. *Glob. Planet. Chang.* **2012**, *98–99*, 6–17. [CrossRef]

138. Woeppelmann, G.; Le Cozannet, G.; de Michele, M.; Raucoules, D.; Cazenave, A.; Garcin, M.; Hanson, S.; Marcos, M.; Santamaria-Gomez, A. Is land subsidence increasing the exposure to sea level rise in Alexandria, Egypt? *Geophys. Res. Lett.* **2013**, *40*, 2953–2957. [CrossRef]

139. Woeppelmann, G.; Marcos, M. Vertical land motion as a key to understanding sea level change and variability. *Rev. Geophys.* **2016**, *54*, 64–92. [CrossRef]

140. Raucoules, D.; Le Cozannet, G.; Woeppelmann, G.; de Michele, M.; Gravelle, M.; Daag, A.; Marcos, M. High nonlinear urban ground motion in manila (philippines) from 1993 to 2010 observed by dinsar: Implications for sea-level measurement. *Remote Sens. Environ.* **2013**, *139*, 386–397. [CrossRef]

141. Teatini, P.; Ferronato, M.; Gambolati, G.; Bertoni, W.; Gonella, M. A century of land subsidence in ravenna, italy. *Environ. Geol.* **2005**, *47*, 831–846. [CrossRef]

142. Ericson, J.P.; Vorosmarty, C.J.; Dingman, S.L.; Ward, L.G.; Meybeck, M. Effective sea-level rise and deltas: Causes of change and human dimension implications. *Glob. Planet. Chang.* **2006**, *50*, 63–82. [CrossRef]

143. Syvitski, J.P.M.; Kettner, A.J.; Overeem, I.; Hutton, E.W.H.; Hannon, M.T.; Brakenridge, G.R.; Day, J.; Vorosmarty, C.; Saito, Y.; Giosan, L.; et al. Sinking deltas due to human activities. *Nat. Geosci.* **2009**, *2*, 681–686. [CrossRef]

144. Tosi, L.; Teatini, P.; Strozzi, T. Natural versus anthropogenic subsidence of venice. *Sci. Rep.* **2013**, *3*, 2710. [CrossRef] [PubMed]

145. World Bank. *Climate Risks and Adaptation in Asian Coastal Megacities: A Synthesis Report*; World Bank Group: Washington, DC, USA, 2010.

146. Erkens, G.; Bucx, T.; Dam, R.; de Lange, G.; Lambert, J. Sinking coastal cities, Prevention and Mitigation of Natural and Anthropogenic Hazards due to Land Subsidence. In Proceedings of the Ninth International Symposium on Land Subsidence (NISOLS), Nagoya, Japan, 15–19 November 2015; pp. 189–198.

147. Kopp, R.; DeConto, R.M.; Bader, D.A.; Hay, C.C.; Horton, R.M.; Kulp, S.; Oppenheimer, M.; Pollard, D.; Strauss, B.H. 2017: Preprint. Available online: https://arxiv.org/abs/1704.05597 (accessed on 10 October 2017).

148. DeConto, R.M.; Pollard, D. Contribution of antarctica to past and future sea-level rise. *Nature* **2016**, *531*, 591–597. [CrossRef] [PubMed]

149. Ritz, C.; Edwards, T.L.; Durand, G.; Payne, A.J.; Peyaud, V.; Hindmarsh, R.C.A. Potential sea-level rise from antarctic ice-sheet instability constrained by observations. *Nature* **2015**, *528*, 115–118. [CrossRef] [PubMed]

150. Bamber, J.L.; Aspinall, W.P. An expert judgement assessment of future sea level rise from the ice sheets. *Nat. Clim. Chang.* **2013**, *3*, 424–427. [CrossRef]

151. Oppenheimer, M.; Little, C.M.; Cooke, R.M. Expert judgement and uncertainty quantification for climate change. *Nat. Clim. Chang.* **2016**, *6*, 445–451. [CrossRef]

152. De Vries, H.; van de Wal, R.S.W. How to interpret expert judgment assessments of the 21st century sea-level rise. *Clim. Chang.* **2015**, *130*, 87–100. [CrossRef]

153. Ben Abdallah, N.; Mouhous-Voyneau, N.; Denoeux, T. Combining statistical and expert evidence using belief functions: Application to centennial sea level estimation taking into account climate change. *Int. J. Approx. Reason.* **2014**, *55*, 341–354. [CrossRef]

154. Le Cozannet, G.; Manceau, J.C.; Rohmer, J. Bounding probabilistic sea-level projections within the framework of the possibility theory. *Environ. Res. Lett.* **2017**, *12*, 014012. [CrossRef]

155. Wong, T.E.; Bakker, A.M.; Keller, K. Impacts of Antarctic fast dynamics on sea-level projections and coastal flood defense. *arXiv*, 2016.

156. De Winter, R.; Reerink, T.J.; Slangen, A.B.; de Vries, H.; Edwards, T.; van de Wal, R.S. Impact of asymmetric uncertainties in ice sheet dynamics on regional sea level projections. *Nat. Hazards Earth Syst. Sci.* **2017**. [CrossRef]

157. Integrated Climate Data Center at the Hambourg University. Available online: http://icdc.cen.uni-hamburg.de/daten/ocean/ar5-slr.html (accessed on 5 May 2017).

158. Becker, M.; Meyssignac, B.; Letetrel, C.; Llovel, W.; Cazenave, A.; Delcroix, T. Sea level variations at tropical pacific islands since 1950. *Glob. Planet. Chang.* **2012**, *80–81*, 85–98. [CrossRef]

159. Meyssignac, B.; Becker, M.; Llovel, W.; Cazenave, A. An assessment of two-dimensional past sea level reconstructions over 1950–2009 based on tide-gauge data and different input sea level grids. *Surv. Geophys.* **2012**, *33*, 945–972. [CrossRef]

160. Melet, A.; Meyssignac, B.; Almar, R.; Le Cozannet, G. Underestimated wave contribution to sea level rise and changes at the coast. 2017, submitted.

161. Woodworth, P.L. A survey of recent changes in the main components of the ocean tide. *Cont. Shelf Res.* **2010**, *30*, 1680–1691. [CrossRef]

162. Pickering, M.D.; Horsburgh, K.J.; Blundell, J.R.; Hirschi, J.M.; Nicholls, R.J.; Verlaan, M.; Wells, N.C. The impact of future sea-level rise on the global tides. *Cont. Shelf Res.* **2017**, *142*, 50–68. [CrossRef]

163. Idier, D.; Paris, F.; Le Cozannet, G.; Boulahya, F.; Dumas, F. Sea-level rise impacts on the tides of the European Shelf. *Cont. Shelf Res.* **2017**, *137*, 56–71. [CrossRef]

164. Albert, S.; Leon, J.X.; Grinham, A.R.; Church, J.A.; Gibbes, B.R.; Woodroffe, C.D. Interactions between sea-level rise and wave exposure on reef island dynamics in the solomon islands. *Environ. Res. Lett.* **2016**, *11*, 054011. [CrossRef]

165. Vafeidis, A.T.; Nicholls, R.J.; McFadden, L.; Tol, R.S.J.; Hinkel, J.; Spencer, T.; Grashoff, P.S.; Boot, G.; Klein, R.J.T. A new global coastal database for impact and vulnerability analysis to sea-level rise. *J. Coast. Res.* **2008**, *24*, 917–924. [CrossRef]

166. Menendez, M.; Woodworth, P.L. Changes in extreme high water levels based on a quasi-global tide-gauge data set. *J. Geophys. Res.-Ocean.* **2010**, *115*. [CrossRef]

167. Holgate, S.J.; Matthews, A.; Woodworth, P.L.; Rickards, L.J.; Tamisiea, M.E.; Bradshaw, E.; Foden, P.R.; Gordon, K.M.; Jevrejeva, S.; Pugh, J. New data systems and products at the permanent service for mean sea level. *J. Coast. Res.* **2013**, *29*, 493–504. [CrossRef]

168. Hemer, M.A.; Fan, Y.L.; Mori, N.; Semedo, A.; Wang, X.L.L. Projected changes in wave climate from a multi-model ensemble. *Nat. Clim. Chang.* **2013**, *3*, 471–476. [CrossRef]

169. Emanuel, K.A. Downscaling cmip5 climate models shows increased tropical cyclone activity over the 21st century. *Proc. Natl. Acad. Sci. USA* **2013**, *110*, 12219–12224. [CrossRef] [PubMed]

170. Little, C.M.; Horton, R.M.; Kopp, R.E.; Oppenheimer, M.; Yip, S. Uncertainty in twenty-first-century CMIP5 sea level projections. *J. Clim.* **2015**, *28*, 838–852. [CrossRef]

171. Giorgi, F.; Jones, C.; Asrar, G.R. Addressing climate information needs at the regional level: The CORDEX framework. *World Meteorol. Organ. (WMO) Bull.* **2009**, *58*, 175. Available online: http://wcrp.ipsl.jussieu.fr/cordex/documents/CORDEX_giorgi_WMO.pdf (accessed on 10 October 2017).

172. Mitrovica, J.X.; Gomez, N.; Morrow, E.; Hay, C.; Latychev, K.; Tamisiea, M.E. On the robustness of predictions of sea level fingerprints. *Geophys. J. Int.* **2011**, *187*, 729–742. [CrossRef]

173. Tamisiea, M.E.; Mitrovica, J.X. The moving boundaries of sea level change understanding the origins of geographic variability. *Oceanography* **2011**, *24*, 24–39. [CrossRef]

174. Spada, G. Glacial isostatic adjustment and contemporary sea level rise: An overview. *Surv. Geophys.* **2017**, *38*, 153–185. [CrossRef]

175. Spada, G.; Bamber, J.L.; Hurkmans, R. The gravitationally consistent sea-level fingerprint of future terrestrial ice loss. *Geophys. Res. Lett.* **2013**, *40*, 482–486. [CrossRef]

176. Lemos, M.C.; Morehouse, B.J. The co-production of science and policy in integrated climate assessments. *Glob. Environ. Chang.-Hum. Policy Dimens.* **2005**, *15*, 57–68. [CrossRef]

177. McNie, E.C. Delivering climate services: Organizational strategies and approaches for producing useful climate-science information. *Weather Clim. Soc.* **2013**, *5*, 14–26. [CrossRef]

178. Weaver, C.P.; Mooney, S.; Allen, D.; Beller-Simms, N.; Fish, T.; Grambsch, A.E.; Hohenstein, W.; Jacobs, K.; Kenney, M.A.; Lane, M.A.; et al. From global change science to action with social sciences. *Nat. Clim. Chang.* **2014**, *4*, 656–659. [CrossRef]

179. Mimura, N.; Pulwarty, R.S.; Duc, D.M.; Elshinnawy, I.; Redsteer, M.H.; Huang, H.-Q.; Nkem, J.N.; Rodriguez, R.A.S. Adaptation planning and implementation. In *Climate Change 2014: Impacts, Adaptation, and Vulnerability*; Part A: Global and Sectoral Aspects; Contribution of Working Group II to the Fifth Assessment Report of the Intergovernmental Panel on Climate Change; Field, B.C., Barros, V.R., Dokken, D.J., Mach, K.J., Mastrandrea, M.D., Bilir, T.E., Chatterjee, M., Ebi, K.L., Estrada, Y.O., Genova, R.C., et al., Eds.; Cambridge University Press: Cambridge, UK; New York, NY, USA, 2014; pp. 869–898.

180. Bisaro, A.; Hinkel, J. Governance of social dilemmas in climate change adaptation. *Nat. Clim. Chang.* **2016**, *6*, 354–359. [CrossRef]

181. Dittrich, R.; Wreford, A.; Moran, D. A survey of decision-making approaches for climate change adaptation: Are robust methods the way forward? *Ecol. Econ.* **2016**, *122*, 79–89. [CrossRef]

182. Hinkel, J.; Bisaro, A. Methodological choices in solution-oriented adaptation research: A diagnostic framework. *Reg. Environ. Chang.* **2016**, *16*, 7–20. [CrossRef]

183. Wong, T.E.; Keller, K. Deep Uncertainty Surrounding Coastal Flood Risk Projections: A Case Study for New Orleans. *arXiv*, 2017.

184. Robinson, A.E.; Ogunyoye, F.; Sayers, P.; van den Brink, T.; Tarrant, O. Accounting for Residual Uncertainty: Updating the Freeboard Guide. Report—SC120014, UK Environmental Agency, Flood and Coastal Erosion Risk Management Research and Development Programme, 2017. Available online: https://www.gov.uk/government/uploads/system/uploads/attachment_data/file/595618/Accounting_for_residual_uncertainty___an_update_to_the_fluvial_freeboard_guide_-_report.pdf (accessed on 6 May 2017).

185. Norton, J. An introduction to sensitivity assessment of simulation models. *Environ. Model. Softw.* **2015**, *69*, 166–174. [CrossRef]

186. Saltelli, A.; Annoni, P. How to avoid a perfunctory sensitivity analysis. *Environ. Model. Softw.* **2010**, *25*, 1508–1517. [CrossRef]

187. Plant, N.G.; Thieler, E.R.; Passeri, D.L. Coupling centennial-scale shoreline change to sea-level rise and coastal morphology in the gulf of mexico using a bayesian network. *Earths Future.* **2016**, *4*, 143–158. [CrossRef]

188. Recommendations of the Task Force for Financial Disclosure of Risk. Available online: http://www.fsb.org/2017/06/recommendations-of-the-task-force-on-climate-related-financial-disclosures-2/ (accessed on 31 August 2017).

189. United Nations Environment Programme (UNEP). *The Adaptation Finance Gap Report 2016*; United Nations Environment Programme (UNEP): Nairobi, Kenya, 2016.

Journal of
*Marine Science
and Engineering*

MDPI

Article

Second-Pass Assessment of Potential Exposure to Shoreline Change in New South Wales, Australia, Using a Sediment Compartments Framework

Michael A. Kinsela [1,2,*], Bradley D. Morris [1], Michelle Linklater [1] and David J. Hanslow [1]

[1] Water, Wetlands & Coasts Science, Office of Environment & Heritage, NSW Government, 59 Goulburn Street, Sydney, NSW 2000, Australia; Bradley.Morris@environment.nsw.gov.au (B.D.M.); Michelle.Linklater@environment.nsw.gov.au (M.L.); David.Hanslow@environment.nsw.gov.au (D.J.H.)
[2] School of Geosciences, Faculty of Science, The University of Sydney, Sydney, NSW 2006, Australia
* Correspondence: Michael.Kinsela@environment.nsw.gov.au; Tel.: +61-2-9995-5661

Received: 30 August 2017; Accepted: 7 December 2017; Published: 20 December 2017

Abstract: The impacts of coastal erosion are expected to increase through the present century, and beyond, as accelerating global mean sea-level rise begins to enhance or dominate local shoreline dynamics. In many cases, beach (and shoreline) response to sea-level rise will not be limited to passive inundation, but may be amplified or moderated by sediment redistribution between the beach and the broader coastal sedimentary system. We describe a simple and scalable approach for estimating the potential for beach erosion and shoreline change on wave-dominated sandy beaches, using a coastal sediment compartments framework to parameterise the geomorphology and connectivity of sediment-sharing coastal systems. We apply the approach at regional and local scales in order to demonstrate the sensitivity of forecasts to the available data. The regional-scale application estimates potential present and future asset exposure to coastal erosion in New South Wales, Australia. The assessment suggests that shoreline recession due to sea-level rise could drive a steep increase in the number and distribution of asset exposure in the present century. The local-scale example demonstrates the potential sensitivity of erosion impacts to the distinctive coastal geomorphology of individual compartments. Our findings highlight that the benefits of applying a coastal sediment compartments framework increase with the coverage and detail of geomorphic data that is available to parameterise sediment-sharing systems and sediment budget principles. Such data is crucial to reducing uncertainty in forecasts by understanding the potential response of key sediment sources and sinks (e.g., the shoreface, estuaries) to sea-level rise in different settings.

Keywords: climate change; coastal barrier; coastal sediment compartment; geomorphology; littoral sediment cell; risk management; sea-level rise; sediment budget; shoreline change; uncertainty

1. Introduction

Beach erosion is a natural process often caused by high waves (and temporarily raised coastal sea levels) during storms that drive the rapid (hours to days) transfer of large volumes of sand from the sub-aerial beach and dunes to the adjacent surf zone and shoreface. Using aerial LiDAR surveys, Harley et al. [1] found that 11.5 million m^3 of sand was temporarily lost to the sea from 177 km of shores in southeast Australia during a single storm in June 2016, with an average of 65 m^3 (and maximum of 228 m^3) being lost from each metre of beach alongshore. While the loss of sand offshore is usually temporary, full beach recovery to the pre-storm state may take several months to several years, depending on the post-storm wave climate and frequency of storms [2–4]. Future change in modal and storm wave climates, due to climate change, may drive changes in shoreline orientation and the range of shoreline variability, as the distinctive morphology of different beaches adjusts to new hydrodynamic conditions [5–7].

Severe beach erosion can generate significant socio-economic and environmental impacts for coastal communities, relating to the damage or loss of: assets (properties, infrastructure, utilities, and public facilities), recreational and commercial beach amenity (including tourism), coastal habitats, and ecosystem services [8–11]. The increasing cost of these impacts has prompted a growing research focus on coupled physical-economic modelling of coastal systems, to examine the value and equity of potential solutions [12–15]. Areas that are subject to beach erosion impacts at present may require careful management and coastal engineering solutions, while future impacts can be minimised by locating new development beyond the future reach of erosion, and by maintaining naturally resilient beaches with sufficient sand supplies to accommodate erosion and recovery cycles.

Global sea-level rise through the present century, and beyond, is expected to drive mean-trend shoreline recession, where the underlying sediment supply to beaches is insufficient to oppose the passive and morphodynamic influences of sea-level rise on shoreline migration [16,17]. In many settings, the rate of shoreline recession will not be simply the rate of passive inundation, but may be enhanced by the cumulative loss of sediment from the beach and dunes, as the beach morphology responds to new boundary conditions [18]. In that case, sand may be progressively lost from the beach and dunes to other depositional environments of the coastal system (e.g., tidal inlets, estuaries, the shoreface), or alongshore. Shoreline recession exposes land (and assets) that has been historically protected from coastal hazards by natural dune buffers to the impacts of beach erosion.

While process-based modelling suggests that global mean sea level may rise by around 0.44 m (RCP2.6 median value) to 0.74 m (RCP8.5 median value) by the end of this century [19], other evidence suggests the potential for global sea-level rise up to 2.5 m by 2100 [20], due to accelerated ice loss from glaciers and ice sheets that is not captured in process-based models. Uncertainty regarding the impact of sea-level rise on global shorelines during the present century and beyond, due to the wide range of sea-level projections, and local controls on sediment redistribution (and thus shoreline response), compels the development of methods for shoreline forecasting that capture the complexity of coastal depositional systems, and communicate the spectrum of potential impacts.

One approach to understanding (and predicting) shoreline change that is caused by a sediment budget imbalance at the beach (whether positive or negative), as may be imposed by sea-level rise, is to model the redistribution of sediments within the coastal depositional system. The approach is founded on the premise that beaches are elements of broader sediment-sharing coastal systems (or *coastal tracts*), which include the key depositional environments of coastal barrier systems (i.e., rivers, estuaries, tidal inlets, dunes, beach, shoreface, continental shelf) [21]. Assuming the perspective of the beach, these depositional environments meet at the littoral sediment transport system, which also connects the coastal tract to adjacent depositional systems (and their beaches) alongshore. Sediment budget principles [22,23] can be applied to map and model sediment exchanges between the various depositional environments, and the resulting shoreline change [24,25].

Coastal sediment compartments (and littoral sediment cells) are spatial tools for understanding (and quantifying) sediment connectivity within and between sediment-sharing coastal systems. Sediment compartments identify more or less contained systems, and are usually based on the broad-scale structure of the coastline and prominent features that impede alongshore transport [26]. Littoral cells are usually finer in scale and identify sectors of uniform alongshore sediment transport, which are separated by convergence and divergence points. Sediment compartments and littoral cells have been mapped in many jurisdictions, including the United States (US) [27], United Kingdom (UK) [28,29], and Australia [30–32], to ensure coastal management and planning initiatives reflect that local beach dynamics are influenced by sediment exchanges with other depositional environments of the sediment-sharing system. They have been applied as qualitative templates simply to identify beaches that are connected by sediment transport, or quantitatively, to parameterise and model sediment budgets and shoreline change. Sediment compartment mapping was recently completed for the entire Australian coastline to inform the national assessment of coastal hazard impacts and risks [33,34]. The benefits of a sediment compartments approach are most fully realised where the framework can be applied to parameterise and model sediment budgets and shoreline change within a sediment-sharing system.

We describe a scalable method for using sediment compartments to assess potential exposure to shoreline change on embayed wave-dominated beaches in New South Wales (NSW), Australia. The sediment compartments framework is used to parameterise coastal morphology and quantify sediment exchanges between sources and sinks. A simple volumetric shoreline encroachment model is applied within the framework to estimate the impact of modelled sediment redistribution on shoreline change. We use a Monte Carlo sampling regime to estimate and communicate uncertainty in shoreline change forecasts statistically, which is necessitated by the intrinsic uncertainty around: environmental change (e.g., sea-level rise), sediment transport processes, and the shoreline model. We demonstrate the approach at a regional scale to develop a second-pass assessment of exposure to shoreline change along the NSW coast this century, and at a local scale to demonstrate the flexibility of the method in capturing the sensitivity of shoreline response to fine-scale geomorphic variability.

The method that we demonstrate is intended to provide a second-pass assessment of potential exposure to coastal erosion and shoreline change. While the approach accounts for the distinctive geomorphology of individual sediment compartments and beaches, we adopt several simplifying assumptions to account for poorly understood or documented processes, with the goal to apply a consistent shoreline response model to all of the NSW beaches. The suitability of those assumptions for individual beaches should be evaluated in any local-scale assessment of shoreline change.

2. Regional Setting

We consider present and future exposure to beach erosion and shoreline change along the coastline of New South Wales (NSW) in southeast Australia. The oceanic coast is roughly 2065 km in length, including 1038 km of sandy shorelines, which are divided into 721 beaches that are primarily influenced by open-coast processes [35,36]. Most beaches are backed by readily erodible Quaternary (Holocene or late Pleistocene) beach and dune deposits [37]. In some locations, emergent bedrock or engineered structures (e.g., sea walls) may restrict the potential extent of beach erosion, or weakly cemented sediments (e.g., indurated sand) may impede the rate of shoreline change. About 45% of all sandy shorelines occur within or are immediately backed by National Park reserves, where infrastructure and development are restricted to minimal recreational amenities, and beach systems generally occur in natural states. Around 15% (150 km) of sandy shorelines along the NSW coast are located within 110 m of existing property lot boundaries.

2.1. Coastal Geomorphology and Processes

The NSW coast is situated on the tectonically passive southeast Australian continental margin (Figure 1A). The geometry of the continental shelf is relatively deep and narrow (30–50 km wide) when compared with classic passive margins (e.g., US Atlantic), meaning that wave energy from the moderate- to high-energy wave climate experiences minimal attenuation before arriving at the coast [38]. Swell and storm wave directions are typically south to south-east [39], driving a northward littoral sediment transport system, which is most effective along the northern third of the coast where the coastal morphology supports relatively continuous alongshore transport [40,41]. The coast is periodically impacted by extratropical cyclones that intensify in the Tasman Sea [42], generating high waves and temporarily raised sea levels (storm surges) that can cause severe beach erosion [1].

The coast features embayed sandy beaches of varying lengths, which are separated alongshore by rocky headlands and cliff coasts that trace the ancient geological framework of the coastline [43]. The northern third of the coast features broader and shallower embayments, low hinterland and broad coastal plains, large rivers, and a more gently sloping inner-continental shelf (Figure 1B). In contrast, the central and southern coasts feature smaller embayments, rugged hinterland and narrow coastal plains, coastal lakes and lagoons with small rivers or streams, and a steeper and narrower inner shelf (Figure 1C). At a finer scale, compartment dimensions and geology, sediment type and availability, and exposure to regional wave climates, all control beach morphodynamics in this wave-dominated and microtidal (approximately 2 m spring tidal range) setting [35,36].

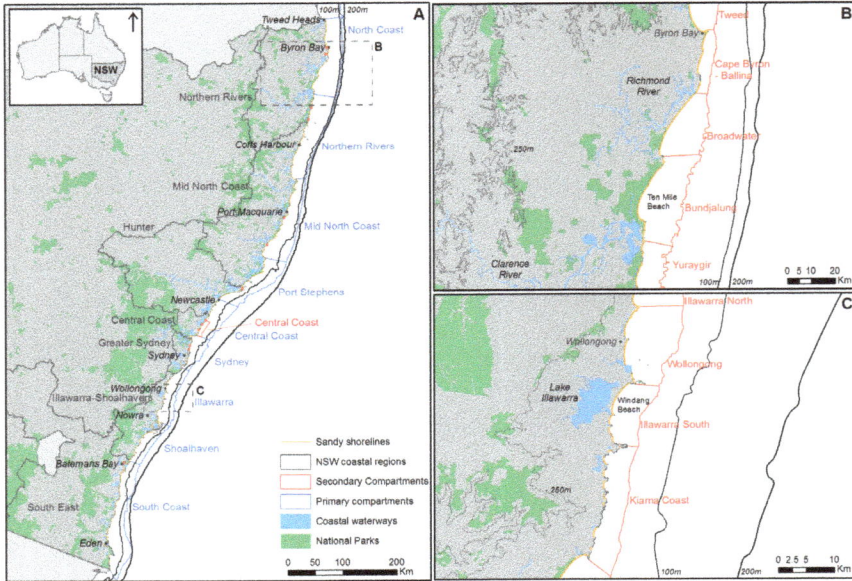

Figure 1. (**A**) Location and orientation of the New South Wales (NSW) coastline, showing: the distribution of sandy shorelines; the variable geometry of the continental shelf (indicated by 100 and 200 m isobaths); the nine primary coastal sediment compartments (including names) [33,34]; and, the seven NSW coastal regions considered in the exposure assessment. The small red dots indicate identified erosion hot spots [44]. Representative secondary sediment compartments [33,34] from the (**B**) North Coast and (**C**) Illawarra primary compartments are also shown. The 250-m elevation contour shown in (**B**) and (**C**) compares coastal relief between the North Coast and Illawarra regions. The location of the Central Coast secondary sediment compartment (Figure 3A) is shown in (**A**).

Sediment cover across the inner shelf is relatively thin, except where relict coastal barriers and shelf sand bodies accumulated during lower sea levels, prior to erosional reworking during the late Holocene post-glacial marine transgression and subsequent sea-level highstand [45,46]. A marine abrasion surface extends along 300 km of the central to southern coast, frequently outcropping across the submerged inner-shelf seabed [46]. The abrasion surface is related to the prominent coastal escarpment that is evident on steeper sectors of the coast, and is thought to have formed by cyclic erosion and planation of the Palaeozoic to Mesozoic bedrock by coastal processes, as the shoreline migrated in and out across the margin during late Quaternary sea level fluctuations [46]. The extent of the marine abrasion surface captured in detailed seabed mapping off the Sydney coastline [47] suggests that many central and southern compartments may be sediment deficient offshore.

Most NSW beaches are elements of coastal sand barriers, which are composite sand bodies that comprise shoreface, beach, dune, and estuarine deposits [48]. The sand barriers are late Pleistocene to Holocene in age, and separate the ocean beaches from back-barrier water bodies (rivers, coastal lakes, and lagoons) and the coastal hinterland [43]. The sub-aerial surfaces of the sand barriers include relatively low-lying coastal plains of varying extents, depending on the compartment dimensions and barrier type [49]. Development is often concentrated on the coastal plains due to their flat and regular morphology. Unlike low-gradient passive margins (e.g., US Atlantic and Gulf coasts), which feature narrow, low-lying barrier islands that are dominated by barrier-bypassing processes (e.g., washover and tidal inlet migration), the bay barriers of NSW are typically higher, wider, and more stable in comparison, and mostly feature well-developed dune morphology formed over hundreds to thousands of years by south easterly wave and wind climates [48,49]. As such, dunes are most developed in the northern ends of embayments, and sediment exchanges between beaches and back-barrier environments are usually via stable tidal inlets (not washover).

2.2. Exposure to Coastal Erosion

Properties and infrastructure along several NSW beaches have been damaged or destroyed by erosion in the past [50–52]. Beach erosion is usually caused by extratropical cyclones (ETCs)—locally termed East Coast Cyclone (ECC) or East Coast Low (ECL) storms. The storms originate from various locations on or adjacent to the Australian continent, depending on the synoptic pattern, and intensify over the NSW coast or in the Tasman Sea, generating high waves (usually with ESE to SE directions) and elevated coastal sea levels [7,42]. An unusual storm featuring a coupled ETC and anticyclonic intensification impacted the entire southeast Australian coast in June 2016, generating high waves with ENE to E directions, and causing severe beach erosion along the NSW coast. The storm coincided with a spring high tide, and the unusual easterly storm-wave direction resulted in minimal wave transformation prior to entering coastal embayments and impacting beaches [1,53]. Considerable damage to properties and infrastructure occurred at several NSW beaches (e.g., Figure 2).

Figure 2. Wamberal Beach, in the Central Coast secondary compartment (Figure 1A), before (**A**), and after (**B–D**), severe erosion caused by a storm during June 2016 that featured an anomalous easterly storm-wave direction. The beach is underlain by a cemented siltstone unit (**D**) that outcrops on the beach face and in the frontal dune along the central part of the beach during severe erosion events.

Because many beaches along the northern NSW coast are connected via a northward-directed littoral sand transport system [40], temporary or persisting divergences in transport rates can also drive periodic beach erosion or ongoing shoreline recession. Periodic erosion may occur where wave climate variability supports only intermittent sediment bypassing of prominent headlands [41,54]. On the other hand, persistent and ongoing shoreline recession may result from a long-term sediment budget imbalance within a coastal sediment compartment. For example, Ten Mile Beach (Figure 1B) is part of a receding sand barrier, where long-term shoreline recession associated with net northward sand transport has exposed indurated sand (coffee rock) along the beach face [55].

Fifteen coastal erosion hot spots [44] have been identified along the NSW coast, where multiple properties are presently threatened by erosion (Figure 1A). However, the total present and future exposure to coastal erosion along the NSW coast is relatively poorly known. A first-pass national assessment of climate change risks to Australia's coasts [56] estimated that 3600 residential buildings in NSW might be at risk from coastal erosion within the present century (i.e., are located within 110 m of erodible shorelines), including 700 buildings that may be presently at risk (i.e., are located within 55 m of erodible shorelines). However, the generalised proximity analysis method that is used, in which all of the properties within fixed distances of erodible sandy shorelines were considered as being potentially exposed, implies that the estimated exposure is only a first approximation.

Coastal erosion hazard zones are defined by local governments in NSW for management and planning purposes, and provide another estimate of potential exposure to coastal erosion impacts at present and in the future. Present-day (immediate) and future (e.g., 2050, 2100) hazard zones are usually defined, accounting for the impacts of storms, as well as historical shoreline trends and projected sea-level rise. However, the existing coverage of erosion hazard zones is incomplete, and the erosion components considered and analysis and modelling methods that are used vary between the 26 local government areas along the open coast of NSW [57,58].

3. Materials and Methods

3.1. Coastal Sediment Compartments Framework

Our approach uses a coastal sediment compartments framework (a hierarchy of compartments and sub-compartments), to parameterise the depositional environments of coastal tract systems [21]. The hierarchy is used to conceptualise and map the cross-shore and alongshore extents of sediment sharing between sources and sinks, for timescales relevant to coastal management and planning (years to centuries) [34]. The framework reflects the natural hierarchy of coastal depositional systems [21], and allows for the model parameterisation to reflect the spatial and temporal scales of the forecast, and the resolution of available geomorphology and process data. The compartments framework is the basis for developing aggregated morphological data models that capture the surface dimensions and relief of depositional landforms within the sediment-sharing systems [24].

The Australian coastline was recently divided into 100 primary sediment compartments that capture the limits of sediment sharing at long timescales (centuries to millennia), as defined by the geological framework of the coast and large coastal landforms [33]. Within the primary sediment compartments, 359 secondary compartments have been identified, representing sediment sharing at intermediate (decadal to centennial) timescales [33,34]. The NSW coast includes 9 primary and 47 secondary sediment compartments (Figure 1A).

As the broad scale of the secondary compartments often includes several embayed beaches (Figure 3A), tertiary compartments, and sub-compartments are often mapped to represent coastal landforms at finer spatial scales, commensurate with sediment sharing at short to intermediate timescales (years to decades). For example, sub-compartment mapping has been carried out in West Australia using high-resolution marine LiDAR data [59], and for the Illawarra-Shoalhaven region of NSW (Figure 1A) using nearshore seabed data and sediment transport modelling [60].

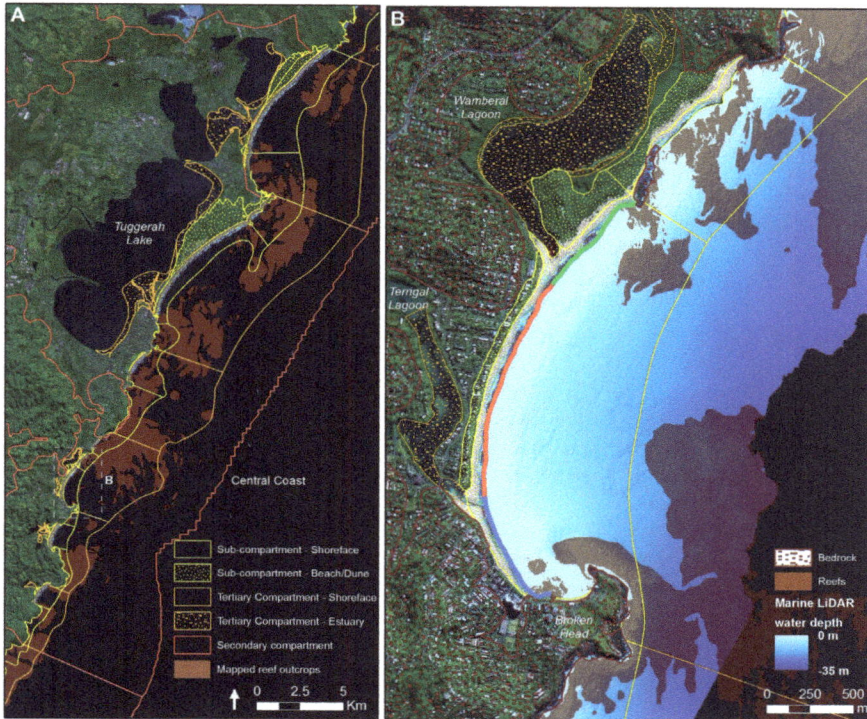

Figure 3. (**A**) Example of sub-compartment mapping for the Central Coast secondary compartment (Figure 1A). Secondary compartments extend to 50 m water depth and capture the Quaternary depositional systems onshore. Tertiary compartments extend to 40 m water depth and include coastal barrier and estuarine flood-tide delta deposits onshore. Sub-compartments extend to 20 m water depth and include beach and dune deposits onshore. (**B**) The Terrigal-Wamberal sub-compartment includes Terrigal (blue), Wamberal (red) and Wamberal North (green) beaches. Marine LiDAR data reveals the distribution of sediments and reef outcrops (shaded brown) across the shoreface seabed.

Within the NSW secondary sediment compartments, we identified 137 tertiary compartments (inter-annual to decadal timescales), most of which contain multiple sub-compartments (annual to inter-annual timescales), to facilitate the application of our shoreline change model to all NSW beaches. In the absence of high-resolution bathymetry and seabed substrate data for the whole NSW coast, our delineation of tertiary compartments and sub-compartments was based on the:

(1) dimensions and average orientation of coastal sectors and embayments;

(2) prominence and alongshore extent of coastal headlands, cliffs and visible nearshore reefs;

(3) extent of tidal inlets and training walls (where present); and,

(4) shoreface geometry depicted in regional-scale bathymetry data.

Figure 3 provides an example of tertiary compartment and sub-compartment mapping for the Central Coast secondary sediment compartment located between Sydney and Newcastle (Figure 1A). The mapping covers both the onshore and offshore depositional environments of the system.

3.2. Simple Shoreline Encroachment Model

We use a simple shoreline encroachment model that scales in complexity with the detail of data available to inform the modelling approach [61]. Encroachment refers to cumulative erosion into a pre-existing beach-dune system, in which sediment is progressively lost from the beach to the littoral zone, where it may be transferred offshore, alongshore, or into tidal inlets [62]. This is different from shoreline migration by barrier rollover, in which barrier-bypassing processes (washover and tidal-inlet sequestration) dominate, transferring sand from the beach face and frontal dune to back-barrier depositional environments [63]. Our model does not support continuous (rollover) or discontinuous (drowning, overstepping) dynamic barrier behaviours [64,65], and is therefore only applicable to simulating shoreline recession on relatively steep, moderate to high energy wave-dominated coasts, where the existing beach and dune morphology is well-developed. The model is not suitable for simulating shoreline change on low-lying barrier island coasts, where dynamic barrier behaviour controls shoreline change [64–66].

The encroachment response is likely to characterise the initial (i.e., present century) response of most NSW beaches to sea-level rise, as dune morphology is typically well developed and continuous alongshore, relative to the predominant wave energy conditions. However, the assumption may be violated where barrier-bypassing processes become dominant as sea level rises—i.e., where existing dune morphology is overwhelmed by combined raised sea level and elevated wave run-up heights. The assumptions and limitations of the simple shoreline encroachment model are considered by the authors to be suitable for the purposes of a second-pass shoreline change exposure assessment, but should be evaluated on a case by case basis for any finer-scale applications.

3.2.1. Volumetric Beach Response

Beach response to changing boundary conditions (e.g., storm wave conditions, sediment supply, sea-level rise) is quantified as the time-averaged sediment-flux change in the littoral transport system, which connects the beach with sources and sinks, both proximal (e.g., surf zone, tidal inlet) and distal (e.g., shoreface, flood-tide delta, up-drift river, down-drift beach). A change in the sediment balance of the littoral transport system is reflected at the beach (in erosion or accretion), and is estimated by solving the sum of the sediment volume redistribution, relative to the initial (present-day) beach volume. For example, a deficit in response to the generation of new sediment *accommodation* across sediment sinks may contribute to shoreline instability and encroachment, but may be offset or moderated by sediment supply from sources. We use the coastal sediment compartments framework (Section 3.1) to identify and (as far as is possible) quantify the sources and sinks, and sediment exchanges between the depositional environments of sediment-sharing coastal systems.

We parameterise sediment redistribution within each compartment and beach response (and thus shoreline change) volumetrically. This means that the predicted shoreline change can closely reflect between-site and alongshore variability in beach and dune morphology, which is now captured in high resolution, and at large spatial scales using remote sensing techniques (e.g., LiDAR).

The method differs from the simple Bruun model [67–69], which is based on idealised beach-dune morphology that is maintained as sea level rises. That is, because the Bruun profile is measured from the dune crest, as the shoreline recedes into the dune system by means of cumulative erosion, the dunes are presumed to aggrade at the same pace as sea level rises. This implies that rates of sand supply and dune aggradation are sufficient to maintain vertical dune growth at the same pace as sea-level rise. Otherwise, the relative dune crest height would decrease and the Bruun profile would flatten as sea level rises. However, we observe that the beach response to sediment deficit conditions on contemporary receding beaches in NSW reflects encroachment, in which cyclic dune scarping and destabilisation manifests as progressive erosion into the relict dune system, with minimal or no dune growth. The prospect of a rapid acceleration in sea-level rise in the present century adds further uncertainty to an assumption of instantaneous dune aggradation during encroachment in this setting.

Adopting a risk-averse approach, our model assumes that the dunes on NSW beaches will not aggrade at the same pace as projected sea-level rise within this century.

The encroachment model calculates the sediment volume (V) lost from the initial beach and dunes at the end of the forecast period, in cubic metres per metre of beach shoreline (m^3/m), due to fluctuating (F) and cumulative (C) erosion processes:

$$V = c_f(F) + c_c(C) \qquad (1)$$

The functions describing F and C may vary in complexity depending on available data and knowledge. We define F using a gamma probability function that approximates the potential range and relative likelihoods of fluctuating erosion on fully exposed open-coast NSW beaches:

$$F \sim \Gamma(k, \theta) \equiv Gamma(v, 3, 30.5) \text{ for } 1 \leq v \leq 350 \qquad (2)$$

The form of F may include multiple components of fluctuating erosion (e.g., erosion due to storms, periodic shoreline rotation) if sufficient data is available. The rationale for the form (i.e., shape, range, and tail) of the gamma distribution in Equation (2) is described in Section 3.4.1. The coefficient c_f (Equation (1)) is used to scale F between beaches, or along a beach (Figure 4). It is used in the regional-scale application (Section 3.4.1) in order to account for variation in exposure to the wave climate between beaches or beach sectors, and in the local-scale application (Section 3.4.2) to account for alongshore variation the resistance of the beach-dune substrate to erosion. Variable substrate resistance may occur, for example, where bedrock or cemented sediments outcrop intermittently along the beach-dune system. Both influences may limit the fluctuating erosion volume along parts of a beach.

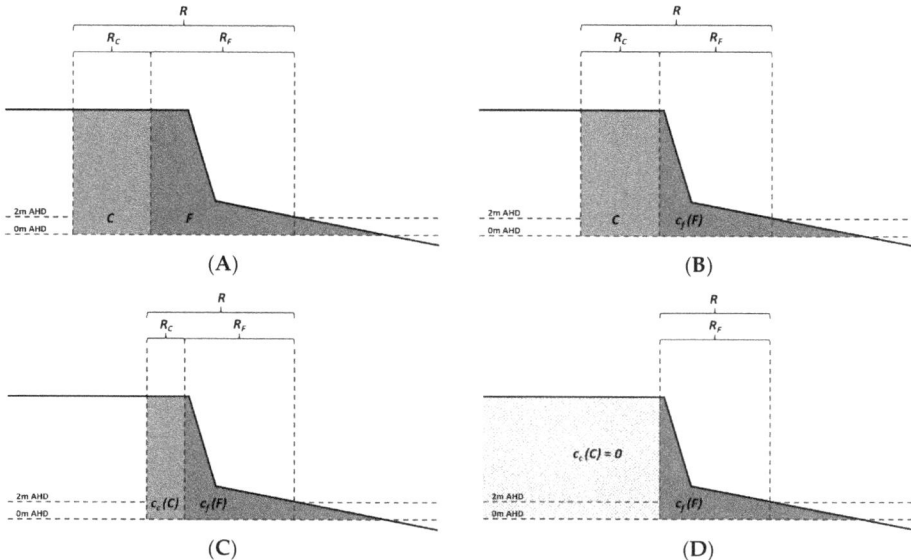

Figure 4. Illustration of the influence of the c_f and c_c scaling coefficients (Equation (1)) on modelled fluctuating (F) and cumulative (C) beach erosion, and the corresponding influence on the shoreline recession distance (R). The parameterisations are (**A**) $c_f = 1$, $c_c = 1$, (**B**) $c_f < 1$, $c_c = 1$, (**C**) $c_f < 1$, $c_c < 1$, and (**D**) $c_f < 1$, $c_c = 0$. Note that erosion is calculated above mean sea level (0 m AHD) and landward from the beach berm position, which is estimated here by the 2-m AHD elevation contour.

Cumulative erosion (C) reflects a long-term imbalance in the littoral sediment budget at the beach, which is driven by persisting sediment redistribution between sources and sinks, in response to ongoing or past change in boundary conditions (e.g., sea-level rise, altered wave climate). We define C as the net sum of interactions between the potential sources and sinks for NSW beaches:

$$C = (q_x + q_y)t + c_s(V_S) + c_e(V_E) + \left(\frac{V_A + V_O + V_M + V_B + V_R}{l} \right) t \tag{3}$$

where t is the forecast period (in years) and l is the total length (m) of sandy shorelines within the tertiary compartment containing the beach system to which the model is applied (Section 3.1).

Where data exists describing mean-trend change in the beach volume due to cross-shore (x) or alongshore (y) sediment transport processes (e.g., photogrammetry analysis, beach survey record, geohistorical data), the q_x and q_y parameters ((m^3/m)/year) may be used to describe underlying change in the littoral sediment budget. The V_S (Equation (5)) and V_E (Equation (6)) variables (m^3/m) capture the response of the shoreface and estuarine flood-tide delta sinks to sea-level rise, respectively, over the forecast period (t). The V_A, V_O, V_M, V_B, and V_R variables (m^3/year) represent annual sediment losses (or gains) within the relevant tertiary compartment, due to: aeolian processes, barrier washover, mega-rips, biogenic sediment production, and river supply, respectively. While such processes may not occur uniformly along the length of sandy shorelines (l), their impacts on shoreline change are distributed alongshore throughout the tertiary compartments by the littoral transport system.

The coefficient c_c (Equation (1)) scales C between or along beaches, which may be desired where complete or partial substrate resistance is anticipated to stop or slow the rate of shoreline recession. For example, well-cemented and alongshore-continuous indurated sands throughout the dune system may slow the overall rate of shoreline recession. Figure 4 illustrates the influence of different parameterisations of c_f and c_c on modelled fluctuating (F) and cumulative (C) beach volume change, and the corresponding effect on the respective components of shoreline change (R).

The beach response to sea-level rise considers the influence of two potential sediment sinks on the beach system: (1) the shoreface adjacent to sandy shorelines; and, (2) the flood-tide (marine origin) deltas of estuaries within each tertiary compartment. Redistribution of sand from the beach to these sinks, to maintain surface morphology in balance with the prevailing geomorphic and hydrodynamic controls under rising sea level, may become a long-term driver of shoreline recession. As sea-level rise exposes dunes to increased erosion during storms, some of the sand that was previously only temporarily lost to the littoral transport system, may be permanently lost to new sediment accommodation that is generated by sea-level rise, rather than returning to the beach during the recovery phase following an erosion event. Successive instances of beach erosion, followed by only partial recovery may contribute to cumulative shoreline recession.

The shoreface component of the response to sea-level rise (V_S) includes the potential sediment accommodation generated across the shoreface by rising sea level. In the modelled response, we assume that sea-level rise drives an upward and landward translation of the shoreface. The potential shoreface sediment accommodation space is estimated using a volumetric implementation of the standard concept of erosion that is caused by sea-level rise, as proposed by Bruun [67–69]. The significance of the modelled shoreface response to sea-level rise depends on the dimensions and geometry of the shoreface, the sampled sea-level rise (S), and the sampled closure depth (h_c), which determines the offshore extent of sediment accommodation and morphologic response across the shoreface.

The geometry of the shoreface is approximated by fitting the following power function through the available shoreface hydrographic data,

$$h(x) = Ax^m \tag{4}$$

where A and m are tuning parameters that are freely determined using linear regression fitting. The sediment accommodation volume generated between the baseline shoreline (x_0), and the offshore position of the sampled shoreface closure depth (x_c), is calculated as the difference between the initial shoreface geometry and the response shoreface geometry, following sea-level rise:

$$V_S = 0.5 \times \left[\int_{x_0}^{x_c} (h(x) + S) dx - \int_{x_0}^{x_c} h(x) dx \right] \tag{5}$$

The potential shoreface sediment accommodation volume is halved to reflect the concept of upward and landward profile translation in response to sea-level rise, as in Bruun's standard model of erosion caused by sea-level rise [67–69]. That is, the shoreface surface does not simply aggrade, the shoreface profile also translates landward with the retreating shoreline.

The scaling coefficient c_s (Equation (3)) allows for the modelled shoreface response to be limited to only the sedimentary portion of a mixed sediment-reef shoreface. Exposed reef outcrops that protrude above an otherwise sedimentary shoreface surface suggest that sediment cannot accumulate there under the prevailing energy conditions. To account for the negative accommodation profile of reef outcrops, we assume that they do not represent potential sediment accommodation. This assumption is not suitable for application in sediment-deficit compartments, where extensive low-profile reef may be exposed, simply due to a lack of sediment in the shoreface environment.

The third term in Equation (3) (V_E) represents the estuarine flood-tide delta component of the response to sea-level rise. This considers the influence of potential sediment accommodation that is generated within tidal inlets and estuaries by rising sea level, which may support the vertical growth (aggradation) of delta deposits with sand sequestered from the adjacent beach and shoreface. The sediment loss from the beach is the product of the sea-level rise (S), and the surface area of submerged active flood-tide delta deposits (A_D). The total response over the forecast period may be moderated using the c_e coefficient (Equation (3)), to simulate a slower (not instantaneous) rate of delta response:

$$V_E = \frac{A_D \times S}{l} \tag{6}$$

We assume the existence of a morphodynamic balance between the hydrodynamic conditions of tidal inlets and delta morphology [70–73], which is supposed to have developed during prolonged sea-level stability that has been experienced in NSW during the late Holocene [45]. The likely rates of flood-tide delta response to sea-level rise are uncertain and may be site specific [72,73]. The potential sand loss to estuaries is distributed along the length (l) of sandy shorelines within the compartment.

3.2.2. Calculating Shoreline Change

We use the Australian beach database [35,36] as a framework for parameterising morphology and response variables for individual NSW beaches, or beach sectors as defined along some longer beaches. Airborne LiDAR topography data is used to calculate shoreline encroachment distances based on the modelled beach volume change. For example, airborne LiDAR covering the entire NSW coast captured between 2009–2014 [74] was used in the region-scale application. Depending on the resolution of the application, the beach face and dune morphology may be characterised by an alongshore-averaged (aggregated) beach-dune terrain profile [21,24] for each beach or beach sector (Section 3.4.1), or by a series of beach-dune profiles that are regularly spaced along the shoreline (Section 3.4.2). In either case, the encroachment distance is calculated by applying the modelled beach change volume (above 0 m AHD) across the beach-dune profile, moving landward from the baseline shoreline (Figure 5). To achieve consistency in defining the present-day baseline at different beaches, we use the 2 m AHD elevation contour as derived from LiDAR topography, which approximates the modal run-up or berm position on NSW beaches.

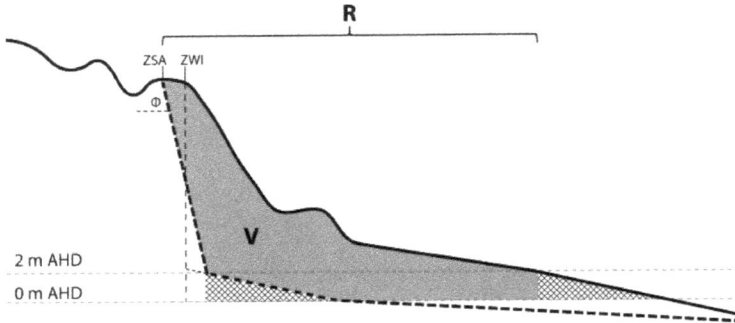

Figure 5. The modelled net sediment volume change (V, Equation (1)) is applied to high-resolution beach-dune terrain profiles to calculate the shoreline change distance (R). The sediment volume is applied above 0 m AHD (present mean sea level), and landward of the 2-m AHD elevation contour baseline (beach berm height on NSW beaches), with an allowance included for dune slumping [75].

To account for dune slumping following encroachment into pre-existing dune morphology, we apply a commonly used allowance that is proportionate to the crest level of the dune erosion escarpment and the angle of repose of the substrate material [75]. Once the net encroachment distance, or the zone of wave impact (ZWI) has been determined using the cumulative volume of the beach-dune profile, the slumping allowance, or zone of slope adjustment (ZSA), is calculated based on the sampled dune height at that location, and the allowance is then added to determine the total encroachment distance (Figure 5). The angle of repose may be a discrete value, or a range of values (using a probability function), depending on confidence in the composition and behaviour of the substrate material. We assume a conservative angle of repose (Φ) for unconsolidated sand of 30°.

3.3. Uncertainty Management

Uncertainty is unavoidable in shoreline change forecasts due to: the potential range of future forcing conditions, the incomplete knowledge-base about beach and shoreline dynamics, and the intrinsic limitations of beach and shoreline response models. Forecasts should be communicated in the context of the uncertainty space to support informed and transparent decision making.

In reference to Earth-surface models, Murray et al. [76] argue that, uncertainty quantification techniques are most suited to simulation models, rather than exploratory models, although most models fall somewhere between clearly defined end members of the two. For process-based simulation models that are used to predict short-term beach response to storms, model boundary conditions (e.g., waves and water levels) are relatively well known to high spatial and temporal resolutions, and so statistical techniques to manage the not insignificant uncertainty that is introduced by the selected parameterisation of complex models are of particular importance [77]. For behaviour-based models that are typically used to predict long-term shoreline change, similar techniques have been demonstrated to manage uncertainty in boundary conditions and model parameterisation [78–81]. Awareness and demand for uncertainty management in beach erosion and shoreline change predictions is growing within the coastal management community [58,81].

While our modelling approach is intended to provide quantitatively reasonable estimates of the potential for shoreline change using simple sediment budget principles, it nonetheless includes many basic assumptions about poorly understood phenomena, such as the nature of the shoreface as a sediment source or sink during sea-level change. When considering the spatial extent and required resolution of forecasts in our regional-scale application, our approach to uncertainty management is designed to support rapid simulation, generating a probability distribution of shoreline change predictions based on 10^6 model iterations, for present-day and future forecasts (2050 and 2100) for all relevant NSW beaches (395 beaches in total).

Based on the general description of uncertainty in modelling provided by Roy and Oberkampf [82], uncertainty in shoreline change forecasts emerges from: (1) the stochastic nature of environmental forcing and coastal processes, such as storms and sea-level rise (*aleatory* uncertainty); (2) a limited understanding of the sediment transport processes that drive beach and shoreline response to changing environmental conditions (*epistemic* uncertainty); and, (3) the simplified representation of complex three-dimensional coastal morphodynamics by aggregated morphology and parameters that describe morphologic response (*model form* uncertainty).

Aleatory uncertainty is typically expressed using a probability distribution that describes the likelihood of occurrence across the feasible range of magnitudes. For example, we use a gamma probability function (Equation (2)) to describe the likelihood of experiencing fluctuating beach erosion, within a feasible range, in the final year of the simulation forecast period. In each model run, the value of F is randomly sampled from Equation (2), and it is combined with the cumulative beach change (C) to calculate the total beach change volume V (Equation (1)). The width and shape of the function is based on available data and knowledge of episodic beach erosion in NSW (Section 3.4.1). Fluctuating beach erosion (F) is described using an asymmetric gamma distribution to reflect the significantly reduced likelihood of experiencing the most severe erosion events in any given year.

We use a Monte Carlo sampling regime to manage *epistemic* uncertainty in cumulative beach change (C) due to sediment redistribution within and between compartments [78]. The feasible range and most likely values for all of the parameters and variables in Equations (3), (5) and (6) (except t, l, x_0), are defined by triangular probability functions, which require the definition of lower (a) and upper (c) bounds, and a modal or most likely (b) value only. This simplistic representation of the uncertainty space is commensurate with the state of knowledge, in that sufficient data or scientific understanding may exist to define the feasible range and best estimate value of a model parameter or variable, although the exact shape of the probability distribution remains largely unknown [78]. The triangular functions capture the estimated uncertainty space around the best estimate value, which might represent an average of measurements or simply the most likely value based on expert knowledge.

The simplicity of the shoreline encroachment model implies that *model form* uncertainty is unavoidable in our findings. However, given the exploratory nature of the long-term forecasts, in particular, few datasets exist with which to calibrate or test model predictions. Furthermore, the volumetric design of the model expresses a direct relationship between the sediment budget principles that control the redistribution of sand from the beach to other depositional features, and the simulated beach-volume and shoreline change. While our assumptions underlying sediment redistribution in response to sea-level rise are founded on geological evidence, and historical observations from naturally evolving or modified systems, the applicability of these assumptions to all NSW beaches requires further scrutiny. In particular, the significance of shoreface and estuarine response in long-term forecasts suggests that a detailed appraisal of the scope for these responses to sea-level rise in different settings would greatly improve our approach. Without such information, we adopt a risk-averse position in developing a second-pass coastal erosion exposure assessment.

3.4. Model Applications

We present regional- and local-scale applications of the modelling approach to demonstrate the utility of the coastal sediment compartments framework for parameterising coastal geomorphology and applying sediment budget principles, and the flexibility of the simple shoreline encroachment model in scaling to suit the available data and required resolution of shoreline change forecasts. The objective of the regional-scale application is to develop a second-pass estimate of potential exposure to coastal erosion and shoreline change in NSW. The local-scale application provides an example of how the approach may be applied to develop more refined shoreline forecasts where more detailed sediment budget data is available. We stress, however, that the local-scale example also applies speculative values for some sediment budget components, due to limited available data. Thus, the example should not be interpreted as a reliable forecast, but serves to highlight key sediment budget components that would benefit from more detailed observations and knowledge.

3.4.1. Regional Scale (NSW Coast)

First, we consider the property and infrastructure exposure to coastal erosion along 395 NSW beaches, comprising 70% of the total length of open-coast sandy shorelines along the NSW coast, and covering the full extent of potential property and infrastructure exposure to coastal erosion. The remaining open-coast beaches, where the modelling approach was not applied, are characterised by non-erodible backshore substrates and thus have limited or no exposure to coastal erosion.

The fluctuating component of shoreline change is described by a gamma probability function (Equation (2)), which is intended to approximate the feasible range and likelihood of fluctuating erosion on fully exposed open-coast NSW beaches. The dominant component of fluctuating change is the so-called "storm demand", which refers to the volume of sand removed from the beach by raised water levels and high waves experienced during an individual or closely grouped series of coastal storms. Gordon [83] presented a relationship for the probability of storm demand on NSW beaches from measured and estimated (i.e., based on beach surveys and photogrammetry analysis) beach erosion volumes (Figure 6A). That relationship, along with more recent storm demand observations in NSW [84], and simulation experiments based on the long-term beach measurement dataset at Collaroy-Narrabeen Beach [85,86], form the basis for our parameterisation of the gamma distribution that is used to describe F (Equation (2)) for fully exposed open-coast NSW beaches (Figure 6B).

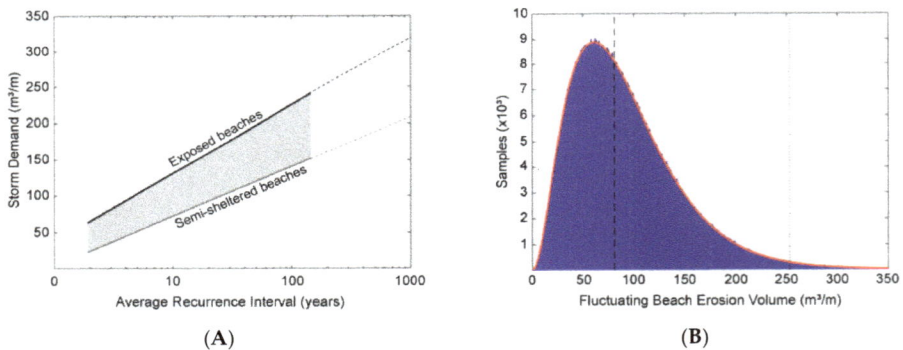

Figure 6. (**A**) Estimated Average Recurrence Interval (ARI) of storm demand on exposed (black line) to semi-sheltered (grey line) NSW beaches [83]. The functions have been extrapolated here (dashed lines) to consider their relationship to more severe erosion events. (**B**) Gamma function (red line) used to describe the probability of fluctuating beach erosion (Equation (2)), with 10^6 random samples (blue columns). The 50th (dashed line) and 99th (dotted line) percentile values are shown.

Consideration was also given to geo-historical evidence including the depositional records from prograded coastal barriers, which store wave climate and shoreline response records for previous centuries. Those records suggest that the recent historical period (including all of the observation data) has been characterised by lower intensity storm conditions relative to previous centuries [87]. As such, it may be imprudent to assume that historical measurements have captured the maximum potential storm demand on NSW beaches. When considering that the upper tail of our F distribution (Figure 6B) also accommodates the possibility of enhanced erosion due to: historically unprecedented storm clustering and wave climate extremes, rips, and other beach processes that are known to enhance the average erosion response, or the coincidence of a severe storm demand with the cyclic differential shoreline oscillation phenomena known as *beach rotation* [6,88,89].

As a modest beach erosion event is likely to occur in any given year, the mode of the *F* distribution is 65 m³/m (Figure 6B), which approximates a two-year average recurrence interval storm demand based on the relationship in Figure 6A. The 50th percentile *F* value is 82 m³/m, and the 99th percentile (1% exceedance level) *F* value is 250 m³/m (Figure 6B), which is about equivalent to the estimated 100-year average recurrence interval storm demand for exposed beaches (Figure 6A). For sheltered beaches, *F* values sampled from the gamma probability function (Figure 6B) were scaled (using c_f, Equation (1)) based on the average shoreline orientation for each beach or sector [36], to account for the effects of enhanced refraction on incident wave energy at the shoreline. The c_f scaling values based on average shoreline orientation are provided in Appendix A.

To evaluate the suitability of Equation (2) (Figure 6B) for describing the range and likelihood of fluctuating erosion on NSW beaches, comparisons were made between modelled erosion on exposed open-coast beaches, and the locations of historical maximum erosion escarpments where available (i.e., mapped dune scarps associated with the most severe historical erosion event at each beach). The comparisons demonstrate that low-probability (e.g., 99th percentiles) modelled beach erosion based on Equation (2) is consistent with mapped historical maximum erosion escarpments (Appendix B).

We applied a limited sediment budget parameterisation to model cumulative erosion (*C*) for the regional-scale application, using only the first three terms in Equation (3). Photogrammetry analysis of historical shoreline change, recorded in aerial photographs captured intermittently since the 1960s, has been carried out for more than 150 NSW beaches. Where a consistent and ongoing long-term shoreline recession trend has been identified in photogrammetry records, the annual average rate of sand volume loss at the beach was applied using the q_y parameter (Equation (3)).

There is the potential for sand supply from the shoreface to NSW beaches. The occurrence of prograded Holocene sand barriers along parts of the central and southern NSW coast, where fluvial sources and the alongshore sand transport system is limited, suggests that shoreface sand supply to beaches was an important process during recent geological time, and may persist today at significant rates along some parts of the coast [24,90,91]. That potential was considered using the regional rates of sand supply to NSW beaches,, which was derived from the analysis of geohistorical records spanning the last several centuries [87], which was applied as annual average rates of sand supply using the q_x parameter (Equation (3)). The shoreface may act as a source or sink, depending on the relationship between the geomorphic setting of the beach, the local wave climate, and the sea-level rise scenario.

When considering Equations (5) and (6), the modelled beach response to sea-level rise is a function of the sampled values of: (1) sea-level rise, (2) compartment-averaged shoreface geometry (Equation (4)), (3) the shoreface closure depth, and (4) the "active" surface area of estuarine flood-tide deltas.

Figure 7A shows the range of sea-level rise projections that were applied in the model to calculate shoreline response, which reflect the Intergovernmental Panel on Climate Change (IPCC) global mean sea-level rise (GMSL) projections from the Fifth Assessment Report [19]. The IPCC projections for southeast Australia suggest a sea-level rise of around 0–10% above the global average [92,93]. As GMSL projections presented in the Fifth Assessment Report (AR5) were restricted to the "likely" range (17th to 83rd percentiles), we used linear extrapolation to extend the distribution tails to cover the 0–100th percentile range. The triangular distributions shown in Figure 7A reflect the combined range of GMSL projections for the three emissions pathways (RCP2.6, RCP4.5, and RCP8.5) considered in AR5. The bounds of the triangular distributions for 2050 (blue) and 2100 (red) reflect the 0th and 100th percentiles of the combined range, while the modes reflect the 50th percentiles of the combined ranges of the three emissions pathways (Figure 7A).

We acknowledge that the range of GMSL projections that were applied in the modelling do not reflect the full uncertainty space for the present century. The AR5 GMSL projections were limited to the consideration of climate process model forecasts only and omitted the potential influence of rapid ice melt this century. Sweet et al. [20] recently reviewed and revised the AR5 GMSL projections in the context of the latest research on upper-end GMSL projections that reflect rapid ice melt processes, and recommended that scenarios covering the range 0.3–2.5 m be considered in assessing the potential impacts of sea-level rise at 2100. However, we limit our consideration of GMSL this century to the AR5 findings, in order to enable comparison between our findings and existing coastal erosion hazard studies that have been developed by local governments in NSW for coastal management and planning.

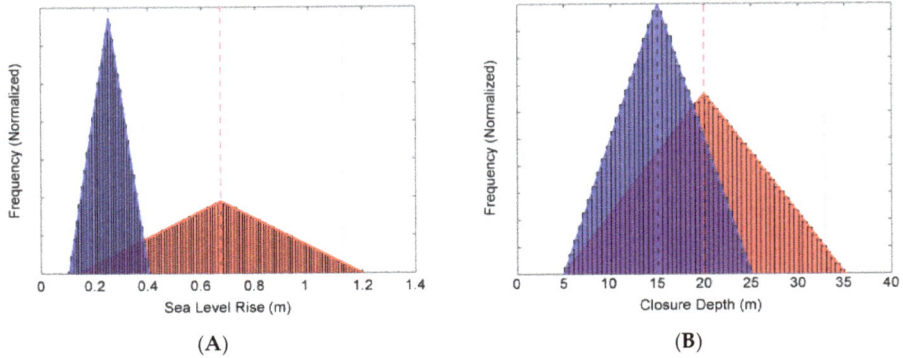

(A)   **(B)**

Figure 7. Input triangular probability distributions applied in the modelling for (**A**) sea-level rise, and (**B**) shoreface closure depth, for the 2050 (blue) and 2100 (red) forecast periods, with 10^6 random samples (columns) also shown. The 50th (dashed line) and 99th (dotted line) percentile values for each forecast period are indicated.

We approximated the shoreface geometry of each beach to calculate the potential shoreface sediment accommodation volume generated by sea-level rise (Figure 8A). The shoreface geometry was generated by fitting Equation (4) to regional-scale bathymetry, averaged alongshore within each sediment compartment. For each beach, the average distance from all of the sandy shorelines within the relevant sub-compartment to the 10 and 20 m isobaths was calculated, while the 30 and 40 m depth coordinates represent the average distance from all of the sandy shorelines within the relevant tertiary compartment to the corresponding isobaths (e.g., Figure 3). The method reflects greater alongshore variability in upper-shoreface (0–20 m water depth) geometry between NSW beaches, relative to lower-shoreface (20–40 m) geometry, which is relatively consistent at the tertiary compartment scale.

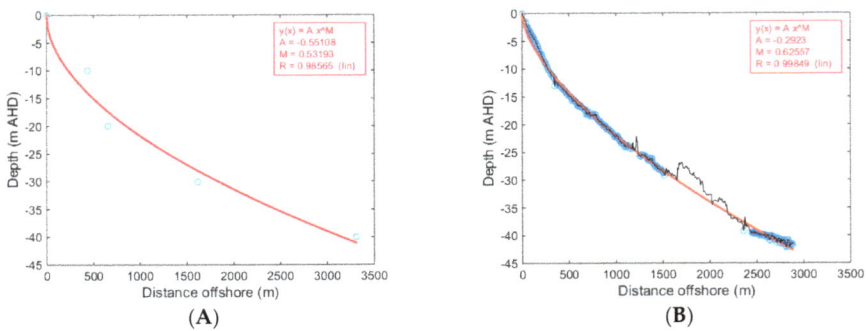

(A)   **(B)**

Figure 8. (**A**) Example of estimated shoreface geometry (red line) fitted to alongshore-averaged regional-scale bathymetric data (blue circles) at Wamberal Beach (Figure 3B). (**B**) Local-scale shoreface geometry (red line) at Wamberal Beach fitted to high-resolution hydrographic data (blue circles) with protruding reef outcrops (black line) omitted. Curve fitting was carried out using the ezyfit.m tool.

The triangular function for shoreface closure depth (h_c) widens from 2050 to 2100 (Figure 7B) to reflect the increasing potential for sediment-accommodation generation and surface response across the lower shoreface at longer timescales [94–96]. The upper limits of shoreface closure depth in Figure 7B reflect representative values of Hallermeier's [97] outer shoal zone limit [98,99], and the only long-term observation dataset [100] for this region. When combined with the widening uncertainty space for accelerating sea-level rise (Figure 7A), and when considering the typical cross-shore extent of shorefaces in this region (Figure 8A), the modelled shoreface sediment-accommodation potential may represent a significant driver of the simulated shoreline recession, particularly for the 2100 forecast period.

The surface area of active submerged flood-tide delta deposits (A_D) was mapped for each NSW estuary using the NSW Coastal Quaternary Geology Data Package [101,102] and recent aerial imagery. The surface area A_D is applied in Equation (6) to calculate the potential sediment accommodation volume generated in estuaries by sea-level rise. Depending on the location of each estuary and connectivity with adjacent beaches, summed values of A_D were applied at the sub-compartment or tertiary compartment resolution, with the influence of estuarine sediment sinks distributed along the total sandy shoreline length (l) corresponding to the relevant compartment. Summary statistics describing the total area of the shoreface and estuarine (flood-tide delta) sediment sinks in each primary sediment compartment (Figure 1A), and the relative difference (i.e., total sink areas divided by the length of sandy shoreline in each compartment), are provided in Appendix C.

3.4.2. Local Scale (Wamberal Beach)

We apply the simple shoreline encroachment model in higher resolution to Wamberal Beach (Figure 3) to investigate sensitivity to local-scale geomorphic complexities. Wamberal Beach is the central sector of the Terrigal-Wamberal sub-compartment, and spans from Terrigal Lagoon inlet in the south to Wamberal Lagoon inlet in the north (Figure 3B). Wamberal Beach fronts a narrow stationary-receded beach barrier comprising a frontal dune only, which is anchored to the bedrock framework in the north, and which separates Terrigal Lagoon from the ocean in the south. Both the frontal dune and back-barrier flat feature moderate density residential development. Historically, Wamberal Beach has been impacted by erosion that is caused by coastal storms, resulting in significant damage and loss of properties [52]. Many beachfront properties were damaged by severe erosion (Figure 2) that was caused by a storm that impacted the entire NSW coast in June 2016 [1]. However, the barrier morphostratigraphy (stationary-receded) suggests that from a geohistorical perspective, Wamberal beach has been relatively stable or very slowly receding during the mid-late Holocene.

In the local-scale example, we use the same alongshore-averaged beach-dune terrain profile to model cumulative shoreline erosion (C) as was used in the regional-scale example (i.e., the red profile in Figure 9). However, we use beach-dune terrain profiles extracted from LiDAR data along regular alongshore-spaced (25 m) transects (Figure 10) to model fluctuating erosion (F). This allows for the modelled R_C (Figure 4) to reflect the distributed influence (i.e., along the length of the beach) of cumulative sand loss from the sub-compartment, while the modelled R_F (Figure 4) reflects alongshore variability in beach-dune geomorphology along Wamberal Beach. Our approach assumes that as the shoreline recedes by encroachment in a time-averaged sense, due to cumulative sand loss from the sub-compartment, nearshore wave processes that are operating at a higher frequency maintain a dynamic or ephemeral beach face with time-averaged sand volume consistent with the present-day setting.

Figure 9A shows the alongshore-averaged beach-dune terrain profiles for the three sectors of the Terrigal-Wamberal sub-compartment (Figure 3B). The three averaged terrain profiles capture the gross alongshore gradient in beach-dune morphology. This emerges from the influence of the prominent Broken Head on transformation of the predominant south to southeast wave climate to the nearshore, which results in greatly reduced exposure along Terrigal Beach relative to Wamberal and Wamberal North beaches (Figure 3B). In the local-scale example, cumulative shoreline change (R_C) is calculated by applying the cumulative beach change volume (*C*) to the alongshore-averaged profile above 0 m AHD and landward of the 4 m AHD elevation contour (not the 2 m AHD contour), which reflects the position of the frontal dune face. This ensures that R_C reflects sediment loss from the dune, not the beach face, which is a transient feature affected by fluctuating processes.

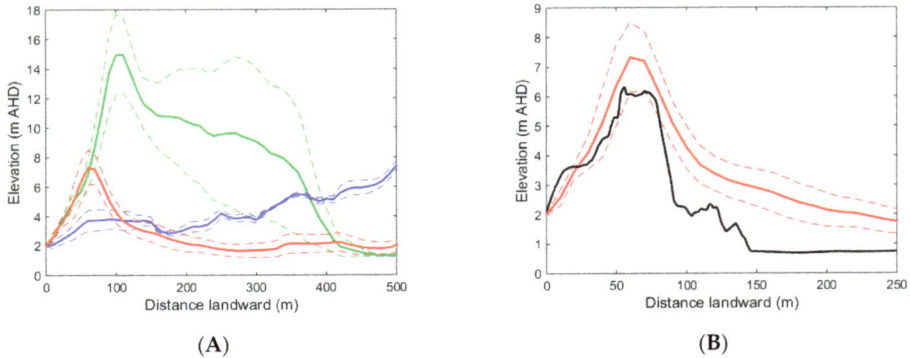

(A)   (B)

Figure 9. (**A**) Alongshore-averaged beach-dune terrain profiles, within envelopes of ± one standard deviation (dashed), for Terrigal (blue), Wamberal (red) and Wamberal North (green) beach sectors (Figure 3B). (**B**) Comparison between the alongshore-averaged terrain profile for Wamberal Beach (red), and a profile from the narrow point (black) of the Wamberal Beach sand barrier (Figure 10).

The modelled fluctuating beach change volume (*F*) is applied to each 25-m alongshore-spaced profile, above 0 m AHD and landward of the 2 m AHD elevation contour, which reflects the beach berm position. Figure 9B compares the alongshore-averaged terrain profile for Wamberal Beach, with a profile from the narrow point of the sand barrier (Figure 10), as an example of terrain variability relative to the alongshore-averaged profile. As the greatest beach erosion volumes are typically achieved when the pre-storm beach state is fully accreted [75], we derive the 25-m alongshore-spaced beach-dune profiles from two LiDAR surveys, which together capture a fully accreted beach state along the northern and southern parts of Wamberal Beach. The 2011 LiDAR survey, from which the alongshore-averaged profile was derived (Figure 9), captured an accreted state along the southern two-thirds of the beach, while a 2016 LiDAR survey [1] captured an accreted state along the northern third of the beach. Similarly, the baseline (2 m AHD elevation contour) from which we measure modelled fluctuating shoreline change (R_F), is a hybrid beach berm position based on both LiDAR surveys, representing a fully accreted beach state along the length of Wamberal Beach (Figure 10).

The local-scale model configuration included the gamma probability function for fluctuating erosion (*F*) for fully exposed open-coast beaches (Equation (2)), as we applied in the regional-scale example (Figure 6B). Measurements of historical beach erosion within the Terrigal-Wamberal sub-compartment suggest that the Wamberal Beach sector is fully exposed to the impacts of storms. For example, Worley Parsons [103] reported that storm-induced beach erosion volumes that were determined using photogrammetry analysis of historical aerial photographs reached 250 m^3/m along parts of the beach following a series of severe storms in 1974, and since then, beach erosion volumes on the order of 200 m^3/m have been measured in response to several other historical storm events.

However, the scaling coefficient for fluctuating erosion (c_f) was used to account for potential alongshore variation in substrate resistance. An investigation of the subsurface geology of Wamberal Beach [104] found that an underlying siltstone deposit approaches the beach face along a 200-m sector of the northern half of the beach. This is indicated in Figure 10 by the bore holes that are marked in red. Severe erosion in June 2016 exposed the siltstone deposit along that sector of beach (Figure 2D).

Additional geotechnical studies that were carried out for private development applications have found that the siltstone deposit rises to 6–8 m AHD within the frontal dune along parts of that same sector.

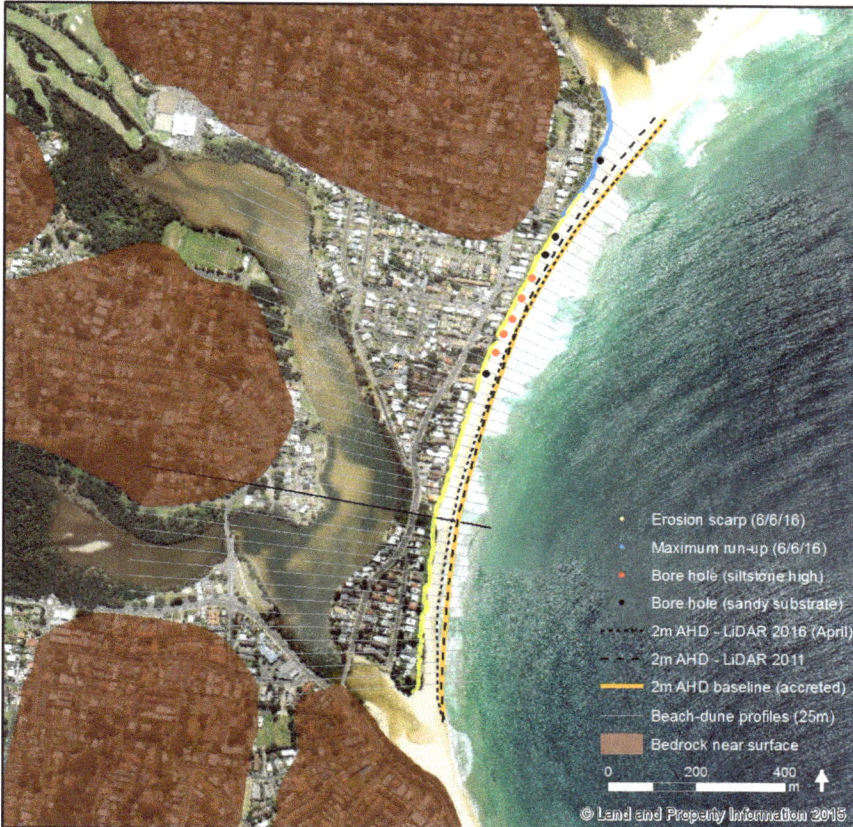

Figure 10. Wamberal Beach, which is the central sector of the Terrigal-Wamberal sub-compartment (Figure 3B), showing the distribution of 25-m alongshore-spaced beach-dune profiles that were used to calculate fluctuating shoreline change. The black profile spans the narrow point of the barrier. Beach berm positions (2 m AHD elevation contour) based on airborne LiDAR surveys carried out in 2011 and 2016 (April) are shown with the hybrid (orange) baseline representing a fully accreted beach. Bore holes sampled by Hudson [104] are also shown, indicating where a buried siltstone deposit approaches the beach face (red). The base of the erosion escarpment (yellow) and the limit of wave run-up (blue) from the June 2016 storm were mapped using RTK-GNSS immediately after the storm.

Given that the siltstone unit is preserved within the dune and beneath the beach, but that it is not exposed on the beach face under modal conditions, and the shoreline curvature appears undisturbed, we consider the material to provide no significant resistance to shoreline recession. This is supported by inspection of the siltstone in June 2016, when the exposed material was found to be easily disturbed. However, an exposure within the dune at 6–8 m AHD elevation may restrict the extent of beach erosion during an individual erosion event. As such, c_f was applied as a triangular probability function ($a = 0.7$, $b = 0.85$, $c = 1$), thereby allowing a maximum 30% reduction in the fluctuating beach erosion volume, due to the presence of the siltstone material in the frontal dune. That is consistent with the reduced response to storms observed at Lake Cathie Beach on the Mid-North Coast (Figure 1A), where indurated sand (coffee rock) lenses within the frontal dune contribute to about a 30% reduction in beach erosion relative to similarly exposed beaches in the region [61].

Photogrammetry analysis for Wamberal Beach does not indicate any steady and persistent trend in shoreline change through the period of historical aerial photographs. Rather, the data suggests variable accretion and erosion along different parts of the beach between aerial photograph captures [103]. This is supported by the variable alignment of beach berm positions (i.e., 2 m AHD elevation contour), which were derived from airborne LiDAR surveys that were carried out in 2011 and 2016 (Figure 10). Thus the photogrammetry data likely reflects short- to medium-term beach oscillation associated with wave climate variability [6,88,89], and the impacts of individual storm events, rather than any underlying shoreline recession or accretion that would be anticipated to persist indefinitely along the sub-compartment. There is limited prospect for alongshore sand transport around Broken Head into the sub-compartment, which is indicated by the prominence and orientation of Broken Head relative to the modal wave direction, and the continuous offshore reefs (Figure 3B). Therefore, no underlying rate of beach volume change was applied using the q_y parameter (Equation (3)).

The shoreface of the Terrigal-Wamberal sub-compartment has been mapped using side-scan sonar [47], single-beam echosounder, and marine LiDAR (Figure 3B), and the substrate has been investigated using seismic reflection and seabed grab samples [105]. The upper- to mid-shoreface is mostly sandy to around 20 m water depth, beyond which extensive reef outcrops protrude through the sandy shoreface (Figures 3B and 8B). Sediment thickness is generally less than 5 m, except where it approaches or slightly exceeds 10 m around a buried palaeo-drainage channel in the northern sector of the sub-compartment [105], which is represented by the only continuous sediment pathway to the lower shoreface (Figure 3B). The stationary-receded barrier morphostratigraphy, and character of the shoreface substrate, both suggest the limited potential for sand supply from the shoreface, and thus no persisting rate of sand supply to the beach was applied using the q_x parameter (Equation (3)).

Rather than using the regional-scale shoreface geometry (Figure 8A) in order to calculate the potential shoreface sediment accommodation volume, more accurate shoreface geometry was determined by fitting Equation (4) to detailed hydrographic survey data (Figure 8B), omitting the reef outcrops to estimate the shoreface geometry if the seabed were entirely composed of unconsolidated sediments. The same triangular probability functions for sea-level rise and shoreface closure depth, as applied in the regional-scale example, were also used in the local-scale application (Figure 7). The shoreface response scaling coefficient (c_S) was applied to account for the mixed sediment-reef substrate within the Terrigal-Wamberal sub-compartment (Figure 3B). That is, the intermittent protruding reefs with relief of several metres represent potential sediment accommodation space that is already filled by reef outcrops, and thus is not available to be filled by sand lost from the beach. The scaling coefficient was defined using a triangular function ($a = 0.65$, $b = 0.7$, $c = 0.75$), reflecting that about 30% of the total shoreface area, extending to the shoreface toe (35 m water depth) within the alongshore extent of sub-compartment, is reef (Figure 3B). That parameterisation of c_S means that only 65–75% of the shoreface represents potential sediment-accommodation space during sea-level rise.

To demonstrate our approach, we also include allowances for other potential influences on the sub-compartment sediment budget, including mega-rips and change in the biogenic (carbonate) sediment component. A previous modelling investigation of future shoreline change at Avoca Beach, immediately south of Broken Head, identified both of the influences as relevant considerations for future sediment budgets within this region [79]. Based on that study, which surveyed a panel of experts to determine appropriate values for sediment budget parameters, we apply V_M as a triangular function ($a = 0$, $b = 815$, $c = 1630$) to consider cumulative sediment loss via mega-rips, which equates to a maximum 0.5 m^3/m sediment volume loss from the beach per year. Application of V_M recognises the potential for sand loss offshore from this sub-compartment during severe storms, with limited potential for sand to return due to the character of the shoreface substrate. Similarly, we apply V_B as a triangular function ($a = -1630$, $b = 815$, $c = 3260$) to consider carbonate sediment dissolution at rates up to 1 (m^3/m)/year, or enhanced carbonate production up to a maximum rate of 0.5 (m^3/m)/year.

3.5. Exposure Assessment

We applied standard spatial-overlay analysis techniques using ArcGIS to identify potentially exposed properties and infrastructure, which are defined as those that are intersected by the modelled coastal erosion hazard zones for each forecast period and exceedance level. We use both the NSW Cadastre dataset and the Geo-coded Urban and Rural Address System (GURAS) to identify addresses that are potentially exposed to coastal erosion. Whereas, the Cadastre describes the location and character of property lots, the GURAS database stores the details of each address within a property lot. While some lots contain multiple addresses, some addresses occupy multiple lots. By jointly querying both databases, all of the valid addresses that were potentially exposed to coastal erosion were identified, including both primary (houses and multi-dwelling buildings) and secondary (individual apartments within multi-dwelling buildings) address types.

Simply counting the number of potentially exposed addresses only partially communicates the relative exposure between beaches or regions. To assess the relative impacts of modelled beach erosion and shoreline change on potentially exposed properties, we also calculate for each potentially exposed property lot, the proportion of land area that is affected by the modelled coastal erosion hazard zones. This is achieved by dividing the land area of the lot that is affected by the modelled coastal erosion hazard zone by the total lot area. Based on that analysis, all of the potentially exposed addresses (including primary and secondary address types) were categorised into five groups, based on the proportion of each property lot that is potentially affected by coastal erosion: <10%, 10–25%, 25–50%, 50–90%, and >90% lot-area identified as exposed.

We also applied spatial-overlay analysis to identify lengths of roadways that are potentially exposed to coastal erosion for each forecast period and exceedance level scenario. Roadways exposure is categorised into five road types: vehicular track, local road, arterial road, primary road, and motorway (by increasing significance, associated infrastructure, and replacement costs). For simplicity, we restrict our assessment of infrastructure exposure to roadways, as many other infrastructure and utilities assets scale in a relatively linear relationship with roadways.

4. Results

4.1. Regional Scale (NSW Coast)

Figure 11 shows examples of the regional-scale coastal erosion hazard modelling for Ten Mile Beach in the Northern Rivers region (Figure 1B), and Windang Beach in the Illawarra-Shoal haven region (Figure 1C). Shaded hazard zones for the present (red) and 2050 (orange) forecast periods represent areas that were exceeded by only 1% of model predictions for each forecast period. Three hazard zones (50%, 10%, 1% exceedance) are shown for the 2100 forecast period (yellow), indicating areas that were exceeded by 50%, 10%, and 1% of model predictions. The forecast extent of coastal erosion increases for longer forecast periods and lower exceedance levels, reflecting increased sand loss from the beach due to simulated sediment redistribution to estuarine and shoreface sinks.

The influence of low back-barrier morphology on modelled shoreline change, once the frontal dune has been breached, is evident in the 2100-10% exceedance and 2100-1% exceedance forecast shoreline positions along the southern end of Ten Mile Beach, at Shark Bay (Figure 11A). The extent of the 2100-10% exceedance and 2100-1% exceedance hazard areas from the baseline shoreline is much greater along that sector, compared with the beach to the north, because of the lower dune and back-barrier morphology along the Shark Bay sector. Along the Shark Bay sector, the encroachment response assumption is violated, and the shoreline change forecast is likely over-estimated. In that location, barrier-rollover processes (washover) would result in back-barrier deposition upon breaching of the frontal dune.

A similar effect is not apparent at Windang Beach, where even the southern end of the beach features relatively well-developed dune morphology that is not breached by the 2100-1% exceedance scenario (Figure 11B). There, most of the coastal development is set back more than 200 m from the shoreline, well behind the frontal dunes, and thus the forecast hazard zones suggest that most development is not likely to be exposed to coastal erosion within the present century, even for low-probability scenarios.

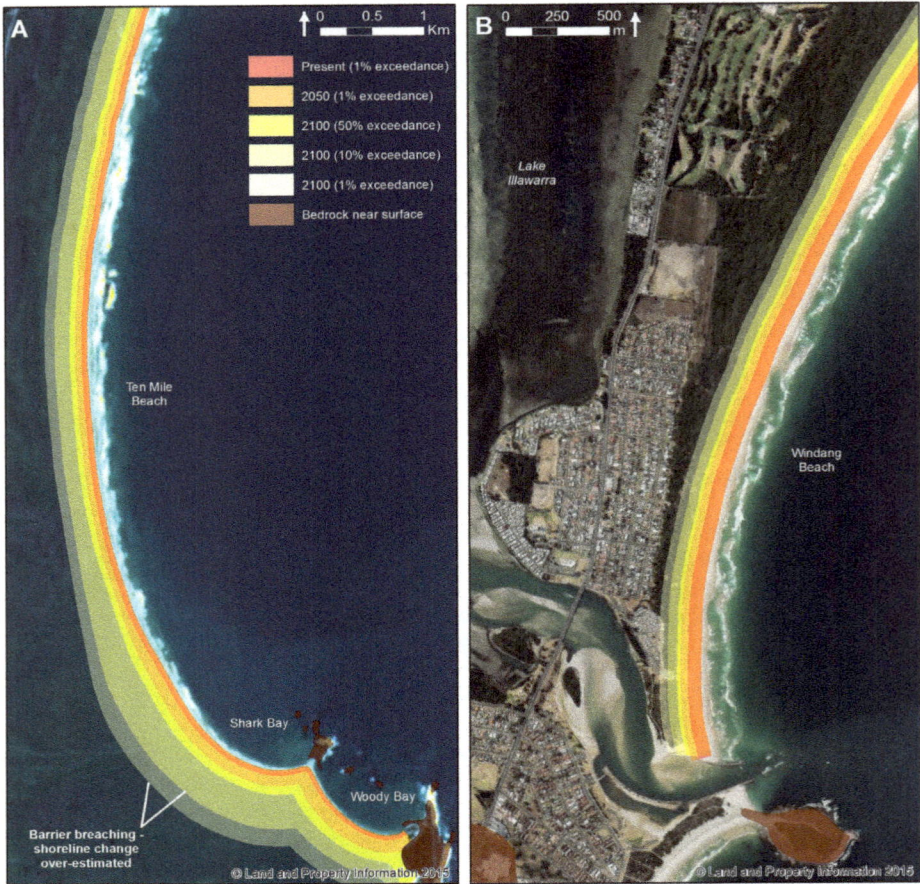

Figure 11. Regional-scale coastal erosion hazard mapping for (**A**) Ten Mile Beach in the Bundjalung compartment (Figure 1B), and (**B**) Windang Beach in the Illawarra South compartment (Figure 1C). The 1% exceedance level is shown for the present and 2050 forecast periods, while 50%, 10% and 1% exceedance levels are shown for the 2100 forecast period. At Shark Bay (**A**), for the 2100-10% and 1% exceedance level scenarios, the encroachment model exceeds the low-volume dune system and erodes into very low-lying back-barrier terrain. In reality, barrier rollover would ensue, and the associated back-barrier deposition suggests that the actual shoreline change would be less than depicted here. In that case, the basic assumptions of the simple shoreline encroachment model are violated.

Figure 12 shows model forecast sample distributions of V for Windang Beach, for the present (i.e., F only) and 2050 (combined F and C) forecast periods. The V distribution for the present scenario (Figure 12A) reflects the input gamma probability function for F (Figure 6B), and the fitted cumulative distribution is a gamma function (Figure 12B). The 50%, 10%, and 1% exceedance beach change (F) volumes were 83, 167, and 255 m^3/m, respectively. The V distribution for the 2050 scenario (Figure 12C) reflects the combined distributions of the gamma probability function for F, and the symmetrical input triangular functions for the various *C* components (e.g., Figure 7). The fitted cumulative distribution is a generalised extreme value (GEV) function (Figure 12D). The 50%, 10%, and 1% exceedance beach change volumes were 234, 342, and 447 m^3/m, respectively.

Figure 13 summarises the forecast shoreline change distances for all 395 modelled beaches, grouped by the nine primary sediment compartments of the NSW coast (Figure 1A). The blue boxes cover the 25th to 75th percentile range of forecast shoreline change (landward from the 2 m AHD beach-berm baselines) for modelled beaches within each primary compartment, with the red lines indicating the median values. The black dashed lines cover approximately ±2.7σ for each sample set, while the red markers indicate values beyond that range that are interpreted as outliers.

The median forecast shoreline change distances are relatively similar along the coast for the present (*F* only) scenario, in the 60–70 m range in all of the primary compartments (Figure 13). The Illawarra compartment features the highest forecast erosion due to fluctuating processes. Variation within compartments is greater for the northern and southern NSW compartments relative to the central NSW coast, due to both larger sample sizes and greater variability in beach-dune morphology. For example, the South Coast compartment includes several very protected beaches within Batemans Bay (a deeply indented open-ocean embayment featuring prominent islands at the bay mouth), which contribute to very high variability in shoreline change forecasts within that compartment.

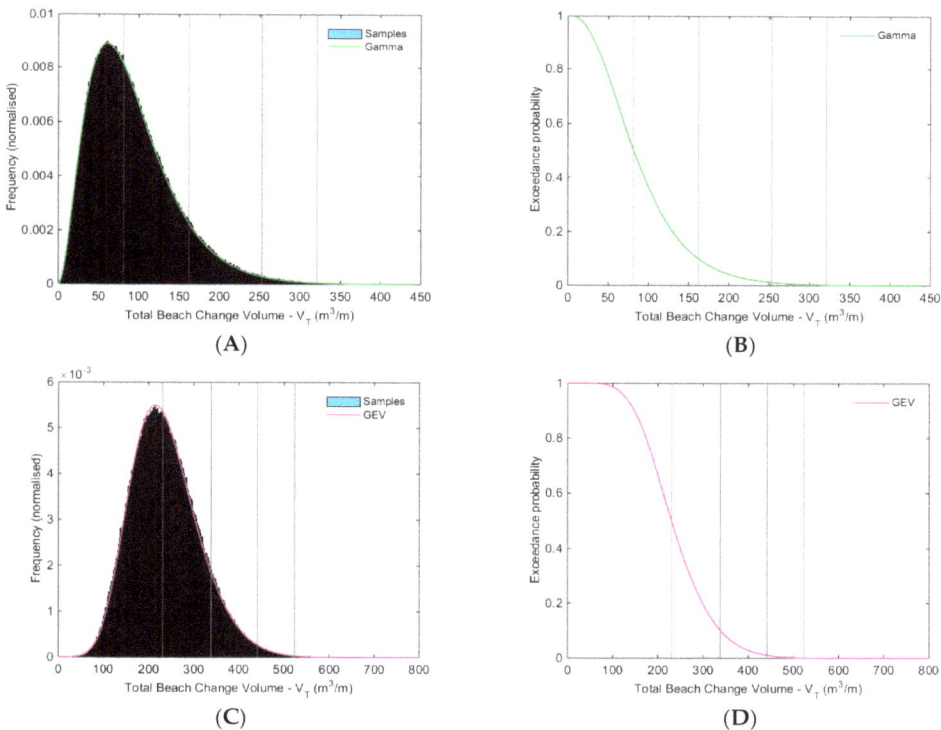

Figure 12. Model forecasts of beach volume change for Windang Beach (Figure 11B) for the present (**A,B**) and 2050 (**C,D**) forecast periods. Frequency distributions of the modelled beach volume change (**A,C**), and the corresponding cumulative probability functions (**B,D**) are shown for the present and 2050 forecast periods, with (left to right) 50th, 90th, 99th and 99.9th sample population percentiles (50%, 10%, 1%, and 0.1% exceedance levels, respectively) indicated on each plot as grey vertical lines.

As expected, given the design of the simple shoreline encroachment model, forecast shoreline change increases in all of the compartments for the 2050 and 2100 (combined *F* and *C*) scenarios (Figure 13). The north coast and south coast compartments are characterised by above average forecast shoreline recession, due to the increasing sediment demand from broader shoreface and estuarine sediment sinks, relative to average beach-dune morphology. Variation between compartments also increases, which is evident in higher deviation about the 25th–75th percentile ranges and increased outlier values for those compartments. For example, the north coast compartments feature large potential shoreface sediment sinks and relatively low dune morphology, particularly in less exposed southern corners of compartments (e.g., Figure 11A), resulting in above average forecast shoreline recession and high variability between individual beaches.

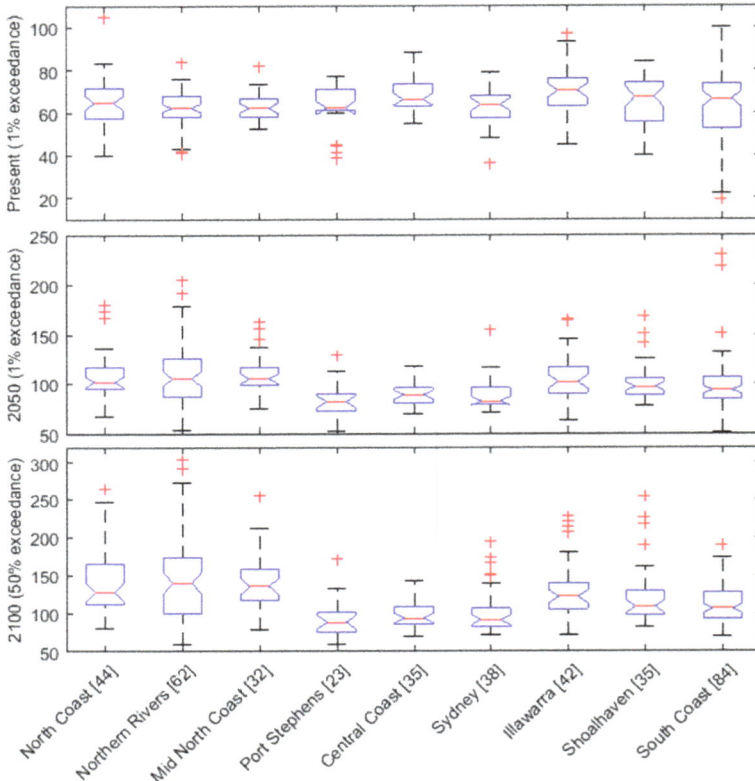

Figure 13. Box plots summarising the modelled shoreline recession distances (m) from the regional-scale application, for each NSW primary compartment (Figure 1A), as measured from the beach berm baseline (2-m AHD elevation contour). The number of beaches modelled in each compartment is indicated in square brackets. Modelled shoreline change distances are shown for the present-1% exceedance (**top**), 2050-1% exceedance (**middle**), and 2100-50% exceedance (**bottom**) scenarios.

The exposure assessment based on the regional-scale modelling identified approximately 1200 property lots (2300 total addresses) in NSW that are potentially affected by coastal erosion at present (1% exceedance level). Property exposure rises to around 3100 lots (5200 total addresses) at 2050 (1% exceedance level), and 4800 lots (8200 total addresses) at 2100 (50% exceedance level).

However, many potentially affected lots may be only partially exposed to coastal erosion, in which case assets on the property may not be affected. When considering only properties for which more than half of the lot area was intersected by the modelled coastal erosion hazard zones (and consequently for which assets are likely to be affected), the property exposure numbers above decrease to 247 property lots (455 total addresses) that may be affected at present, rising to around 1862 lots (2718 total addresses) at 2050, and 3300 lots (5076 total addresses) at 2100.

Figure 14 shows the distribution of property exposure to coastal erosion between the seven NSW regions (Figure 1A), based on the regional-scale modelling. The exposure assessment findings are organised by the NSW planning regions, rather than the primary sediment compartments, reflecting the intended application of our investigation. All of the regions except for Hunter and South East feature significant property exposure to coastal erosion at present (based on the 1% exceedance scenario). Property exposure at 2050 (1% exceedance) and 2100 (50% exceedance) increases most significantly in the Northern Rivers, Mid North Coast, and Illawarra-Shoalhaven regions. Large increases in addresses for which more than half of the property lot is affected by modelled coastal erosion (orange and red shading) highlights regions that may experience a very significant increase in exposure to coastal erosion during the present century.

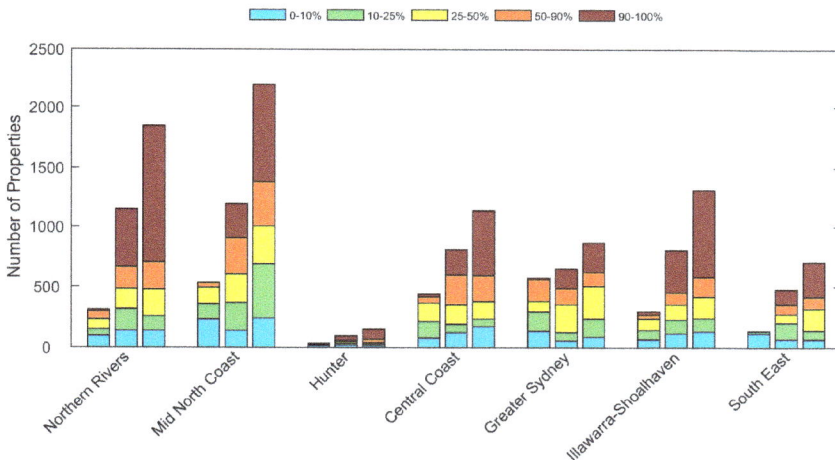

Figure 14. Total address exposure by NSW regions (Figure 1A) for the present-1% exceedance (left column in each group), 2050-1% exceedance (middle columns) and 2100-50% exceedance (right columns) scenarios. The colouring categorises exposed addresses by the proportion of each associated property lot that was intersected by the modelled erosion hazard zones for each region and scenario.

Based on the regional-scale modelling, about 70 km of NSW roadways may be exposed to coastal erosion at present (1% exceedance), increasing to 196 km at 2050 (1% exceedance), and 311 km at 2100 (50% exceedance). However, Figure 15 shows that, at present, vehicular tracks account for most of the roadway exposure behind NSW beaches. The analysis shows that increasing lengths of local and arterial roads may be exposed to coastal erosion by 2050 and 2100. No primary roads or motorways were affected by the three forecast period/exceedance level scenarios that were considered in the regional-scale exposure assessment. Roadway exposure was found to be relatively low in the Hunter, Central Coast, and Greater Sydney regions, moderate in the Illawarra-Shoalhaven and South East regions, and highest in the Northern Rivers and Mid North Coast regions (Figure 15).

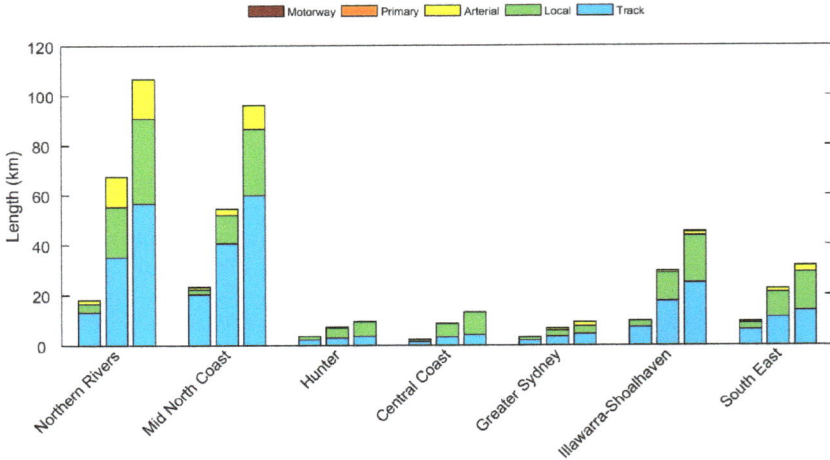

Figure 15. Roadway exposure by NSW regions (Figure 1A) for present-1% exceedance (left columns), 2050-1% exceedance (middle columns) and 2100-50% exceedance (right columns) scenarios. The colouring indicates the types of roadways that were identified as potentially exposed to coastal erosion based on each scenario.

4.2. Local Scale (Wamberal Beach)

In the local-scale application, we consider the sensitivity of fluctuating erosion to variation in geomorphology along the beach, using 25-m alongshore-spaced profiles to calculate R_F. (Figure 10). Figure 16A shows the 50%, 10%, and 1% exceedance level present-day (F only) erosion hazard zones for Wamberal Beach. The impact of the 50% exceedance erosion event (82 m^3/m) is limited to the beach face, while the 10% (160 m^3/m) and 1% (250 m^3/m) exceedance erosion events also impact the frontal dune. The effect of applying c_f along the 200 m stretch of the beach where the buried siltstone unit rises to the surface (indicated by the red drill hole markers) is evident in the closer spacing between the forecast shoreline positions for the three exceedance levels, and the relative proximity of the forecast shoreline positions to the model baseline in that location. Nonetheless, the modelling suggests that almost all of the properties immediately fronting Wamberal beach may be exposed to severe beach erosion caused by extreme coastal storms (e.g., a 1% exceedance level erosion event).

Figure 16B adds cumulative erosion (C) by 2050, based on the alongshore-averaged beach-dune profile, to the fluctuating erosion (F), as shown in Figure 16A, for the corresponding exceedance levels. At 2050, the modelling suggests that the combined impacts of F and C at the 50% exceedance level for each component would be limited to the beach and the face of the frontal dune. For the combined 10% and 1% exceedance levels, the impacts of coastal erosion extend across the frontal dune. The results suggest that the combined 1% scenario (i.e., a 1% exceedance level fluctuating beach erosion event with a 1% exceedance shoreline recession) may lead to breaching of the dune at the narrow point of the barrier. While the effect of scaling fluctuating erosion (using c_f) along the 200-m stretch of the beach where the buried siltstone unit rises to the surface is preserved in the combined model forecast, as seen in closer spacing of hazard zones along that section of the beach (Figure 16B), the influence becomes less prominent as the relative magnitude of cumulative erosion increases.

Figure 16. Local-scale coastal erosion hazard zones for Wamberal Beach, the central sector of the Terrigal-Wamberal sub-compartment (Figure 3B). (**A**) 50%, 10% and 1% exceedance level forecasts for fluctuating erosion only (at present). (**B**) 2050-50% exceedance, 2050-10% exceedance, and 2050-1% exceedance forecasts (combined fluctuating and cumulative erosion). Bore holes sampled by Hudson [104] are shown, with red markers indicating where a buried siltstone deposit approaches the beach face, and rises up to between 6–8 m AHD elevation within some parts of the adjacent frontal dune.

5. Discussion

5.1. Modelling Approach and Limitations

Application of the simple shoreline encroachment model at a regional-scale demonstrates the potential variation in the sensitivity of NSW beaches to projected sea-level rise, primarily relating to the influence of differences in the dimensions of estuarine and shoreface sediment sinks between sediment compartments along the coast. The volumetric design of the model allows for the consideration of the impacts of sediment redistribution between all of the potential sources and sinks on shoreline change. Because the modelling approach does not assume that existing dune morphology will aggrade at the same pace as sea-level rise (e.g., in contrast to Bruun's model), the impacts of sediment redistribution at the shoreline reflect erosion into the contemporary morphology of each NSW beach, as measured by airborne LiDAR surveys. In that way, the modelled shoreline recession reflects the behaviour of presently receding beaches on this coastline, in which the beach and shoreline encroach into the dune system. This is a risk-averse position, consistent with other assumptions in our approach, as dune deposition during sea-level rise may slow the rate of shoreline recession. Our modelling approach and applications are subject to many limitations arising from the model design, and the datasets available to inform the model parameterisation.

Regarding the model design, a key limitation is the assumption of an encroachment response [62], and the lack of support for shoreline recession by barrier roll-over [63]. This may lead to an inaccurate estimation of the rate of shoreline retreat where the frontal dune is breached by fluctuating and/or cumulative erosion, and the remaining dune or back-barrier morphology is low enough to support washover processes. In that case, the assumption of an encroachment response is no longer valid, and the model will likely over-estimate potential shoreline recession. When considering the well-developed dune morphology of most NSW beaches, relative to low-relief barrier island coasts e.g., [64], and following manual review of the regional-scale model predictions, we are confident that this limitation does not affect the model scenarios that are considered in our exposure assessment (i.e., present-1% exceedance, 2050-1% exceedance, and 2100-50% exceedance forecasts). In some settings, however, low-exceedance (e.g., 1%) and long-term (e.g., 2100) forecasts may over-estimate potential shoreline change (Figure 11A). Around estuary entrances, and for very narrow coastal barriers or typically sheltered NSW beaches, low dune and back-barrier morphology means that the assumption of an encroachment response may be invalid for model predictions that include high sampled sea-level rise and broad sediment sinks. Because we use alongshore-averaged beach-dune profiles to model cumulative erosion in both examples, low and/or narrow barrier morphology must be consistent along the length of the beach for it to influence shoreline change forecasts.

Another important limitation of the model design is the absence of dune growth during sea-level rise. Although we suggest that dune aggradation at the same pace as projected accelerating sea-level rise seems unlikely in this setting, aeolian deposition and dune growth is likely to play some role in shoreline response. While the washover parameter (V_O) could be used to simulate the effect of washover deposition or dune growth in slowing the rate of shoreline recession, we neglect its use due to lack of data or previous examples. Nonetheless, a more rigorous dynamic barrier model [64,65,78] could be implemented within the framework to address these processes in more detail. We emphasise again that the model as applied here is intended to provide a risk-averse mid-resolution forecast of potential shoreline change. Future applications should consider more refined methods to describe the sediment transport processes that drive sediment redistribution and shoreline change.

Regarding our model parameterisation, the most significant limitation is the assumption that the shoreface will act as a sediment sink during sea-level rise in all settings along the NSW coast. Where this is not the case, the model is likely to over-predict shoreline recession (e.g., Figure 11A). While a sampled closure depth at the low end of the uncertainty space, such as 5–10 m (Figure 7B), assumes that only the nearshore surf zone acts as a sediment sink during sea-level rise, which is reasonable in any setting to maintain surf zone morphodynamics, a deeper closure depth imposes a more extensive sediment sink across the mid to lower shoreface. In reality, whether the shoreface represents a source or sink at each NSW beach will depend on the relationship between present shoreface geomorphology and the prevailing depositional controls (e.g., sediment distribution, type, and availability, and the energy regime). The presence of Holocene prograded barriers along parts of the central and southern NSW coasts is evidence that some of the shorefaces have acted as a source of sediments for adjacent beaches during the mid to late Holocene [24,90,91]. Whether or not the shoreface remains a significant source of sand supply for these NSW beaches, particularly under conditions of accelerating sea-level rise, remains an area of ongoing research [91]. Along the lower gradient northern NSW coast and shelf (Figure 1), the complete filling of many embayments with Pleistocene barrier deposits (and consequent absence of Holocene barrier deposits) [43], and the strong northward alongshore transport system, together make it more difficult to determine the current depositional relationship between beaches and the shoreface. Even where the shoreface does act as a sediment sink, the extent and timescale of shoreface response will likely depend on the rate and ultimate magnitude of future sea-level rise [96]. Similarly, the timescales and extent of estuarine response to sea-level rise [70–73] also remains largely unstudied and unknown on this coast, and for the larger north coast systems (Figure 1A) in particular, the volume and rate of river sediment supply is crucial to understanding the relative contribution from fluvial and littoral sources.

Comparison between the model parameterisations of shoreface geometry and composition, between the regional- and local-scale applications (Figure 8), demonstrates the importance of high-resolution coastal seabed mapping and sampling for understanding and quantifying coastal sediment budgets, and forecasting the potential for sediment redistribution and future shoreline change. Beyond the relationship between shoreface geometry and the prevailing depositional controls, the structure and composition of the seabed (Figure 3B) provides a direct insight to the potential for sediment accommodation across the shoreface—or the potential for ongoing shoreface sediment supply. For example, the presence of extensive low-relief reefs may be evidence of a sediment-deficient compartment, or the occurrence of protruding reef outcrops on an otherwise sedimentary shoreface suggests that sediment cannot accumulate in such areas under the prevailing energy conditions, thereby reducing the potential shoreface accommodation space. Understanding the balance of supply and accommodation in each sediment compartment is critical for interpreting if shoreface reefs represent negative sediment accommodation, an insufficient sediment supply, or both, in the context of sediment redistribution in response to sea-level rise.

Our approach provides a simple and scalable method to model potential shoreline change that considers distinctive beach morphology and compartment-based sediment budget principles, which we demonstrate through regional- and local-scale applications. However, we acknowledge that the approach remains a framework that would benefit from site-specific investigations and observation data to determine the applicability of our assumptions, and more rigorous methods of simulating shoreline response to sediment redistribution. While such data and methods already exist for some settings, and could be applied within our framework, our assumptions and the limitations of our approach highlight focus areas for future research. As a first step, an improved understanding of the probability distributions that describe key model variables (including sea-level rise) would help.

5.2. Exposure to Beach Erosion and Shoreline Change

Despite the limitations described above, regional-scale application of our modelling approach has enabled a second-pass assessment of property and infrastructure exposure to coastal erosion in NSW, which accounts for variation in the response of different beaches to both fluctuating and cumulative erosion—the first analysis and dataset of its kind. This is a considerable improvement on an earlier national-scale first-pass exposure assessment, which identified exposed assets simply by applying a uniform buffer distance (110 m) around all of the sandy and potentially erodible shorelines [56]. In comparison, our approach considers the distinctive morphology of individual NSW beaches (using LiDAR topography), in the context of the distinctive characteristics of the 47 secondary sediment compartments of the NSW coast. The regional-scale coastal erosion hazard mapping and exposure data that we present is designed for state-wide applications, such as to guide strategic planning along the NSW coast, and does not negate the need for more detailed investigations to inform local-scale coastal management and planning initiatives [106].

In NSW, assets that fall within the immediate (present-day) erosion hazard management zones, as defined by local governments, are considered to be potentially exposed to fluctuating beach erosion at present. Erosion hazard management zones are also defined for land and asset planning periods (often to 2050 and 2100), which consider the impacts of fluctuating beach erosion, the persistence of historical trends in shoreline change, and the potential response to sea-level rise. While the existing erosion hazard management zones may provide some indication of state-wide exposure, the coverage of NSW beaches is incomplete, and often focusses on locations that have been historically impacted by coastal erosion [57,58]. Therefore, while an analysis of the existing erosion hazard management zones may provide some indication of present exposure, it is anticipated that the potential future exposure is currently under-estimated.

To investigate this, we compiled the existing coastal erosion hazard mapping that is used by local governments to compare the identified exposure with our regional-scale modelling. We apply the same exposure assessment method (Section 3.5) to the incomplete coverage of local government erosion hazard management zones to determine the current definition of exposure. The results suggest that more than 2000 addresses are identified as being potentially exposed to coastal erosion at present, while that number increases to over 3700 and 6800 addresses by 2050 and 2100, respectively (Figure 17). When considering only properties for which more than half of the lot area may be affected, the exposure numbers reduce to 210 property lots (321 total addresses) potentially affected at present, rising to around 731 lots (1134 total addresses) at 2050, and 2040 lots (3315 total addresses) at 2100. The exposure figures that are based on existing coastal erosion hazard mapping compare with 247 property lots (455 total addresses) at present, 1862 lots (2718 total addresses) at 2050, and 3300 lots (5076 total addresses) at 2100, based on our regional-scale modelling (Figure 14).

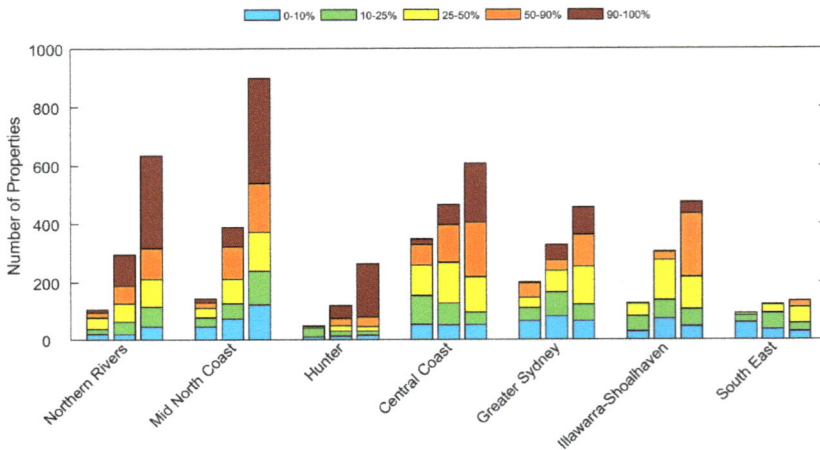

Figure 17. Address exposure by NSW regions (Figure 1A) for present (left column), at 2050 (middle column) and at 2100 (right column), based on existing coastal erosion hazard management zones prepared for local governments. Colouring categorises the exposed addresses by the proportion of each associated property lot that was intersected by local government erosion hazard zones.

As anticipated, the comparison suggests that the existing mapping likely captures most of the present-day exposure, but may under-estimate future exposure, due to the coverage bias toward locations at present risk from erosion. Comparison between Figures 14 and 17 shows that the distribution of exposure between the NSW regions is similar based on our regional-scale modelling and the existing erosion hazard management zones, respectively. Present exposure is highest in the relatively small area of central NSW (Central Coast and Greater Sydney regions), relative to northern NSW (Northern Rivers, Mid North Coast, and Hunter regions), and southern NSW (Illawarra-Shoalhaven and South East regions). This is reflected by the distribution of identified coastal erosion hot spots (Figure 1A). Exposure to coastal erosion increases in all of the regions by 2050 and 2100, with the majority of the increase representing addresses for which more than 25% of the property lot could be affected by erosion. The relative increase in exposure within and between regions is consistent with the patterns seen in Figure 14.

Differences in the rate and degree (as quantified by the affected land-area of exposed properties) of increasing exposure between regions reflects both regional coastal geomorphology, and the distribution of coastal development within each region. In central NSW, and southern NSW in particular, steeper and more rugged coastal geomorphology, coupled with higher exposure to the predominantly S-SE wave climate, contribute to well-developed dune morphology and relatively reduced dimensions of potential sediment sinks, such as the shoreface and estuaries (Appendix C). Combined with the restricted dimensions of low-lying coastal plains (for exposed developments to occur), and historically low development in southern NSW, these factors contribute to a lower potential increase in future exposure to coastal erosion relative to northern NSW. In contrast, the broad dimensions of potential sediment sinks in northern NSW suggests the potential for a greater increase in future exposure, based on our approach (Figure 14) and current practice (Figure 17).

The comparison between exposure figures based on our regional-scale modelling (Figure 14), and existing erosion hazard management zones that are used by local governments (Figure 17), suggests that: (1) the distribution and coverage of existing erosion hazard management zones along the NSW coast may be biased towards present exposure; and, (2) allowances for cumulative erosion through the present century in existing erosion hazard management zones may not reflect the full potential for sediment redistribution within coastal systems. This is evident in comparable present exposure being identified using both datasets, but considerably different exposure numbers for the 2050 and 2100 forecast periods. Kinsela and Hanslow [57] reviewed the various methods used to define erosion hazard management zones in NSW, and found that consideration of the potential response to sea-level rise, in particular, was often limited in scope by the application of a profile closure depth restricted to the upper shoreface (10–12 m water depth) only (c.f., Figure 7B), and the lack of consideration for the influence of sea-level rise on other sediment sinks, such as flood-tide delta environments within estuaries and tidal inlets. In contrast, our second-pass assessment and regional-scale mapping may over-estimate exposure where the shoreface and estuaries do not represent sediment sinks during sea-level rise (Section 5.1), again highlighting the need for improved data and methods to understand and model coastal sediment budgets.

5.3. Improving the Sediment Compartments Approach

Our approach demonstrates the utility of the sediment compartments framework, coupled with a statistical (Monte Carlo) model input sampling regime, for investigating the sensitivity of beach erosion and future shoreline change to depositional controls and coastal geomorphology, as defined at regional or local scales. Although we use the simple shoreline encroachment model to demonstrate the potential sensitivity of future shoreline change to the distribution, dimensions, and responses of sources and sinks within sediment-sharing coastal systems, more complex beach response and shoreline change models could be applied within our modelling approach. In any application, the model design and parameterisation should reflect the coverage and detail of available input data, and the intended purpose and required resolution of the model forecasts.

Our regional-scale application (Section 3.4.1) was limited by the availability of data for all NSW beaches to inform the parameterisation of various sediment budget components that may influence shoreline response on NSW beaches (Equation (3)). However, the resolution of the morphological data models captured the distinctive beach and dune topography of individual beaches, and was commensurate with the scale and intended application of the model predictions. The local-scale application (Section 3.4.2) demonstrates the scope for improvement in application of the approach where high-resolution data describing the coastal geomorphology (and processes) is available. For example, the use of regularly spaced (25-m alongshore) beach-dune profiles to capture the impact of fluctuating beach erosion, and alongshore-averaged beach-dune profiles to capture the distributed impact of cumulative sediment loss on shoreline change, is an important consideration where alongshore variability in the substrate and beach-barrier morphology presents the prospect of complex shoreline responses (Figure 10).

Beyond the simple assumption that the shoreface and estuaries act as sediment sinks during sea-level rise, uncertainty in both the second and third terms in Equation (3) also stems from limited knowledge regarding the response timescales of the shoreface [95,96] and estuarine depositional environments [72,73]. The fourth term in Equation (3) contains several volumetric sediment budget components that are potentially relevant to the future shoreline change on NSW beaches. However, those components remain difficult to quantify due to a limited understanding of the associated sediment transport processes, and were omitted from the regional-scale application. In our local-scale example (Section 3.4.2), we relied on expert opinion from a nearby beach to estimate the potential contribution of V_M and V_B to apply in our approach. As such, the second, third, and fourth terms of Equation (3) highlight focus areas for future research to better understand sediment dynamics on the NSW coast (and elsewhere), particularly considering the projected effects of global climate change within the present century.

A rigorous sub-compartment classification, and the mapping and quantification of sediment sources, sinks, and pathways, relies on an adequate understanding of the depositional environments of coastal sediment-sharing systems. Undoubtedly, the potential for the sediment compartments framework to improve the reliability of shoreline change forecasts depends on the coverage and detail of geomorphic data describing the distribution, dimensions, and connectivity of sources and sinks within coastal systems. Our understanding of sediment transport processes then determines uncertainty in the rate and volume of sediment redistribution in response to environmental change (e.g., sea-level rise). Confidence in our tertiary- and sub-compartment classification (Section 3.1) was limited by the coverage and resolution of bathymetry and seabed substrate data that is available along the NSW coastline. However, our examples suggest that increased effort in geomorphic data collection should lead to more reliable forecasts of future shoreline change, by enabling much refined sediment budget parameterisations, and providing new insights to the likely response of key sources and sinks to sea-level rise. While detailed geomorphic mapping has been completed for the coastal plains and valleys of the NSW coast [74,101,102], equivalent mapping describing the geomorphology of the inner-continental shelf, in particular, is essential to developing more refined shoreline change forecasts using the sediment compartments framework and sediment budget principles.

6. Conclusions

1. Coastal sediment compartments provide a hierarchical framework to conceptualise and quantify potential sediment redistribution between the various depositional environments (sources and sinks) of sediment-sharing coastal systems. Sub-compartment classifications allow for sediment transport processes, which accumulate into meaningful sediment exchanges between sources and sinks across varying time scales, to be connected with the spatial scales of their impact on beach fluctuation and cumulative shoreline change.

2. Volumetric approaches to modelling fluctuating and cumulative erosion provide a means to forecast the impacts of compartment-based sediment redistribution on beach and shoreline response, which reflects both compartment sediment budgets and transport pathways, and the distinctive beach-face and dune morphology of individual beaches.

3. Based on our simplistic modelling approach and assumptions, exposure to coastal erosion is expected to increase into the future on open-coast NSW beaches, primarily due to the influence of sea-level rise on shoreline recession, driven by the redistribution of beach and dune sand to adjacent depositional environments (sediment sinks). The increase in exposure will vary between NSW beaches, reflecting regional- and local-scale variation in coastal geomorphology, and the present (and future) distribution of coastal development within each region.

4. Assumptions regarding the response of key depositional environments (e.g., the shoreface and estuaries) to sea-level rise remains the most significant limitation to the reliability of long-term shoreline change forecasts, because of the overwhelming potential sediment demand that is imposed on littoral sediment budgets. Site-specific data and investigation is necessary to determine the likely roles and morphological response rates of these depositional environments, as sources or sinks within sediment-sharing systems.

5. Opportunities to improve shoreline change forecasting based on the sediment compartments framework increase with the coverage and resolution of geomorphic data that is available to describe the distribution, dimensions, and depositional histories of sediment sources and sinks.

For example, detailed seabed mapping and sampling covering the inner-continental shelf and estuary inlets is critical to reducing uncertainty in the future responses of shoreface and flood-tide delta depositional environments to sea-level rise.

Acknowledgments: This study was funded by research grants from the New South Wales (NSW) Environment Trust, Climate Change Fund, and Office of Emergency Management (Natural Disasters Resilience Program). The authors thank the School of Aviation and the Water Research Laboratory at UNSW Sydney for the use of airborne LiDAR data collected at Wamberal Beach in 2016, funded by the Australian Research Council (DP150101339) with assistance from the NSW Office of Environment and Heritage Coastal Processes and Responses Node. The authors thank Bruce Thom, Andrew Short, Angus Gordon and Marc Daley for helpful comments on the design of the modelling approach. The authors also thank two anonymous journal reviewers for detailed, insightful and constructive comments that greatly improved the manuscript.

Author Contributions: M.A.K. and D.J.H. conceived the overall approach. M.A.K. lead the design and development of the erosion modelling and exposure assessment methods, and carried out the modelling and analyses. B.D.M. contributed to the design of the erosion modelling and exposure assessment methods and lead the development of the modelling and analysis tools. M.L. contributed to the sediment sub-compartment classification methods and lead the development of the sediment compartment datasets. M.A.K. prepared the manuscript with input from all co-authors. D.J.H. provided input and review on all aspects of the study.

Conflicts of Interest: The authors declare no conflict of interest. The funding bodies had no role in the design of the study; in the collection, analyses, or interpretation of data; in the writing of the manuscript, and in the decision to publish the results.

Appendix A

The shoreline orientation of each beach, as recorded in the Australian beach database, was used as a pragmatic means to scale fluctuating erosion (F) for sheltered NSW beaches. The scaling coefficient for fluctuating beach erosion (c_f in Equation (1)) was adjusted based on the average shoreline orientation of each beach or beach sector, as described in Table A1. The a, b and c parameters describe the lower bound, mode, and upper bound of triangular probability functions respectively. A beach with shoreline orientation of 90° faces due east on average, while a beach with shoreline orientation of 180° faces due south. The relatively higher exposure of north coast beaches to easterly storm wave conditions, which are more common in northern NSW, was accounted for by scaling fluctuating erosion for beaches with shoreline orientation <80°. In contrast, for central and southern NSW beaches, scaling was applied for shoreline orientations <90°. Considering the south to southeast wave climate, fluctuating erosion was also scaled for all beaches with shoreline orientation >180°.

Table A1. Scaling values for fluctuating erosion (F) applied in the regional-scale example, to account for shoreline exposure to wave climate. Exposure varies between beaches based on the average shoreline orientation for each beach or sector. The a, b and c and values correspond to the lower bound, mode and upper bound, respectively, of the triangular probability functions used to represent c_f.

Shoreline	South/Central			North		
(°)	a	b	c	a	b	c
0–29	0.5	0.55	0.6	0.5	0.55	0.6
30–59	0.6	0.65	0.7	0.6	0.65	0.7
60–70	0.7	0.75	0.8	0.7	0.75	0.8
70–74	0.7	0.75	0.8	0.8	0.85	0.9
75–79	0.7	0.75	0.8	0.9	0.95	1
80–84	0.8	0.85	0.9	1	1	1
85–89	0.9	0.95	1	1	1	1
90–179	1	1	1	1	1	1
180–189	0.9	0.95	1	0.9	0.95	1
190–199	0.8	0.85	0.9	0.8	0.85	0.9
200–219	0.7	0.75	0.8	0.7	0.75	0.8
220–239	0.6	0.65	0.7	0.6	0.65	0.7
240–299	0.5	0.55	0.6	0.5	0.55	0.6
300–360	0.4	0.45	0.5	0.4	0.45	0.5

Appendix B

To evaluate the suitability of the gamma function (Figure 6B) for describing the probability of fluctuating erosion (F) on fully exposed NSW beaches, we compared model predictions from our present-1% exceedance level scenario, with historical maximum erosion escarpments where they have been mapped on NSW beaches. A recent investigation of the geohistorical record of beach response to severe storms on this coastline compiled historical maximum erosion escarpments from 10 NSW beaches, which were identified using photogrammetry analysis of aerial photographs [87]. Figure A1 shows the result of the comparison for three representative fully exposed NSW beaches from that dataset. The comparison indicates that the 1% exceedance level fluctuating erosion volume in our approach is consistent with the impacts of some of the most severe coastal storms that have occurred during the recent historical period.

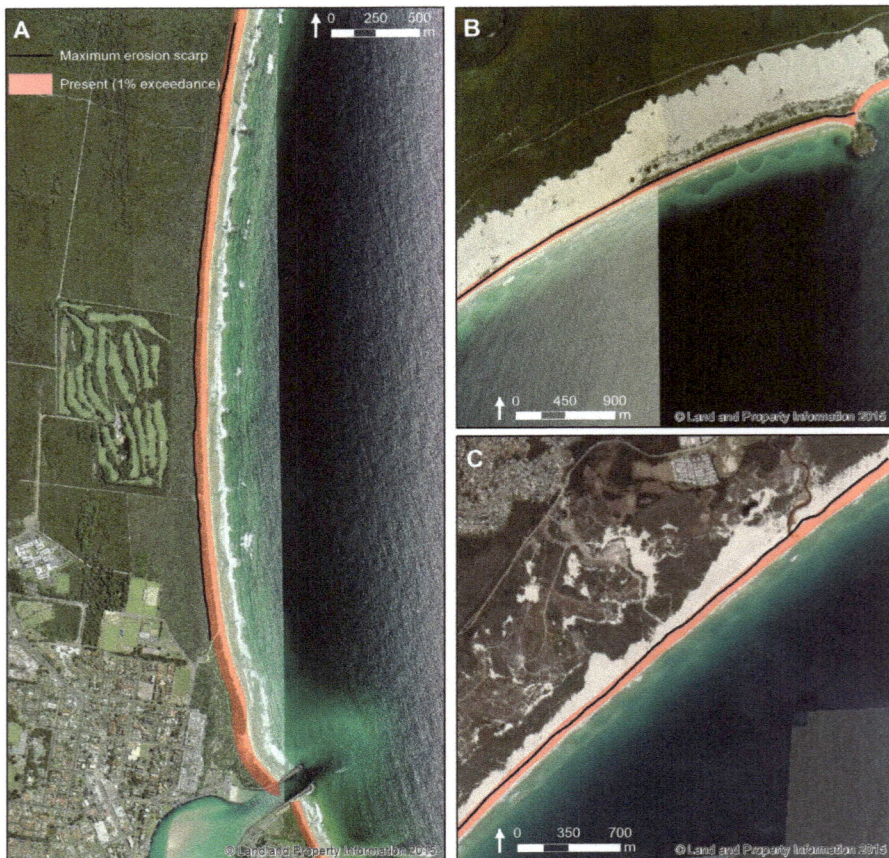

Figure A1. Comparison between mapped historical maximum erosion escarpments and the present-1% exceedance erosion scenario from our regional-scale modelling. Examples are provided for the following exposed NSW beaches where historical maximum erosion escarpments have been mapped (year of maximum erosion event is indicated): (**A**) Nine Mile Beach at Tuncurry, Mid North Coast compartment (1963); (**B**) Bennetts Beach, Port Stephens compartment (1974); and (**C**) Redhead Beach, Central Coast compartment (1974). See Figure 1A for compartment locations.

J. Mar. Sci. Eng. **2017**, *5*, 61

Appendix C

The shoreface (V_S) and estuarine flood-tide delta (V_E) depositional environments represent prominent sediment sinks in the cumulative erosion function of the simple shoreline encroachment model (Equation (3)). The degree of influence that they have on forecast shoreline change varies within each model simulation, depending on the input values randomly sampled from the relevant probability functions (e.g., Figure 7). However, low-exceedance forecasts (e.g., 1% exceedance level) are representative of input values sampled from the upper tails of the input probability distributions.

To interpret the relative influence of the shoreface and estuary sediment sinks on variability in forecast shoreline change along the NSW coast, we provide their total summed surface areas for each primary compartment in Table A2. The surface areas are also normalised by dividing by the length of sandy shorelines within each compartment, to account for the varying alongshore extents of the primary compartments (Figure 1A). Generally speaking, the simple shoreline encroachment model assumes that beaches within compartments with higher normalised shoreface and estuary delta surface areas will experience greater sediment loss to these sinks during sea-level rise.

Table A2. Summary statistics by primary sediment compartment (Figure 1A) describing the total length of sandy shorelines, total shoreface surface area (0–40 m water depth) and total surface area of estuarine flood-tide deltas. The shoreface and estuary delta areas are also expressed as area per metre of sandy shoreline to account for the varying alongshore extents of the compartments.

Primary Compartment	Shoreline Length [1] (km)	Total Shoreface Area (km^2)	Normalised Shoreface Area [2] (m^2/m)	Total Estuary Delta Area (km^2)	Normalised Estuary Delta Area [3] (m^2/m)
North Coast	152.1	841.7	5534.5	15.0	98.6
Northern Rivers	164.0	1080.6	6587.7	23.8	144.9
Mid North Coast	157.4	875.3	5562.4	7.3	46.5
Port Stephens	118.4	502.7	4246.0	3.2	27.2
Central Coast	66.0	355.3	5383.7	4.8	73.0
Sydney	66.0	174.8	2647.5	0.76	11.5
Illawarra	79.7	474.1	5946.7	3.5	44.3
Shoalhaven	93.3	369.3	3956.3	2.2	23.7
South Coast	167.3	635.4	3797.2	8.8	52.5

[1] Sandy shorelines primarily influenced by open-coast processes as defined by Smartline [37]. [2] Total shoreface (0–40 m water depth) surface area divided by sandy shoreline length. [3] Total estuarine flood-tide delta surface area divided by sandy shoreline length.

References

1. Harley, M.D.; Turner, I.L.; Kinsela, M.A.; Middleton, J.H.; Mumford, P.J.; Splinter, K.D.; Phillips, M.S.; Simmons, J.A.; Hanslow, D.J.; Short, A.D. Extreme coastal erosion enhanced by anomalous extratropical storm wave direction. *Sci. Rep.* **2017**, *7*, 6033. [CrossRef] [PubMed]

2. Thom, B.G.; Hall, W. Behavior of beach profiles during accretion and erosion dominated periods. *Earth Surf. Process. Landf.* **1991**, *16*, 113–127. [CrossRef]

3. Phillips, M.S.; Harley, M.D.; Turner, I.L.; Splinter, K.D.; Cox, R.J. Shoreline recovery on wave-dominated sandy coastlines: The role of sandbar morphodynamics and nearshore wave parameters. *Mar. Geol.* **2017**, *385*, 146–159. [CrossRef]

4. Harley, M.D.; Turner, I.L.; Middleton, J.H.; Kinsela, M.A.; Hanslow, D.J.; Splinter, K.D.; Mumford, P.J. Observations of beach recovery in SE Australia following the June 2016 east coast low. In Proceedings of the Coasts & Ports 2017, Cairns, Australia, 21–23 June 2017.

5. Barnard, P.L.; Short, A.D.; Harley, M.D.; Splinter, K.D.; Vitousek, S.; Turner, I.L.; Allan, J.; Banno, M.; Bryan, K.R.; Doria, A.; et al. Coastal vulnerability across the Pacific dominated by El Niño/Southern Oscillation. *Nat. Geosci.* **2015**, *8*, 801–807. [CrossRef]

6. Mortlock, T.R.; Goodwin, I.D. Impacts of enhanced central Pacific ENSO on wave climate and headland-bay beach morphology. *Cont. Shelf Res.* **2016**, *120*, 14–25. [CrossRef]

7. Goodwin, I.D.; Mortlock, T.R.; Browning, S. Tropical and extratropical-origin storm wave types and their influence on the East Australian longshore sand transport system under a changing climate. *J. Geophys. Res. Oceans* **2016**, *121*, 4833–4853. [CrossRef]
8. Leatherman, S.P. Social and economic costs of sea level rise. *Int. Geophys.* **2001**, *75*, 181–223.
9. Phillips, M.R.; Jones, A.L. Erosion and tourism infrastructure in the coastal zone: Problems, consequences and management. *Tour. Manag.* **2006**, *27*, 517–524. [CrossRef]
10. Barbier, E.B.; Hacker, S.D.; Kennedy, C.; Kock, E.W.; Stier, A.C. The value of estuarine and coastal ecosystem services. *Ecol. Monogr.* **2011**, *81*, 169–193. [CrossRef]
11. Gopalakrishnan, S.; Smith, M.D.; Slott, J.M.; Murray, A.B. The value of disappearing beaches: A hedonic pricing model with endogenous beach width. *J. Environ. Econ. Manag.* **2011**, *61*, 297–310. [CrossRef]
12. Murray, A.B.; Gopalakrishnan, S.; McNamara, D.E.; Smith, M.D. Progress in coupling models of human and coastal landscape change. *Comput. Geosci.* **2013**, *53*, 30–38. [CrossRef]
13. Williams, Z.C.; McNamara, D.E.; Smith, M.D.; Murray, A.B.; Gopalakrishnan, S. Coupled economic-coastline modeling with suckers and free riders. *J. Geophys. Res. Earth Surf.* **2013**, *118*, 887–899. [CrossRef]
14. Lazarus, E.D.; Ellis, M.A.; Murray, A.B.; Hall, D.M. An evolving research agenda for human-coastal systems. *Geomorphology* **2016**, *256*, 81–90. [CrossRef]
15. Jongejan, R.; Ranasinghe, R.; Wainwright, D.; Callaghan, D.P.; Reyns, J. Drawing the line on coastline recession risk. *Ocean Coast. Manag.* **2016**, *122*, 87–94. [CrossRef]
16. FitzGerald, D.M.; Fenster, M.S.; Argow, B.A.; Buynevich, I.V. Coastal impacts due to sea-level rise. In *Annual Review of Earth and Planetary Sciences*; Annual Reviews: Palo Alto, CA, USA, 2008; pp. 601–647.
17. Stive, M.J.F.; Cowell, P.J.; Nicholls, R.J. Impacts of Global Environmental Change on Beaches, Cliffs and Deltas. In *Geomorphology and Global Environmental Change*; Slaymaker, O., Spencer, T., Embleton-Hamann, C., Eds.; International Association of Geomorphologists, Cambridge University Press: Cambridge, UK, 2009; pp. 158–179.
18. Zhang, K.; Douglas, B.C.; Leatherman, S.P. Global warming and coastal erosion. *Clim. Chang.* **2004**, *64*, 41–58. [CrossRef]
19. Church, J.A.; Clark, P.U.; Cazenave, A.; Gregory, J.M.; Jevrejeva, S.; Levermann, A.; Merrifield, M.A.; Milne, G.A.; Nerem, R.S.; Nunn, P.D.; et al. 2013: Sea Level Change. In *Climate Change 2013: The Physical Science Basis. Contribution of Working Group I to the Fifth Assessment Report of the Intergovernmental Panel on Climate Change*; Stocker, T.F., Qin, D., Plattner, G.-K., Tignor, M., Allen, S.K., Boschung, J., Nauels, A., Xia, Y., Bex, V., Midgley, P.M., Eds.; Cambridge University Press: Cambridge, UK, 2013.
20. Sweet, W.V.; Kopp, R.E.; Weaver, C.P.; Obeysekera, J.; Horton, R.M.; Thieler, E.R.; Zervas, C. *Global and Regional Sea Level Rise Scenarios for the United States*; National Oceanic and Atmospheric Administration (NOAA): Silver Spring, MD, USA, 2017.
21. Cowell, P.J.; Stive, M.J.F.; Niedoroda, A.W.; de Vriend, H.J.; Swift, D.J.P.; Kaminsky, G.M.; Capobianco, M. The coastal-tract (part 1): A conceptual approach to aggregated modeling of low-order coastal change. *J. Coast. Res.* **2003**, *19*, 812–827.
22. Komar, P.D. The budget of littoral sediments: Concepts and applications. *Shore Beach* **1996**, *64*, 18–26.
23. Rosati, J.D. Concepts in sediment budgets. *J. Coast. Res.* **2005**, *21*, 307–322. [CrossRef]
24. Cowell, P.J.; Stive, M.J.F.; Niedoroda, A.W.; Swift, D.J.P.; de Vriend, H.J.; Buijsman, M.C.; Nicholls, R.J.; Roy, P.S.; Kaminsky, G.M.; Cleveringa, J.; et al. The coastal-tract (part 2): Applications of aggregated modeling of lower-order coastal change. *J. Coast. Res.* **2003**, *19*, 828–848.
25. French, J.; Burningham, H.; Thornhill, G.; Whitehouse, R.; Nicholls, R.J. Conceptualising and mapping coupled estuary, coast and inner shelf sediment systems. *Geomorphology* **2016**, *256*, 17–35. [CrossRef]
26. Davies, J.L. The coastal sediment compartment. *Aust. Geogr. Stud.* **1974**, *12*, 139–151. [CrossRef]
27. Patsh, K.; Griggs, G. *Development of Sand Budgets for California's Major Littoral Cells: Eureka, Santa Cruz, Southern Monterey Bay, Santa Barbara, Santa Monica (Including Zuma), San Pedro, Laguna, Oceanside, Mission Bay, and Silver Strand Littoral Cells*; Institute of Marine Sciences, University of California: Santa Cruz, CA, USA, 2007.
28. Bray, M.J.; Carter, D.J.; Hooke, J.M. Littoral cell definition and budgets for central southern England. *J. Coast. Res.* **1995**, *11*, 381–400.

29. Cooper, N.J.; Pontee, N.I. Appraisal and evolution of the littoral "sediment cell" concept in applied coastal management: Experiences from England and Wales. *Ocean Coast. Manag. Coast. Manag.* **2006**, *49*, 498–510. [CrossRef]

30. Sanderson, P.G.; Eliot, I. Compartmentalisation of beachface sediments along the southwestern coast of Australia. *Mar. Geol.* **1999**, *162*, 145–164. [CrossRef]

31. Eliot, I.; Gozzard, B.; Nutt, C. Geologic frameworks for coastal planning and management. In Proceedings of the Australasian Coasts & Ports Conference 2011, Perth, Australia, 28–30 September 2011.

32. Eliot, I.; Nutt, C.; Gozzard, J.; Higgins, M.; Buckley, E.; Bowyer, J. *Coastal Compartments of Western Australia: A Physical Framework for Marine and Coastal Planning*; Damara WA Pty. Ltd.: Perth, Australia, 2011.

33. McPherson, A.; Hazelwood, M.; Moore, D.; Owen, K.; Nichol, S.; Howard, F.J.F. *The Australian Coastal Sediment Compartments Project: Methodology and Product Development*; Record 2015/25; Geoscience Australia: Canberra, Australia, 2015.

34. Thom, B.G.; Eliot, I.; Eliot, M.; Harvey, N.; Rissik, D.; Sharples, C.; Short, A.D.; Woodroffe, C.D. National sediment compartment framework for Australian coastal management. *Ocean Coast. Manag.* **2017**, in press.

35. Short, A.D. Australian beach systems—Nature and distribution. *J. Coast. Res.* **2006**, *22*, 11–27. [CrossRef]

36. Short, A.D. *Beaches of the New South Wales Coast*, 2nd ed.; Sydney University Press: Sydney, Australia, 2007.

37. Sharples, C.; Mount, R.; Pedersen, T.; Lacey, M.; Newton, J.; Jaskierniak, D.; Wallace, L. *The Australian Coastal Smartilne Geomorphic and Stability Map Version 1: Project Report*; School of Geography and Environmental Studies (Spatial Sciences), University of Tasmania: Hobart, Tasmania, 2009.

38. Wright, L.D. Nearshore wave-power dissipation and coastal energy regime of Sydney-Jervis Bay region, New-South-Wales—Comparison. *Aust. J. Mar. Freshw. Res.* **1976**, *27*, 633–640. [CrossRef]

39. Short, A.D.; Trenaman, N.L. Wave climate of the Sydney region, an energetic and highly variable ocean wave regime. *Aust. J. Mar. Freshw. Res.* **1992**, *43*, 765–791. [CrossRef]

40. Boyd, R.; Ruming, K.; Goodwin, I.; Sandstrom, M.; Schroder-Adams, C. Highstand transport of coastal sand to the deep ocean: A case study from Fraser Island, southeast Australia. *Geology* **2008**, *36*, 15–18. [CrossRef]

41. Goodwin, I.D.; Freeman, R.; Blackmore, K. An insight into headland sand bypassing and wave climate variability from shoreface bathymetric change at Byron Bay, New South Wales, Australia. *Mar. Geol.* **2013**, *341*, 29–45. [CrossRef]

42. Browning, S.A.; Goodwin, I.D. Large-scale influences on the evolution of winter subtropical maritime cyclones affecting Australia's east coast. *Mon. Weather Rev.* **2013**, *141*, 2416–2431. [CrossRef]

43. Roy, P.S.; Thom, B.G. Late Quaternary marine deposition in New South Wales and southern Queensland—An evolutionary model. *J. Geol. Soc. Aust.* **1981**, *28*, 471–489. [CrossRef]

44. NSW Coastal Erosion Hot Spots. Office of Environment and Heritage (OEH). Available online: http://www.environment.nsw.gov.au/coasts/coasthotspots.htm (accessed on 8 August 2017).

45. Thom, B.G.; Roy, P.S. Relative sea levels and coastal sedimentation in southeast Australia in the Holocene. *J. Sediment. Petrol.* **1985**, *55*, 257–264.

46. Thom, B.G.; Keene, J.B.; Cowell, P.J.; Daley, M. East Australian marine abrasion surface. In *Australian Landscapes*; Bishop, P., Pillans, B., Eds.; Geological Society: London, UK, 2010; pp. 57–59.

47. Gordon, A.D.; Hoffman, J.G. Sediment features and processes of the Sydney continental shelf. In *Recent Sediments in Eastern Australia—Marine Through Terrestrial*; Frankel, E., Keene, J.B., Waltho, A.E., Eds.; Geological Society of Australia Special Publication: Sydney, Australia, 1986; pp. 29–51.

48. Roy, P.S.; Cowell, P.J.; Ferland, M.A.; Thom, B.G. Wave dominated coasts. In *Coastal Evolution: Late Quaternary Shoreline Morphodynamics*; Carter, R.W.G., Woodroffe, C.D., Eds.; Cambridge University Press: Cambridge, UK, 1994; pp. 121–186.

49. Thom, B.G. Transgressive and regressive stratigraphies of coastal sand barriers in southeast Australia. *Mar. Geol.* **1984**, *56*, 137–158. [CrossRef]

50. Thom, B.G. Coastal erosion in eastern Australia. *Search* **1974**, *5*, 198–209. [CrossRef]

51. Chapman, D.M.; Geary, M.; Roy, P.S.; Thom, B.G. *Coastal Evolution and Coastal Erosion in New South Wales*; Coastal Council of New South Wales: Sydney, Austrilia, 1982.

52. Hanslow, D.J.; Howard, M. Emergency Management of Coastal Erosion in NSW. In *Planning for Natural Hazards—How Can We Mitigate the Impacts?* Morrison, R.J., Quin, S., Bryant, E.A., Eds.; Proceedings of the Symposium on Natural Hazards: Wollongong, Austrilia, 2005; pp. 103–116.

53. Mortlock, T.; Goodwin, I.; McAneney, J.; Roche, K. The June 2016 Australian East Coast Low: Importance of Wave Direction for Coastal Erosion Assessment. *Water* **2017**, *9*, 121. [CrossRef]
54. Proudfoot, M.; Petersen, L.S. Positive SOI, negative PDO and spring tides as simple indicators of the potential for extreme coastal erosion in northern NSW. *Aust. J. Environ. Manag.* **2011**, *18*, 170–181. [CrossRef]
55. Chapman, D.M. Coastal erosion and the sediment budget, with special reference to the Gold Coast, Australia. *Coast. Eng.* **1981**, *4*, 207–227. [CrossRef]
56. *Climate Change Risks to Australia's Coasts: A First Pass National Assessment*; Australian Government Department of Climate Change: Canberra, Australia, 2009.
57. Kinsela, M.A.; Hanslow, D.J. Coastal erosion risk assessment in New South Wales: Limitations and future directions. In Proceedings of the 22nd NSW Coastal Conference, Port Macquarie, Australia, 12–15 November 2013.
58. Wainwright, D.J.; Ranasinghe, R.; Callaghan, D.P.; Woodroffe, C.D.; Cowell, P.J.; Rogers, K. An argument for probabilistic coastal hazard assessment: Retrospective examination of practice in New South Wales, Australia. *Ocean Coast. Manag.* **2014**, *95*, 147–155. [CrossRef]
59. Stul, T.; Gozzard, J.R.; Eliot, I.G.; Eliot, M.J. *Coastal Sediment Cells between Cape Naturaliste and the Moore River, Western Australia*; Geological Survey of Western Australia for the Western Australian Department of Transpo: Perth, Australia, 2012.
60. Carvalho, R.C.; Woodroffe, C.D. From catchment to inner shelf: Insights into NSW coastal compartments. In Proceedings of the 24th NSW Coastal Conference, Forster, Australia, 11–13 November 2015.
61. Kinsela, M.A.; Morris, B.D.; Daley, M.J.A.; Hanslow, D.J. A flexible approach to forecasting coastline change on wave-dominated beaches. *J. Coast. Res.* **2016**, *SI75*, 952–956. [CrossRef]
62. Cowell, P.J.; Roy, P.S.; Jones, R.A. Simulation of large-scale coastal change using a morphological behavior model. *Mar. Geol.* **1995**, *126*, 45–61. [CrossRef]
63. Leatherman, S.P. Barrier dynamics and landward migration with Holocene sea-level rise. *Nature* **1983**, *301*, 415–417. [CrossRef]
64. Moore, L.J.; List, J.H.; Williams, S.J.; Stolper, D. Complexities in barrier island response to sea level rise: Insights from numerical model experiments, North Carolina Outer Banks. *J. Geophys. Res.* **2010**, *115*, F03004. [CrossRef]
65. Lorenzo-Trueba, J.; Ashton, A.D. Rollover, drowning, and discontinuous retreat: Distinct modes of barrier response to sea-level rise arising from a simple morphodynamic model. *J. Geophys. Res. Earth Surf.* **2014**, *119*, 779–801. [CrossRef]
66. Walters, D.; Moore, L.J.; Vinent, O.D.; Fagherazzi, S.; Mariotti, G. Interactions between barrier islands and backbarrier marshes affect island system response to sea level rise: Insights form a coupled model. *J. Geophys. Res. Earth Sci.* **2014**, *119*, 2013–2031. [CrossRef]
67. Bruun, P. Sea level rise as a cause of shore erosion. *J. Waterw. Harb. Div. ASCE* **1962**, *88*, 117–130.
68. Bruun, P. Review of conditions for uses of the Bruun Rule of erosion. *Coast. Eng.* **1983**, *7*, 77–89. [CrossRef]
69. Bruun, P. The Bruun Rule of erosion by sea-level rise—A discussion on large-scale two- and three-dimensional usages. *J. Coast. Res.* **1988**, *4*, 627–648.
70. Eysink, W.D. Morphologic response of tidal basins to changes. In *22nd International Conference on Coastal Engineering*; American Society of Civil Engineers: Reston, VA, USA, 1990; pp. 1948–1961.
71. Stive, M.J.F.; Capobianco, M.; Wang, Z.B.; Ruol, P.; Buijsman, M.C. Morphodynamics of a tidal lagoon and the adjacent coasts. In *8th International Biennial Conference on Pysics of Estuaries and Coastal Seas*; Dronkers, J., Scheffers, M., Eds.; Elsevier: Roterdam, The Netherland, 1998; pp. 397–407.
72. Van Goor, M.A.; Zitman, T.J.; Wang, Z.B.; Stive, M.J.F. Impact of sea-level rise on the morphological equilibrium state of tidal inlets. *Mar. Geol.* **2003**, *202*, 211–227. [CrossRef]
73. Kragtwijk, N.G.; Zitman, T.J.; Stive, M.J.F.; Wang, Z.B. Morphological response of tidal basins to human interventions. *Coast. Eng.* **2004**, *51*, 207–221. [CrossRef]
74. Spatial Services Imagery and Elevation Programs: NSW Digital Elevation Data Set. Available online: http://spatialservices.finance.nsw.gov.au/mapping_and_imagery/imagery_programs (accessed on 5 August 2017).
75. Nielsen, A.F.; Lord, D.B.; Poulos, H.G. Dune stability considerations for building foundations. *Civil Eng. Trans. Inst. Eng. Aust.* **1992**, *CE34*, 167–174.

76. Murray, A.B.; Gasparini, N.M.; Goldstein, E.B.; van der Wegen, M. Uncertainty quantification in modeling earth surface processes: More applicable for some types of models than for others. *Comput. Geosci.* **2016**, *90*, 6–16. [CrossRef]

77. Simmons, J.A.; Harley, M.D.; Marshall, L.A.; Turner, I.L.; Splinter, K.D.; Cox, R.J. Calibrating and assessing uncertainty in coastal numerical models. *Coast. Eng.* **2017**, *125*, 28–41. [CrossRef]

78. Cowell, P.J.; Thom, B.G.; Jones, R.A.; Everts, C.H.; Simanovic, D. Management of uncertainty in predicting climate-change impacts on beaches. *J. Coast. Res.* **2006**, *22*, 232–245. [CrossRef]

79. Mariani, A.; Flocard, F.; Carley, J.T.; Drummond, C.D.; Guerry, N.; Gordon, A.D.; Cox, R.J.; Turner, I.L. *East Coast Study Project—National Geomorphoc Framework for the Management and Prediction of Coastal Erosion*; Water Research Laboratory, University of New South Wales: Sydney, Australia, 2013.

80. Anderson, T.R.; Fletcher, C.H.; Barbee, M.M.; Frazer, N.; Romine, B. Doubling of coastal erosion under rising sea level by mid-century in Hawaii. *Nat. Hazards* **2015**, *78*, 75–103. [CrossRef]

81. Wainwright, D.J.; Ranasinghe, R.; Callaghan, D.P.; Woodroffe, C.D.; Jongejan, R.; Dougherty, A.J.; Rogers, K.; Cowell, P.J. Moving from deterministic towards probabilistic coastal hazard and risk assessment: Development of a modelling framework and application to Narrabeen Beach, New South Wales, Australia. *Coast. Eng.* **2015**, *96*, 92–99. [CrossRef]

82. Roy, C.J.; Oberkampf, W.L. A comprehensive framework for verification, validation, and uncertainty quantification in scientific computing. *Comput. Methods Appl. Mech. Eng.* **2011**, *200*, 2131–2144. [CrossRef]

83. Gordon, A.D. Beach fluctuations and shoreline change—NSW. In Proceedings of the 8th Australasian Conference on Coastal and Ocean Engineering, Launceston, Australia, 1987.

84. Rollason, V.; Gordon, A. Back to the future of beach fluctuations and shoreline change. In Proceedings of the 24th NSW Coastal Conference, Forster, Australia, 11–13 November 2015.

85. Callaghan, D.P.; Nielsen, P.; Short, A.; Ranasinghe, R. Statistical simulation of wave climate and extreme beach erosion. *Coast. Eng.* **2008**, *55*, 375–390. [CrossRef]

86. Callaghan, D.P.; Ranasinghe, R.; Roelvink, D. Probabilistic estimation of storm erosion using analytical, semi-empirical, and process based storm erosion models. *Coast. Eng.* **2013**, *82*, 64–75. [CrossRef]

87. Goodwin, I.D.; Burke, A.; Mortlock, T.; Freeman, R.; Browning, S.A. *Technical Report of the Eastern Seaboard Climate Change Initiative on East Coast Lows (ESCCI-ECLs) Project 4: Coastal System Response to Extreme East Coast Low Clusters in the Geohistorical Archive*; Marine Climate Risk Group, Climate Futures, Macquarie University: Sydney, Australia, 2015.

88. Harley, M.D.; Turner, I.L.; Short, A.D.; Ranasinghe, R. A reevaluation of coastal embayment rotation: The dominance of cross-shore versus alongshore sediment transport processes, Collaroy-Narrabeen Beach, southeast Australia. *J. Geophys. Res. Earth Surf.* **2011**, *116*. [CrossRef]

89. Davies, G.; Callaghan, D.P.; Gravios, U.; Jiang, W.; Hanslow, D.; Nichol, S.; Baldock, T. Improved treatment of non-stationary conditions and uncertainties in probabilistic models of storm wave climate. *Coast. Eng.* **2017**, *127*, 1–19. [CrossRef]

90. Cowell, P.J.; Stive, M.J.F.; Roy, P.S.; Kaminsky, G.M.; Buijsman, M.C.; Thom, B.G.; Wright, L.D. Shoreface sand supply to beaches. In Proceedings of the 27th International Coastal Engineering Conference, Sydney, Australia, 16–21 July 2001; pp. 2495–2508.

91. Kinsela, M.A.; Daley, M.J.A.; Cowell, P.J. Origins of Holocene coastal strandplains in Southeast Australia: Shoreface sand supply driven by disequilibrium morphology. *Mar. Geol.* **2016**, *374*, 14–30. [CrossRef]

92. Church, J.A.; White, N.J.; Domingues, C.M.; Monselesan, D.P.; Miles, E.R. *Sea-Level and Ocean Heat-Content Change*, 2nd ed.; Elsevier: Amsterdam, The Netherlands, 2013; p. 103.

93. Church, J.A.; McInnes, K.L.; Monselesan, D.; O'Grady, J. *Sea-Level Rise and Allowances for Coastal Councils around Australia—Guidance Material*; Commonwealth Scientific and Industrial Research Organisation (CSIRO): Canberra, Australia, 2016.

94. Niedoroda, A.W.; Swift, D.J.P.; Hopkins, T.S.; Ma, C.M. Shoreface morphodynamics on wave-dominated coasts. *Mar. Geol.* **1984**, *60*, 331–354. [CrossRef]

95. Stive, M.J.F.; de Vriend, H.J. Modeling shoreface profile evolution. *Mar. Geol.* **1995**, *126*, 235–248. [CrossRef]

96. Kinsela, M.A.; Cowell, P.J. Controls on shoreface response to sea level change. In Proceedings of the Coastal Sediments '15, San Diego, CA, USA, 11–15 May 2015.

97. Hallermeier, R.J. A profile zonation for seasonal sand beaches from wave climate. *Coast. Eng.* **1981**, *4*, 253–277. [CrossRef]

98. Meleo, J.F. *Shoreface Variability in Southeastern Australia*; The University of Sydney: Sydney, Australia, 1994.

99. Cowell, P.J.; Hanslow, D.J.; Meleo, J.F. The shoreface. In *Handbook of Beach and Shoreface Morphodynamics*; Short, A.D., Ed.; John Wiley: Hoboken, NJ, USA, 1999; p. 392.

100. Patterson, D.C. Shoreward sand transport outside the surf zone, northern Gold Coast, Australia. In Proceedings of the 33rd International Conference on Coastal Engineering, Santander, Spain, 1–6 July 2012.

101. Troedson, A.L.; Hashimoto, T.R. *Coastal Quaternary Geology—North and South Coast of NSW*; Geological Survey of New South Wales: Sydney, Australia, 2008.

102. NSW Coastal Quaternary Geology Data Package (Version 3)—Geological Survey of New South Wales. Available online: https://search.geoscience.nsw.gov.au/product/40 (accessed on 5 August 2017).

103. *Worley Parsons Open Coast and Broken Bay Beaches Coastal Processes and Hazard Definition Study*; Worley Parsons for Gosford City Council: Sydney, Australia, 2014.

104. Hudson, J.P. *Gosford City Council Open Ocean Beaches Geotechnical Investigations*; Coastal & Marine Geosciences: Sydney, Australia, 1997.

105. Hudson, J.P. *Gosford City Beach Nourishment Feasibility Study—Stage 2—Investigation of Marine Sand Resources*; Coastal & Marine Geosciences: Sydney, Australia, 1999.

106. Hanslow, D.J.; Dela-Cruz, J.; Morris, B.D.; Kinsela, M.A.; Foulsham, E.; Linklater, M.; Pritchard, T.R. Regional scale coastal mapping to underpin strategic land use planning in south east Australia. *J. Coast. Res.* **2016**, *75*, 987–991. [CrossRef]

MDPI AG
St. Alban-Anlage 66
4052 Basel, Switzerland
Tel. +41 61 683 77 34
Fax +41 61 302 89 18
http://www.mdpi.com

Journal of Marine Science and Engineering Editorial Office
E-mail: jmse@mdpi.com
http://www.mdpi.com/journal/jmse

www.ingramcontent.com/pod-product-compliance
Lightning Source LLC
Chambersburg PA
CBHW051712210326
41597CB00032B/5456